All About Growing Fruits, Berries & Nuts

Created an
the editori
ORTHO BO(

Editor
Barbara Ferguson

Illustrator
Ron Hildebrand

Designer
Gary Hespenheide

Ortho Books

Production Director
Ernie S. Tasaki

Managing Editors
Karin Shakery
Michael D. Smith
Sally W. Smith

System Manager
Leonard D. Grotta

National Sales Manager
Charles H. Aydelotte

Marketing Specialist
Susan B. Boyle

Operations Coordinator
Georgiann Wright

Administrative Assistant
Deborah Tibbetts

Senior Technical Analyst
J. A. Crozier, Jr.

Address all inquiries to
Ortho Books
Chevron Chemical Company
Consumer Products Division
Box 5047
San Ramon, CA 94583

Copyright © 1982, 1987
Chevron Chemical Company
All rights reserved under international and
Pan-American copyright conventions.

	6	7	8	9	
		91	92		

ISBN 0-89721-096-4
Library of Congress Catalog Card
Number 87-070194

Chevron Chemical Company
6001 Bollinger Canyon Road, San Ramon, CA 94583

Acknowledgments

Manuscript Researcher
Philip Edinger

Manuscript Consultants
James Beutel, University of California
Dr. Joseph D. Norton, Auburn University, Alabama
Dr. Lee Reich, Ulster County Community College, New York

Art Director
Craig Bergquist

Copy Chief
Melinda Levine

Copyeditors
Andrea Y. Connolly
Loralee Windsor

Editorial Coordinator
Kate Rider

Pagination By
Linda M. Bouchard
Bob Miller

Proofreader
Judy Bess

Indexer
Frances Bowles

Production By
Studio 165

Separations By
Creative Color

Lithographed By
Webcrafters, Inc.

Consultants
Dr. Harry J. Amling, Auburn University, Alabama
Dr. Claron O. Hesse, University of California
Dr. Alec Hutchinson, Horticultural Research Institute, Ontario, Canada
Dr. W. L. Mellenthin, University of Oregon
Dr. Robert A. Norton, Northwest Washington Research and Extension Unit
Dr. H. P. Olmo, University of California
Dr. Henry P. Orr, Auburn University, Alabama
Fay Paquette, Camarillo, California
Perley Payne, University of California
Dr. Robert G. Platt, University of California
Dr. Raymond L. Self, Auburn University, Alabama
Dr. Warren C. Stiles, Cornell University, New York
Dr. Beth L. Teviotdale, University of California
Dr. R. A. Van Steenwyk, University of California
Dr. Robert D. Way, New York Experiment Station at Geneva

Photographers
William C. Aplin: 75
Heidi Bishop: 79, 84TL, Back cover BR
Josephine Coatsworth: 4, 7B, 7T, 14T, 14BL, 14BR, 16–17
Derek Fell: 18, 64, 65, 66R, 68L, 68R, 81, 82C, 82L, 82R, 83BL, 83TL, 90BR, 95, 100L, 100R, 105L, 105R, 106
Saxon Holt: Front cover
Michael Lamotte: 58–59
Michael Landis: 9, 11B, 18, 70R, 92BR, 101
John Lund: Back cover BL
Michael McKinley: 25
Pam Peirce: 102R
C. R. Reasons: 55, 63C, 85, 86, 87L, 92TL, 94L
Susan Roth: 1, 6L, 8, 26, 38–39, 70BL, 72L, 72R, 74, 84BL, 84R, 88R, 89L, 96BL, 96R, 96TL, 99L, 102L, 103L, 103R, 104, 107L, 107R, 108
D. Van Cleveland: 69TR
Wolf Von dem Bussche: 70TL, 73R

Special Thanks To
Apple Blossom Ranch, Sebastopol, California

Front Cover
The harvest is the orchardist's reward.

Back Cover
Four delicious reasons to raise your own fruit and vegetables: clockwise from the top left photo, these are 'Steuben' grapes, 'Bosc' pears, 'Royal Anne' cherries, and 'Barcelona' filberts.

All About Growing Fruits, Berries & Nuts

The Home Fruit Garden

Let your garden do more for you. When you select dwarf varieties and plan carefully, you can grow a wide range of fruits and berries for your family in a small space.

Strictly speaking a fruit is the seed-bearing portion of any plant; but the term commonly suggests the delicious edible fruits that grace our tables—apples, pears, peaches, plums, and tangy sweet berries. In this book you'll find all the most popular fruits, berries, and nuts suitable for growing by the home gardener. The book also presents many of the best varieties for specific parts of the country with descriptions of their important characteristics and their cultural requirements.

Many gardeners share the misconception that producing a good fruit crop requires the knowledge and skill of the expert orchardist. In fact extensive maintenance and culture are necessary only for maximum commercial production. The home gardener can get by with less complicated spraying, feeding, and pruning programs and still harvest a decent-sized crop. In addition we now have improved varieties with better fruiting, disease resistance, and tolerance of special soil and climate conditions. Your chances of growing good fruit are greater today than ever before. If you learn about your growing conditions, choose the varieties that fit them, and give your plants the attention and care outlined in this book, you are bound to succeed.

Information on planting, general care, and maintenance begins on page 17; pruning and training instructions begin on page 39; and descriptions of specific varieties, maps indicating plant adaptations, and a discussion of special needs for each plant can be found under the entries in the "Encyclopedia of Fruits, Berries & Nuts", beginning on page 59.

Modern dwarf fruit trees can be tucked into a vegetable garden without sacrificing vegetable space.

Fruit trees don't need to be relegated to the orchard. This quince makes an interesting and decorative landscape feature.

FRUITS IN THE LANDSCAPE

Gardeners who grow fruits, berries, and nuts at home will tell you that the fresh-picked taste is more than enough reason to grow them; but fruit plants also enhance the landscape—even in a small yard or garden.

Not long ago a gardener with an average-sized lot had to be content with very little in the way of fruit—perhaps a single apple tree in the center of the lawn and a grapevine growing over an outbuilding or arbor. As the average lot has grown even smaller, modern horticulture has met the challenge. Fruit trees are now available in a range of sizes that permit using them almost anywhere—spotted about the smallest yards and gardens, as borders and hedges, as groundcovers, or even as small shade trees. Modern dwarfing techniques and simplified training methods allow you to grow as many as a dozen fruit trees in the same small garden and still have plenty of sunny space available for vegetables or flowers.

Nut trees are the exception to this trend. In most cases plenty of space is still necessary for the average nut tree, so they are best on large lots. Because it is so important to many homeowners, small-space gardening is addressed whenever possible throughout the

book. This first chapter focuses on dwarf fruit trees and growing fruits in containers, another effective means of keeping fruit plants small. On pages 39–57 we offer two techniques for keeping fruit plants compact and productive: pruning and training. The "Encyclopedia of Fruits, Berries & Nuts" beginning on page 59, includes many dwarf varieties well suited to your garden.

Above: With dwarf trees and careful training, fruit can be fit into any landscape, even the smallest. By growing several tiny apple trees, a family can enjoy fresh apples over a long season. Left: A grape trellis sets off a gladiolus in the border.

The Edible Landscape

When you landscape with fruit, you combine beauty with practicality. Fruit can serve many functions. For example, apple trees make superb shade trees anywhere in the

Pomegranates have beautiful flowers in the spring, followed by colorful fruit in the summer, which will stay on the tree all winter.

yard if you prune them to a branch high enough to allow passage underneath. A large crabapple tree or a spreading cherry will also provide good shade.

Any fruit tree you like can be used as a focal point or accent in the yard or garden. The most striking trees in bloom are apples, cherries, quince, and some of the peaches. Crabapples are especially effective. These hardy trees are the most widely adapted of all flowering trees and offer abundant displays of red to pink to white blossoms followed by brilliantly colored fruit. As a rule they require some winter chilling, but there are varieties that bloom beautifully even in the mild

Pacific Coast climates. Some crabapples have fragrant blossoms and some have red to purple foliage.

Shrub fruits can also play an important role in the landscape, either as individual accents or as hedges or shrub borders. Try blueberries for their subtle colors or currants for their beautiful flower clusters and brilliant scarlet fruit.

Genetic dwarf peach trees make splendid flowering hedges, and showy-flowered dwarf or standard peaches can be trained in the same way. Espaliered apples or pears can also form attractive hedges or borders.

Genetic dwarf peach 'Bonanza' adds to the lush look of this patio. It also fits comfortably into Japanese settings.

strawberries, for about 5 years, and reseed themselves freely.

These are just a few ideas for planting with fruit. Other possibilities depend on climate, soil, available varieties, and your own taste.

DWARF TREES

The key to good fruit in the small home garden lies in the effective use of dwarf fruit trees. They provide good fruit and attractive shape, foliage, and flowers.

You will readily appreciate the development of dwarf trees when you consider what would happen to your gardening space if you planted a standard-sized apple tree, which can grow from 20 to 40 feet high and spread 30 to 40 feet. The tree would take up a considerable amount of space and block out so much sun that the ground beneath it would be too dark to grow most other plants. Compare this with the dwarf apple varieties that can be held to a height of about 10 feet with a 10-foot spread. Standard apricots, peaches, and plums can grow to 30 feet tall with a spread of 30 feet; sweet oranges to 20 to 30 feet tall with a spread of 15 to 20 feet; and pears to a towering 45 feet with a spread as wide as 30 feet. Most dwarf varieties of these trees can be kept to a height of about 10 feet—in certain cases they can be kept to even less.

Genetic Dwarfs

Dwarf trees are produced in nature or through horticultural practices. Natural dwarfs are called *genetic dwarfs.* Among apples, the most common genetic dwarfs are spur apples, so named because on a given amount of wood they produce more fruiting spurs than ordinary apple trees. Fruit production can have a dwarfing effect because it uses energy that would otherwise go into the growth of the tree. The heavier crops of spur apples mean slower tree growth.

These natural mutations occur with many varieties of apple trees; for example, there are several spur varieties of the popular 'Red Delicious' apple available to gardeners under such names as 'Redspur' and 'Starkrimson'. Spur apples grow more slowly but eventually reach about three-quarters normal size. This may be as much as 15 to 30 feet tall with a similar

You can even use fruit as a ground cover. Strawberry plants are effective, especially in smaller areas, but plan to replace them every three years with new plants if you want a heavy fruit crop. For larger areas, use the low-bush blueberry.

The European alpine strawberry ('Alexandria' is one variety) doesn't form runners, so it makes a neat border or ground cover for a small area. The fruit is edible and delicious, with an intense "wild" flavor, but it isn't very large or prolific. Children love to hunt and pick the berries. These strawberries are grown from seed. They last longer than other

spread—still a large tree for many home gardens. However, all genetic dwarfs can be made even smaller by grafting them onto dwarfing rootstocks.

An example of a genetic dwarf apple that is not a spur type is the variety sold as 'Garden Delicious'. It bears fruit resembling the popular 'Golden Delicious', but grows to only 6 or 8 feet, 3 feet when grown in a container.

Apricots, sweet cherries, sour cherries, peaches, nectarines, and plums all have genetic dwarf varieties. There are several good, extremely cold-tolerant, genetic dwarf sour cherries that grow to perhaps 10 feet under ideal conditions but most often stay at 6 or 7 feet. There is a wide range of genetic dwarf peaches, all of which grow slowly to 8 or 9 feet. As these have almost no stem between the leaves, they have a typically dense or lumpy look. Dwarf peaches and nectarines (the nectarine being a sport, or mutant, of the peach) are quite decorative, with large, showy flowers. They are also fairly tender, however, and tolerate little cold.

There are many natural dwarf plums, but they are not the popular varieties; rather, they are crosses of the western sand cherry, or true cherry, and the shrub plum. Most are

This apple has been dwarfed by grafting it onto dwarfing rootstock.

produced commercially for very cold climates and are readily available in the North.

Man-Made Dwarfs

There are several horticultural methods of producing dwarf trees. All work by limiting the supply of nutrients to the tree.

Pruning   If it is done while the plant is actively growing, pruning can have a dwarfing effect, because it removes foliage that is photosynthesizing, enabling the plant to grow. Heavy summer pruning increases dwarfing.

Root pruning   Another effective dwarfing procedure is root pruning. In this technique a sharp spade is sunk into the ground around the perimeter of the plant to sever some of the feeder roots.

Girdling and scoring   Tree growth can also be reduced by girdling and scoring. Girdling removes a strip of bark, about ¼-inch wide, from around the tree. Scoring makes a single cut around the tree. These cuts are small enough not to kill the tree, but interfere with the supply of nutrients to the roots, slowing their growth and dwarfing the tree.

Above: These trees have been dwarfed by training.
Left: This apple has been dwarfed by a combination of training and grafting.

Training   Bending or twisting branches to direct growth can slow growth, and also redirect it to make the tree smaller.

All these techniques produce only a temporary dwarfing effect and must be repeated periodically. Growing in containers can produce permanent dwarfing by confining the roots of the tree to a small growing space. You must be sure, however, that the roots do not grow out of the bottom of the container and become established in the soil below it.

The easiest and most effective way to produce permanent dwarfing is by grafting *scions* (shoots or buds of a desired variety) to dwarfing rootstocks. This method offers many advantages to growers, horticulturists, and gardeners. Grafting is one easy way to produce large numbers of plants in a relatively short time. It ensures true reproduction of a desired variety, which is important because seed does not always breed true. And grafted dwarf trees remain uniformly smaller and tend to bear fruit at a younger age than standard trees, sometimes as early as their second year of growth.

Grafted dwarfs are readily available at most nurseries and garden centers. More curious, adventuresome, or enterprising gardeners may want to try their own grafting.

Dwarfing Rootstocks

The key to producing grafted dwarf trees is in growth-limiting rootstocks. The most extensive research on such plants has been undertaken with apples and has resulted in the development, at the Malling Research Station in England, of the numbered Malling rootstocks. There are several Malling rootstocks available, each having a different degree of dwarfing effect on the apple variety that is grafted onto it. Michigan State University has also developed a dwarfing rootstock for apples called 'Mark'.

Other major fruit trees can also be dwarfed by grafting, but far less research has been done on these. As a result fewer kinds of dwarfing rootstocks are known, and these are not always as effective as the ones for apples. Apricots, peaches, nectarines, and plums can be dwarfed on 'Nanking' cherry rootstock and

cherries on 'St. Lucie' or 'Vladimir' cherry. Quince is a reasonably satisfactory rootstock for dwarfing pears and there is also a dwarfing rootstock made from the cross of 'Old Home' pear with 'Farmingdale' pear.

One of the advantages of growing dwarf trees is easy care. When the distance from roots to treetop is 10 feet or less, sprays for pests and diseases are easier to apply and pruning is less difficult and time-consuming. These advantages increase if you further reduce size by applying the pruning and training techniques offered in this book.

CLIMATE

Most of the fruits we discuss in this book are referred to as temperate-climate fruits. (For information about tropical-climate fruits, see Ortho's book *All About Citrus & Subtropical Fruits*.) Temperate climates are characterized by hot summers and cold winters. Most of the United States lies in the Northern Temperate Zone. Within this zone, however, climates vary greatly. Furthermore, they vary not only over large distances, but also within cities from district to district, and even from place to place within a garden. This is why gardeners must plan in terms of microclimates.

Microclimates

Your own garden may have several microclimates. For example, a spot protected from the wind will have a warmer microclimate than will a spot out in the open. If the sheltered spot is backed by a wall that reflects heat, the area will be warmer still. In a northern garden such a location might be ideal for helping a tree bear better fruit; in a southern garden the location might be too hot.

The extreme diversity of climate in the United States makes it impossible to provide an exact guide to climatic conditions. The

Apples can stand the coldest northern winters, but a late spring frost may burn the blossoms, destroying the crop.

Above: The dwarf peach in front is in its third year of training. The apple behind it has reached maturity.
Right: This espaliered apple has had a different variety grafted on top.
Far right: The garden has room for salad vegetables and flowers as well as fruit, and yet doesn't appear crowded.

small maps with each entry in the encyclopedia (beginning on page 59) will serve as a guide to where the fruit can be grown in the United States and parts of Canada. The information in the individual varietal descriptions will help you choose the plants most likely to do well in your garden. For more detailed information, check with local nurseries, garden centers, agricultural extension agents, and especially gardening friends and neighbors who can tell you about their own successes and failures.

A SAMPLE GARDEN

The plan to the right illustrates some of the remarkable space-saving possibilities in gardening with fruit. Notice the orientation of the garden. Fruit needs sun to set a crop, and the illustrated arrangement provides maximum exposure to the sun as it passes from east to west. If your location does not have sun all day, at least plan to give your fruit plants southern or western exposures.

Notice also how this plan makes use of espaliered apple trees and grapevines to provide an attractive border and plenty of fruit and still allow space for many other plants. The raspberries and currants are trained on trellises and oriented approximately north and south. They take up little space and bear heavily. The grapes are planted mainly on the north side of the garden to give them full southern exposure to the sun, which is necessary in order to develop good sugar content in the fruit. Note that there is still room in the center of the garden for dwarf fruit trees, raised beds in which to grow vegetables or flowers, and a plot of annuals.

This plan is based on a real fruit garden that is situated beside a home in the California wine country. The photographs on the opposite page, as well as several of the photographs on other pages in this chapter, were taken in this garden. It measures only 15 by 50 feet but contains 17 fruit trees, several grape varieties, cane berries, ornamental plants, and vegetables. The trained plants are key features in the space-saving aspect of this garden. This requires a little effort, but once established, espaliered trees can be grown closer together and still receive good light, perfect air circulation, and plenty of root space.

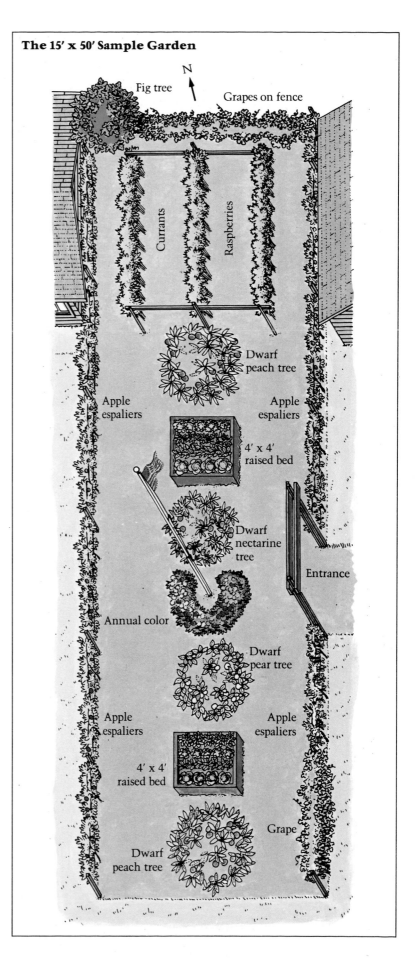

The 15' x 50' Sample Garden

N

Fig tree

Grapes on fence

Currants

Raspberries

Dwarf peach tree

Apple espaliers

Apple espaliers

4' x 4' raised bed

Dwarf nectarine tree

Entrance

Annual color

Dwarf pear tree

Apple espaliers

Apple espaliers

4' x 4' raised bed

Grape

Dwarf peach tree

Planting and Caring for The Home Orchard

Learn the principles of fruit growth and apply them when caring for your home orchard. If you follow the planting, feeding, and watering instructions in this chapter, you'll get maximum production from your garden.

The art of growing abundant fruit lies in selecting the right varieties for your region, then growing them with informed skill. This chapter discusses how fruit plants produce fruit, how they grow, and tells you everything you need to know to get them started and keep them healthy and productive.

Every time you bite into an apple or pear, you're tasting the results of plant breeding, particularly the act of pollination. With a few exceptions (certain figs, for example), fruit will not form unless pollen from the male parts of a flower is transferred to the female parts of a flower. The pollinating insects for most of the fruits in this book are bees. The presence of bees around your plants, however, does not necessarily mean you'll get a crop. The pollen the bees carry must be of the right sort. Most of us know that apple pollen, for example, will never pollinate a pear blossom; it's also true that apple pollen will not always pollinate an apple blossom.

The same sweet goodness that attracts us to fruit attracts a wide variety of insects. A few well-timed sprays, however, protect your fruit against unwelcome guests.

To get a crop from most fruit trees, you need a separate source of pollen and a bee to carry it.

POLLINATION REQUIREMENTS

Some plants are called *self-pollinating* or *self-fertile*. This means that their flowers can be fertilized by pollen from flowers either on the same plant or another plant of the same kind. Self-fertile plants will produce fruit even if they are planted far away from any other plant of their kind. Among the self-fertile plants are a few types of apples, pears, and plums; most peaches, apricots, and crabapples; and all sour cherries.

Other plants set fruit only when they receive pollen from a plant of a different variety. When a plant's pollen is ineffective on its own flowers, it is called *self-sterile*. This group includes some peaches, apricots, and crabapples; most apples, pears, and plums; and all sweet cherries. The plant that can fertilize a sterile plant is called a *pollinator.*

Never assume that because you have a bearing fruit tree you can plant a new tree of a different variety nearby and be sure of a crop. Plants must bloom at about the same time for successful cross-pollination to occur; for example, an early self-sterile apple will not bear fruit unless the pollinator is another early apple variety.

Planting for Pollination

A fruit plant that needs a pollinator needs it close by. The maximum distance is 100 feet, but the closer the better. The bees that carry the pollen are unlikely to fly back and forth if the distance between the trees is any greater.

If your neighbor has a pollinating variety across the back fence, you're in good shape; if not, do one of the following.

☐ Plant two trees fairly close together.

☐ Graft a branch of another variety onto a tree that needs pollination.

☐ Place a bouquet of flowers from a pollinating tree in a vase or jar of water and lodge the container in the branches of a second tree.

The "Encyclopedia of Fruits, Berries & Nuts," beginning on page 59, will tell you which varieties need pollinators and which varieties act as pollinators.

HOW FRUIT PLANTS GROW

All plants must have sugar to produce energy and grow. They make the sugar through photosynthesis. You can stimulate this process by planting them in a sunny spot; pruning and training them for good leaf exposure; keeping the soil properly watered; and keeping leaves free of dust, pests, and disease. Each piece of growing fruit needs some 30 leaves working for it, not including the leaves that supply nourishment to roots and branches.

The illustration on the opposite page gives you some idea of the day-to-day workings of a fruit plant. While the leaves are busy above ground, the roots spread out underground searching for water, oxygen, and mineral nutrients. These essential elements are then transported to the green leaf tissues where photosynthesis is carried out using the energy supplied by sunlight to manufacture needed sugars. The sugar not immediately converted to usable energy for the plant's growth is stored throughout the plant, including the fruit. It's easy to understand that the greater the supply of factors that produce sugar—sunlight, water, and carbon dioxide—the more abundant and sweet the fruit.

SOIL

Roots depend on the soil for a good supply of air and consistent moisture. The best soil for fruit trees allows air into the soil quickly after a rain or irrigation and holds much water.

Heavy soils are soils that drain slowly. You can improve them for fruit trees by adding plentiful organic matter, and by planting high, so the tree sits on a low mound.

Some fruits, such as pears, will tolerate dense, airless, heavy soil. Apples and crab-apples will take short periods of airless soil, but apricots, cherries, figs, plums, grapes, and currants all need fair drainage. Strawberries, cane berries, and peaches need good drainage, and blueberries must have perfect drainage.

In gardens with extremely heavy soil, you can still plant fruits that prefer porous soils by using containers or raised beds. A raised bed for a standard fruit tree should be 3 feet deep and 6 feet square. Soils for containers are discussed on page 26.

Soils that don't hold much water are called *droughty*. They are most easily improved by adding large amounts of organic matter. The organic matter acts like a sponge, holding water until the tree needs it. Most fruit trees can be grown on droughty soil, but they should be watered and fertilized more frequently than trees on better soil.

You can supply these needs best if you first examine your soil. If it is rock hard when dry and gummy when wet, you have the very fine-textured soil called clay. Clay holds moisture so well that there is little or no room for air. To correct this, aerate clay soil by adding organic matter such as peat moss or compost. Spread 4 or 5 inches of organic matter over the soil and mix it in evenly. Ideally you should add organic material wherever the plant's roots might spread at maturity; the roots spread more widely than the branches.

If water soaks directly into your soil without significant spreading and the soil dries up a few days after watering, your soil is sandy. Sandy soils contain a great deal of air, but moisture and nutrients wash away quickly. Additional organic matter helps here, too, by filling in spaces between the coarse soil particles and retaining the water. Peat moss, compost, and manures are especially beneficial to sandy soils. You can also use sawdust or ground bark, but with these you must add extra nitrogen—¼ pound ammonium nitrate, or equivalent, per bushel—or the soil microbes will rob nitrogen from plants as they convert the sawdust or bark into humus.

If you have soil that feels moist for days after watering, but still crumbles easily when you pick up a handful and squeeze it, you are blessed with loamy soil and shouldn't need to amend it.

PLANTING

Nurseries and garden centers sell plants in three ways: bareroot, balled and burlapped (with the rootball wrapped in burlap), or in containers.

Most deciduous fruit plants are sold bare root. The leafless plant is taken from the ground in late fall or winter after it has gone dormant and it is then shipped to the nursery where it is held in moist sand or wood shavings. Sometimes the roots are enclosed in a plastic bag full of damp shavings. Bare-root plants are fragile and must be kept cool and moist. Plant them as soon as possible.

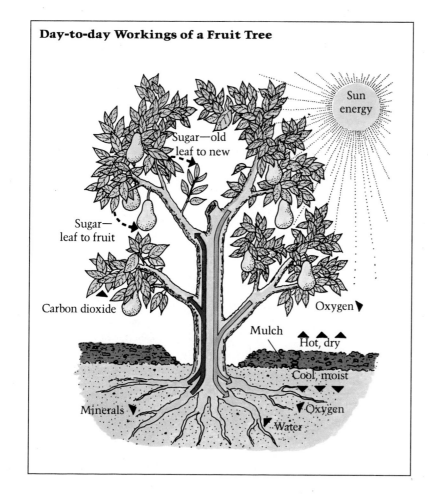

Day-to-day Workings of a Fruit Tree

Sun energy

Sugar—old leaf to new

Sugar—leaf to fruit

Carbon dioxide

Oxygen

Mulch

Hot, dry

Cool, moist

Minerals

Oxygen

Water

Ball and Burlap Planting

Do not jar the rootball or you may damage the tree. After the first layer of fill is pressed down, lay back the burlap and fill again. Soak, make mound, soak again. Protect all newly planted trees from winds or strong sun.

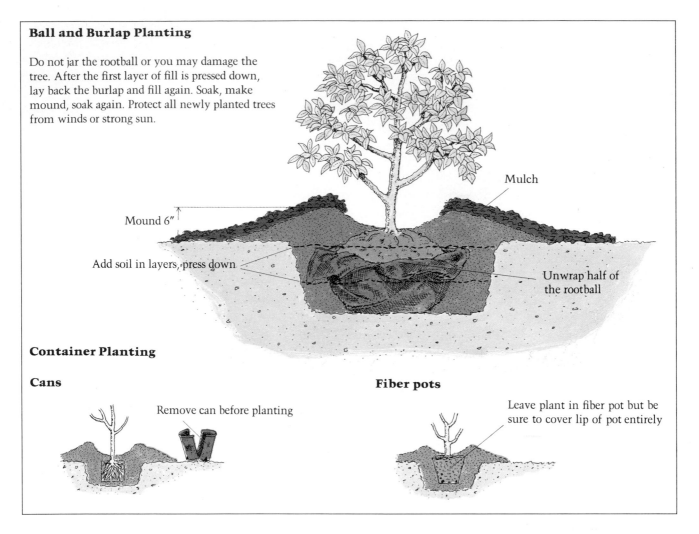

Mulch

Mound 6″

Add soil in layers, press down

Unwrap half of the rootball

Container Planting

Cans

Remove can before planting

Fiber pots

Leave plant in fiber pot but be sure to cover lip of pot entirely

Bare-root plants are sometimes put into containers at the nursery. If you buy them in winter or while they're still dormant, you can bare the roots again to plant them. If they have already leafed out, keep them in their containers until May or June so the root system has time to knit the container soil.

Fruit trees are seldom sold balled and burlapped, but they are frequently sold in containers made of plastic, pulp, or metal. Balled and burlapped plants are sold at the same time as bare-root plants and should go into the ground or their permanent containers quickly. Trees sold in containers are available the year around and may be held until time to plant, as long as you don't cut the container.

Never let bare-root or balled and burlapped plants lie around unprotected. If you must keep bare-root plants for a time before you can plant, dig a shallow trench, lay the plants on their sides with the roots in the trench, and cover the roots with moist soil. This is called "heeling in." Wrap balled and burlapped plants in a sheet of plastic so the rootball stays moist.

Planting Trees and Shrubs

The illustrations here will give you some idea about how to plant a tree from the nursery. Remember never to plant if the soil is very wet. Working wet soil packs it, driving out the air and trapping the roots. In rainy climates you can dig the hole for the plant in the fall and keep the soil mound dry by covering it with a weighted plastic sheet. The soil will then be workable any time.

A good rule of thumb is to dig the planting hole twice the width of the rootball. Another good rule of thumb is always to plant high. Notice in the illustrations that the planting soil is mounded above the normal soil line. The most fragile part of a woody plant is the crown, that section where the roots branch and the soil touches the trunk. The crown

Planting Bare-root Trees

Clip off broken roots. The hole must be wide enough for the roots to spread. Soak the soil after the hole is refilled; make a volcano-shaped mound, then soak again from the top, running water slowly so it sinks in. Mound higher in dense soil, lower in good soil.

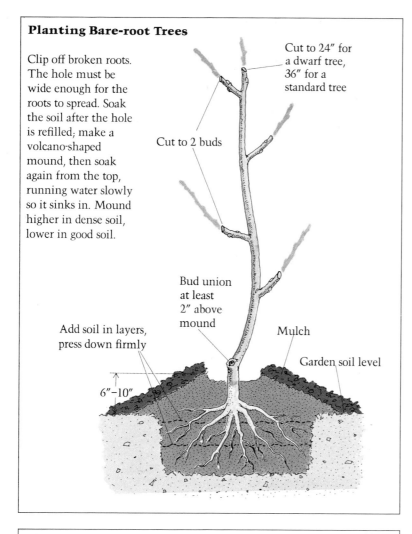

Cut to 24" for a dwarf tree, 36" for a standard tree

Cut to 2 buds

Bud union at least 2" above mound

Add soil in layers, press down firmly

Mulch

Garden soil level

6"–10"

Planting Cane Berries and Grapes

Plant a rooted cutting with two or three buds above the soil. Cover these buds with mulch.

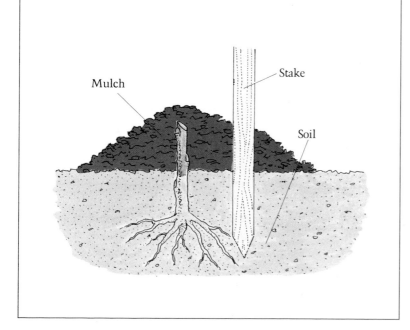

Mulch

Stake

Soil

must be dry most of the time, especially in spring and fall. Raised planting minimizes crown rot fungus (which can be fatal to the plant) by making it impossible for water to puddle near the trunk. If you plant at soil level, the soil in the planting hole might settle, causing your plants to sink.

Be especially careful when planting grafted dwarf fruit trees. Dwarfing rootstocks may blow over unless they have support and most apples on dwarf rootstocks require staking. Many growers now place the bud of the fruiting variety high on the rootstock, up to 6 or 8 inches above the roots. This bud union shows later as a bulge with a healed scar on one side. It's tempting to plant the tree close to this union for stability, but don't plant the tree deeper than it was at the nursery, or you may have problems with crown rot later.

Be careful not to bury the bud union in soil or mulch at any time during the life of the tree (see illustration, at left). If moist material touches the union, the upper fruiting part will root, and its vigorous root system will produce a full-sized tree instead of the dwarf you bought. Check the bud union frequently for signs of rooting and keep mulches a few inches away from it.

Planting Cane Berries and Grapes

Cane berries and grapes are usually sold bare root in the spring. Plant at the same depth as they were growing before being harvested for sale. There is usually a line on a bare root plant that indicates the previous soil level. Cut back canes to two or three buds. If you are staking the plant, place the stake at planting time to avoid injuring the roots.

Planting Blueberries, Currants, and Gooseberries

These plants are often sold in containers and should be planted slightly deeper in the ground than they were in the container. For currants and gooseberries, leave only the three strongest branches and cut these back to eight inches. Space plants 3 to 4 feet apart.

Plant blueberries about an inch deeper than they grew in the container. Plants should be spaced about 4 to 6 feet apart. Rabbiteye blueberries grow larger and should be spaced a bit farther apart.

Planting Strawberries

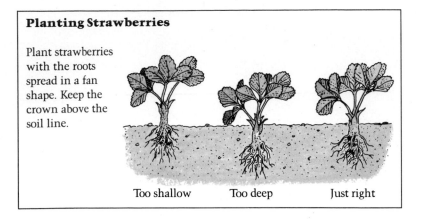

Plant strawberries with the roots spread in a fan shape. Keep the crown above the soil line.

Too shallow   Too deep   Just right

Planting Strawberries

Strawberries are sold either bare root in early spring or planted in flats, six-packs, or 4-inch containers. Plant strawberries with the roots spread in a fan shape. Be sure to plant so that the crown is above the soil line.

Strawberries can be grown in the matted row or double row hill systems. In the matted row system, all runners are allowed to grow which gives larger yields the first bearing season. The plants should be spaced 12 inches apart and rows should be spaced 3 feet apart.

In the double-row hill system, plants are 12 inches apart in the row and rows are 12 inches apart. All runners are picked off. Each double row is raised and separated by a 24-inch trough that allows you to walk among the plants to pick or care for them. The system lends itself to everbearing strawberries or those that don't send out many runners.

FEEDING

Plants "eat" sunlight; what we feed them is actually mineral nutrients. The three primary plant nutrients are nitrogen, phosphorous, and potassium. Plants also need three secondary nutrients—calcium, magnesium, and sulfur—and small quantities of trace nutrients, including iron, manganese, and zinc. Nitrogen in some usable form is the only element that is always in short supply. You can add it as ammonium nitrate, ammonium sulphate, calcium nitrate, complete fertilizers, or manure.

Fruits rarely need extra phosphorus, but they will occasionally need potassium and the other nutrients. If growth is slow or leaves and fruit look unnatural or unhealthy, check with your nursery or agricultural extension agent to find out what should be added.

Feeding is less a matter of exact measurement than of how the plant responds. Nitrogen forces leafy growth, but too much growth at the wrong time can harm your fruit crop. In most cases feed the plant less from mid-June through leaf fall. If it produces only a few inches of growth one season, step up feeding. If it sprouts up like a geyser, feed less. Slow growth and pale or yellowish leaves mean that the tree is getting too little nitrogen.

The following are general guidelines for the amount of 10-10-10 fertilizer needed for fruiting plants each year. These three numbers, found on all fertilizer labels, refer to the percentages of nitrogen, phosphorus, and potassium (always in that order) contained within that particular product. If a different analysis fertilizer is used, adjust with respect to the percent nitrogen (that is, double the amount of fertilizer if the analysis is 5-10-10 but don't change the amount if the label reads 10-5-5).

Although one or two feedings are often recommended, we suggest feeding equal amounts of chemical fertilizers four times at evenly spaced intervals between early spring and late June unless otherwise noted. Keep fertilizers away from the trunks of trees and shrubs and water very deeply after feeding.

Fruit and Nut Trees

Use the following schedules for standard-sized fruit trees. Dwarf trees will require proportionately less, and very large nut trees will require more.

First and second season   Four tablespoons 10-10-10 fertilizer per year; one at each feeding, scattered evenly.

Third to seventh season   Double the amount each year: 8 tablespoons or 2 tablespoons per feeding in the third year, 16 tablespoons or 4 tablespoons per feeding in the fourth year, 32 tablespoons or 8 tablespoons per feeding in the fifth year, 64 tablespoons or 16 tablespoons (one cup) per feeding in the sixth year, and 128 tablespoons or two cups per feeding in the seventh year.

Mature tree   Continue feeding with 2 cups per feeding, for a total of 8 cups per year.

Grapes

Apply 4 tablespoons 10-10-10 fertilizer per plant per feeding four times a year for a total of one cup per plant per year.

Strawberries

Apply 2 cups 10-10-10 fertilizer per 100 feet of row at the end of June and again in August.

Brambles

Apply 2 cups 10-10-10 fertilizer per 100 feet of row per year.

Blueberries

Apply 2 tablespoons to 1 cup of ammonium sulfate (which makes the soil more acid) per plant, depending on plant vigor and size.

Feeding With Animal Manure

Animal manure improves the soil texture as well as adding nutrients but it is lower in nitrogen than chemical fertilizers, and it may contain salts, which can be harmful in dry climates. Be especially careful of bird, rabbit, and feed-lot manures in dry climates. If leaves show brown edges from salt burn, soak the root area for several hours and change to another feeding method. Well-rotted barn or stable manure is safer.

Since manures contain less nitrogen per pound than chemical fertilizers, you can use relatively more, and because they release nitrogen slowly through bacterial action, you can put the whole amount around the tree at one time. For young trees begin with a little less than ½ pound of dry bird manure, or about 1 pound of dry cattle manure, and double each year. For mature trees, use 50 to 70 pounds of well-rotted bird or rabbit manure, spreading it under the outer branches in fall. For the same trees, use 100 to 200 pounds of well-rotted cattle manure.

WATERING

Standard fruit trees need a lot of water. If this is not supplied by spring and summer rains, deep watering is necessary. Dwarf trees on shallow-rooted stocks may not need as much, but they must have a constant moisture supply. At planting time, water each layer of soil in the planting hole. If the garden soil is dry, soak the hole itself before you put in the plant. Finish by creating a depression in the planting mound to hold the water, then soaking the tree thoroughly. Take care that water does not run out of the depression.

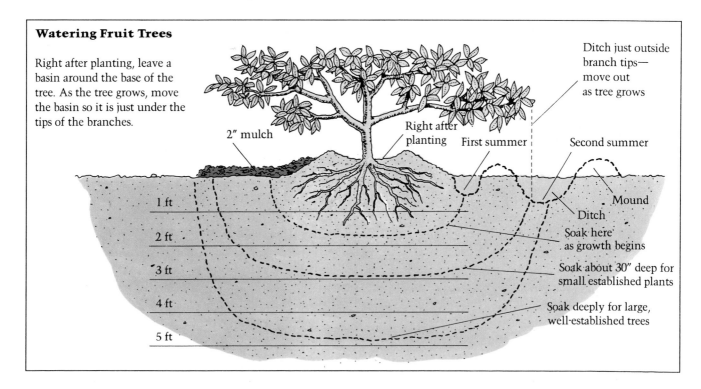

Watering Fruit Trees

Right after planting, leave a basin around the base of the tree. As the tree grows, move the basin so it is just under the tips of the branches.

2″ mulch

Right after planting   First summer

Ditch just outside branch tips—move out as tree grows

Second summer

Mound
Ditch
Soak here as growth begins
Soak about 30″ deep for small established plants
Soak deeply for large, well-established trees

1 ft
2 ft
3 ft
4 ft
5 ft

Do not water again before new growth begins unless the soil seems dry. The roots are not growing actively at this time, and soggy soil will invite rot. When new growth begins, let the top inch of soil dry and then give the plant a thorough soaking. Be sure to water at the top of the planting mound. This is especially important with balled and burlapped plants since the soil in the rootball may not take up water unless it is applied directly overhead.

When first-season growth is abundant and plants are growing well in midsummer, build a shallow basin around the base of the planting mound to direct water to the plant's roots. Water by filling the basin. This basin should be expanded in size each year so that it is just outside the tips of the branches.

Plants that are actively growing generally need 1 inch of water once a week, or about 2 gallons of water per square foot of root spread once a week. (The roots generally spread out somewhat farther than the top canopy of the tree.) A newly planted tree would have a root spread of up to 2 square feet and, therefore, would need 2 to 4 gallons of water a week. Adjustments can be made for rainfall and soil type. Your tree may need water quite often in very sandy soil, less often in heavier soil. Always dig down a few inches into the soil first to see if watering is necessary.

Trees in a lawn area should have a deep soaking about twice a summer in addition to normal lawn watering.

FRUIT IN CONTAINERS

Containers make plants mobile. Moving fruit plants to shelter when cold weather comes or to a shady spot if excess heat is the problem, makes it possible to grow them outside their normal climate ranges. With containers you can relocate plants to find out where they do best and even try varieties not usually recommended for your climate, such as peaches in North Dakota. Deciduous trees can survive a winter season in a garage. One caution, however: plants in containers are not as hardy as those in the ground. You must give container plants winter protection in climates where the ground freezes to any depth at all.

What fruits can you plant in containers? Virtually any you like. Trees, of course, must be grafted or genetic dwarfs. The "Encyclopedia of Fruits, Berries & Nuts," beginning on page 59, lists a great many dwarf varieties of apples, apricots, cherries, nectarines, peaches, pears, and plums, all of which are suitable for container culture. Any fig can be grown in a container. Strawberries can be planted in large or small containers, and blueberries and currants make excellent container plants. You can even plant grapes, providing you give them a trellis or other support during the growing season.

Choosing Containers

What kinds of containers are suitable for growing fruit plants? Again, virtually anything you like, as long as it will hold the plant and a sufficient amount of soil, is nontoxic, and contains holes for adequate drainage. Half-barrels, for example, make excellent containers for fruit trees, as do wooden and ceramic planters.

Whatever type of container you choose, the size should be just 2 or 3 inches wider than the roots of your plants. The right size container allows roots easy access to water and nutrients without giving them so much space that root growth occurs at the expense of top growth. If you start with a bare-root apple, a pear, or one of the genetic dwarf fruits, your first container should be about the size of a 5-gallon can. Let the young tree grow for a season and then repot it the following spring in a larger container.

If you plan to relocate the plant and its container, carefully consider container size in advance. The maximum for mobile containers should be about bushel size. Anything bigger will be too bulky to handle or move. Half-barrels, or any boxes or pots that hold about an equal volume of soil, are roughly the right size. The smaller the container, the easier it is to move. However, keep in mind that the plant must have sufficient room for its roots and that more work is involved in feeding, watering, and root pruning with smaller containers. All these factors must be balanced when choosing the best container.

Strawberries are the ideal container plant; they are attractive, tasty, and small enough to be easily moved. Bring them up front when they are flowering and fruiting, then move them to an out-of-sight corner until the following spring.

You can grow what would otherwise be a large tree—such as this fig—in a container, but use a light mix or it will be difficult to move. Substitute perlite for some of the sand in the recipe on this page to make it lighter.

Container Soils

Because containerized soil loses moisture easily, the soil must hold moisture well. To prevent soggy roots and the possibility of disease, it must also have good drainage. Commercial mixes, often referred to as soilless mixes or synthetic soils, fit the bill.

Synthetic mixes offer several advantages. They are free of disease organisms, insects, and weed seeds. They are lightweight—half the weight of garden soil when both are wet—which is an advantage both in relocating container plants and in growing them on roofs or balconies. In addition they can be used just as they come from the bag, often without needing to be moistened for planting.

If you plan to fill a number of containers, you may want to save money by mixing your own planting medium. Here is a basic recipe:

> 9 cubic feet of fine sand designated as 30-270 (numbers refer to screen sizes the sand will pass through)
> 9 cubic feet of ground bark
> 5 pounds 5-10-10 fertilizer
> 5 pounds ground limestone
> 1 pound iron sulphate

Some gardeners like to add a little rich loam to the mix of sand and organic material. Add up to one-third loam if you like, but be careful not to include clay soil, which holds too much water for a container mix. Also remember that if you add topsoil to the mix, you may give it good physical properties but you will also increase the risk of introducing soil pests and diseases that can harm your plants.

Potting and Repotting

Before you pot your fruit plant, make sure your container has drainage holes. Cover the holes with screen or broken pieces of crockery but don't cover the holes tightly or you'll retard drainage. Do not fill the bottom with rocks or coarse gravel; contrary to popular opinion these do not improve drainage, they simply take up space in the container.

To plant a bare-root plant, place enough tamped-down soil mix in the bottom of the pot so that the plant crown is slightly below the container rim when the roots are touching the mix. Hold the plant at that level and toss in enough mix to support it, tamping lightly as you go, and filling the container to about an inch below the rim. The soil will settle, leaving room to water.

When planting fruits grown in nursery containers, remove the container and scratch the rootball all around with a fork to rough up the roots and direct them outward. Be sure to cut off long roots spiraling at the bottom of the rootball. The plant should be planted just as deep as it was in the nursery container—and no deeper—with at least an inch between the pot rim and the soil level.

Repotting is similar, as shown in the illustrations on the opposite page. Repotting is necessary because plants tend to bunch feeder roots at the wall of a container where they dry out and die more easily. This creates shortages of water and nutrients even when you are providing proper care. When you shave off an inch of root and add fresh soil, the plant will grow healthy young roots in the new reservoir of moisture and nutrients. New top growth will soon follow new root growth. After potting or repotting, give the plant a good deep watering.

Steps in Repotting

1. Remove plant in early spring. Shave 1″ from sides of roots and 1½″ from the bottom, using a large knife.

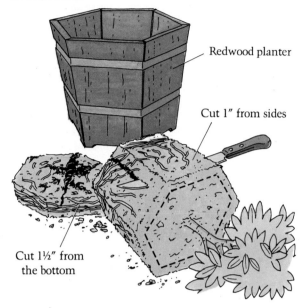

Redwood planter

Cut 1″ from sides

Cut 1½″ from the bottom

2. Reset the plant on fresh soil with the crown just below the rim of the container.

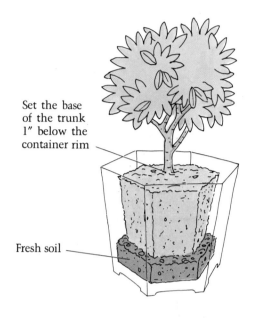

Set the base of the trunk 1″ below the container rim

Fresh soil

3. Fill in around side, tamping lightly with a stick as you go.

4. Water slowly until water flows freely from the bottom. Prune lightly, removing tangles and extra-long branches.

Feeding Container Fruit Plants

If you are using a purely synthetic mix, you must be careful about feeding. The nutrients you add often wash through the soil when you water, so you'll have to feed more often. It's best to keep to a regular schedule.

One good feeding method is to give each plant about half the recommended quantity of complete fertilizer (one that contains nitrogen, phosphorus, and potassium) every two or three weeks. If the label recommends 1 tablespoon per gallon of water each month, use ½ tablespoon per gallon instead and feed every two weeks. A liquid fertilizer is easier to measure in exact proportions and is also less likely to burn roots.

It can be difficult to know when to water hanging plants because you can't see—and sometimes can't feel—the soil surface. Try lifting the container slightly to test its weight. With some practice you will be able to gauge its dryness accurately.

Another good method is to use slow-release fertilizer pellets. These dissolve slowly over a period of time, releasing nutrients with every watering.

Feed from the beginning of the growing season until the end of summer if the plant is to receive winter protection. Stop about mid-July if the plant is to stay outdoors. This will give both a chance to harden new growth.

Watering Container Plants

If you check the soil occasionally by digging down an inch or two, you'll soon learn when to water. Water whenever the soil just under the surface begins to dry. Judge when to water by the behavior of your plant. It should never wilt, but it shouldn't stand in soggy soil either. The top inch may stay moist for a week in fairly cool weather, but in hot, windy weather you'll need to water more often, perhaps even every day for a plant that needs repotting. (This is why well-drained soil is important. You can water liberally without drowning the roots.) Water enough with each irrigation so that a good amount drains from the bottom of the container. Don't count on rain to do all of your watering. The foliage of plants in containers can act as an umbrella, shedding most of the rainfall. Check the soil even when rain has been abundant.

Vacation watering   When you leave home for a long period, group your containers near a water source and away from the afternoon sun. Grouping them will help keep them moist, shade will further cut the need for water, and if they are located near a hose, your vacation waterer won't miss any of them by accident. For large numbers of containers, you can hook up a permanent system of small hoses and add a timer that turns water on at regular intervals. Drip systems are particularly effective.

Leaching   It is important to leach container soil occasionally to remove built-up mineral salts that can burn leaves. Salts accumulate from fertilizers and from hard water. (Any water that won't produce good soapsuds or leaves bathtub rings is hard water and has a high salt content.) You'll know you have a salt problem when you see brown leaf edges. Leaching is running enough water through the soil to wash away the salts.

Every couple of months, leach your soil. Put your garden hose in each container and let it run slowly for about 20 minutes. The water should flow just fast enough that it soaks through the soil and out the drainage holes of the container. It is a good idea to fill the container until water runs freely from the bottom, go on to other containers, then return and repeat the process. This will keep salts to a minimum. Avoid using softened water on your plants because it contains harmful chemicals.

Winter Protection

If you live in a climate where your garden soil freezes in winter, then container soil is likely to freeze all the way through. Gardeners in the coldest northern zones should plan to protect even hardy deciduous plants during the coldest months of the year when these plants are in containers.

The easiest way to protect a container plant from freezing winter temperatures is to move it into an unheated cellar, garage, porch, or room. You can also bury the pot in the ground, making sure to place it deep enough that the soil covers the rim. Heaping sawdust, leaves, or woodchips around the pot will help protect the roots if temperatures do not drop too far below freezing.

PESTS AND DISEASES OF FRUIT

The more energy your plants expend recovering from the effects of pests and diseases, the less fruit they will bear. Here are some tips on giving them a helping hand that will bring ample rewards at harvest time.

One of the best defenses against pests and disease is a vigorous plant. Healthy fruit and nut trees and berry plants resist infection and can overcome insect attacks, but sickly, weak plants will succumb to these problems.

Sanitation is another good method of disease and insect prevention. Be sure to remove all remaining fruits, berries, and nuts at the end of the season and clean up the ground below the plants. This material can provide a home for overwintering pests; either burn it or seal it in a bag to be discarded. Your neighbors' trees and shrubs may be the source of some of your trouble, but there is no way to make others keep their plants pest-free. Just be sure to prepare the soil properly and keep the tree, shrub, or vine watered and fed. Beyond that you must accept the fact that you may lose some fruit each season: In most cases there will be plenty left for a good harvest.

When you use chemical control measures, read the label carefully and follow all directions exactly.

Fruit Pests

The pests listed here are among the most common, although you may encounter others. A few are confined to specific regions of the United States.

Aphids   These are soft-bodied insects that damage leaves and fruit by sucking plant sap. A dormant oil spray kills overwintering eggs, and malathion or diazinon contact sprays help control the insects during the growing season if sprayed once a week.

Apple maggots (railroad worms)   This pest is found primarily east of the Rockies, but it is also becoming a problem in the Pacific Northwest and northern California. The apple

Rosy apple aphid

Apple maggot

Birds

maggot has adult flies that lay eggs under the skin of the fruit. When the eggs hatch the larvae tunnel through the flesh. The flies are active from July through harvest. Keep trees clean and remove damaged fruit. Spray with diazinon or carbaryl (Sevin®) products, following label directions. Do not use diazinon within 14 days of harvest. These pests can also be controlled effectively by hanging "sticky balls"—red spheres coated with auto grease or latex, which traps the insects—in your trees.

Birds   The biggest problem with cherries, blueberries, and other small fruits is that birds can remove the entire fruit. When fruit begins to ripen, cover the whole plant with plastic netting, which is available from nurseries and hardware dealers. Throw the net directly over the plant, or build simple wood frames to support netting over dwarfs and bushes. For larger trees, strands of cotton twine will annoy the birds when they try to land and may be sufficient. Throw a ball of twine over the tree repeatedly from different sides. The twine usually rots away over the next winter.

Cherry fruit fly

Flatheaded borer

Cherry (pear) slug

Leaf roller

Grape berry moth on grape

Codling moth

Cherry fruit flies   This fruit fly starts as a white larva that burrows through the cherries, leaving a hole. Spray at seven-day intervals with diazinon insecticide. Do not spray diazinon within 14 days of harvest.

Cherry (pear) slugs   These small, wet-looking green worms are the larvae of a wasp. They skeletonize leaves, leaving lacy patches. When you notice them (probably in June and again in August) spray with malathion insecticide or a contact spray registered for control of the pest. Follow label directions.

Codling moths   The major pest of apples and pears, these moths lay eggs in the blossoms and their larvae tunnel in the fruit leaving holes and droppings (frass). After petals fall spray with diazinon insecticide or malathion and methoxychlor insecticide and continue to spray every two to four weeks as directed.

Flatheaded borers   This western pest is the larva of a beetle. The borer burrows into bark that has been damaged or sunburned. To avoid the pest, avoid the damage. Paint or wrap young trunks or those exposed by heavy

pruning, and be careful not to cut them with tools or machinery. When you find tunnels and droppings, cut away bark and wood and dig out the borers, then paint the wound with tree seal or asphalt emulsion. For apples, cherries, and pears, use a lindane product registered for control of this pest.

Grape berry moths   This insect damages grapes in the northeastern part of the country. Since the insect overwinters on pieces of grape leaves on the ground, it can be controlled by tilling the leaves into the soil. It can also be controlled with malathion insecticide applied just before bloom, just after bloom, and two months later.

Leaf rollers   This moth larva hides in rolled leaves and feeds on both foliage and fruit. Once established it is protected from spray because the spray cannot reach it. Infested leaves must be picked off. Spray with diazinon, carbaryl (Sevin®), or malathion insecticides when pests first appear and continue spray treatments according to the intervals recommended on the label. Leaf rollers can also be eradicated using *Bacillus thuringiensis* (a parasitic bacteria).

Mites Well-watered, vigorous plants are much less susceptible to spider mite infestations than plants subject to drought, dust, and dirt, so keep plants both clean and well-watered. Over-spraying for other insects may trigger a mite attack because spraying kills the mites' enemies. You'll know mites are present if there is a silvery webbing under the leaves, or if the leaves are curled, stippled, or bronzed. Kill overwintering mites with dormant oil spray. During the growing season, use dicofol miticide or other products registered for mite control on fruit trees. In summer oil sprays are safe only for pears.

Oriental fruit moths The larvae of this pest of peaches, plums, nectarines, and apricots burrows into the twigs, which causes the tips to wilt. The most serious damage occurs later in the season when the larvae feed on the fruit. The worms cannot be killed with pesticides alone, but you can control the adult moth by spraying with malathion insecticide and cleaning up infested fruit.

Peach tree borers There are two kinds of peach borer: One bores into twigs and fruit (peach twig and fruit borer); the other, very common borer attacks the trunk at the soil line (trunk borer). To check for the latter, dig soil away from the trunk and check for tunnels and droppings. Kill the worm by pushing a bit of wire down its tunnel. For trunk borer, spray trees with diazinon, carbaryl (Sevin®), or lindane products according to the directions on the label. Check cherry and plum trees for infestations of the same pest.

To control the twig and fruit borer, spray with diazanon insecticide at the following times: in winter, in May, and before harvest (during the safe period indicated on the product label). Trunk borers can also be controlled by placing a ring of mothballs on the ground close to, but not touching the trunk. Mound soil 2 to 3 inches deep over the ring for 6 weeks and then uncover it.

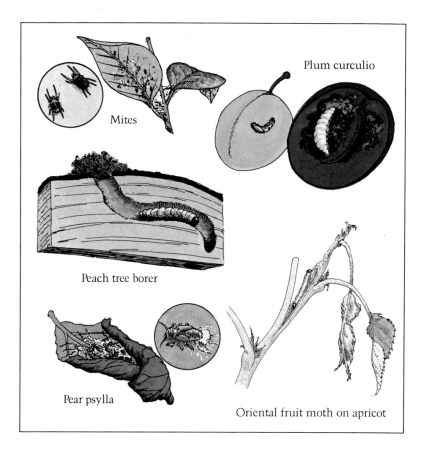

Mites

Plum curculio

Peach tree borer

Pear psylla

Oriental fruit moth on apricot

Trees damaged by lawn mowers, frost cracking, or other borers are more susceptible to peach tree borer.

Pear psyllas These insects are related to aphids. Like aphids, the larvae cluster on leaves and suck plant juices, and they excrete a sticky, sweet honeydew that coats the leaves and fruits. A black, sooty fungus may grow on the honeydew, reducing photosynthesis and weakening the tree. Dormant oil spray, applied just before the buds swell, is an effective treatment against pear psylla.

Plum curculios This pest, which belongs to the beetle family, is a serious problem for apples, peaches, cherries, and other wild and cultivated fruits east of the Rockies. Both the adults and larvae damage the fruit. Look for the crescent-shaped scars on fruit made by the female plum curculio when laying eggs. The pests are active for 3 to 4 weeks starting at petal fall. Spray with a malathion and methoxychlor insecticide or other products labeled for control of this pest.

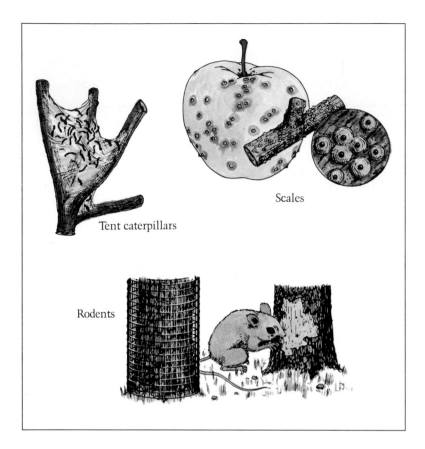

Tent caterpillars

Scales

Rodents

Fruit Diseases

The following are the most common of the many fruit tree diseases. Some are easily controlled with proper sprays (timing is very important); others are best fought by choosing resistant plants, and some require the removal and burning of infected plant parts.

Apple and pear scab   These diseases overwinter on apple and pear leaf debris, so be sure to rake up under trees and destroy the material. The fungus infects foliage and fruit. The disease is severest in wet weather and is less of a problem in dry-summer areas. Crabapples also are affected and need spraying. Choose resistant plants when possible.

To control scab, apply captan, dodine, lime sulfur, or benomyl to apples at regular intervals before and after the tree blooms.

Bacterial leaf spot   Primarily attacking cherries, peaches, and plums, this bacterial disease lives through winter in leaf debris. In the East and South, the growing of Japanese plums is severely limited due to this problem. The infection, which occurs during rainy periods in the spring, causes brown spots that form in the leaves and develop into widening holes. It may attack fruit spurs and cause fruit drop.

This disease cannot be eliminated, but it can be suppressed by spraying with basic copper sulfate at petal fall or by using streptomycin and terramycin antibiotics later when trees are in bloom and again one month later when rains may occur. For peaches another alternative is to plant only resistant varieties such as 'Redhaven', 'Sunhaven', 'Madison', or 'Loring'.

Bacterial canker of cherries   This disease causes long, narrow, damp-looking, gum-edged patches on the trunk or branches. Branches die as they are girdled by these patches. In wet climates avoid 'Bing', 'Lambert', 'Royal Ann', and 'Van' cherries. Resistant varieties are 'Corum' and 'Sam'. The disease can also affect apricots, blueberries, peaches, and prune plums. Do not use susceptible peaches in damp climates.

Rodents   Mice, voles, and rabbits all eat the bark of young trees, especially when the ground is covered with mulch or snow in winter and better food is unavailable. If enough bark is removed, the tree will die after the first growth surge of spring. Protect the lower trunk in winter or the year around with a cylinder of hardware cloth. Check it occasionally during the growing season and loosen or replace it as necessary.

Scales   This pest causes spots to develop on infested fruit. Scales appear in masses when infestations are heavy. Use a delayed dormant oil spray to control mature scales before crawlers (the immature stage) hatch. Crawlers can be controlled with malathion or diazinon sprays applied in May or June.

Tent caterpillars   You probably won't see this pest if you have sprayed early for others. Tent caterpillars build large webs among branches; these webs contain hundreds of hairy caterpillars that emerge to eat leaves. Spray with diazinon or malathion insecticides. Tent caterpillars can also be eradicated by spraying *Bacillus thuringiensis*.

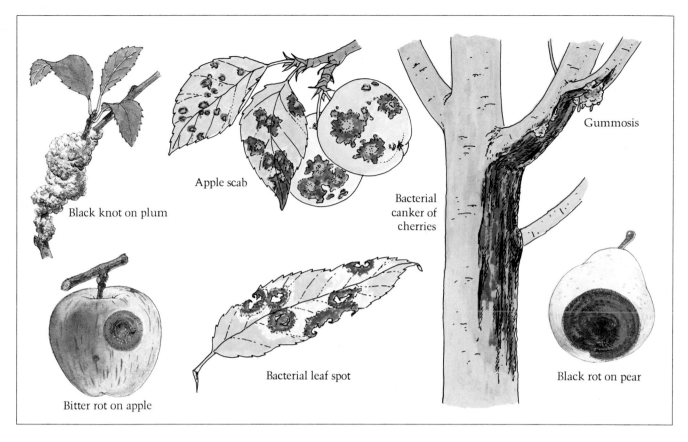

Black knot on plum

Apple scab

Bacterial canker of cherries

Gummosis

Bitter rot on apple

Bacterial leaf spot

Black rot on pear

Bitter rot   This is the most serious fruit rot of apples in the Southeast. Dark round lesions appear on the fruits and irregular brown spots on the leaves. Remove mummified and newly infected fruit. Captan fungicide sprays will control the disease.

Black knot   This disease causes black tar-like swellings on plum branches. The infection can spread from infected wild cherries and plums to cultivated trees, so plant at least 600 feet from any infected trees. Prune out infection. Moderately resistant varieties include 'Methley' and 'Italian Prune'. Resistant varieties are 'Shiro' and 'Santa Rosa'. Some control is possible using a captan or benomyl spray, but spraying alone will not control this disease. Prune out and destroy infected twigs and branches in the fall and winter, making sure to cut at least 4 inches below visible signs of infection.

Black rots   There are two major black rot diseases. The first is black rot on apples and is sometimes called frog-eye leaf spot because it causes brown spots surrounded by purple margins on the leaves. It is one of the three fruit rots important on apples in the Southeast. (The others are white rot and bitter rot.) The rotted area is characterized by concentric, alternating black and brown rings. Control by pruning out diseased branches, removing mummified fruit, and applying captan fungicides.

The other black rot is the most serious disease of grapes in the East. Attacking all young, growing parts of the plant, the disease first appears as soft, round, tan spots on the berries. Eventually the berries shrivel to hard, dry mummies resembling raisins.

Plants that dry off quickly are less likely to get black rot, so choose a site in the sun with good air circulation and prune the plants annually. Captan fungicide sprays control black rot and are applied when new shoots are 6 to 10 inches long, just before bloom, just after bloom, and then every 10 days until berries reach full size. A moderately resistant variety is 'Delaware'. Except in unusually wet seasons, muscadine varieties are not attacked by black rot.

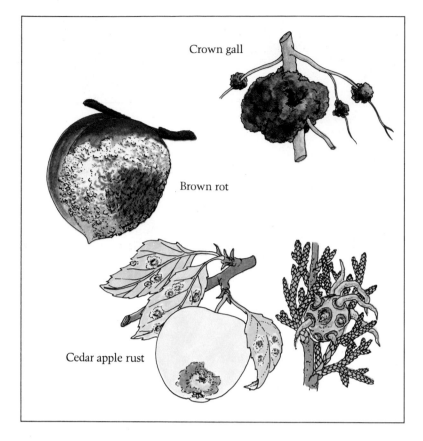

Crown gall

Brown rot

Cedar apple rust

cup-shaped structures; then they turn yellow and fall. The galls form on the juniper or red cedar plants. Remove junipers and red cedars, or avoid planting them. If you have ornamental cedars, remove the galls in summer. Galls are brownish and globe shaped and look like part of the tree. Spray the apples with a product registered specifically for control of cedar apple rust (ferbam or zineb products), following directions carefully. The fungicides in multipurpose sprays usually do not control rust. Select resistant cultivars where cedar apple rust is a big problem.

Crown gall　Of bacterial origin, this disease occurs in many soils. It attacks young trees, producing soft, corky galls, or swellings, on the crown and roots. The galls often grow until they girdle and stunt the tree.

Avoid buying young trees that show galls, and plant young trees carefully to avoid injury, as injury allows bacteria to enter the plant. Older trees with galls can be treated with a product called Gallex®.

Recent experimental work has indicated that young plants can be successfully inoculated against the disease.

Crown rot　Almost any plant growing in a soil that is frequently or constantly wet may be subject to this serious disease. Crown rot is caused by a fungus. Infected branches redden and foliage yellows or discolors. Look at the bark below the soil line to see whether it is dead; if it is, scrape the bark away and pull back the soil so that air can reach the infection.

Avoid crown rot on established trees by planting high and watering away from the trunk. The soil at the crown should dry as quickly as possible.

Brown rot　Serious on all stone fruits, but especially on apricots, peaches, and nectarines, this disease, in some regions, makes it nearly impossible for fruit to grow to the edible stage. The disease causes blossoms to brown, turn wet looking, and drop. Brown rot eventually causes fruit rot on the tree. Infected fruit must be removed by hand to prevent reinfection.

To control the blossom-blight phase, spray as the first pink petals show but before the flowers open, using a fungicide such as captan, benomyl, dodine, ferbam, or ziram.

To control attacks on fruit, spray as the fruit begins to ripen (green fruit is rarely attacked) and repeat if there is a period of wet weather. On peaches and nectarines, the disease may attack twigs and overwinter on them. Pruning out dead twigs and removing overwintering fruit "mummies" controls the disease in the following year.

Cedar apple rust　This disease appears only where the alternate host, certain species of juniper or red cedar, grows near apples. The apple leaves first show orange spots and odd

Downy mildew　Grapes in the Northeast may be attacked by this disease when the weather is cool and moist, first in June and then again toward late summer. A white cottony growth appears, usually on the oldest leaves, but it can also attack young shoots, tendrils, or berries. Control with captan fungicide, and clean up dead leaves where the fungus overwinters.

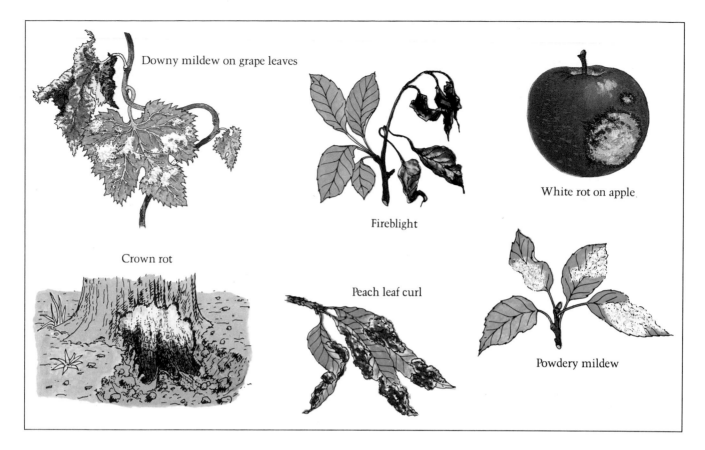

Downy mildew on grape leaves

Fireblight

White rot on apple

Crown rot

Peach leaf curl

Powdery mildew

Fireblight   This disease is spread by insects during the bloom period and shows in spring as new growth wilts, turns dark, and finally blackens as if burned.

Only severe pruning and burning of infected wood has proven effective for the home gardener. Choose resistant pears such as 'Lincoln' or 'Magness'. Where the disease is severe, plant resistant apple cultivars such as 'Cortland' or 'Delicious'.

To control infection, cut off any blighted branches several inches below the infection as soon as you notice an attack. Sterilize pruning tools between each cut by dipping them in a 10-percent bleach solution or full-strength rubbing alchohol.

Gummosis   Deposits of gum on branches of stone fruits (peaches, nectarines, apricots, plums, and cherries) are fairly common and may be caused by mechanical damage, insect damage, or a number of diseases. Several serious bacterial diseases with this symptom almost rule out the planting of certain fruit varieties in some areas. There is no spray treatment. A disease that often causes gummosis is bacterial canker of cherries, described above.

Peach leaf curl   This disease first shows up as a reddening of leaves; leaves then turn pale green or yellow, curl, blister, and may have a powdery look; and finally they fall. A second crop of leaves may grow, which may or may not be affected. No spraying is effective once the disease has appeared.

To control curl you should spray immediately after leaf drop, or just before the buds break, with a fixed copper spray or lime sulfur, wetting every twig and branch completely.

Powdery mildew   This is a fungus that causes a grayish, powdery coating to form over young shoots, leaves, and flower buds, often deforming or killing them. It thrives in still, shady spots.

When an infection begins, clip off severely mildewed twigs and spray with benomyl. A cold winter will eliminate much of the fungus.

White rot   This is an important fruit rot of apples in the Southeast. Papery orange cankers form on branches. Prune out infected branches and spray with captan fungicide.

Spray Schedules

To get the kind of unblemished fruit you are used to buying in grocery stores, it is usually necessary to spray fruit trees according to a regular schedule. This chart shows the approximate timing for controls for some of the major problems of fruit trees. See the text for recommended sprays for each problem.

Recommended Spray Schedules

Timing	Apple	Pear
Dormant (Late Winter)	Dormant oil*	Dormant oil* Fireblight
Swollen Bud (all buds closed—no green showing)	Scab Powdery mildew Fireblight	Fireblight
Prebloom (petal color just beginning to show)	Scab Cedar apple rust Black rot Powdery mildew	Scab Leaf roller
Bloom	Fireblight Powdery mildew Scab	Scab—spray at 10-day intervals until rains stop Fireblight
Petal fall	Scab Cedar apple rust Plum curculio Codling moth Fruit rots Aphid Leaf roller	Codling moth Scab Scale Aphid Fireblight Plum curculio Leaf roller
7 days after petal fall	Scab Rust Plum curculio Codling moth Fruit rots Aphid	Codling moth Mites Scab Scale Fireblight Fruit rots
14 days after previous spray	Scab Plum curculio Codling moth Fruit rots Aphid	Codling moth Mites Scab Fireblight Fruit rots
14 days after previous spray	Aphids Plum curculio Codling moth Fruit rots Apple maggot	Pear psylla Codling moth Mites Fireblight Fruit rots
Thereafter: every 14 days until 2 to 4 weeks before harvest, according to label directions of material used.	Scab Plum curcuiio Codling moth Fruit rots Apple maggots	Pear psylla Codling moth Mites Fruit rots
After leaf fall		

*Do not spray if temperatures are expected to drop below 32° F within 24 hours.

Not all of these problems will occur where you live, and they will probably not take as many sprays as shown. For example, apple maggots don't occur in much of the country, and fireblight can be a problem all summer in humid areas, but is only a spring disease in dry-summer regions.

Apricot, Peach, Nectarine	Plum	Cherry
Dormant oil* Peach leaf curl	Dormant oil* Peach twig borer	Dormant oil* Gummosis
Peach twig borer	Black knot	
Brown rot Leaf roller Bud moth	Brown rot Leaf roller Black knot Bud moth	Brown rot Leaf roller Bud moth Aphid
Brown rot Powdery mildew	Brown rot Powdery mildew Black knot	Brown rot Powdery mildew
Brown rot Plum curculio Leaf roller Scales Oriental fruit moth Peach twig borer Powdery mildew	Brown rot Plum curculio Leaf roller Black knot	Brown rot Plum curculio Leaf roller Aphid Scales
Brown rot Plum curculio Mites Oriental fruit moth Peach twig borer Scales	Brown rot Plum curculio Scales Peach twig borer Mites Codling moth Aphid Black knot	Brown rot Plum curculio
Brown rot Plum curculio Oriental fruit moth Mites Peach twig borer	Brown rot Plum curculio Codling moth Mites Black knot	Brown rot Plum curculio Powdery mildew
Plum curculio Oriental fruit moth Mites	Plum curculio Mites Black knot	Powdery mildew Cherry fruit fly
Plum curculio Mites Peach tree borer	Plum curculio Mites Black knot	Gummosis Mites Powdery mildew
Peach leaf curl		

Pruning and Training

Pruning encourages healthy growth and larger fruit; training conserves space and makes harvesting easier. With just a bit of practice at both techniques, you'll feel like a pro.

Pruning and training are two operations of extreme importance to the gardener who grows fruit. Pruning is simply cutting out branches. Most training involves a lot of careful pruning and a little bit of actual training, that is, tying or propping branches to create desired shapes. First we give you instructions for pruning fruit plants correctly, and then we show you how these techniques apply to training plants into functional shapes. Once you have mastered the basics, you can experiment with techniques for training into more formal shapes.

Is it really necessary for the home gardener to learn how to prune and train? Most of us have seen long-neglected apple or pear trees or unpruned tangles of blackberry vines that still bear delicious fruit. This suggests that these practices are not really essential and that we can get by without them.

Plants will live, grow, and bear fruit without ever being pruned, but experience has shown that good pruning and some training can prevent or remedy many of the problems that arise in most plants. Pruning is probably best viewed as the most effective means to head off trouble, improve your plants' performance, and keep them in excellent condition. Also, training is the area where the craft of raising fruit trees can be raised to the level of an art. Trees can be trained to be beautiful, functional, or even whimsical.

This apple tree has been espaliered against a south-facing wall for protection. Espaliering is the most highly developed form of training.

PRUNING PROCEDURES

In pruning, you remove part of a plant to benefit the whole. When you cut away any part of a plant, it directly affects the plant's growth. Depending on how and when it is done, pruning can be used to achieve the following results:

- ☐ Produce new growth where it is desired
- ☐ Help control growth
- ☐ Shape a young plant
- ☐ Correct or repair damage
- ☐ Help control insects and diseases
- ☐ Rejuvenate or reshape an older plant
- ☐ Bring about earlier blooming
- ☐ Increase the production, size, and quality of fruit

These advantages make pruning well worth undertaking, even if you are inexperienced and, like many, timid about cutting your plants.

Good pruning requires knowledge, foresight, and care. As a rule of thumb, never make a cut without a clear idea of its probable effect on the plant. On the other hand, don't be so fearful of cutting that you can't get the job done. If you keep in mind that proper pruning is beneficial to plants and then proceed carefully, you'll get good results.

When to Prune

In freezing climates, pruning should be done in spring just as the buds begin to swell. Avoid pruning early in the dormant season, November or December, in climates with severe weather. Freezing injury can occur to plants if they are pruned in fall or early winter.

In more moderate climates you can prune anytime during the dormant period. That is, the period between leaf fall and the beginning of bud swell in the spring.

Getting Started

No job can be done well without the right tools. To start you'll need a good pair of pruning shears, and if you plan to be making cuts larger than shears can handle, you'll need more tools such as those shown in the illustration below. It's worth noting that the best-quality pruning tools are generally more expensive, but are worth the extra cost in the long run, because they will last much longer than inexpensive tools.

Pruning Tools

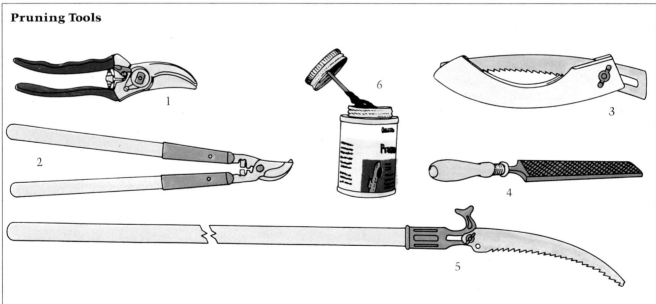

A basic set of pruning tools includes:

1. Hand pruners, for stems up to ¾" in diameter. There are two main types: scissor-style pruners have sharpened blades that overlap in making the cut, and anvil-style pruners, which have a sharpened top blade that snaps onto a flat plate of softer metal.

2. Loppers, for stems up to 1¼". Heavy duty loppers are available for cutting through wood 1¾" thick.

3. Hand pruning saw, for larger stems. Cuts on the pull stroke. Works best on green wood.

4. Wood rasp, for smoothing rough edges of large cuts.

5. Pole saw cuts high branches without climbing. The curved blade operates like the narrow curved pruning saw.

6. Pruning compound seals large cuts.

If you find yourself in a quandary about where to begin, remember that you can't hurt a plant by cutting out dead, diseased, or damaged wood or wood that crosses and rubs against other wood (which can cause wounds that become susceptible to infection). On the contrary, you'll be doing a great deal of good. Eliminating these problems is the place to start for inexperienced and experienced pruners alike. It will provide a clearer view of the tree and of the remaining work to be done and will open up the tree to more air and light.

There are two basic types of pruning cuts: heading and thinning cuts. Thinning removes wood and puts an end to growth. All thinning cuts are to the base of the branch or sucker so that there are no buds left to sprout new growth. Heading is the process of shortening a branch, not removing it entirely. Buds on the remaining portion of wood are stimulated to grow when growth above them is removed.

Fruit trees vary in their needs for pruning. Some need a good deal of pruning every year, some need only a little in an entire lifetime, and some never need pruning unless injured. No other plants in the garden depend on pruning as much as fruit trees. Unfortunately no trees vary as widely in the most effective means of pruning. If you want to grow beautiful fruit, you must learn the difference between pruning an apple and a peach.

The objective in pruning a fruit tree is to produce an abundance of good-quality fruit throughout the branches, including the lower and interior ones. An unpruned peach or nectarine tree will bear mostly on branch tips, and the leaves that produce the sugars that nourish the plant and accumulate in the fruit will grow only where they receive enough light for photosynthesis.

An unpruned tree with heavy growth in its interior restricting light and air circulation is more susceptible to fungus diseases. In addition heavy loads of fruit at branch tips can make harvest difficult and cause splitting and breaking in branches, wounds that invite pests and disease. To head off these problems and produce a more even distribution of leaves and fruit, you need to thin out the tree

Pruning Basics

A tree's response to a pruning cut depends on where on the branch the cut is made. Both types of cuts are used in pruning fruit trees and grapes.

Heading Cuts

Make a slanting cut just beyond a bud.

Several buds left on the cut branch grow, making denser, more compact foliage on more branches.

Thinning Cuts

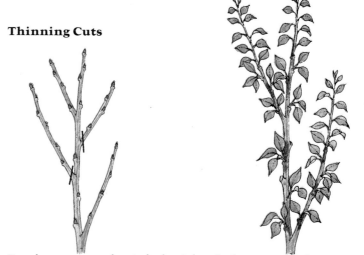

Branches are removed entirely, leaving no buds to grow. Their energy is diverted into remaining branches, which grow more vigorously.

by removing selected branches. The increased light and air circulation achieved by this pruning will help prevent fungus diseases, and the open tree will be much more accessible to thorough coverage with any necessary sprays. The tree will also bear more uniformly and will be easier to harvest. Clearly, then, one of the main purposes in fruit tree pruning is to allow light and air into interior leaves and fruit.

The illustrations on the next page will help you recognize the various parts of a plant that you need to know when you get to work.

Anatomy of a Fruit Tree

Crotch: The angle where branches fork, or where a main limb joins the trunk. Strong crotches are wide angled—45 degrees or more; weak crotches are narrow.

Scaffold: The main limbs branching from the trunk.

Watersprout: A very vigorous shoot from a dormant bud on an old branch. Remove by cutting at the base.

Sucker: A vigorous shoot from the roots or from below the bud union. Cut off at the base.

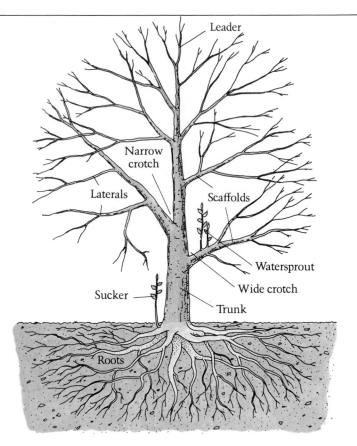

Choosing the Right Bud

Prune to the lateral bud that will produce the branch you want. An outside bud will usually produce an outside branch. The placement of that bud on the stem points the direction of the new branch.

Parts of the Branch

Terminal Bud: The fat bud at a branch tip will always grow first and fastest if you leave it. Cut it, and several buds will grow behind it.

Leaf Bud: Flattish triangle on the side of a branch. To make one grow, cut just above it. Choose buds pointing outward from the trunk so the growing branch will have space and light.

Flower Bud: Plump compared to leaf buds and first to swell in spring. On stone fruits they grow alone or beside leaf buds. On apples and pears they grow with a few leaves.

Spurs: Twiglets on apples, pears, plums, and apricots. They grow on older branches, produce fat flower buds, then fruit. Don't remove them.

Bud Scar: A ring on a branch that marks the point where the terminal bud began growing after the dormant season. The line marks the origin of this year's growth.

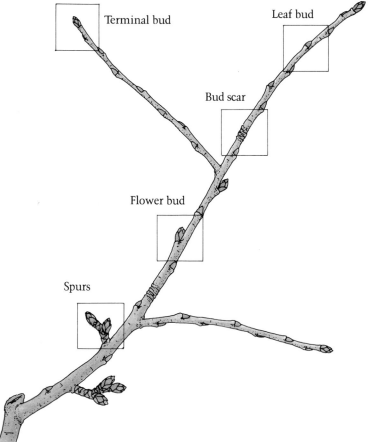

Making a Cut

When you cut away part of a plant, a wound is left, susceptible to pests and diseases. To avoid trouble always try to make wounds as small as possible.

The smallest possible wound is made by removing a bud or twig. If a new sprout is growing in toward the center of the tree or toward the trunk or threatening to tangle with another branch when it grows longer, pinch it off now to save pruning later. If you see the bud of a sucker down near the soil, rub it off with your thumb.

Always make cuts close to a node. Branches grow only at these nodes, and if you leave too long a stub beyond the node the stub will die and rot. Be sure to cut at a slight angle so that there is no straight "shoulder" left to attract disease or burrowing pests.

Summer Pruning

If a tree grows too vigorously while it is young, fruiting will be delayed several years. Also, overly-vigorous trees produce many *water-sprouts*—vertical shoots that spoil the shape of the tree.

Overly vigorous growth can be controlled by summer pruning, which weakens a plant by removing leaves that manufacture nourishment. Where space is limited this pruning is the main means of confining trees, but too much summer pruning can damage a tree. Experience will teach you how much pruning is necessary, but here are some guidelines.

In early summer remove only water-sprouts. These may suddenly shoot out much farther than any other growth. Cut them off at the base. Also remove any suckers from below the bud union, cutting to the base.

When the new growth matures and slows its pace, begin snipping it back. The season will vary depending on weather, feeding, and watering, but you can begin to prune some branches in July and finish up by early September. Cut off all but about four leaves of the current season's growth on each new branch. Don't thin out branches. You can do that during winter if necessary.

Thinning Fruit

Thinning fruit is a form of pruning. It is almost always necessary to thin if you want a large, sweet, top-quality fruit, a return crop, and less limb breakage. Without thinning nature will produce the most fruit in order to get the largest number of seeds to perpetuate the species. The tree has only so much food and energy, and you have a choice of one large fruit or two small ones. To see how much difference thinning makes, leave a branch unthinned and compare its fruit at harvest time to that of a properly thinned branch.

A good rule of thumb is to thin to distances of twice the expected diameter of the fruit. For example, if you have an apple tree that should produce 3-inch fruits, leave 6 inches between apples after thinning.

Although each variety has a best time for thinning, generally it is best to thin before the fruit has gone through half its growing season. Usually there is a natural fruit drop about 3 to 6 weeks after bloom, sometime in early June. You can do a light thinning before this time but wait until after the fruit drop to do a final thinning. If you lose a lot of fruit you may not want to thin at all. In this case, do not worry if fruits are bunched together, rather than evenly spaced according to the above rule. It is the overall fruit load, not the spacing, that is most critical.

More information on thinning specific fruits can be found under individual fruit entries in the "Encyclopedia of Fruits, Berries & Nuts," beginning on page 59.

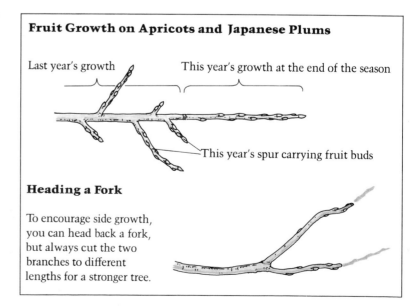

Fruit Growth on Apricots and Japanese Plums

Last year's growth   This year's growth at the end of the season

This year's spur carrying fruit buds

Heading a Fork

To encourage side growth, you can head back a fork, but always cut the two branches to different lengths for a stronger tree.

Three Methods of Training

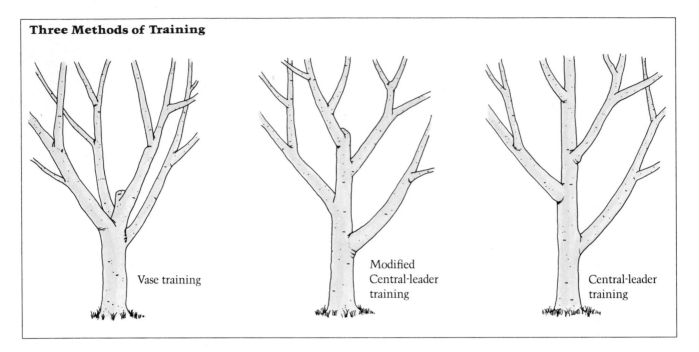

Vase training

Modified
Central-leader
training

Central-leader
training

TRAINING METHODS

Commercial fruit growers train fruit trees to one of three forms, and each has its own advantages. Most growers prune their trees into a vase shape; the central leader form is sometimes used; and the modified central leader combines both shapes. Remember that dwarf trees will require less severe pruning because they are smaller.

Productive fruit trees should be trained to one of these forms (vase, central leader, or modified central leader) starting when the trees are planted. Such pruning and training will keep the tree balanced in form and—more importantly—balanced in new and young wood. Left unpruned the tree will become dense with weak, twiggy growth and overloaded with small, less healthy fruit.

Once fruit trees begin to produce a crop each type should be pruned differently. Pruning methods depend on the part of the tree that bears fruit and this differs from one fruit type to another. All bearing fruit trees can be pruned annually, with additional light pruning in the summer to expose fruit spurs.

Vase Training

In this method the tree is shaped to a short trunk of about 3 feet with three or four main limbs, each of which has fully filled-out secondary branches. This shape creates an open center allowing light to reach all branches.

Vase pruning is the training method most often used. It is always used with apricots, plums, and peaches and often used with pears and apples.

Central-Leader Training

In this technique the tree is shaped to one tall trunk that extends upward through the tree, with tiers of branches along the trunk like a Christmas tree. Since branches at the top are shorter than those lower down, light penetrates throughout the tree and fruit is produced from top to bottom. This shape is very strong, but makes it difficult to reach the highest branches.

The smallest dwarf apples are pruned in this shape. Because the trees are so small, shade and height are not problems. Large apples are occasionally trained this way, but this is not suggested for home gardens.

Modified Central-Leader Training

This method produces both the strength of a central trunk and the sun-filled center of a vase-shaped tree. A single trunk is allowed to grow vertically with whorls of branches in the same manner as the central-leader form. At the end of the third or fourth season, when the tree is from 6 to 10 feet tall, the main

Training for a Vase Shape

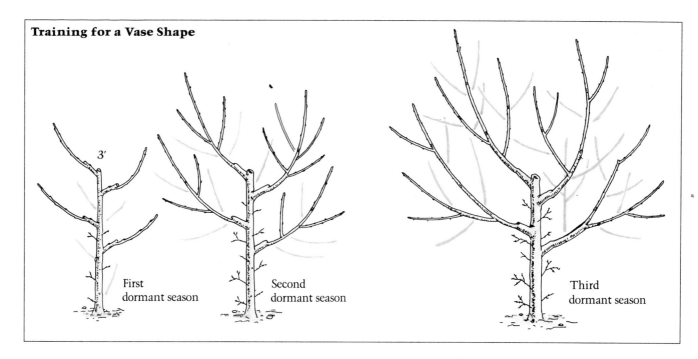

First dormant season

Second dormant season

Third dormant season

leader is cut off at about 3 or 4 feet. Main scaffold branches are then selected and subsequently pruned to form a vase shape.

HOW TO PRUNE FOR A VASE SHAPE

The vase shape is probably the most common shape for a fruit tree. The general requirements for establishing a vase shape are given here, but be sure to consult the specific procedures in the following section for each type of fruit tree.

When you plant a young bare-root tree, it will normally consist of a thin vertical shoot, called a *whip*, and some twiggy side branches. To start the tree on a good course, cut the whip back to about 2 feet above the ground for a dwarf, 2½ feet for a standard tree. Cut just above a bud and then prune any side branches back to two buds.

First Dormant Season

After the new tree has grown through the spring, summer, and fall into its first winter dormancy, choose three or four branches with wide crotches, as shown in the illustration above. Examining the tree from above, look for branches that radiate evenly around the

trunk with almost equal distance between them. You should also try to have at least 6 inches vertical distance between branches, with the lowest branch about 15 inches above the ground. If there are three such branches, cut off the vertical stem just above the top one. If there are fewer than three good branches, leave the vertical stem and choose the remaining scaffold branches during the next dormant season.

Second Dormant Season

If necessary choose the remaining scaffold branches and cut off the vertical stem just above the highest selected scaffold branch. The scaffold branches you chose during the first dormant season will have grown side branches. Remove the weakest of these, leaving the main stem and laterals on each branch. Do not prune twiggy growth.

Third Dormant Season

Now is the time to thin surplus shoots and branches. Select the strongest and best-placed terminal shoot near the tip of each scaffold branch, as well as one or two other side shoots on each branch. Remove all other shoots on the branch. Leave the short weak shoots that grow straight from the trunk to shade it and help produce food for the tree.

Pruning Apples, Pears, and European Plums

Trim lightly to remove tangled branches or damaged wood. Cut dangling limbs or vertical watersprouts at the base. Head back branch tips to maintain size of older trees. Leave twiggy spurs for fruit production.

Bearing spur

Young spur

SPECIFIC TRAINING PROCEDURES

The following are training methods for each kind of fruit and nut tree and berry bush listed in the "Encyclopedia of Fruits, Berries & Nuts" beginning on page 59. Most of the training involves careful pruning procedures.

Training Almond Trees

Like peaches, almonds are borne on one-year-old shoots, but the wood continues to bear up to 5 years. Pruning is similar for both trees although almond trees are pruned less severely.

Train young almond trees to a vase shape as described on page 45. Your ultimate goal is to have a wide tree with an open top. Keep pruning the branches that are growing upward back to laterals that are growing outward. This creates a wider branching habit.

Training Apple Trees

Apples bear on long-lived spurs. The fruit forms at the tip of last year's spur growth, and the spur itself then grows a bit more, off to the side of the fruit. Each spur bears for 10 years or more, so don't tear it off when you pick.

Any training method in this book will suit any apple, but vase pruning has been the method commonly used by orchardists to train standard apple trees. After you have picked your scaffold branches, as described earlier, cut them back one-third to encourage a strong branch system near the trunk.

In the second and third dormant season, reduce the length of all new growth by one-third, and thin out to create a strong, evenly spaced framework of branches. These secondary scaffold branches are the ones that will develop fruit spurs on their lateral branches. The pruning during this period should always be to a bud on the top of a branch that points outward. This will develop the vase shape.

With semidwarf and dwarf apple trees modified central-leader or vase pruning can be used, but the central-leader system makes these trees stronger and earlier bearing. When planting the bare-root dwarf, cut back all branches, including the top, about one-quarter or about 5 to 10 inches. Make cuts to a strong outside bud.

For the second and third year, repeat the process to train the central leader up and the scaffold branches out parallel to the ground. Most dwarfs will begin to bear the second and third years and will bear heavily thereafter. It is important to maintain the single, upright central leader throughout the life of the dwarf tree. Be sure to remove any fruit that forms on this leader because fruit formation will stunt the leader and another branch may become dominant.

Training Apricot Trees

Apricots appear on the previous season's shoots and on short-lived spurs on older wood. Pruning is essential to apricot production for several reasons: It stimulates a certain amount of new growth for next year's crop; it keeps the tree open; and it prevents fruit from being borne only high in the tree.

Pruning Apricots and Japanese Plums

Most fruit is formed on short spurs growing on two- or three-year-old wood. Head back new whips by a half. The half you leave will form fruit spurs the following summer and produce a good crop of fruit the year after that. Trim away tangles.

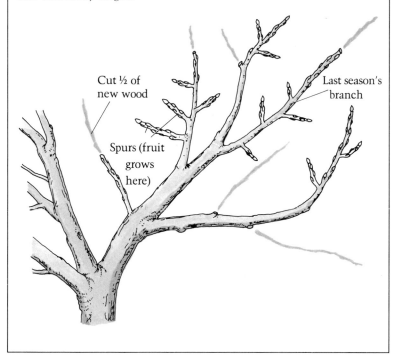

Cut ½ of new wood

Last season's branch

Spurs (fruit grows here)

Pruning a Young Sweet Cherry

Cut (head) back polelike growth to an outside bud to encourage more open branching habit.

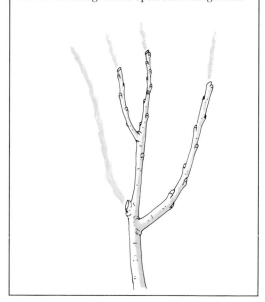

Like plums, apricots bear on spurs that produce for two to four years and then need to be pruned out and replaced with younger wood. In pruning apricots you need to head back long new whips by one-half and remove the oldest fruiting wood. See the sketch on this page. Fruit may form in the second year, but don't expect a heavy crop until the third or fourth year.

Training Chestnut Trees

Chestnuts bear on new wood. Pruning is not necessary for good fruit set.

Train young trees to a central leader (see page 44). Once that framework is established, trees need only occasional pruning to remove dead, weak, or poorly placed branches.

Training Cherry Trees

All cherries bear on long-lived spurs. Spurs on tree cherries begin to bear along two-year-old branches and can produce for 10 years and more. Count on the first crops in the third or fourth year after planting. Bush cherries may bear sooner.

As the sweet cherry grows, it should be pruned to the central-leader system described earlier. Make sure that the leader or upper scaffold branches are not crowded and choked by lower scaffold branches that grow upward. After the tree begins to bear, prune out weak branches, those that develop at odd angles, and those that cross other branches. Be sure to head back the leader and upright side branches to no more than 12 to 15 feet so that the mature tree can be kept at about 20 feet.

Sour cherry trees differ from the sweet cherries in that they tend to spread wider and are considerably smaller. In fact, some varieties resemble large bushes. The sour cherry can be pruned in a central-leader shape, or—if you prefer to keep the tree smaller—prune it to a vase shape (see page 45). It is quite easy to keep the sour cherry under 12 feet with either system.

Cherries need no thinning and little pruning after the first two seasons of growth. Sweet cherries may need heading back in the first years of growth to encourage branching.

Pruning Peaches and Nectarines

All fruit grows on wood that grew during the previous season. Long branches produce more fruit than short ones. Trim away about half the new growth, removing shorter pieces completely, heading back long shoots by about a quarter to a third. The heavy pruning encourages abundant summer growth for a good crop the following year. Cut out tangled center growth for more light penetration.

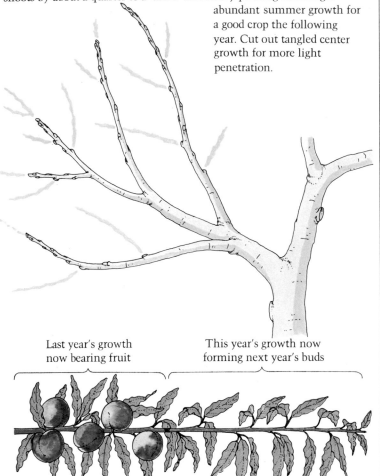

Last year's growth now bearing fruit

This year's growth now forming next year's buds

Training Crabapples

Crabapples fruit on long-lived spurs, generally producing clusters of several fruits on each spur. Since crops are heavy, you can cut back new wood without losing anything.

Train the young trees to a vase shape with three or four scaffolds (see page 45). After the second year you can leave these scaffolds alone or cut them back to maintain size.

Training Fig Trees

Because fig trees bear fruit on new and one-year-old wood, pruning is necessary to stimulate new growth each year for next year's crop. Prune figs to suit the growing situation. Different varieties grow in different ways. 'Adriatic' grows like a spreading shrub. Do not head this tree back, for it will never grow tall or wide. Select scaffold branches at the first dormant period and prune to keep future branches off the ground. Each year remove low branches that touch the ground or interfere with picking.

The 'Kadota' fig is a vigorous grower that should be kept low and spreading. Head new growth short in the middle of a tree and longer on the outside. When a tree reaches its mature shape, head the new growth back 1 or 2 feet each year.

Figs have two crops per year. The first crop blooms on new wood of the previous season, and the second crop appears on new wood of the current season. When a tree is cut back to confine it, you usually lose most of the first crop. You also lose it in cold regions where winter does your pruning. Prune the young tree to an open shape in the first two years and remove any suckers at the base. Pull—don't cut—the suckers. Shrub forms need no attention except for removing dead wood.

Training Filbert Trees

Filberts form on one-year-old wood, so as plantings age, remove older, played out stems. Head back remaining branches to encourage new growth as you would for peaches only not as heavily; nuts are small and a large bearing surface is necessary for a good crop. Filbert trees usually start to bear at four years.

If you want a single-trunked filbert, train to the central-leader system as shown on page 45. Remove all suckers as they appear. To train as a shrub, let some suckers develop.

Training Peach Trees

Peaches fruit on one-year-old wood but unlike almonds, once a peach is harvested the section of branch on which it grew will never fruit again. Encourage new growth for replacement branches by pruning heavily every winter.

Peaches are twiggy trees, but the greatest number of flower buds form on sturdy new branches that made more than a foot of new growth the previous summer. Keep these and thin the more anemic twigs. You can head the strong new branches back by one-third to one-half if you want to keep the tree small. The tree will bloom on the remaining branches.

Most newly planted young peach trees are pruned to the vase shape. See page 45. Young peach trees should be pruned moderately. Your ultimate goal is to have a wide tree with an open top 12 or 13 feet high. As the tree grows, cut the branches that are growing upward back to laterals that are growing outward. This creates a wider branching habit.

When the tree is maturing (about 6 to 8 feet tall in the West and Midwest and 10 to 12 feet tall in the East and South), start severely cutting back the new growth on the top of the tree, being sure to maintain the open center that will admit light to the lower inside parts of the tree. In general pruning should be lighter on young bearing trees than on older peach trees.

Training Pear Trees

Pears bear on long-lived spurs, much as apples do. These spurs last a long time if you're careful not to damage them when picking fruit.

You should train the young pear tree to the central-leader system by selecting five or six scaffold branches over a two-year period. Since it is characteristic of pear varieties to grow upright, be careful not to have too many heading-back cuts for they will promote too many upright shoots. If you want a small pear tree, buy a dwarf; don't try to make a standard tree smaller by heavy pruning.

The pear is very susceptible to fireblight, especially in the soft succulent growth that results from heavy pruning, so be careful about heading back or thinning shoots on mature trees. Once fireblight takes hold, there is very little that can be done except to remove infected growth.

A pear is as trainable as an apple, and a trained tree can last 75 years. Plant pears as espaliers, train them to 45-degree angles for an informal hedge, or try them in tubs as a single cordon or on a trellis.

Training Pecan Trees

Pecans bear on new wood, and little pruning is needed other than topping to limit tree size.

You can train young pecan trees to a central-leader framework (see page 44), heading back any overvigorous laterals that would divert energy from upward growth. Trees will start to bear five to eight years after planting.

Training Persimmons

Fruit is borne on new wood. On a naturally shaped tree it will set on the outer portion. Thinning fruit is unnecessary but will help keep lawn trees neater.

Persimmons are often allowed to grow naturally, forming globe-shaped trees up to 25 or 30 feet high. They can be pruned back in spring to keep them smaller. Little pruning is necessary, however. Train the young tree to three widely spaced scaffolds and leave it alone thereafter, or control it by cutting each year to strong lateral branches, removing as much growth as necessary to maintain the size you want.

Persimmons can be trained as espaliers or into hedges. In pruning an espalier, cut off enough of the previous year's growth to expose the most interesting lines of the plant.

Training Plum Trees

Plums fruit on spurs on older branches with the heaviest production on wood that is from two to four years old.

There are many plum varieties, but they fall into two main groups: European and Japanese. European plums need only occasional thinning and heading once the tree shape has been formed. Japanese plums overgrow and overbear. They are particularly prone to branch splitting when mature and bearing heavy crops. Cut back the long whips as discussed under "Training Apricot Trees," and thin fruit when it reaches thumbnail size, leaving about 4 to 6 inches between the remaining fruits.

Remove one-third of the new wood on Japanese plums each year by thinning and heading. This heavy pruning is necessary to produce larger fruit. Keep long, thin branches headed to give the tree a stubby, wide shape. When the fruit spurs on a branch have borne for five to six years, select a new branch from lateral shoots on this branch. The next year, remove most of the old branch, cutting it off just above the selected lateral.

Pruning Rigid and Trailing Blackberries and Raspberries

Last year's growth is blooming and bearing fruit as new shoots emerge from the crown (those shown near the ground). Remove all but five of the new shoots. Let them continue to grow on the ground.

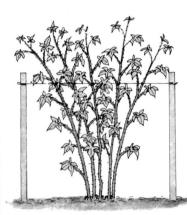

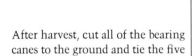

Head back the new canes at a point a few inches above the wire to encourage lateral growth along the wire.

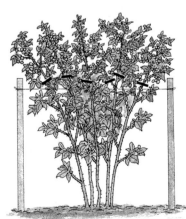

In winter, cut the laterals back to 18″. They will bear the next summer and continue the cycle.

After harvest, cut all of the bearing canes to the ground and tie the five new canes to the wire.

After the fall harvest of everbearing red and yellow raspberries, cut back the portion that fruited.

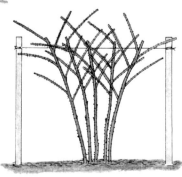

Training Pomegranate Trees

Pomegranate blossoms and fruit form on the current year's growth, so you can prune them back heavily without loss of flowers or fruit.

Although they do not require pruning for good fruit set, you can prune pomegranates to fountain-shaped shrubs or as single- or multiple-trunk trees.

Training Quince Trees

Quince fruits on new wood and because the large, heavy fruits are borne singly at branch ends, it is wise to head back any overlong branches to prevent limbs from breaking or sprawling under the weight of the crop.

To grow as a small tree, train to a vase shape as described on page 45. Otherwise let multiple stems grow from ground level and select the strongest, best-placed ones for framework. Roots are shallow and will sucker if damaged by cultivation. Little pruning is needed in subsequent years other than to maintain a good shape.

Training Walnut Trees

Like pecans, walnuts bear on new wood and need a minimum of annual pruning after the initial shape is established. Some heavy-bearing varieties require pruning to thin them out.

Walnuts are usually trained to a modified central-leader with five or six main laterals.

Training Cane Berries

All cane berries produce fruit on canes sprouted the previous year.

Cane berries have a variety of growing habits. They have either erect canes or trailing canes that tend to creep, and they either grow in clumps or have runners that cause them to spread. Blackberries can have erect or trailing canes, depending on the variety, but all red and yellow raspberries have erect canes and all black and purple raspberries have trailing canes.

Blackberries, black raspberries, and purple raspberries   All blackberries and black and purple raspberries have the clumping habit that allows them to be planted in hills (slightly raised groupings). These berries fruit on twiggy side branches, called laterals, that grow on canes of the previous season.

Training Rigid Berries on a Double Wire

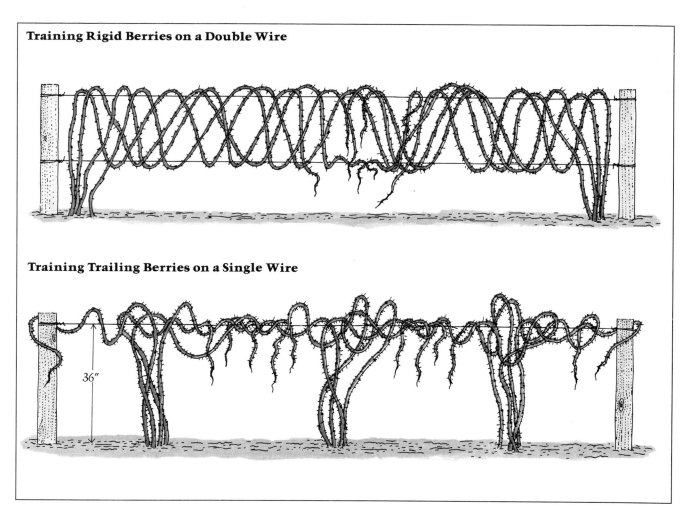

Training Trailing Berries on a Single Wire

36"

The canes fruit only once and must be removed entirely each year. Cut bearing canes to the ground after you have harvested the crop (or during the following winter). If you are growing berries in hills, leave about five of the best new canes per hill and cut out the rest. Leave about 6 inches between plants if they are growing in rows. You can paint the bases of the canes white to enable you to distinguish them later from new summer growth.

In winter cut all canes back to about 45 inches in height. Cut laterals on blackberries to 15 to 18 inches, and laterals on black and purple raspberries to 10 to 12 inches.

In summer, as the new canes grow, gather them in bunches, tie them very loosely, and lay them along the ground until it's time for training. This will keep them out of the way of the year-old canes that are bearing. Pinch back the *new* young blackberry canes when they reach 36 inches; pinch the new black and purple raspberries at 24 to 30 inches. Pinched canes will send out lateral growth that should

be cut back in winter as described above. If you wish tie erect canes to a wire 15 inches above ground.

Berries that trail can be trained to a wire stretched 36 inches above the row of plants. After harvest remove old canes and cut off new ones at 45 inches. Canes that sprawl left are then pulled back to the right side of the wire. Canes that grow right are pulled to the left side. Tie if necessary. Extremely long laterals will grow outward, knitting together. Train them as needed along the wire.

Red and yellow raspberries   These berries always grow on erect stems with no laterals and do not clump but spread by underground runners. They are often grown in rows. These raspberries are either one-crop (also called single-crop), like the berries described above, or two-crop (also called ever-bearing), which means they produce a second crop of berries in the fall on the new canes that grew in the summer.

For one-crop berries, cut out bearing canes as soon as the harvest is over. Train new canes to a post or to one or two horizontal wires. If your berries are growing in a row, prune out new canes to leave 6 inches between remaining plants. Those canes that grew in the summer will produce a crop next year.

Two-crop raspberries produce a crop at the tops of new canes in fall. Cut out old canes that fruited early in the season, and train young canes. In the fall these young canes will fruit at the top. Cut this upper fruiting portion off the canes after harvest; the lower portion will bear the following summer. In winter cut canes back to about 45 inches in height.

Training Bush Berries

Bush berries such as blueberries, currants, and gooseberries need some annual pruning. Blueberries are borne on buds of last year's wood so light pruning is done to remove older and weaker wood. Currants and gooseberries bear on one-, two-, and three-year-old wood. Four-year-old and older wood produces poor berries and should be removed. Clean up bushes by removing the oldest shoots (four years or older) in winter, thinning out the new growth, and cutting out dead wood.

If berries are very small one year, thin the following winter.

Training Grapes

Grapes fruit on lateral shoots on year-old canes. All grapes require heavy pruning to produce fruit, but after the first three growing seasons, different types of grapes need different pruning. Wine grapes and muscadines usually need *spur pruning* and American grapes and 'Thompson Seedless' require *cane pruning*. Both spurs and canes grow from a permanent trunk with arms (cordons) that you train on a trellis or arbor.

In spur pruning, you cut all side branches on the lateral arms or cordons to two buds in fall or winter. Two new shoots grow on the spur you leave, and each produces a cluster or bunch of fruit.

Cane pruning is for grapes that do not produce fruit on shoots growing too near the main scaffold. 'Thompson Seedless' and many American grapes such as 'Concord' are among these. Instead of cutting to a short spur in

Grape Spurs

Grape spurs are created by cutting back old canes each year to two buds. The resulting jagged small branch is called a spur. Not all grapes are pruned this way; some are cane pruned.

winter, leave two whole canes from the previous growing season and two canes cut back to two buds. The short canes will form replacements for the others when they are through fruiting. When fruit forms from side growth along this cane, clip the cane off beyond the next set of leaves. You thereby encourage two new canes that will bear fruit the following year.

The grape variety list in the "Encyclopedia of Fruits, Berries & Nuts" (page 91) indicates whether you should practice spur or cane pruning on a particular variety.

Grapes can also be trained to a head, that is, freestanding. The fruiting stubs, or spurs, are selected early—about four per vine—then clipped back each year until the leafless plant looks like a caveman's club.

Stake the young trunk and allow up to four shoots to grow, beginning about 24 or more inches above the ground. To spur prune cut each shoot to two buds in winter each year. To cane prune gather the fruiting canes upward and tie them together toward the tip. Let growth from renewal buds trail.

You can also train grape vines onto an arbor. An arbor is an overhead frame at least 5 to 10 feet long on a side and supported by posts. Train a vine up each vertical post with two branches crossing the top horizontally. It may take more than two seasons for the vines to reach the top. Then begin training the horizontals according to the method suggested for that variety.

Training Spur and Cane Grapes for the First Three Seasons

When you plant: Plant a rooted cutting with two or three buds above the soil, then bury those in light mulch.

First growing season: Leave the plant alone. It will grow a number of shoots.

First dormant season: Choose the best shoot and cut others to the base. Head remaining shoot to three or four strong buds.

Second growing season: When new shoots reach about 12" long, select the most vigorous and pinch off others at the trunk. Tie the remaining shoot to a support (arbor post, trellis post). When the shoot reaches branching point at arbor top or trellis wire, pinch it to force branching. Let two strong branches grow. Pinch any others at 8" to 10" long.

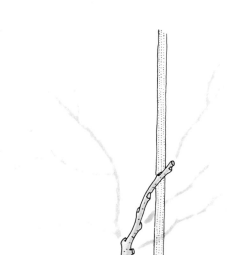

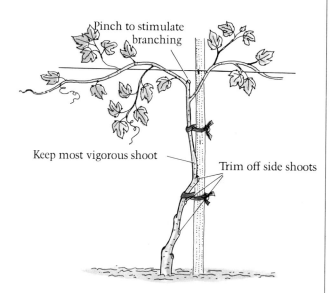

Pinch to stimulate branching

Keep most vigorous shoot

Trim off side shoots

Second dormant season: Cut away side shoots, leaving only the trunk and two major branches. Tie these to the arbor top or the trellis wire.

Third growing season: Let the vine grow. Pinch tips of sprouts on trunk. After this, spur and cane pruning differ.

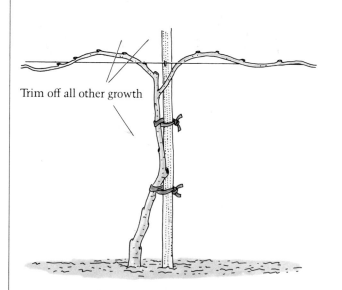

Trim off all other growth

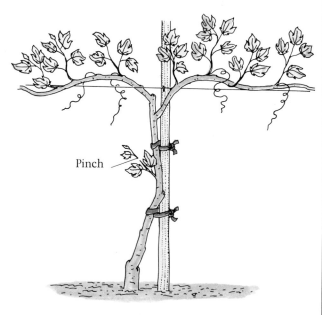

Pinch

Spur Training of Grapes

Third dormant season: Remove all shoots from the vertical trunk. Choose the strongest side shoots on horizontal branches and cut to two buds. Remove weak shoots at the base, spacing a spur, cut to two buds, every 6″ to 10″.

Annually: Every dormant season after this, each spur will have two shoots that produced fruit during the summer. Cut off weak spurs. Cut the stronger spurs to two or three buds. These buds will produce fruit-bearing shoots in summer. Repeat each year. Always keep the trunk clear of growth.

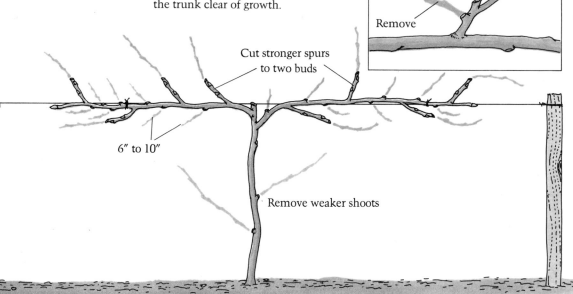

Cut to two buds

Remove

Cut stronger spurs to two buds

6″ to 10″

Remove weaker shoots

Cane Training of Grapes

Third dormant season: Remove shoots from the trunk. Cut horizontal branches back so that two long shoots remain on each. On a two-wire trellis, you can leave up to eight shoots per vine. Tie the shoot farthest from the trunk to the trellis. Cut the other to two or three buds. The tied shoot will fruit the following summer. The clipped shoot will produce growth to replace it the next winter, and fruit the year after.

Annually: When the outside cane has borne fruit, cut it back to the inside stub, now holding two or three new canes. Select the best and tie it to the trellis for fruit. Cut the next cane to two or three buds. Remove the weakest at the base. Repeat each year.

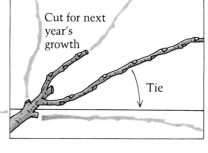

Cut for next year's growth

Tie

Cut renewal spur to two or three buds

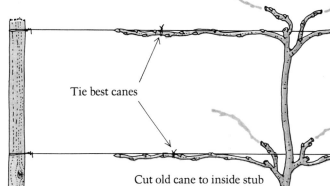

Tie best canes

Cut weakest canes at base

Cut old cane to inside stub

SPECIAL TRAINING PROCEDURES

Fruit trees can be grown successfully as espaliers or hedges that create garden dividers and boundary plantings. These special training techniques save space and allow the home gardener to grow more varieties than would otherwise be possible. This section details the use of dwarfing and training to confine fruit trees within small spaces.

Growing fruit in tight spaces is really no harder than maintaining a healthy rosebush, but keep the following points in mind.

☐ Be especially careful about planting and general maintenance.

☐ Prepare your soil well, and where drainage is a problem, use low raised beds.

☐ Feed and water on a regular schedule, and keep a careful watch for signs of any insects or diseases.

☐ Don't let new growth escape from you and spoil the pattern.

☐ Inspect your plants frequently.

In limited-space planting, training continues all seasons throughout the life of the plant. Be ready to pinch or snip at any time. Major pruning is still a winter task, but in summer you will need to head, or cut away wild growth and suckers, and you may need to loosen or renew ties or add new ones.

Be sure you understand the normal growth patterns of the plants you intend to train. For example, apples and pears bear fruit in the same places for years. Although fruiting spurs may need to be renewed over the years, the growth pattern means that you can confine these trees to formal shapes and keep them that way.

On the other hand, peaches and nectarines fruit on branches that grew the previous year. Old branches will not bear, so they should be cut away like berry canes and replaced with new growth from the base of the tree. This heavy pruning makes rigid training patterns impossible. Peaches and nectarines can be fanned out over walls or grown as hedges, but they cannot be held to strict geometric shapes.

Grapes are vines and therefore are almost always trained on a wall, wire, or fence. See Training Grapes on page 52 for instructions. A variety that requires a little more heat than your region normally offers may produce good fruit when grown on a south or west wall.

You can train cane berries flat against fences or walls, and treat them something like peaches, since you must replace all canes that have fruited with canes of the current season.

The poorest subjects for limited space training are the quince and cherry. The quince fruits at the tips of new twigs, and the cherry is normally too large to confine and will not fruit at all without a pollinator close by. Both of these plants can be trained, but your efforts would be better spent on something more rewarding.

Espaliers

Technically speaking an espalier is a plant pruned to grow all in one plane. Most often, a symmetrical pattern is established through careful pruning and training. Supple young branches are fastened to a fence, a wall, or wires with soft string or bands of rubber. Plants trained against a wall should be at least 6 inches away from the wall to allow good air circulation and room to grow.

Espaliers usually require the formation of *cordons*—side arms—off a main trunk. Cordons should be spaced no less than 16 inches apart to allow for optimal growth.

A word of caution: Where summer temperatures reach 90° F or more, heat from the wall will cook the fruit. In this situation, espalier a tree onto a free-standing trellis for better air circulation.

These apples are in their third year of training as a hedge. Because hedge training is less structured and formal than is espalier training, it is a little simpler.

Training Apples and Pears

Both apples and pears can be grown as hedges.

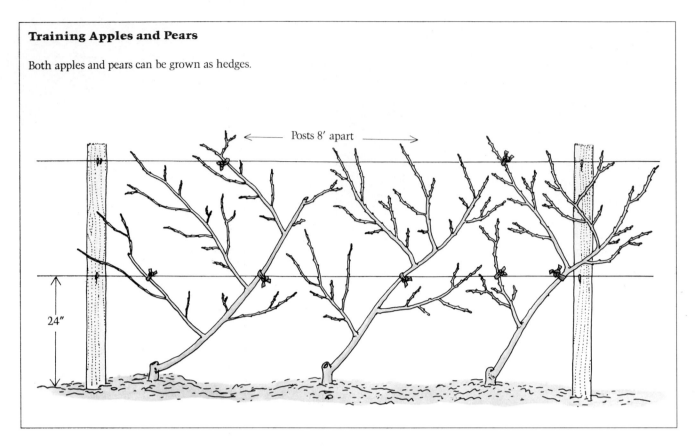

Posts 8′ apart

24″

Apples and pears as espaliers Of all the fruit trees, apples and pears are really the best suited for this specialized training. Run horizontal wires 18 inches across a wall or between two posts. Plant a bare-root whip, then cut it off at 18 inches, at or just below the height of the first wire. This will activate the buds just below the cut.

The first summer, train two side buds onto the wire, letting a third bud develop as a trunk. Tie the two side branches onto the wire so that the tips are lower than the branches. Rub off all growth from the trunk and prune the tips and the branches.

Next winter, cut the trunk off a little below the second 36-inch-high wire. This will activate another set of buds. Again, keep two buds for side branches and one for the trunk extension. Train these as you did the first set of buds. Cut the laterals on the branches on the first wire back to three buds. These will develop into fruiting spurs.

Continue training until three wires (or as many as you wish) are covered with branches. On the top wire there will be no trunk extension, just the two side branches.

Hedges

Home gardeners are not the only ones concerned with limited space planting. Commercial growers are experimenting with training methods that let them grow fruit in hedgerows and harvest their crops without ladders.

Apples and pears as hedges Both dwarf apples and pears grow and fruit well when trained as hedges against horizontal wires. Set posts about 8 feet apart. Stretch a bottom wire between the posts 24 inches above the ground. For very small trees, place the upper wire at 4 or 5 feet. For larger trees, place a third wire at 6 or 7 feet.

Plant the young bare-root trees about 3 feet apart, beginning next to an end post. The last tree should be placed about 2 to 3 feet short of the final post. If you buy unbranched trees, bend the trunk at a 45-degree angle and tie it to the wire. If there are any branches with wide crotches, cut them so only two leaf buds remain. Clip off branches with narrow crotches at the trunk.

During the first season, train the trunk and any new branches at about 45 degrees, tying loosely where they touch the wires. Pinch off at the tip any branches that seem badly spaced or that extend from the fence at right angles. The first winter remove badly placed branches at the trunk. Remove the tips from well-placed branches, cutting to a healthy bud on the top of each branch.

The second summer continue training the shoots at the ends of branches upward at a 45-degree angle. Cut side growth to four buds beginning in July.

Each winter thereafter remove tangled or damaged growth and cut remaining long shoots to four leaf buds. Each summer cut out suckers and excessively vigorous sprouts as they appear. Shorten new growth to four leaves after July.

This training method allows side branches to grow outward away from the fence. Your hedge will eventually become 3 to 4 feet wide. You can hold it at that width by pulling some of the outward growth back toward the fence with string, but check ties frequently or they will cut the branches. If parts of your hedge begin to escape and grow too far outward, trim them back to healthy side branches in May. To maintain the proper height of 5 to 8 feet, cut top growth back to a healthy side shoot in May. Make the cut close to the top wire.

Peaches and nectarines as hedges   Since a peach hedge must have its fruiting wood renewed annually, you will need long replacement branches each year. Plant your hedge as described for apples, using wires at 2, 4, and 6 feet. Cut the whips to about 24 inches tall, and shorten side branches that point along the fence to two buds each. Cut off other branches at the trunk. Train all new growth at 45-degree angles in both directions. Remove any suckers from below the bud union, cutting to the trunk.

The first winter cut out about half the new growth at the base, choosing the weakest branches for removal. Cut off the tops of branches you retain if they have grown beyond the hedge limits.

Fruit will form on the branches that grew the previous summer. The original trunk and the lowest branch will form an approximate V shape at or below the lowest wire. During

Training Peaches and Nectarines

Each year cut out branches that have fruited, letting new branches replace them to bear next year's fruit.

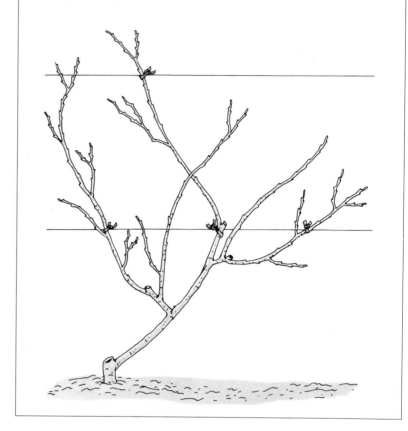

the second summer, choose the healthiest shoots from the lower portions of these main branches, and pinch back all other growth—especially above the second wire—after it produces six to eight leaves. The lower shoots will replace the entire upper structure and should be tied back loosely to the fence. Continue to remove suckers below the bud union.

When leaves drop in fall, cut out all branches that have fruited, and head back the V-shaped main structure to the middle wire. Train the new growth to the fence. During the summer again encourage the lower shoots and pinch back the upper growth. Always be sure that there is new growth above the bud union.

Apricots and plums as hedges   Use approximately the same technique as described for peaches, but instead of replacing all growth each year replace about one-third and head back new growth on the remaining branches to four to six leaves.

Encyclopedia of Fruits, Berries & Nuts

A cornucopia of photographs follows. Flavor and color descriptions are included for hundreds of fruit, berry, and nut varieties. Hardiness, time of fruiting, and special care requirements are discussed.

The information in this encyclopedia is designed to help you choose the best possible fruits for your garden. It describes the best and most popular fruits. Some varieties are known by more than one name. In these cases the most common name is used with the less well known alternatives following in parentheses.

Many varieties have chilling requirements that must be met for the fruit to develop properly. The chilling requirements are described in terms of the number of hours of exposure to winter temperatures below 45° F. Fruits and berries that require chilling fall into one of three general categories:

Low chill   300 up to 400 hours below 45° F
Moderate chill   400 up to 700 hours below 45° F
High chill   700 up to 1000 hours below 45° F

The maps accompanying each fruit show where it is best adapted. The darkest part of the map shows where most varieties are well-adapted; the lighter part shows where you will need to use adapted varieties. You may still be able to raise the fruit in parts of the country uncolored on the map if special varieties are available for your region, or if you use special techniques, such as bringing the plants into a greenhouse for the winter.

Just because a variety is recommended for one area does not necessarily mean that it can't do well in others. Local climate, as well as special treatment from the gardener, can support plants that generally are not expected to do well in a given region.

The home orchardist's reward—a bountiful harvest.

Apples

Well over a thousand apple varieties are available today. Many of these are sports, or accidental mutations of another variety. Others, especially the more recent introductions, are the result of

painstaking crossing and selection by apple breeders. Each parent plant supplies half the heritage of seedlings, but that half may be a set of characteristics either partly or completely hidden in the parent. The seedlings are an unknown mixture until breeders grow them to fruiting size to see what characteristics they have. This work takes time, and many seedlings prove to be inferior to their parents.

Sports, or mutations, may occur at any time, often without apparent reason: Suddenly one branch of a tree is different. Occasionally the odd branch results from mechanical damage, such as pruning; sometimes experimenters purposely change genetic structure with chemicals or radiation. Most sports are worthless, but now and then one turns out to have characteristics that make it worth propagating to create a new strain.

'Delicious', which is by far the most popular and economically important apple in America, first sprouted in an Iowa orchard in 1872. Its parentage is uncertain, but one parent may have been a nearby 'Yellow Bellflower' apple. That 'Delicious' exists at all today is almost a miracle. The owner, Jesse Hiatt, cut the seedling down twice, but it resprouted each time, so finally he let it grow. In about 1880 it bore fruit that Hiatt thought was the best he'd ever tasted. The name 'Delicious' was given at

a fruit show by C. M. Stark of Stark Nurseries. Stark didn't learn the name of the grower until 1894, and by then the apple had already begun its rise to fame.

'Delicious' has produced a number of sports, including the original red sport, 'Starking'; the redder 'Richard', 'Royal Red', 'Hi Early', 'Chelan Red', and 'Red Queen'; and the spur-type 'Starkrimson', 'Redspur', 'Wellspur', 'Hardispur', and 'Oregon Spur'. 'Delicious' is also a parent of 'Melrose'.

The first seedling of 'Jonathan' sprouted in Woodstock, New York, apparently from the fruit of an 'Esopus Spitzenburg'. A Judge Buel of Albany found the apple so good that he presented specimens to the Massachusetts Horticultural Society, naming it for the man who first showed it to him. 'Jonathan' was the most important commercial variety before 'Delicious' took over.

Red sports of 'Jonathan' include 'Jon-A-Red' and 'Jonnee'. Hybrid descendants include 'Jonagold', 'Jonamac', 'Idared', 'Melrose', 'Minjon', and 'Monroe'.

The 'McIntosh' apple came from the McIntosh Nursery in Ontario, Canada. John McIntosh discovered it about 1811 but did not propagate grafted stock until 1835, when the grafting technique was perfected. Well-known descendants of 'McIntosh' include 'Summerred', 'Niagara', 'Early McIntosh', 'Puritan', 'Tydeman's Red', 'Jonamac', 'Macoun', 'Empire', 'Cortland', 'Spartan', and the spur variety, 'Macspur'.

Other apples with long lines of descendants include 'Rome', 'Golden Delicious', 'Northern Spy', and 'Winesap'.

The extensive work on dwarfing rootstocks for apples has produced plant sizes ranging from a 4-foot bush to a 30-foot spreading tree. There is

An informal espaliered apple

even a true, or genetic, dwarf that stays small on any rootstock.

Spur-type apple varieties are sports of standard varieties. They grow more slowly than other plants, and their spurs are packed closer together on the branch. This less vigorous growth means that they are a kind of genetic dwarf, but they are still good-sized trees unless grafted to dwarfing roots. Spur varieties are difficult to train formally. If you buy spur varieties on dwarfing roots, use a training method that doesn't call for any particular form.

Pruning methods depend on how you grow the tree. For general pruning of the larger dwarfed trees or standard trees, see pages 40–43. For special training, turn to page 46.

Thinning is crucial with apple varieties. If left alone the trees set too much fruit, and the heavy crop can snap branches. Even more important, many apple varieties tend to bear every other year. If you leave too much fruit you encourage this alternate bear-

ing: The following year you may find that your tree bears only a handful of apples because the large crop of the previous year has depleted the tree's reserves. Most important of all, the quality of the remaining fruit is better after thinning.

There are many thinning methods, but the best method is to make a light first thinning by the time the fruit is pea size. After this, wait for the natural drop of young fruit in June, then thin the remaining fruit so that there is a single apple every 6 inches along the branches. Each spur may have a cluster of fruit. A single fruit is less likely to become diseased, so leave only the largest fruit on each spur.

Thin carefully or you will damage the spurs or even pull them off with the young fruit. If the apples are small one year, thin more heavily the next year. If the fruit set is light but the fruit is large, thin less next season.

Most apples are self-infertile, so for a good crop most varieties need a pollinator.

A formal espalier

A "spur" type tree

'McIntosh'

Almost any two kinds that bloom together offer good cross-pollination. The following varieties produce poor pollen so cannot pollinate other varieties: 'Jonagold', 'Spigold', 'Mutsu', 'Gravenstein', 'Winesap', 'Stayman', and 'Stayman' sports such as 'Blaxstayman' and 'Staymared'. If you plant one of these varieties, you will need to plant three different varieties in total to get fruit from all of them. Also, if you plant only a very early and a very late variety, they will not cross-pollinate.

All apples need some cool winter weather, but there is an enormous range in this requirement, so varieties are available for any climate except tropical and low desert regions.

Apples are subject to attack by many organisms, but the gardener will have most trouble with codling moth and other fruit-spoiling pests and with the usual aphids, mites, and scales. See the pest and disease section on pages 29–37 for further details. A regular spray schedule is best. Repeated sprays can control diseases such as mildew.

Early Season Varieties

'Akane' A hybrid of 'Worcester Pearmain' and 'Jonathan', 'Akane' has bright red skin; crisp, juicy white flesh; and 'Jonathan'-like flavor. The moderately productive tree is less susceptible to fireblight than the 'Jonathan' parent is. Origin: Japan.

'Jerseymac' A 'McIntosh' cross that ripens in August, this red fruit is medium firm, juicy, and of good quality. The tree produces a crop every year and is generally available. Origin: New Jersey.

'Liberty' The medium-sized fruit is sweet and juicy, somewhat coarse-grained, and abundantly produced; the skin is almost entirely red. The tree's greatest virtue is extreme resistance to rust and scab and high resistance to fireblight and mildew. Origin: New York.

'Lodi' The fruit is up to 3 inches in diameter, with light green skin, sometimes with a slight orange blush. The flesh is nearly white with a greenish tinge; fine grained, tender, and juicy; but sour. The eating quality is only fair, but 'Lodi' is excellent in sauce and pies. The tree tends to overset fruit and must be thinned. Widely available. Origin: New York.

'Tydeman's Early' A 'McIntosh' type, similar in shape and ripening four weeks earlier, this apple is almost entirely red from a very early stage. Fruit drops quickly at maturity and should all be picked within a few days for optimum quality and flavor. The eating quality is good and the fruit keeps much longer than most early varieties. When few other varieties are being harvested, early ripening is a virtue. One drawback is growth habit: The branches are undesirably long and lanky and need to be controlled by pruning. For best results grow this one on dwarf or semidwarf rootstocks. Widely available. Origin: England.

Early to Midseason Varieties

'Gravenstein' Fruit is large but not uniform, with skin that's red against light green. The greenish yellow flesh is moderately fine textured, crisp, firm, and juicy. It is excellent for eating fresh, in sauce, and in pies. The trees are strong, very vigorous, upright, and spreading. Widely available along with 'Red Gravenstein'. Origin: Germany.

'Jonamac' This 'McIntosh' type dessert apple is of very good eating quality, milder in flavor than the 'McIntosh'. Origin: New York.

'McIntosh' If you write down the attributes of a great apple—medium-to-large fruit with sweet, tender, juicy white flesh; very good fresh or in sauce, pies, or cider—you are describing 'McIntosh'. The skin is yellow with a bright red blush. The tree is strong and very vigorous. Widely available. Origin: Ontario.

'Golden Delicious'

'Spartan'

'Jonagold'

'Red Delicious'

'Paulared'   This apple rates high on several counts. It has an attractive solid red blush with a bright yellow ground color. The flesh is white to cream and nonbrowning. Its excellent, slightly tart flavor makes it good both for eating fresh and in sauce and pies. Although it colors early, for quality apples it should not be picked until nearly mature. Fruit holds well on the tree and is harvested in two pickings; it has a long storage life. The tree is everything an attractive tree should be—strong and upright, with good branch structure. Origin: Michigan.

'Prima'   This juicy red apple has fair quality, but its main feature is its resistance to scab, mildew, and fireblight. Origin: Illinois.

Midseason Varieties

'Cortland'   According to many apple growers, this is excellent—even better than 'McIntosh'—as a dual-purpose apple, for eating and cooking. The tree bears heavy crops of large, red-striped fruit with white flesh that is slow to turn brown when exposed to air, making it especially suited for use in salads. The tree is strong and very vigorous, with a spreading, drooping growth habit. Widely available. Origin: New York.

'Empire'   This cross between 'McIntosh' and 'Delicious' has medium, uniform fruit with dark red striped skin and whitish cream flesh that is firm, medium textured, crisp, very juicy and of excellent eating quality. A major fault is that it develops full color long before maturity, tempting the grower to harvest too early. The trees are moderately vigorous and of spreading form. Origin: New York.

'Gala'   This variety gains high marks for quality as a fresh fruit with the advantage of long storage life. The medium-sized fruit is yellow brightly striped with red and borne on a large, upright tree. Origin: New Zealand.

'Jonathan'   The standard 'Jonathan' is one of the top varieties grown in commercial orchards in the Central States. The fruit is medium sized and uniform; the skin is washed red and pale yellow; and the flesh is firm, crisp, and juicy. Rich flavor makes it a good choice for snacks, salads, and all culinary uses. Trees bear heavily. Widely available. Origin: New York.

'Spartan'   A cross between 'McIntosh' and 'Yellow Newtown', the fruit is medium sized, uniform, and symmetrical. It has solid dark red skin and light yellow, firm, tender, crisp, and juicy flesh. The tree is strong, moderately vigorous, and well shaped. It must be thinned to assure good size and annual bearing. Widely available. Origin: British Columbia, Canada.

Midseason to Late Varieties

'Golden Delicious'   For a great eating and cooking apple, 'Golden Delicious' ranks as high as any. The fruit is medium to large and uniform in size. The skin is greenish yellow with a bright pink blush. The flesh is firm, crisp, juicy, and sweet—excellent fresh and in desserts and salads, and very good for sauce. The tree is of medium height, moderately vigorous, upright, and round, with wide-angled crotches. It bears very young and continues to bear annually if thinned. This is an excellent pollinator and will set some crop without cross-pollination. Widely available. Origin: West Virginia.

'Jonagold'   A cross of 'Jonathan' and 'Golden Delicious', this is a beautiful large apple with a lively yellow-green ground color and bright red blushes. The cream-colored flesh is crisp and juicy and has good flavor. It is good for cooking, is among the very best apples for fresh eating, and stores well. The trees are vigorous with wide-angled branches. Origin: New York.

'Granny Smith'

'Idared'

'Mutsu'

'Northern Spy'

'Red Delicious' The number one supermarket apple. there is no question about its dessert and fresh-eating quality. The fruit is medium to large with striped to solid red skin. The flesh is moderately firm in texture and very sweet and juicy. Your best choices are the red sports such as 'Wellspur' or 'Royal Red'. The tree tends to produce full crops every other year unless properly thinned for annual bearing. Widely available. Origin: Iowa.

'Yellow Newtown' The medium-sized fruit has greenish yellow skin and crisp, firm flesh. It is good for eating fresh and excellent for sauce and pies. The trees are strong and vigorous. Widely available. Origin: New York.

Late Varieties

'Fuji' This variety is later to ripen than 'Granny Smith' and, like that variety, needs a long growing season (at least 200 days). Origin: Japan.

'Granny Smith' The fruit is medium to large and bright glossy green. The flesh resembles 'Golden Delicious' but is more tart. It is very good eaten fresh or in desserts, salads, sauce, and pies. The tree is strong, vigorous, upright, and spreading, but it can only be grown in areas with a very long growing season. It has recently become the favorite tart apple in groceries. Widely available. Origin: Australia.

'Idared' A cross of 'Jonathan' and 'Wagener', this hybrid has an attractive, nearly solid red skin with a smooth finish. The large, uniform fruit has white, firm, smooth-textured flesh that is excellent for eating fresh and for cooking. It has a long storage life. The tree is vigorous, upright, and productive. Widely available. Origin: Idaho.

'Mutsu' A cross of 'Golden Delicious' and the Japanese 'Indo', this relative newcomer has gained the approval of both growers and consumers. Large, oblong, greenish fruit develops some yellow color when mature. The flesh is coarse, firm, and crisp. The flavor is excellent (tarter than 'Golden Delicious') when eaten fresh, and it is good for sauce, pies, and baking. Unlike 'Golden Delicious' it does not shrivel in storage. The tree is vigorous and spreading. Origin: Japan.

'Northern Spy' Trees of this variety are very slow to begin bearing; sometimes 14 years elapse before they produce their first bushel (but they bear much sooner on dwarf rootstock). The fruit is large, with yellow and red stripes, and the flesh is yellowish, firm, and crisp. The quality is excellent fresh and for pies. The fruit bruises easily, but has a long storage life. Trees are vigorous and bear in alternate years. Widely available. Origin: New York.

'Rome Beauty' This variety and its sports are the world's best baking apples. Many red sports (such as 'Red Rome') are available in a beautiful, solid medium-dark red. The fruit is large and round, and the flesh is medium in texture, firm, and crisp. The tree is moderately vigorous, starts to produce at an early age, and is a heavy producer. The fruit has a long storage life. Widely available. Origin: Ohio.

'Stayman' This variety is a very late ripener. Where it can be grown, it is good for cooking or eating fresh. The fruit is juicy with a moderately tart, rich, winelike flavor. The skin is bright red and has a tendency to crack. The flesh is fine-textured, firm, and crisp. The tree is medium sized and moderately vigorous. Widely available. Origin: Kansas.

Extrahardy Varieties

In cold-winter areas where some of the favorite apple varieties are subject to winter damage, gardeners may choose one of three hardy varieties developed by the University of Minnesota.

'Honeygold'   Midseason to Late. This apple boasts a 'Golden Delicious' flavor. The fruit is medium to large with golden to yellowish green skin and yellow flesh that is crisp, smooth, tender, and juicy. It is good for eating fresh and in sauce and pies. The tree is moderately vigorous. Origin: Minnesota.

'Red Baron'   Midseason. This cross of 'Golden Delicious' and 'Red Duchess' has round, medium-sized fruit with cherry-red skin. The flesh is crisp and juicy with a pleasantly tart flavor. It is good eaten fresh or in sauce and pies. Origin: Minnesota.

'Regent'   Late. This variety is recommended for a long-keeping red winter apple. The fruit is medium sized, with bright red skin and crisp-textured, creamy white, juicy flesh. Rated excellent for cooking or eating fresh, it retains its fine dessert quality late into winter. The tree is vigorous. Origin: Minnesota.

Low-Chill Varieties

'Anna'   Early. This apple flowers and fruits in Florida and Southern California. The apple is green with a red blush and fair quality. It is normally harvested in July but sometimes sets another late bloom that produces apples for the fall. Use an early blooming variety such as 'Dorsett Golden' or 'Ein Shemer' as a pollinator. Origin: Israel.

'Beverly Hills'   Early. This is a small to medium-sized apple, striped or splashed with red over a pale yellow skin. The flesh is tender, juicy, and

tart. Overall the apple resembles 'McIntosh'. Use it fresh or cook it in sauce or in pies. The tree is suited mainly to cooler coastal areas, since heat spoils the fruit. Locally available. Origin: California.

'Dorsett Golden'   Early. This large 'Golden Delicious'–type fruit requires no frost or significant winter chill and performs well in coastal Southern California and the hot-summer regions of the Deep South. Use it for eating fresh or for cooking. A good pollinator for 'Anna' and 'Ein Shemer'. Origin: Bahamas.

'Ein Shemer'   Early. This is another 'Golden Delicious'–type fruit that is well-adapted to the Deep South, Texas, and Southern California. The tree begins bearing at an early age. Makes a good copollinator for 'Dorsett Golden'. Origin: Israel.

'Gordon'   Early to midseason. The crisp flesh is enclosed in red-striped green skin. The blooming and bearing period is unusually prolonged—August to October in California. It performs particularly well in coastal Southern California. The fruit is good both for eating fresh and for cooking. Self-fruitful. Origin: California.

'Winter Banana'   Midseason. The large fruit is strikingly beautiful. The skin color is pale and waxy with a spreading pink blush. The flesh is tender, with a wonderful aroma and tangy flavor. 'Winter Banana' requires a pollinator such as 'Red Astrachan' in order to set a good crop. Locally available. Origin: Indiana.

'Winter Pearmain'   Midseason. This large green apple has moderately firm flesh of excellent quality. It is a consistent producer in Southern California. Origin: Unknown.

'Sungold'

Apricots

In the colder regions of the country, the selection of apricot varieties is limited because apricots bloom early and may suffer frost damage. In recent years, however, breeders have produced a number of hybrids

with hardy Manchurian apricots, and now varieties such as 'Chinese' will fruit fairly regularly even in the northern plains. The choice of varieties widens in milder regions, and more tender varieties such as 'Moorpark' will bear even in the eastern states.

Dwarfed apricots on special rootstocks produce fair-sized trees, and a full-sized tree will fill a 25-foot-square site, but you can train the tree to branch high and use it in the landscape as a shade tree. Trees are fairly long-lived and

may last from 15 to 30 years, depending on care.

Many apricots are self-fertile, but in colder regions it is usually best to plant a second variety for pollination to encourage the heaviest fruit set possible. Frost may thin much of the young fruit.

Thinning is generally natural, either from frost or from natural drop in early summer. If your tree sets heavily, you will get larger apricots by thinning to 2 inches between each fruit. For pruning and training details, see page 46.

Apricots can also be used as stock plants for grafts. Plums do well on apricot stock, and peaches may take, although the union is weak. Your apricot tree can bear several different fruits over a long season.

Brown rot and bacterial canker are serious pests.

'Moorpark'

Varieties

Check for climate adaptability and pollinating requirements, and be sure to buy hardy trees in the colder regions.

'Blenheim' ('Royal') This is the best eating, drying, and canning apricot in California. The fruit is medium sized and flat orange with some tendency to have green shoulders. It requires moderate chilling and will not tolerate excessive heat (over 90° F) at harvest time. Origin: England.

'Chinese' ('Mormon') Its Utah birthplace marks this variety as a good choice for the coldest regions of the West's apricot climates. Late flowering gives blossoms a chance to escape late frosts. Trees bear heavy crops of small, sweet, juicy fruit at an early age. Origin: Utah.

'Flora Gold' This genetic dwarf apricot reaches about half the size of a full-sized tree. Its small to medium-sized fruit is of high quality—best for eating fresh and for canning. The heavy crop ripens early, about a month before 'Blenheim'. Moderate-chill requirement. Origin: California.

'Goldcot' Late flowering, late bearing, and hardiness to -20°F recommend this variety to midwestern and eastern growers. The medium-sized to large fruit is tough-skinned and flavorful, good for eating fresh and for canning. Self-fruitful. Origin: Michigan.

'Harcot' Another cold hardy variety with late flowering but early ripening. Fruit is medium to large and flavorful. Heavy-bearing, compact trees resist brown rot and are somewhat resistant to bacterial spot. Origin: Ontario, Canada.

'Harogem' Small to medium-sized fruit is blushed bright red over orange; the flesh is firm. This variety ripens in midseason and the fruit is especially long-lasting when picked. The tree is resistant to perennial canker and brown spot. Origin: Ontario, Canada.

'Moorpark' This variety, dating from 1760, is considered by many to be the standard of excellence among apricots. The large fruit is orange with a deep blush, sometimes overlaid with dots of brown and red. The flesh is orange and has excellent flavor and a pronounced and agreeable perfume. Ripening is uneven, with half the fruit still green when the first half is already ripe. This is an advantage in the home garden, since the gardener does not have to use the fruit all at once. The tree does well in all but the most extreme climates. Widely available. Origin: England.

'Perfection' ('Goldbeck') The fruit is very large, oval and blocky, and light orange-yellow without a blush. The flesh is bright orange and of fair quality. The tree is vigorous and hardy but blooms early and so is uncertain in late-frost areas. Since it requires little winter chill it will grow in mild-winter areas. It needs a separate pollinator and sets a light crop. Good for the South and West. Origin: Washington.

'Rival' Its northwestern origins make 'Rival' especially well adapted there. Large, heavily blushed fruit is firm, mild flavored, and particularly good for canning. The tree is large and rangy, blooms early, and needs another early-flowering pollinator such as 'Perfection'. Origin: Washington.

'Royal Rosa' This is a good choice for fresh-off-the-tree eating. The bright yellow fruit is firm fleshed and aromatic with a tart tang to its sweetness. The compact, medium-sized tree bears heavy crops early in the season. Origin: California.

'Scout' This variety originally came from a Manchurian fruit experiment station. The flat, bronzy fruit is medium to large with deep yellow flesh. It is good fresh and can also be canned or used in jams. The tree is tall, upright, vigorous, and hardy. The fruit ripens in late July. Good for the Midwest. Origin: Manitoba.

'Sungold' This is a selection from the same cross as 'Moongold', and the two must be planted together for pollination. The fruit is rounded and of medium size, with a tender, golden skin blushed orange. The flavor is mild and sweet, and the fruit is good fresh or preserved. The tree is upright, vigorous, and of medium size. The fruit ripens somewhat later than 'Moongold'. Good for all zones. Origin: Minnesota.

'Tilton' The vigorous tree bears heavily most years. The fruit is yellow-orange and tolerates heat when ripening. It has a high-chill requirement (over 1,000 hours below 45° F) but performs well in hot summer climates. Origin: California.

'Wenatchee' The fruit is a large, flattened oval with orange-yellow skin and flesh. Trees are fairly long lived and may last from 15 to 30 years, depending on location and care. The tree does well in the Pacific Northwest and the West. Origin: Washington.

An espaliered cherry tree

'Black Tartarian'

Cherries

Cherries come in three distinct forms with many varieties in each category. The *sweet cherry* sold in markets is planted commercially in the coastal valleys of California and in the Northwest, espe-

cially Oregon. There are also extensive commercial plantings near the Great Lakes.

All cherries require considerable winter chilling, which rules out planting in the mildest coastal and Gulf climates, but they are also damaged by early intense cold in fall and by heavy rainfall during ripening. Sweet cherries are especially tricky for the home gardener, but try them where summer heat and winter cold are not too intense.

Sour, or *pie, cherries* are more widely adaptable and are good for cooking and canning. These are the most reliable for home gardeners, and there are

many varieties developed for special conditions. The dwarf 'Meteor' and 'Northstar' pie cherries were developed for

Minnesota winters. These and 'Early Richmond' and 'Montmorency' can all withstand both cold and poor spring weather better than sweet cherries.

Sour cherries are all self-fertile and there are two types: the amarelle, with clear juice and yellow flesh; and the morello, with red juice and flesh. In the coldest northern climates, the amarelle is the commercial cherry.

Duke cherries are hybrids with the shape and color of sweet cherries and the hardiness, flavor, and tartness of sour cherries.

Standard sour cherries and sweet cherries on dwarfing roots both reach 15 to 20 feet. A standard sweet cherry with-

out a dwarfing rootstock is one of the largest fruit trees and can equal a small oak in size if the climate permits. Such cherries can serve as major shade trees.

See page 47 for pruning and training information.

All sweet cherries, with the exception of 'Stella', need a pollinator. 'Windsor', 'Van', and 'Black Tartarian' are good pollinators and bear well, but always plant at least two varieties or use a graft on a single tree. Sour cherries are self-fertile.

Dwarf pie cherries have lovely flowers and make fine hedges and screens. They produce good crops and larger cherries can be grafted onto them for a choice of fruit and good pollination.

Birds are the major pests, but cherries also need protection from fruit flies, pear slugs (actually an insect larva), and bacterial leaf spot.

For any cherry, check the recommended climate. If you try a cherry outside its growing zone, offer protection in fall and winter.

Early Season Varieties

'Black Tartarian' Medium-sized, this sweet black cherry is fairly firm when picked but softens quickly. It is widely planted because it is one of the earliest cherries and an excellent pollinator. The trees are erect and vigorous. Use any sweet cherry as a pollinator. Good for all zones. Widely available. Origin: California.

'May Duke' This duke cherry produces medium-sized, dark red fruit of excellent flavor for cooking or preserves. In cold climates use an early sweet cherry for pollination. In mild climates it is self-fertile. Good for the West. Origin: France.

'Northstar' This is a genetic dwarf sour morello, excellent for the home garden. It has red fruit and flesh and resists cracking. The tree is small, attractive, vigorous, and hardy and resists brown rot. The fruit ripens early but will hang on the tree for up to two weeks. Good for all zones. Widely available. Origin: Minnesota.

'Bing'

'Montmorency'

'Royal Ann'

'Sam' This medium to large, black-fruited sweet cherry is firm, juicy, and of good quality. The fruit resists cracking, and the tree is very vigorous, bearing heavy crops. Use 'Bing', 'Lambert', or 'Van' as a pollinator. Good for the North and West. Widely available. Origin: British Columbia.

Midseason Varieties

'Bing' This variety is the standard for black sweet cherries. The fruit is deep mahogany red, firm, and very juicy. It is subject to cracking and doubling. The tree is spreading and produces heavy crops but suffers from bacterial leaf spot attack in humid climates. It is not easy to grow, although it is quite popular. Use 'Sam', 'Van', or 'Black Tartarian' as a pollinator (not 'Royal Ann' or 'Lambert'). Good for the West. Widely available. Origin: Oregon.

'Chinook' Like 'Bing' this variety has large, heart-shaped, sweet fruit with mahogany skin and deep red flesh. The tree is spreading,

vigorous, and a good producer. It is slightly hardier than 'Bing'. Use 'Bing', 'Sam', or 'Van' as a pollinator. Good for the West. Origin: Washington.

'Corum' This sweet variety is the recommended pollinator for 'Royal Ann' in the Pacific Northwest. The fruit is yellow with a blush and thick, sweet, firm flesh. It is moderately resistant to cracking and is a good canning cherry. The tree is fairly vigorous. Use 'Royal Ann', 'Sam', or 'Van' as a pollinator. Good for the West. Locally available in the Pacific Northwest. Origin: Oregon.

'Emperor Francis' This large, yellow, blushed cherry resembles 'Royal Ann' but is redder and more resistant to cracking. The sweet flesh is very firm. The tree is very productive and hardier than 'Royal Ann'. Use 'Rainier' or 'Hedelfingen' as a pollinator (not 'Windsor' or 'Royal Ann'). Good for the North. Origin: unknown—European.

'Garden Bing' This genetic dwarf plant remains only a few feet high in a container but grows to perhaps 8 feet in the ground. It is self-pollinat-

ing and bears sweet, dark-red fruit like 'Bing'. Good for the West. Origin: California.

'Kansas Sweet' ('Hansen Sweet') This is not really a sweet cherry but a fairly sweet form of the pie cherry group. The fruit is red and has firm flesh that is palatable fresh as well as in pies. The tree and blossoms are hardy in Kansas. It is self-fertile. Good for the North. Origin: Kansas.

'Meteor' This amarelle sour cherry is a genetic dwarf that reaches only about 10 feet tall. The fruit is bright red and large for a pie cherry, with clear yellow flesh. The tree is especially hardy but also does well in milder climates and is an ideal home garden tree for all cherry climates. Good for all zones. Widely available. Origin: Minnesota.

'Montmorency' This amarelle is the standard sour cherry for commercial and home planting. The large, brilliant red fruit has firm yellow flesh and is strongly crack resistant. The tree is medium to large, vigorous, and spreading. Various strains have slightly

different ripening times and fruit characteristics. Good for all zones. Widely available. Origin: France.

'Rainier' In shape this sweet cherry resembles 'Bing', but it is a very attractive blushed yellow like 'Royal Ann' with firm, juicy flesh. The tree is vigorous, productive, and spreading to upright spreading. It is particularly hardy. Use 'Bing', 'Sam', or 'Van' as a pollinator. Good for the South and West. Origin: Washington.

'Royal Ann' ('Napoleon') This very old French sweet variety is the standard for blushed yellow cherries. It is the major cherry used in commercial candies and maraschino cherries. The firm, juicy fruit is excellent fresh and good for canning. The tree is very large, extremely productive, and upright, spreading widely with age. The tree is moderately hardy. Use 'Corum', 'Windsor', or 'Hedelfingen' as a pollinator (not 'Bing' or 'Lambert'). Good for all zones. Widely available. Origin: France.

'Stella'

'Lambert'

'Schmidt'   'Bing' is being replaced by 'Schmidt' as a major commercial black cherry in the East. The fruit is large and mahogany colored with thick skin. The wine-red flesh is sweet but somewhat astringent. The large vigorous tree is upright and spreading. It is hardy, but the fruit buds are fairly tender. Use 'Bing', 'Lambert', or 'Royal Ann' as a pollinator. Good for the North and South. Widely available. Origin: Germany.

'Stella'   This is the first true sweet cherry that is self-fertile (requiring no pollinator). The fruit is large, dark in color, and moderately firm. The tree is vigorous and fairly hardy and bears early. It can be used as a pollinator for any other sweet cherry. Good for the South and West. Origin: British Columbia, Canada.

'Utah Giant'   This new sweet variety produces large, dark red fruit that has been compared to the quality of 'Bing' and 'Lambert'. The fresh fruit is excellent, and in canning it retains its color, firmness, and flavor. Good for the West. Origin: Utah.

'Van'   Large and dark, this sweet cherry has some resistance to cracking. The tree is very hardy and especially good in borderline areas, since it has a strong tendency to overset and therefore may produce a crop when other cherries fail. It bears from one to three years earlier than 'Bing'. Use 'Bing', 'Lambert', or 'Royal Ann' as a pollinator. Good for all zones. Widely available. Origin: British Columbia, Canada.

Late Varieties

'Angela'   This large, dark cherry is comparable to 'Lambert' but is hardier and late flowering, so its blossoms are not likely to be frost damaged. The sweet fruits are more resistant to cracking than those of 'Lambert'; the tree is easier to manage, vigorous, and very productive. For pollinators, use 'Emperor Francis' or 'Lambert'. Origin: Utah.

'Black Republican' ('Black Oregon')   This sweet cherry is firm and very dark with slightly astringent flesh. The tree is quite hardy but tends to overbear heavily and produce small fruit. Use any sweet cherry as a pollinator. Origin: Oregon.

'English Morello'   This late-ripening morello sour cherry is medium sized, dark red, and crack resistant. The tart, firm flesh is good for cooking and canning. The tree has drooping branches and is small and hardy but only moderately vigorous and productive. Good for the North. Origin: Unknown.

'Hedelfingen'   The sweet variety bears dark, medium-sized fruit with meaty, firm flesh. One strain resists cracking, but some trees sold under this name do not. The tree is winter hardy, has a spreading and drooping form, and bears heavily. Use any sweet cherry listed here as a pollinator. Good for the North and South. Origin: Germany.

'Lambert'   This large, dark, sweet cherry is similar to 'Bing' but ripens later. The tree is more widely adapted than 'Bing' but bears erratically in many eastern areas and is more difficult to train and prune. The strongly upright growth produces weak crotches if left untrained. Use 'Van' or 'Rainier' as a pollinator (not 'Bing', 'Royal Ann', or 'Emperor Francis'). Good for all zones. Widely available. Origin: British Columbia.

'Late Duke'   This large, light red duke cherry ripens in late July. Use it for cooking or preserves. In cold climates it requires a sour cherry pollinator. In mild climates it is self-fertile. Good for the West. Origin: France.

'Windsor'   This is the standard late, dark, commercial sweet cherry in the East. The fruit is fairly small and not as firm as 'Bing' or 'Lambert'. Its buds are very hardy, and it can be counted on to bear a heavy crop. A fine choice for difficult borderline areas where others may fail, the tree is medium sized and vigorous with a good spread. For a pollinator, use any sweet cherry except 'Van' and 'Emperor Francis'. Good for the North and South. Widely available. Origin: Unknown.

'Siberian Crab'

'Hyslop' blossom

'Young America'

Crabapples

Fine for jellies or pickled whole fruit, crabapples are also the most decorative of fruit trees. Flowers range from red to pink to white. The fruits are of many sizes, from tiny cherrylike varieties to large,

yellow, pink-cheeked kinds. The varieties sold for their flowers also have edible fruit, but the large-fruited varieties are better if your aim is to grow the fruit for jelly.

Crabapples range from small, 10-foot trees to spreading trees 25 feet tall. The large-fruited kinds are larger trees. If you have no space but want a light crop for jelly, graft a branch to an existing apple tree. All types are self-fertile, but you can graft several kinds that bloom at different times onto one tree to extend the flowering season.

Crabapples are subject to the same diseases as apples, and scab is a major problem for some varieties. Choose resistant kinds.

Varieties

The following includes both large-fruiting kinds and those that are mainly ornamental, but all offer a good crop of smaller fruits. Use red- or pink-fruited varieties if you want pink jelly.

'Barbara Ann' This ornamental offers dark, reddish purple, ½-inch fruit with reddish pulp. The tree produces a profusion of 2-inch, purple-pink, full, double flowers. It grows to about 25 feet tall and is reasonably disease resistant. Origin: Massachusetts.

'Chestnut' This very large, bronze-red crabapple is big enough to make a good dessert or lunchbox fruit, and can also be used to make a deep pink jelly. Its flavor is especially pleasing. The tree is very hardy, medium sized, and reasonably disease resistant. Origin: Minnesota.

'Dolgo' The smallish, oblong red fruit is juicy and, if picked before fully ripened, gels easily into a ruby-red jelly. The tree is hardy, vigorous, and productive. The fruit ripens in September. Widely available. Origin: Russia.

'Florence' The large yellow fruit has an attractive red blush. Use it for pale pink jelly or for pickling whole. The tree is medium sized and somewhat tender, so it is best planted in warmer regions. It ranges from fairly to very productive. Widely available. Origin: Minnesota.

'Hyslop' This medium-sized fruit is yellow blushed with red. Use it whole for relishes or for pale pink jelly. The tree is fairly hardy and ornamental, with single pink flowers. Origin: Unknown.

'Katherine' The tiny fruit of this variety is an attractive yellow with a heavy red blush. It can be made into a pink jelly. The tree is small, slow growing, and fairly hardy, but it flowers and fruits only every other year. It grows about 15 feet tall and is reasonably disease resistant. It has double

flowers that open pink and then fade to white. Origin: New York.

'Montreal Beauty' This medium-sized green crabapple with red striping makes good jelly on its own or is a good base for mint or rose geranium jellies. The tree is medium to large, hardy, and fairly disease resistant. Locally available. Origin: Quebec, Canada.

'Profusion' The tiny scarlet fruit of this variety is good in jellies. The tree spreads only slightly, is small (about 15 feet), and produces small single flowers that are deep red in bud and open to purplish red to blue pink. It is moderately susceptible to mildew. Origin: Holland.

'Siberian Crab' This variety bears an abundance of clear scarlet, medium-sized fruit that can be jellied or pickled whole. The tree is vase-shaped and reaches 15 to 30 feet tall, depending on climate and soil. The 1-inch-wide white flowers are fragrant. Some strains are disease resistant, and some are not. Origin: Russia.

'Transcendent' These large yellow crabapples are blushed with pink on one side. Use them for clear jellies or eat them fresh if you like the wild, astringent flavor. The tree is medium to large and somewhat disease resistant, but not very hardy. Origin: Unknown.

'Whitney' An old favorite, this variety has very large fruit, good for fresh eating, jelly, preserves, and apple butter. The fruit is yellow with red stripes. The tree is hardy, medium to large, and reasonably disease resistant. Widely available. Origin: Illinois.

'Young America' The large and abundant red fruit on this variety makes a clear red jelly with splendid flavor. The tree is especially vigorous, and the fruit ripens about mid-September. Origin: New York.

Figs

Although the fig is generally thought of as a subtropical fruit suited mainly to the warmer parts of the country, some varieties will bear in the milder parts of the

A heavily pruned old fig tree

'Brown Turkey'

'Mission'

Northwest and Northeast. If a freeze knocks the plant down, it will sprout again quickly.

In warm regions the fig bears big, juicy fruit in early summer, then sets a heavier crop of small fruit, perfect for drying, in the fall. It lives for many years, loves clay soil if drainage is good, and needs next to no attention. You have a choice of dark fruit with red flesh or greenish-yellow fruit with bright pink flesh.

In cold-winter regions fig shrubs reach 10 feet tall and spread that much or more. In warm regions trees reach 15 to 30 feet and spread wide and low, but you can easily cut them back or confine them. Figs can also be grown as container plants for use on a patio, allowing you to protect them in winter by moving the container to a garage or storage area.

Figs are not really fruit in the botanical sense. They are flowers, borne on the inside of a balloonlike stem and accessible to the outside world only through a hole at the base.

Most home garden varieties of figs need no pollination. The California commercial fig, 'Calimyrna', does need a special type of pollinating and is not recommended for home use.

No fruit thinning is necessary and figs need no attention to pests or disease. See page 48 for pruning and training information.

Varieties

'Adriatic' The fruit is green skinned with a strawberry-pink pulp. In hot areas the second crop has a paler pulp, and in cool areas the fruit of both crops is larger. The tree is vigorous and large. This fig is used principally for drying and for processing into figbars in California. Locally available. Origin: Italy.

'Brown Turkey' There are two 'Brown Turkey' varieties.

'Brown Turkey' of California ('Black Spanish', 'Negro Largo', 'San Piero') A good variety for fresh use, the large fruit is violet-brown to purplish black on the outside and strawberry-pink inside. Prune heavily. Origin: Italy.

'Brown Turkey' of the eastern United States The medium-sized fruit is coppery brown with strawberry-pink pulp. The flavor is very good. Good for container culture.

'Celeste' ('Blue Celeste', 'Celestial', 'Malta', 'Sugar') The bronzy fruit has a violet tinge and the pulp is amber with rose tones. 'Celeste' is the most widely recommended fig

in the Southeast but is also grown in the West. A hardy plant. Origin: Malta.

'Conadria' One parent is 'Adriatic'. The fruit is thin skinned and white with a violet blush. The red flesh resists spoilage. The tree is vigorous and precocious, producing two crops. Recommended for the hot valleys of California. Locally available. Origin: California.

'Kadota' ('Florentine') The fruit is tough skinned and greenish yellow, and the first crop has a richer flavor. This is principally a canning and drying variety. Recommended for hot California valleys. Origin: Italy.

'King' ('Desert King') The fruit is green with flecks of white; the pulp is violet-pink. The tree comes back from the roots after a freeze and bears in fall. Recommended for Oregon fig climates. Locally available. Origin: California.

'Latterula' ('White Italian Honey Fig') This large greenish yellow fig with honey-colored pulp grows on a very hardy tree that bears

two crops. Recommended for Oregon fig climates. Locally available. Origin: Italy.

'Magnolia' ('Brunswick', 'Madonna') This is a large straw-colored fig on a fairly hardy tree. Recommended for the Southeast. Origin: England.

'Mission' ('Black Mission') This variety bears two heavy crops of black fruit with deep red pulp. The first crop has larger fruit; the second crop can be dried. The tree is large and vigorous. Recommended for California and desert regions but also grown in warmer southeastern zones. Origin: Spain—but came to California via Mexico.

'Texas Everbearing' ('Dwarf Everbearing') The fruit and tree resemble 'Brown Turkey'. This variety will resprout after a freeze kills the top. Recommended for the South.

Also check the local availability of 'Granata', 'Negronne', and 'Neveralla' in Oregon; 'Genoa' ('White Genoa') and 'Osborne Prolific' on the California coast; and 'Green Ischia' and 'Hunt' in the Southeast. Origin: Texas.

Peaches

The peach is one of the most popular of homegrown fruits. Both peaches and their close relatives, the nectarines, are at their best when tree ripened, so a home gardener's time and effort are rewarded

by a product that money can't buy. Probably the biggest drawback is that peaches are susceptible to many pests and diseases and spraying is often necessary for a good harvest. If you choose not to spray your peach trees you will have to be prepared to lose some or all of your crop in a bad year.

Peaches cannot tolerate extreme winter cold or late frost, so in the northern Plains States and northern New England, peaches are purely experimental. The hardiest, such as 'Reliance', may survive and bear in a protected spot, but you can't be sure. Peaches do well in the more temperate climates near the Great Lakes, but choose the warmest site available for planting. A protected sunny spot where cold air can't collect and sit is the right place for your tree.

Some of the greatest peach-growing country in the world is in the West: California alone produces 50 percent of the commercial peaches in the United States. Peaches also do well in South Carolina, Georgia, semicoastal areas of the East, and dry areas of Washington.

To produce great peaches the climate must fulfill high-chill requirements (700 to 1,000 hours of cold winter weather at 45° F or below) unless otherwise stated. This should be followed by warm dry spring weather and hot summers. Gardeners not blessed with this prime cli-mate can grow satisfactory fruit by selecting the right varieties for their own gardens. Selected low-chill varieties can fruit well in all subtropical climates but southern Florida.

The standard tree on a peach rootstock grows to about 15 feet tall and 15 feet wide. It could grow larger if left alone, but it is best pruned heavily each year to maintain that size and to encourage lots of new growth along the branches.

A totally satisfactory dwarfing rootstock for peaches has yet to be found. However, the genetic dwarf peaches grow in bush shape to about 4 to 7 feet tall and require no pruning to maintain size or force growth. At most you will clip out tangles and remove broken twigs. Genetic dwarfs are a good choice for patio containers or small yards. You won't have any trouble fitting a genetic dwarf peach into whatever space is available to you.

In the home garden peach trees can be planted two or three to a hole if the varieties are well chosen. This will spread the harvest over several weeks. Dig an extra large planting hole and set two to three varieties together with their roots almost touching. You can also graft different limbs to different varieties and have three varieties on a single tree.

Once a crop sets on a peach tree, you may not even see the branches through all the fruit. You can't leave it all on the tree because it will be small and of poor quality, it slows branch growth, and it may snap branches. Thin it out when it reaches thumbnail size. For early-season peaches, leave 6 to 8 inches of space between fruit; for late season peaches, leave 4 to 5 inches between fruit.

If a frost knocks off much of your crop, leave all the remaining fruit, even if it is clustered. What is important is the ratio of leaf surface to peaches, so a sparse crop will do equally well in singles or bunches.

Only a few peach varieties need a pollinator. Normally the trees are self-fertile, although bees are a big help in pollen transfer.

All peaches like a winter rest. Without it they bloom late, open their leaves erratically, and finally die. Be sure to choose varieties that suit your climate. Low-chill peaches have been bred for short, mild winters and may bloom too early or freeze in the North. Be sure to buy hardy, high-chill varieties for the North. A high-chill peach will leaf out and flower erratically in southern Mississippi, while a low-chill peach may try to bloom before the last frost in Tennessee.

The universal peach ailment in the West is leaf curl, but you can control it easily with a copper spray. You will also probably encounter the major insect pest, the peach tree borer, gnawing the trunk at ground level. Brown rot attacks fruit but is controllable with sprays. In the East and South, check the list for varieties resistant to bacterial spot. Brown rot and plum curculio are the chief pests in the North and South. See pages 29–37 for control methods.

Very Early Varieties

'Desert Gold' This medium-sized, round fruit has yellow skin with a red blush. The flesh is yellow, firm, and semifreestone. The tree is fairly vigorous and productive and requires heavy thinning. The chilling requirement is very low: 200–300 hours. Good for the desert and coastal areas of the West. Origin: California.

'Springtime' The small to medium-sized fruit has yellow skin with a high blush and abundant short fuzz. The flesh is white and semifreestone. Good for the West. Origin: California.

'Tejon' The medium-sized fruit is yellow with a red blush over half its surface; light fuzz. The yellow flesh is semifreestone. The tree bears very well. Good for the West, particularly Southern California. Locally available. Origin: California.

Early Varieties

'Fairhaven' This large peach is bright yellow with an attractive red cheek and light fuzz. The firm freestone flesh is yellow with red at the pit. The fruit freezes well. The tree has showy flowers. Good for the West. Origin: Michigan.

'Flavorcrest' A large, firm, yellow freestone with good flavor, its skin is blushed red. Good for California. Origin: California.

'Garnet Beauty' This variety is an early sport of 'Redhaven'. Medium to large semifreestone fruit hangs on the tree until overripe. The firm flesh is yellow streaked with red and is slightly fibrous. The tree is vigorous and hardy and produces heavy crops that achieve good size and color even inside the tree. It is susceptible to bacterial leaf spot. Good for the North. Widely available. Origin: Ontario, Canada.

'Golden Jubilee' An old standby, this medium to large freestone has skin mottled bright red. The flesh is yellow, firm, and coarse. The tree is hardy and sets heavy crops but is self-thinning. Good for all zones. Widely available. Origin: New Jersey.

'Redhaven'

'Sunhaven'

'Redhaven'　One of the finest early peaches, this medium-sized freestone is widely recommended. The skin is deep red over a yellow ground. The flesh is yellow, firm, and non-browning. Fruit sets heavily and is good for freezing. This tree needs heavy thinning but rewards with outstanding fruit. The tree is spreading, vigorous, and highly productive and resists bacterial leaf spot. Good for all zones. Widely available. Origin: Michigan.

Note: **'Early Redhaven'** is nearly identical, but two weeks earlier.

'Redtop'　The large fruit is nearly covered with an attractive blush and light fuzz. The yellow freestone flesh is unusually firm and good for canning or freezing. The tree is moderately vigorous and somewhat susceptible to bacterial leaf spot. The flowers are showy. Good for the West. Origin: California.

'Reliance'　A promising home garden variety, this tree is very winter hardy. It will withstand -20° to -25° F in January and February and will still produce a crop that same year. The large freestone fruit has dark red skin over a yellow background. The flesh is bright yellow, medium firm, and slightly stringy. The flowers are showy. Good for the North and West. Widely available. Origin: New Hampshire.

'Springcrest'　This is a medium-sized, flavorful, yellow freestone variety. The tree is vigorous and productive and has showy flowers. The fruit matures in late May. Good for the West. Origin: California.

'Sunhaven'　The skin of this medium to large peach is bright red over a golden ground and has short, soft fuzz. The firm, fine-textured, and nonbrowning flesh is yellow flecked with red. The tree is vigorous and consistently productive. Recommended for all zones. Widely available. Origin: Michigan.

'Ventura'　This is a good low-chill (400 hours below 45° F), yellow-fleshed free-stone for Southern California. The tree has average vigor and productiveness. The fruit has yellow skin with a red blush, good flavor, and firm flesh. Available in Southern California. Origin: California.

'Veteran'　A favorite in western Washington and Oregon, this medium to large fruit is yellow splashed with red and has medium fuzz. The nearly freestone flesh is yellow and soft. The tree is vigorous and highly productive—one of the very best in cool Pacific climates. Good for the West. Origin: Ontario, Canada.

Midseason Varieties

'Babcock'　The small to medium-sized fruit is light pink blushed red with little fuzz. The skin peels easily. The white flesh is red near the pit, tender, juicy and has a mild flavor. The medium to large tree is spreading and vigorous, but needs heavy thinning early in the season to produce large fruit. Good for the West, particularly Southern California. Origin: California.

'Early Elberta' (Gleason Strain)　This large freestone matures 3 to 10 days before 'Elberta'. The flesh is yellow and is of better flavor than 'Elberta'. It is good for canning and freezing. The tree is hardy and consistently productive. Good for the South and West. Widely available. Origin: Utah.

'J. H. Hale'　The skin of this extralarge freestone is deep crimson over a yellow background and nearly fuzzless. The flesh is golden yellow and firm. This variety needs cross-pollination for best production. Good for all zones. Widely available. Origin: Connecticut.

'July Elberta' ('Kim Elberta')　This variety is best suited for the Willamette Valley in Oregon. The medium-sized fruit is greenish yellow blushed and streaked with dull red and very fuzzy. The yellow flesh is of high quality. The tree is vigorous and bears heavily but is susceptible to bacterial leaf spot. Good for the West. Origin: California.

'Veteran'

'Early Elberta'

'Loring' This medium-sized freestone has a slight fuzz and is blushed red over a yellow ground. The flesh is yellow, firm, and medium textured. It resists bacterial leaf spot. Good for the North and South. Widely available. Origin: Missouri.

'Suncrest' This firm, large freestone has a red blush over yellow skin. It is susceptible to bacterial leaf spot and should be grown in the West and other areas without this disease. It is hardy in cold sections of the North. Widely available. Origin: California.

Late Varieties

'Belle of Georgia' ('Georgia Belle') The skin of this outstanding white peach is red blushed over creamy white. The flesh is white and firm, has excellent flavor, and is fair for freezing but poor for canning. The tree is vigorous, very winter hardy, and productive but very susceptible to brown rot. Good for the North and South. Widely available. Origin: Georgia.

'Blake' This large freestone has red, slightly fuzzy skin. The flesh is yellow and firm. It is good for freezing and excellent for canning. It is susceptible to bacterial canker. Good for the North and South. Origin: New Jersey.

'Cresthaven' The skin of this medium to large freestone is bright red over a gold ground and almost fuzzless. The flesh is yellow and nonbrowning. The tree is hardy. It is good for canning and freezing. Good for the North and South. Widely available. Origin: Michigan.

'Elberta' This large freestone is the old favorite for a midseason crop. The skin is red blushed over a deep golden yellow. The fruit tends to drop at maturity. It is resistant to brown rot. Good for all zones. Origin: Georgia.

'Fay Elberta' In California this ranks as the most popular all-purpose freestone peach. It equals 'Elberta' for eating fresh, cooking, and canning, and excels it for freezing. It ranks below 'Elberta' in adaptability, growing where winters fall to 20° F. The color is yellow heavily blushed with red. This one may require considerable thinning for large fruit. The blossoms are especially showy. Origin: California.

'Jefferson' Especially suited to localities where late spring frosts are a problem, this peach is noted for its fine texture and flavor. The skin is bright red over a bright orange background. The flesh is yellow and firm. It is a reliable producer that cans and freezes well. It has some resistance to brown rot. Good for the North and South. Origin: Virginia.

'Madison' Adapted to the mountain areas of Virginia, this variety has exceptional tolerance to frosts during the blossoming season, setting crops where others fail. The skin of this medium-sized freestone fruit is bright red over a bright orange-yellow ground. The flesh is orange-yellow, very firm, and fine in texture. The growth of the tree is average to vigorous. Good for the North and South. Widely available. Origin: Virginia.

'Raritan Rose' This vigorous, winter hardy tree produces delicious white-fleshed freestone peaches. The skin is red. Available in the East and North. Origin: New Jersey.

'Redskin' This popular peach ripens after 'Elberta'. It has good red color and handles well. It is excellent for freezing, canning, and eating fresh. Widely available in the East and North. Origin: Maryland.

'Rio Oso Gem' The skin of this large freestone is red over a yellow ground; the flesh is yellow, firm, fine in texture, and nonbrowning. It is good both fresh and for freezing. The blossoms are light pink, very large, and showy, and appear later than most peach blossoms. The tree is productive but not vigorous. Good for the South and West. Widely available. Origin: California.

'Sunhigh' This is a very good medium to large freestone. The skin is bright red over a yellow ground. The flesh is yellow and firm. The tree is vigorous and spreading. It is very susceptible to bacterial leaf spot and requires thorough summer spraying. Good for the North and South. Origin: New Jersey.

Nectarines

The nectarine is simply a fuzzless peach. Peach trees sometimes produce nectarines as sports, and nectarine trees will produce fuzzy peach sports. The two plants are nearly identical, but nectarines are generally more susceptible to brown rot. Gardeners in the South may have trouble with the disease because hot humid weather encourages it. You will have to spray regularly to control it. Otherwise, nectarines require the same care as peaches.

Early Varieties

'Earliblaze' This is a medium-sized, clingstone, yellow-fleshed fruit that ripens ahead of the 'Redhaven' peach. It has red skin and a prominent suture (seam down the length of the fruit). Good for the North and South. Origin: California.

'Independence' This medium-sized, oval fruit has brilliant cherry-red skin. The freestone flesh is yellow and firm. The tree is productive and moderately vigorous, with showy flowers. It will take warm winters. Good for the South and West. Origin: California.

'Pocahontas' The medium to large oval fruit is bright red. The semifreestone flesh is yellow, slightly stringy, and of good quality. This variety resists brown rot and frost during the blossoming season. The flowers are not showy. Good for the North and South. Origin: Virginia.

'Silver Lode' The skin of this fruit is red. The freestone flesh is white and sweet and of good texture. The tree requires little chilling. Good for the South and West. Origin: California.

'Earliblaze'

'Sungold' This medium-sized freestone has red skin and firm yellow flesh. It is a moderate-chill (555 hours below 45° F) variety and has some resistance to brown rot. Good for the South. Origin: Florida.

'Sunred' This low-chill nectarine (300 hours below 45° F) is adapted to Florida and ripens there in May. It is a small, yellow-fleshed clingstone with red skin. Origin: Florida.

Midseason Varieties

'Fantasia' This fairly large fruit has bright yellow skin covered up to two thirds with a red blush. The freestone flesh is yellow, firm, and smooth. The tree is vigorous and productive with showy flowers. It requires moderate chilling (500–600 hours below 45° F) and is susceptible to brown rot and bacterial leaf spot. Good for the South and West. Origin: California.

'Flavortop' The large oval fruit is mostly red, with firm, smooth, freestone, yellow flesh. The tree is vigorous and productive with showy flowers and needs moderate winter cold. It is susceptible to brown rot and bacterial leaf spot. Good for the South and West. Origin: California.

'Mericrest' An extremely winter hardy, yellow-fleshed, red-skinned nectarine with a large suture. It resists bacterial leaf spot and brown rot. Good for the North. Origin: New Hampshire.

'Nectared 4' The fairly large fruit is yellow with a red blush over much of the surface. The semifreestone flesh is yellow. The tree is productive with showy flowers. Good for the North and South. Origin: New Jersey.

'Nectared 5' The large fruit is smooth with a blush covering most of the yellow skin. The yellow flesh is semifreestone until fully ripe, then freestone. The tree is productive. Good for the South. Origin: New Jersey.

'Panamint' The fruit has a red skin, and the freestone flesh is yellow. The tree is vigorous and productive and needs little winter chilling. Good for the South and West. Origin: California.

'Pioneer' The fruit has a thin red skin. The freestone yellow flesh is red near the pit and has a rich, distinctive flavor. The tree requires little chilling and has large, showy blossoms. Good for the West. Locally available. Origin: California.

'Redchief' This medium fruit is bright red and attractive. The flesh is freestone, white, and fairly firm. The tree is vigorous and productive, has showy flowers, and is very resistant to brown rot. Good for the South. Origin: Virginia.

'Redgold' A hardy, firm, freestone nectarine with glossy red skin, resists brown rot and cracking, but is susceptible to mildew. Good for the North and South. Origin: California.

Late Varieties

'Cavalier' The medium fruit is orange-yellow with splashes and mottles of red. The yellow freestone flesh is firm, aromatic, and slightly bitter. The vigorous and productive tree has showy flowers and resists brown rot. Good for the North and South. Origin: Virginia.

'Fairlane' This very late ripening, red-skinned, yellow clingstone is good for the West. Origin: California.

'Flamekist' This is a large, red-skinned, yellow-fleshed, clingstone nectarine. It has moderate-chill requirements (500–600 hours below 45° F). It is unfortunately susceptible to brown rot and bacterial leaf spot. Good for the West. Origin: California.

'Gold Mine' The large fruit of this favorite old variety is white blushed with red. The juicy white freestone flesh has a sweet aroma and excellent flavor. It is a moderate-chill variety (500 hours below 45° F). Good for the West. Origin: New Zealand.

'Late Le Grand' The large clingstone fruit has yellow skin with a light red blush. This was the first large, firm, yellow commercial nectarine. The spreading tree is productive and has large, showy flowers. It is susceptible to brown rot and bacterial leaf spot. Good for the West. Origin: California.

'Bonanza'

Genetic Dwarf Peaches and Nectarines

The genetic dwarf peaches and nectarines form dense bushes, with long leaves trailing in tiers from the branches. In spring the branches are entirely hidden by flowers that are usually semidouble and always very showy. In winter the bare plants are also visually interesting. The fruit is of normal size.

Most require moderate winter chilling (400–600 hours below 45° F) for good bloom. None are blossom hardy in really cold places, but they can be grown in containers and protected until the warm season. If you try this method in the coldest northern regions, you may have to pollinate the flowers yourself with a pencil eraser, touching it first to pollen, then to the stigma of a different flower.

The plants can be kept in containers until about 5 feet tall, but in the ground they will eventually reach 6 to 8 feet and spread 6 to 9 feet. They can be used as ornamentals and require minimal pruning. Their fruit flavor and texture are not as good as those of standard-sized varieties, so they are not used commercially. The fruit must be thinned, and the trees need normal spraying for all the peach diseases and pests. These dwarfs are more susceptible to mites than normal-sized peach trees.

These dwarf plants were created by breeding numerous varieties, but they all probably share the common heritage of the 'Swataw' peach or the 'Flory' peach, both Chinese genetic dwarf varieties.

Genetic Dwarf Peaches

'Bonanza' A medium-sized, yellow-fleshed freestone with a red blush, this was the original genetic dwarf peach developed for the home gardener from earlier dwarfs like 'Flory'. It has a moderate-chill requirement (about 500 hours below 45° F), and the fruit ripens in mid-June in California. Good for the West and South. Origin: California.

'Compact Redhaven' This tree is larger than other dwarfs (up to 10 feet), and its leaves and growth habit resemble those of standard trees more than genetic dwarfs. The fruit resemble 'Redhaven' in size, quality, and color but are borne on a more compact tree. It tolerates cold better than other genetic dwarfs. Good for all zones, especially the North, Midwest, and East. Origin: Washington.

'Empress' This medium-sized, yellow-fleshed clingstone with glowing pink skin has a sweet flavor and juicy texture. It has a moderate-chill requirement (500–600 hours below 45° F) and ripens in early August in California. Good for the West and South. Origin: California.

'Garden Gold' This is a large, yellow-fleshed freestone with red skin and cavity. A moderate-chill variety (500–600 hours below 45° F) with showy flowers, the fruit ripens in mid-August in California. Good for the West and South. Origin: California.

'Garden Sun'   This large, yellow-fleshed freestone has red skin and cavity. A moderate-chill variety (500–600 hours below 45° F), the fruit ripens in early August in California. Good for the West and South. Origin: California.

'Honey Babe'   A large, firm, orange-fleshed freestone with red skin, this fruit rates high for flavor and sweetness. It is a moderate-chill variety (500–600 hours below 45° F), ripening before 'Redhaven'— mid-June in California. Good for the West and South and worth trying in the East with protection. Origin: California.

'Southern Flame'   A large, yellow freestone with red skin and cavity, this is a good eating fruit that ripens in late July in California. It has low-chill requirements (about 400 hours below 45° F). Good for the West and South. Origin: California.

'Southern Rose'   This is a large, firm, yellow-fleshed freestone with red blush. Rated as a low-chill (300–400 hours below 45° F) variety, the fruit ripens in early August in California. Good for low-chill areas of the West and South. Origin: California.

'Southern Sweet'   This medium-sized, yellow-fleshed freestone has a red blush and good flavor. This moderate-chill variety (500–600 hours below 45° F) matures in mid-June in California, ahead of 'Redhaven'. Origin: California.

'Sunburst'   A large, firm, yellow-fleshed clingstone with a red blush, the fruit is juicy with a red cavity, has good flavor, and ripens in mid-July. It is a high-chill variety (900 hours below 45° F) suggested for warm areas of the East and South and colder areas of the West. Origin: California.

'Nectarina'

Genetic Dwarf Nectarines

'Garden Beauty'   This yellow-fleshed clingstone with red skin has a low-chill requirement (about 400 hours below 45° F). It has large double flowers and ripens in late August in California. Good for the South and West. Origin: California.

'Garden Delight'   A yellow-fleshed freestone, this has a low-chill requirement and red skin. The fruit ripens in mid-August in California. Good for the South and West. Origin: California.

'Garden King'   This yellow-fleshed clingstone with red skin has a low-chill requirement and ripens in mid-August in California. Good for the South and West. Origin: California.

'Golden Prolific'   This large, yellow-fleshed freestone with orange skin and a red center has a high-chill requirement (900 hours below 45° F). The fruit ripens in late August. Good only for high-chill areas in the West but worth trying in the East and North if given winter protection. Origin: California.

'Nectarina'   A medium-sized, yellow-fleshed freestone with a red blush and cavity, the fruit of this low-chill variety (300–400 hours below 45° F) ripens in mid-July. Good for the South and West. Origin: California.

'Southern Belle'   This is a large, yellow-fleshed freestone with red blush. The fruit of this low-chill variety (300–400 hours below 45° F) ripens in early August in California. Good for the South and West. Origin: California.

'Sunbonnet'   This is a large, firm, yellow-fleshed clingstone with a red blush. The fruit of this moderate-chill variety (about 500 hours below 45° F) ripens in mid-July in California. Origin: California.

Pears

Pears, especially dwarf pears, are a fine choice for the home gardener. The trees are attractive even in winter; they require little pruning after they begin to bear; they begin to bear early; and the fruit stores

fairly well without any special requirements. The plants take well to formal or informal training so space is not a problem.

Standard pears will spread 25 feet across and grow as tall or taller. A dwarf in natural shape needs a space about 15 feet square, but with the pruning and training methods described on pages 39–57 you can grow a pear flat against a fence or wall using very little space.

You don't need to thin the fruit, but if a very heavy crop sets, remove fruit that is damaged or very undersized. Thin a few weeks before harvest.

'Bartlett'

All pears need a pollinator. Use almost any other pear. 'Bartlett' is a poor pollinator for 'Seckel', however, and 'Magness' does not pollinate anything.

The one real drawback with pears is fireblight, but a home gardener can work around it by choosing varieties wisely and diligently pruning off diseased wood. Fireblight is at its worst in spring, when insects carry it from tree to tree. Resistant plants are the best answer. Cut off any infected tissue well below the infection and burn it. Other pests are codling moth, mites, pear psylla, and pear slug. See pages 29–37.

Most fruits are best when picked ripe or nearly so. Pears are the exception. A tree-ripened pear breaks down and turns soft and brown at the core. Always harvest pears when they have reached full size but are still green and firm. Hold them in a cool, dark place if you intend to eat them within a few weeks. For longer storage refrigerate the harvested fruit and remove it from cold storage about a week before you want to use it. Pears ripen faster if they are held with other pears in a poorly ventilated spot. For fast ripening place several in a plastic container.

Early Varieties

'Clapp's Favorite' This large yellow fruit with red cheeks resembles 'Bartlett'. The flesh is soft, sweet, and good both for eating and canning. The tree is attractively shaped and very productive but highly susceptible to fireblight. Since it is hardy, this variety is best in cold, late-spring zones. Good for the North and West. Widely available. Origin: Massachusetts.

'Moonglow' This large attractive fruit is soft and juicy with a mild flavor. Use it for canning or eating fresh. The tree is upright, vigorous, and heavily spurred and begins bearing a good crop when quite young. It resists fireblight, so it is good wherever the disease is a severe problem. Good for all zones. Widely available. Origin: Maryland.

'Orient' This nearly round fruit has firm flesh that makes it a good canner; however, the flavor is too mild for a good fresh pear. The tree produces moderate crops and resists fireblight. Good for the South. Origin: California.

'Red Clapp' ('Starkrimson') An attractive red-skinned sport of 'Clapp's Favorite', it does well in the West or North but is susceptible to fireblight. It has good quality fruit. Good for the West and North. Origin: Michigan.

Midseason Varieties

'Bartlett' This familiar commercial pear is yellow, medium to large, and thin skinned. The flesh is very sweet and tender, fine for eating, and good for canning as well. The tree does not have especially good form and is subject to fireblight. It takes summer heat, provided there is adequate cold in winter. In cool climates it needs a pollinator (any variety but 'Seckel' or 'Magness') to set fruit well. Good for all zones. Widely available. Origin: England.

'Lincoln' Called by some "the most dependable pear for the Midwest," this variety bears large fruit abundantly. The tree is extremely hardy and blight resistant. Good for the North and South. Origin: unknown—Midwest.

'Magness' The medium-sized oval fruit has a slightly russet color. The flesh is highly perfumed. The tree is vigorous and spreads widely even for a pear. This variety produces small amounts of good quality fruit. It will not pollinate any other pear varieties. It is highly resistant to fireblight. Good for the South and West. Origin: Maryland.

'Maxine' ('Starking Delicious') This large and attractive fruit has firm, juicy, sweet white flesh. The tree is somewhat blight resistant. Good for the North and South. Origin: Ohio.

'Parker' This medium to large pear is yellow with a red blush. The flesh is white, juicy, and pleasantly sweet. The tree is upright, vigorous, and fairly hardy but susceptible to fireblight. Good for the North. Origin: Minnesota.

'Anjou'

'Bosc'

'Comice'

'Seckel'

'Sensation Red Bartlett'
('Sensation')   Juicy, white,
'Bartlett'-flavored flesh is covered by yellow skin heavily
blushed red. The tree form resembles 'Bartlett' but is smaller; leaves and shoots have a
reddish tinge. It is susceptible
to blight, and in cool climates
it needs a pollinator. Good for
the West. Origin: Australia.

Late Varieties

'Anjou'   The fruit is large
and green, with a stocky neck.
The firm flesh has a mild flavor and is not especially juicy.
It stores well and is good for
eating fresh or for canning.
The tree is upright and vigorous but susceptible to fireblight. Originating in the mild
area near the Loire, it is not
recommended for hot-summer
areas. Good for the North and
West. Widely available. There
are also red 'Anjou' selections
available. Origin: France.

'Bosc'   This long, narrow
fruit has a heavy russet color.
The flesh is firm, even crisp,

with a heavy perfume that
makes some people consider it
among the very finest pears.
It is good fresh or canned and
is especially fine for cooking.
The tree is large and susceptible to fireblight. Good for
the North and West. Widely
available. Origin: France.

'Comice'   The large, round
fruit is green to yellow-green
with a tough skin. This sweet,
aromatic, and juicy pear is the
finest for eating but is not recommended for canning. The
large vigorous tree is slow to
bear and moderately susceptible to fireblight. It sets fruit
better with a pollinator and
should be grown on dwarfiing
quince rootstock. This is the
specialty of the Medford region in Oregon, but it does well
in home gardens along the
California coast. Good for the
West. Origin: France.

'Duchess'   This pear is
greenish yellow and very
large. The flesh is fine textured and of good flavor. The
tree is symmetrical and bears
annually. Good for the North.
Origin: France.

'Gorham'   Of excellent quality, this fruit strongly resembles 'Bartlett' but ripens
later and can be stored longer.
The tree is dense, upright, vigorous, and productive. Good
for the North and South.
Origin: New York.

'Kieffer'   This sand pear
hybrid has large yellow fruit
that is often gritty and therefore poor for fresh use, but it
keeps well in storage and is
excellent for cooking and canning. The tree is especially
recommended because of a
high resistance to fireblight
amounting to near immunity.
It needs little winter chill but
stands both cold and heat well,
so its range is wide. Good for
the East, North, South, and
Midwest. Widely available.
Origin: Pennsylvania.

'Mericourt'   This pear is
green to yellow-green, sometimes blushed deep red and
flecked with brown. The
creamy white flesh is nearly
grit free and is good fresh or
for canning. A vigorous tree,
it will withstand -23° F during
full dormancy. It resists both
fireblight and leaf spot.
Good for the South. Origin:
Tennessee.

'Patten'   This large, juicy
pear is particularly good fresh
and fair for canning. Since the
tree is especially hardy, it
should be considered for the
northern Mississippi valley
where 'Bartlett' and 'Anjou'
fail. Good for the North.
Origin: Louisiana.

'Seckel'   This is a small,
yellow-brown fruit that is not
especially attractive but has
the finest aroma and flavor of
any home garden pear. Eat it
fresh or use it whole for spiced
preserves. The tree is highly
productive and very fireblight
resistant. It sets fruit best
with a pollinator (any pear
but 'Bartlett' or 'Magness').
Good for all zones. Widely
available. Origin: New York.

'Shenseiki'

Asian Pears (Apple Pears)

Asian pears are true pears, but are a different species than the common pear. The common name "apple pear" has probably been given them because their texture is crisp like an apple, and some of

them are shaped like apples. But they are not crosses between apples and pears. They are a distinctly different fruit, with their own unique flavor. This group of pears is native to Japan and China. They were selected for size, shape, flavor, and lack of grittiness. The fruit is eaten firm like an apple, and it will keep in the refrigerator for four to eight months without getting soft like a 'Bartlett' pear. They bloom and ripen just like a 'Bartlett' pear and the trees are pruned like ordinary pear trees. Like other pears, they espalier well. The blossom is white and attractive.

The fruit has its own characteristic flavor, texture, and juiciness. All are susceptible to fireblight and need cross-pollination with any other pear that flowers at the same time.

All Asian pears grow well on the West Coast and they may grow in the South or the East if adequate fireblight protection can be provided. They are best grown on Asian rootstocks, although they are very dwarf on quince rootstock. Fruit should be thinned to one fruit per spur and this is best done when fruit is ¾ inch in diameter, six to seven weeks after bloom.

The fruit should be picked when it is ripe, not picked early and ripened indoors as with other pears.

Varieties

'Chojuro' A flat, russet-skinned variety with a strong flavor, this fruit is very firm, stores a long time, and bears as regularly as clockwork every year. Origin: Japan.

'Hosui' Golden brown skin covers a large, apple-shaped fruit with notably fine-textured, juicy, sweet flesh. Fruit lasts up to six months after picking. Origin: Japan.

'Kikusui' This flat, yellow pear has good texture and is a very juicy, mild-flavored variety. Pick when the skin begins to turn yellow. Origin: Japan.

'Shenseiki' A flat, yellow pear with good texture and flavor, this is the earliest maturing quality Asian pear and should be picked when the skin is yellow. Origin: Japan.

'Shinko' Perhaps the heaviest bearer of Asian pears, it is also fine textured and rich flavored. The medium-sized applelike fruit is a glowing golden russet color; lasting time after harvest is two to three months. Origin: Japan.

'Twentieth Century' ('Nijisseiki') A flat, green pear with fine flavor, this is the most popular Asian pear in California. It tends to bear in alternate years since it crops very heavily. Thin to one fruit per cluster. Origin: Japan.

'Ya Li' A pear-shaped fruit with fine texture and flavor. 'Ya Li' is partially self-fertile so it will set fruit without a pollinator but will produce a heavier crop with a pollinator. Because it blooms earlier than most other varieties, it must have an early pollinator such as 'Tsu Li' or 'Seuri'. Thin for best size and annual bearing. An extremely low chill requirement (300 hours below 45° F) means this apple pear will set fruit in warm southern areas. Origin: China.

'Hachiya'

'Fuyu'

Persimmons

The persimmon belongs to the same family of plants as the ebony tree of southern Asia. The American persimmon, *Diospyros virginiana,* grows as a native from Connecticut to Kansas and southward, but

it won't take the extreme cold of the northern plains or northern New England. It has small, edible fruit up to 2½ inches in diameter.

The large persimmon found in the market is the Oriental persimmon, *Diospyros kaki,*

and its many varieties. It could be far more popular than it is if more gardeners realized the great value of both tree and fruit. The tree grows

well in any well-drained soil and makes a fine medium-sized shade tree with large leaves that turn a rich gold to orange-red in the fall. A heavy crop of orange fruit decorates the bare branches until winter. It can be grown in the southern states and on the West Coast.

Persimmon foliage is large and glossy, with leaves reaching 4 to 6 inches in length. The new spring leaves are bronze or reddish, and in fall they turn to shades of yellow, pink, and red. The fruit hangs on into the first frosts and is orange with a red blush.

Store persimmons in the refrigerator and use only after they soften. Placing them in a bag with an apple will hasten the ripening process. Eat them when they soften, or use the flesh as you would applesauce or bananas. If you want to store it, mash the soft pulp out of the skin for freezing, and discard the tough skin. 'Fuyu' is the one persimmon that is

not astringent when firm. You do not need to ripen and soften it before eating.

Use a persimmon tree as an attractive background plant in a shrub border, or in front of evergreens (where it shows off its leaves and fruit best). Since the persimmon grows slowly, it takes well to espalier training. Train it informally against a flat surface, or use a trellis to form a persimmon hedge. It will also grow well as a single lawn tree, but you'll have a problem in late fall when the soft fruit drops and squashes.

American persimmons are normally dioecious, meaning that some trees are male, producing pollen but no fruit, while others are female. You will need a female tree for fruit and a male close by for pollen. Plant both unless you have wild trees near your garden. Some improved varieties bear fruit without requiring a separate pollinator, but these are not yet generally available.

Oriental persimmons set fruit without pollination. The large fruit, 3 to 4 inches in diameter, are usually picked in October before the first frost. Oriental persimmons stand winter temperatures to about 0° F, but they need only a short chill period (100–200 hours below 45° F) to fruit well in southern locations.

In the West the persimmon has no serious pests. In the East a flat-headed borer may attack the trunk, but it can be removed by hand.

Varieties

American Persimmons
Good varieties include 'Early Golden', 'Garretson', 'Hicks', 'John Rick', 'Juhl', and 'Meader', which sets seedless fruit and needs no pollinator.

Oriental Persimmons   Good varieties include 'Chocolate', 'Fuyu', 'Hachiya', and 'Tanenashi'. 'Hachiya' is the popular large fruit sold commercially. 'Chocolate' has dark flesh around its seeds and is a type of persimmon rather than a variety.

Plums

Of all the stone fruits, plums are the most varied. They range from hardy little cherry plums and sand cherries to hybrids with the hardiness of natives, sweet European plums (and the prunes made from them), and sweet or tart Japanese plums.

European plums tend to be small, and most varieties are egg-shaped. The flesh is rather dry and very sweet. Prunes from these plums are the sweetest and easiest to dry. The plants are fairly

hardy, but some varieties do well in mild-winter areas. All varieties are self-pollinating, except for those noted.

Japanese plums have relatively large, soft, and juicy fruit, sometimes with tart flesh near the pit. The plants are the least hardy of the various kinds of plum, although selected varieties are grown in

the milder northern regions. Taste one to test for ripeness before you harvest. Most Japanese plums need cross-pollination. Exceptions include 'Santa Rosa', 'Methley', 'Beauty', and 'Climax', but all plums set fruit better with a pollinator. Most are very susceptible to bacterial leaf spot in the South and East. Some resistant varieties are listed and rated by local extension agencies. Check with your farm advisor or nursery.

Plum trees bear for 10 to 15 years or more, and standard plum trees take space. Expect your tree to fill an area 15 to 20 feet square. Bush and cherry plums reach 6 feet or so and may spread as wide or wider. A dwarfed European plum on Nanking cherry roots will get as tall as 10 to 12 feet in height.

All the large-fruit Japanese plums must be thinned five to eight weeks after bloom. Thin fruit to 4 to 6 inches apart. European plums should have clusters thinned to two or three fruit per spur. The young trees should be pruned as discussed on page 49. Bush varieties need their oldest shoots trimmed off at ground level after about four years of bearing to encourage new growth.

Tree plums don't lend themselves to confinement, so use bush types if your space is limited. Use bush types as shrubby screens or try them in containers.

Brown rot is a major concern and requires summer spraying. See page 34. Bacterial leaf spot is a serious problem for most Japanese plums in the South and the East, but it is not a problem on other types of plums. Japanese types do best in the West; European types are best in the East; and bush types grow well in the South, Midwest, and North.

Early Varieties

'Bruce'   This large Japanese plum has red skin, red flesh, and good flavor. The fruit matures early, ripening in June. The tree bears young and heavily. Use 'Santa Rosa' as a pollinator. Good for the North and South. Origin: Texas.

'Early Golden'

'Earliblue'   This European blue plum has tender, green-yellow flesh resembling 'Stanley' but softer. Production is moderate, but fine for the home garden and the tree is hardy. It is best planted in the North, and ripens in mid- to late July in Michigan. Origin: Unknown.

'Early Golden'   A round, medium-sized Japanese plum, it is yellow and of fair quality. The stone is small and free. The tree is vigorous, outgrowing other varieties, but it has a tendency to bear in alternate years. Thin carefully. Pollinate with 'Shiro' or 'Burbank'. The fruit ripens in Michigan in mid-July. Good for the North. Origin: Canada.

'Mariposa'   The large, heart-shaped, Japanese fruit has mottled red and yellow skin enclosing sweet, red, freestone flesh. Fruit is good both for eating fresh and for cooking. The ripening time is mid-July. Because of its low-chill requirement (only 400 hours below 45° F), this is good choice for the mildest winter climates. For pollinators use 'Late Santa Rosa', 'Santa Rosa', or 'Wickson'. Origin: California.

'Methley'   This small to medium-sized Japanese fruit is reddish purple with red flesh and excellent flavor. It ripens over a long period, requiring several pickings. The tree is upright with hardy flower buds. For better crops pollinate with 'Shiro' or 'Burbank'. The fruit ripens in Michigan in mid-July, earlier in the South. Good for the North. Widely available. Origin: South Africa.

'Burbank'

'Santa Rosa'

'Green Gage'

'Santa Rosa'   This popular large Japanese plum has deep crimson skin. Its flesh is purplish near the skin and yellow streaked with pink near the pit. It is good for dessert or canning. Use any early or midseason plum for improved pollination. The fruit ripens in California in mid-June, later in the North. Good for all zones. Widely available. Origin: California.

Early Midseason

'Abundance'   This purple-red Japanese plum has tender yellow flesh that softens quickly. It is good for dessert or cooking. The tree tends to bear every other year. Use 'Methley' or 'Shiro' as a pollinator. The fruit ripens in Michigan in late July. Good for the North. Origin: California.

'Satsuma'   This is a Japanese plum with red juice. The meaty fruit is small to medium, with a dull, dark red skin, mild red flesh, and a small pit. Use it for dessert or preserves. Use 'Santa Rosa' or 'Wickson' as a pollinator. Good for all zones. Widely available. Origin: California.

'Shiro'   This medium to large Japanese plum is round and yellow and has a good flavor. Use it fresh or for cooking. The tree produces heavily. Use 'Early Golden', 'Methley', or 'Santa Rosa' as a pollinator. The fruit ripens in early July in California and the South, in late July in Michigan. Good for all zones. Widely available. Origin: California.

Midseason Varieties

'Burbank'   This large red Japanese plum has amber flesh of excellent flavor. The trees are fairly small and somewhat drooping. Use the fruit for canning or dessert. Use 'Early Golden' or 'Santa Rosa' as a pollinator. The fruit ripens in early August in the Northwest and in mid-July in California. Good for all zones. Widely available. Origin: California.

'Damson'   This old European plum is derived from a different species than other European plums. The smallish blue fruit is best for jam, jelly, and preserves. Improved varieties include 'Blue Damson', 'French Damson', and 'Shropshire Damson'. The trees are small and self-pollinating, and the fruit ripens at the end of August or in September. It is a late plum in the North. Good for all zones. Widely available. Origin: England.

'Green Gage' ('Reine Claude')   The greenish yellow European fruit has amber flesh and is good fresh, cooked, or preserved. The trees are medium sized and self-pollinating. The fruit ripens in mid-July, later in the North. Good for all zones because it has a low-chill requirement and is cold hardy. Widely available. Origin: Unknown.

'Ozark Premier'   This extremely large, red Japanese plum has yellow flesh. The trees are disease resistant, hardy, and productive. The fruit ripens early in August.

Good for the North, Midwest, and South. Widely available. Origin: Missouri.

'Queen Ann'   The large, free-stone purple fruit has golden orange flesh. The combined qualities of juiciness, rich flavor, and no tartness at the pit make it an esteemed Japanese dessert plum. The fruit ripens in mid-July. The tree is less vigorous than other Japanese plums. Use 'Santa Rosa' as a pollinator. Origin: California.

'Stanley'   The most widely planted European plum in the East, Midwest, and South, this tree has large, dark blue fruit with firm, richly flavored yellow flesh. It bears heavily every year, is hardy into central Iowa, and is self-pollinating. The fruit ripens after mid-August, into September in northern regions. Good for the North. Widely available. Origin: New York.

'Stanley'

'Yellow Egg'

'Italian Prune'

'Sugar' This very sweet, dark blue European plum is fairly large and excellent for home drying and canning. The trees are self-pollinating and bear in alternate years, with light crops in off years. The fruit ripens after July 15. Good for all zones. Origin: California.

'Yellow Egg' This golden yellow European plum has a thick skin and yellow flesh. The round-topped, vigorous tree is hardy and productive. In the West the tree is planted in Washington. It is self-pollinating, and the fruit ripens in late August. Good for the North and West. Origin: Unknown.

Late Varieties

'Bluefre' This large blue European freestone has yellow flesh. The trees are vigorous and self-pollinating and bear young. The fruit ripens early in September and hangs on well after ripening. It has some sensitivity to brown rot. Good for the North. Origin: Missouri.

'French Prune' The small European fruit is red to purplish black and very sweet with a mild flavor. This is the main prune variety in California. The tree is large and long-lived, often surviving even after orchards have become housing developments. It is self-pollinating, and the fruit ripens in late August to September. Good for the South and West. Widely available in California. Origin: France.

'Italian Prune' ('Fellenberg') This dark blue European plum is very sweet and good for dessert, canning, or drying. It has been the major plum of the Washington-Oregon area. The fruit ripens in late August and September. Good for the South and West. Widely available. Origin: Germany.

'Late Casselman' and 'Late Santa Rosa' These firm, late-ripening Japanese plums resemble regular 'Santa Rosa' in tree shape and appearance of fruit, but the fruit is sweeter and much firmer. They mature six weeks later than 'Santa Rosa'. Origin: California.

'President' This large, dark blue European fruit has amber flesh and ripens very late, after other plums. It lacks outstanding flavor, but use it for winter cooking or canning. Use another late European plum as a pollinator. The fruit ripens in Michigan at the end of September. Good for the North. Origin: England.

Hardy Plums

These plums were especially selected and bred for the coldest northern and Great Plains climates.

'Pipestone' This large red fruit has tough skin that is easy to peel. The flesh is yellow and of excellent quality but somewhat stringy. The tree is vigorous and hardy, performing reliably in cold regions. Use 'Toka' or 'Superior' as a pollinator. Origin: Minnesota.

'Superior' This large, conical red fruit with russet dots and heavy bloom has yellow, firm flesh that is excellent for eating fresh. The tree bears

very young and prolifically. Use 'Toka' as a pollinator. Origin: Minnesota.

'Toka' This large, pointed fruit is medium red, and often described as apricot colored. The flesh is firm and yellow with a rich spicy flavor. The tree is a spreading, medium-sized heavy producer, but it may be short-lived. Use 'Superior' as a pollinator. Origin: Minnesota.

'Underwood' This very large, red, freestone plum has golden yellow flesh that is somewhat stringy but of good dessert quality. Ripening extends over a long season beginning in July. The tree is vigorous and among the hardiest. Use 'Superior' as a pollinator. Origin: Minnesota.

'Waneta' This is a large, tasty, reddish purple plum with yellow flesh. Use 'Superior' as a pollinator. Origin: South Dakota.

Pomegranates

With its shiny leaves, fleshy orange flowers, and bright red fruit, the pomegranate is one of the most beautiful fruiting plants. The leaves have a reddish tint in spring and are bright yellow in fall, providing

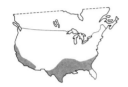

Pomegranate seeds

'Wonderful'

'Smyrna'

a background that makes the fruit especially attractive.

Pomegranates are often thought of as a tropical or desert fruit, but in fact they withstand winter temperatures down to about 10° F. While they do ripen their fruit best in very hot, arid climates, you can harvest edible fruit in cooler areas. They are ideal plants for the desert Southwest because they tolerate drought. Rain or irrigation close to harvest can cause pomegranates to split.

The edible portion of a pomegranate is the juicy scarlet flesh around the abundant seeds. If you score the skin just down to these seeds in about six places, cutting from stem to flower, you can open the fruit and expose all the seeds at once. Pomegranates are good in fruit salads and make an excellent syrup when cooked with sugar and a little water. This syrup is sold commercially as grenadine.

You can grow pomegranates as fountain-shaped shrubs or single- or multiple-trunk trees. They reach about 10 to 12 feet tall under ideal conditions but often remain smaller. A shrub can spread from 6 to 8 feet across.

Blossoms form on the current year's growth, and as the fruit grows heavier it pulls down the slender new branches, making a decorative weeping effect. The plant stands drought well, but keep the moisture level even. Thin-

ning is not necessary. An excess crop is very decorative if left on the tree. The trees are self-fertile, so even a single specimen will bear fruit.

Normally pomegranates have no pest or disease problems, but the leaves can develop fungus diseases in humid climates.

Variety

'Wonderful'   This is the most common pomegranate and the only one you're likely to see in nurseries.

Quince

The quince is an underdog among fruits, perhaps because it must be cooked before it is edible. Since this processing is no more difficult than stewing tomatoes, quinces deserve to

be more widely grown both for their distinctive fruit and their ornamental value.

The plants are 15 to 20 feet tall and slow growing so they can be trained as trees or as multiple-stemmed shrubs. White or pale pink 2-inch blossoms appear at the tips of new growth in spring; flowering is therefore late enough to escape frost damage. The 2- to 4-inch dark green leaves have white, feltlike undersides; foliage rather sparsely covers attractively angled or gnarled branches. The large fruit is decorative, and after harvest the foliage turns yellow before dropping.

Wherever winter temperatures remain above -15° F—except in the low deserts of the Southwest and West—quinces stand a good chance of success. They prefer a heavy but well-drained soil but will tolerate damp soil as well as light soils and even some drought.

Fireblight is the one serious disease of quince, especially in humid regions. Since fireblight attacks new growth, avoid fertilizing and heavy pruning, both of which will stimulate vulnerable shoots. Codling moth is the principal insect pest.

The trees are self-fertile and begin bearing at four to five years. They continue to produce for 35 years. The fruit is rounded or somewhat elongated, depending on the variety. Ripening occurs in late summer or early fall, when fruit color changes from green to yellow. The fruit will be hard when ripe but needs careful handling because it bruises easily; storage life is only a few weeks.

Only a few cultivars are available, with little difference among them.

Varieties

'Orange'   This is sometimes sold as 'Apple' quince and is apple-shaped with orange flesh. Origin: Southern Europe.

'Pineapple'   A rounded fruit with white flesh, this quince can be cooked without the addition of water. Origin: California.

'Smyrna'   This variety is larger than the others with elongated fruit and yellow flesh. Origin: Turkey.

BERRIES

Berries are tempting to grow because most offer rich rewards for a small investment of time and space. A little sunlight and a pot, for example, are all you need to grow a crop of luscious strawberries.

The small-fruited plants can return bumper crops with minimal effort on your part, and several of the shrubby or vining plants can also add beauty to your landscape.

In considering berries you must work out the space and number of plants needed for a reasonable supply of fruit. If the plants are right for your climate and are given excellent care, the number of plants necessary to supply a family of five would be something like this:

Strawberries:
    25 (20–30 quarts)
Raspberries: 24 (20–30 quarts)
Blackberries:
    12 (10–15 quarts)
Blueberries: 4 (15–17 quarts)
Currants: 3 (10–12 quarts)
Gooseberries: 3 (10–12 quarts)

Strawberries are without question the easiest plants to work into any space you may have available. On a south-facing apartment terrace you can produce a crop in containers such as strawberry jars or moss-lined wire strawberry trees. An ideal plant for containers, where you can find it, is the European wild strawberry, or *fraise de bois*. This plant won't make runners. It grows in a clump, so a container planting stays compact.

The cane berries—blackberries and raspberries—take more space, although you can grow a few in large containers. If you train them carefully along a fence or trellis and keep them pruned, they won't take very much space, but they will produce heavy crops of fruit that you just can't buy since the finest flavor disappears during transportation to your grocery store.

Trailing blackberries tied to a trellis

Blueberries and currants make extremely ornamental shrubs, covered with bloom in spring, and with decorative fruit in later seasons. Blueberries require light, acid soil, and constant moisture, so try them where you would grow azaleas. Currants and gooseberries are an interim host to a serious disease of five-needle pines, so in some areas you're not allowed to plant them. Where they are permitted nothing takes less care, is more decorative, or gives a tastier crop.

Grapes, with their lush foliage, fall color, and interesting vines, are among the best landscaping plants. Use them on arbors, against walls, as fences, or as freestanding shrubs on a pole or trellis. Choose varieties recommended for your climate since grapes are especially sensitive to summer heat.

Blackberries

Blackberries and raspberries are closely related and have similar growing requirements, but blackberries are larger and more vigorous, and some varieties are less hardy. Blackberries come in two fairly

distinct forms—erect and trailing—and have a number of different names.

The ordinary blackberry is a stiff-caned, fairly hardy plant that can stand by itself if properly pruned. The trailing kind, generally called *dewberries,* are tender and grown mainly in the South. Trailing plants from the Pacific Coast are sold under their variety names—for example, 'Boysen' and 'Logan'—and are not referred to as dewberries. These varieties will freeze in the East and the North without winter protection.

Blackberries like a light, well-drained soil with a high moisture-holding capacity. Do not plant them where tomatoes, potatoes, or eggplants have grown previously, since the site may be infected with verticillium wilt and the berries cannot grow there.

Plant in early spring a month before the last frost. Set plants 4 to 6 feet apart in rows 6 to 9 feet apart. Before planting, clip canes to 6-inch stubs and plant at the depth they grew in the nursery. As soon as new growth begins, cut any stubs that do not sprout and burn them to protect plants from anthracnose, a fungal leaf-spot disease that can infect bramble plants. It is a problem in moist, warm climates, especially in the South.

Several inches of mulch will help keep soil moist, prevent weed growth, and help prevent suckers. Mulches such as fresh straw or sawdust use up nitrogen in the soil and you must supply extra nitrogen, but in general, don't fertilize too heavily or you'll get lush plant growth at the expense of a fruit crop.

The stiff-caned berries need no support, but can be confined between two wires to save space. Trailing blackberries should be cut to the ground after fruiting and the clippings destroyed to reduce the chances of spreading disease. New growth that sprouts during the last part of summer will fruit the following year.

If you disturb or cut roots of blackberries they will sucker badly. If you want more plants, chop off pieces of root beside the parent plants and set them in the new planting site like seed. If you don't want more plants, mulch the planting instead of cultivating for weed control. Blackberries can be more invasive than any other cultivated plant and, if abandoned, can quickly grow out of control.

Blackberries are subject to many pests and diseases. Save yourself trouble by buying certified plants and keeping them away from any wild plants. Some varieties resist a few diseases. Spray for blackberry mite and don't worry too much about the rest.

Either dewberries or erect blackberries can be planted in much of the South. In colder parts of the South, choose only the erect blackberry or be prepared to offer winter protection by burying canes under 2 inches of soil after the first frost, and then digging them out just as buds begin to swell.

Erect Blackberries

Erect blackberries are not recommended for the very coldest northern regions of the country but may succeed if you bundle up the canes in straw and burlap for the winter.

'Alfred'   This plant produces large, firm berries early. Locally available. Good for the North. Origin: Michigan.

'Bailey'   The fruit is large, medium firm, and of good quality. The bush is reliably productive. Good for the North and parts of the Pacific Northwest. Origin: New York.

'Black Satin'   These vigorous, thornless vines are semierect, producing heavy crops of large, elongated, dark berries equally good for fresh eating or for cooking. Ripening time coincides with 'Eldorado'. Good for the South. Origin: Maryland.

'Brainerd'   This large, highquality fruit is excellent for processing. The plant is productive, vigorous, and hardy. Locally available. Good for the South. Origin: Georgia.

'Brazos'   This is a popular variety in Texas, Arkansas, and Louisiana. The large fruit matures early and bears over a long period. The plant is vigorous and resistant to disease. Locally available. Good for the South. Origin: Texas.

'Cherokee'   This vigorous upright plant produces moderately thorny canes. Heavy crops of good-quality, mediumsized berries come in midseason. Widely available. Good for the South. Origin: Arkansas.

'Comanche'   The plant is similar to 'Cherokee' but the very large berries are better for eating fresh (also good for cooking). The crop ripens two weeks earlier than 'Cherokee'. Widely available. Good for the South. Origin: Arkansas.

'Darrow'   The berries are large and irregular, with firm flesh. They ripen over a very long season, sometimes into fall. The bush is hardy and reliable. Grows wherever the cold is not too intense. Origin: New York.

'Ebony King'   The large fruit is glossy black, sweet, and tangy. It ripens early and resists orange rust. Widely available. Good for the South, North, and Pacific Northwest. Origin: Michigan.

'Marion'

'Eldorado'   This very hardy and productive old variety resembles 'Ebony King' and is totally immune to orange rust. Good for the South and North. Origin: Ohio.

'Flint'   This blackberry needs only moderate winter chill. The berries are fairly large in clusters of 8 to 15, and the plant is highly resistant to leaf spot and anthracnose. Locally available. Good for the South. Origin: Georgia.

'Hendrick'   The fruit is large, medium firm, and tart. The bush is productive. Locally available. Good for the North. Origin: New York.

'Humble'   This low-chill Texas variety has large, somewhat soft berries and comparatively few thorns. Locally available. Good for the South. Origin: Texas.

'Jerseyblack'   This vigorous, semitrailing plant is notably rust resistant. It produces a midseason crop of large fruit that is similar to 'Eldorado' in appearance and flavor. Good for the South. Origin: New Jersey.

'Ranger'   This large, firm berry is best when fully ripe. It is especially recommended for Virginia and similar climates. Origin: Maryland.

'Raven'   This large berry is of high quality fresh or processed. The plant is erect, vigorous, and productive but rather tender. Origin: Maryland.

'Smoothstem'   The berries ripen late and are rather soft. Production is quite heavy in large clusters. The plant is thornless and hardy from Maryland southward. Origin: Maryland.

'Thornfree'   The medium to large fruit is tart and good. The semi-erect canes reach 8 feet, with up to 30 berries on each fruiting twig. The plant is rather tender. Widely available. Origin: Maryland.

'Williams'   The mediumsized fruit ripens in late June and is very good fresh. The bush is semi-erect, vigorous, and thorny. It resists most cane and leaf diseases. Locally available. Good for the South. Origin: North Carolina.

'Olallie'

'Thornless Logan'

Trailing Blackberries

All of these berries are tender and need protection from cold.

'Aurora' This very early fruit is large, firm, and of excellent flavor. The canes are most productive on the bottom 5 feet, so they do well planted close together and cut back heavily. Locally available. Origin: Oregon.

'Boysen' ('Nectar') A Pacific Coast variety with large and aromatic fruit produced over a long season, this plant is vigorous and fairly thorny. In California this variety provides an early crop from May 20 to June 20, depending on the area, and a second crop may extend the harvest through August. Also good for the South and Pacific Northwest. Origin: California.

'Carolina' This dewberry is vigorous and productive with very large fruits. It resists leaf spot diseases. Locally available. Good for the South. Origin: North Carolina.

'Cascade' Fresh or preserved, the flavor of this berry is unsurpassed. The plant is productive but tender. Good in milder parts of the Pacific Northwest. Origin: Oregon.

'Early June' The large, round fruit has excellent flavor and is acid enough for jam, jelly, and pies. These dewberries ripen in early June. The plant is semithornless and somewhat resistant to anthracnose and leaf spot. Locally available. Good for the South. Origin: Georgia.

'Flordagrand' The large fruit is very soft and tart, good for cooking and preserves. It ripens very early. Canes are evergreen. This dewberry must be planted with 'Oklawaha' for pollination. Locally available. Good for the South. Origin: Florida.

'Lavaca' This plant is a seedling of 'Boysen' that is hardier than the parent and more resistant to disease. The fruit is firmer and less acid. Locally available. Good for the South. Origin: Unknown.

'Lucretia' This hardy old favorite is a vigorous and productive dewberry with very large, long, soft fruits that ripen early. It needs winter protection in the North. Origin: North Carolina.

'Marion' The fruit of this midseason variety is medium to large, long, good quality, and excellent in flavor. The plants send out a few vigorous canes that are up to 20 feet long and very thorny. Good in milder parts of the Pacific Northwest. Origin: Oregon.

'Oklawaha' This dewberry resembles 'Flordagrand' and should be planted with it for pollination. Locally available. Good for the South.

'Olallie' This is the prime California variety, with large, firm, high-quality berries that are shiny black, firm, and sweet. The canes are thorny and very productive. The plant has a low-chill requirement and resists verticillium wilt and mildew. It is especially good for Southern California. Origin: Oregon.

'Thornless Boysen' This summer-bearing Pacific Coast berry is flavorful with a fine aroma, and grows on tender plants that must be trained. Bury the canes for the winter in colder climates. Widely available. Origin: California.

'Thornless Evergreen' A top commercial berry in Oregon, this variety produces large, firm, sweet fruit. Plants are vigorous and produce heavily, but are very tender. Pinching canes at 24 inches encourages more canes and laterals. Canes sometimes revert to a thorny type. Origin: Oregon.

'Thornless Logan' This large, reddish, tangy Pacific Coast berry is good for jam, pies, and a syrup base for drinks. Bury the canes in winter in colder climates. A thorny form, 'Logan', is not grown as widely. 'Thornless Logan' can revert to the thorny type. Origin: California.

'Young' This large, purplish black dewberry of excellent flavor is easy to pick. The plant produces few long canes. Anthracnose is a serious threat. Good for the South. Origin: Louisiana.

Keep birds away from berries with plastic netting.

'Bluecrop'

Blueberries

Blueberries demand the right climate and planting soil but take very little care if you provide suitable conditions. They are about as hardy as a peach but need a fair amount of win-

ter chill and will not grow well in mild winter climates.

Blueberries belong to the heath family and count azaleas, rhododendrons, mountain laurel, and huckleberries among their cousins. If any of these grow naturally near your garden, or if you have prepared an artificial site that suits them, then blueberries will also do well.

Blueberries like soil rich in organic material such as peat—very acid, but extremely well drained. Such soils are found in areas of high rainfall, which is lucky, since the berries need constant moisture,

even though they cannot tolerate standing water.

There are major commercial plantings of blueberries in sandy soils in New Jersey, especially Burlington and Atlantic Counties; in certain areas of Michigan; in Washington and Oregon; and to a certain extent in New York, Massachusetts, and Indiana.

Southern gardeners have a choice of two kinds of blueberries, depending on climate. The high-bush blueberry grows commercially in large plantings in southeastern and western North Carolina. A home gardener who hopes to succeed with the plant should live in, or north of, that area. If you know of native blueberries near your home, nursery plants should do well.

The rabbiteye blueberry, or southern high-bush blueberry, grows wild along streambeds in Georgia and northern Florida. With proper care it thrives where muscadine grapes succeed.

Soil must be extremely well-drained and acid. Plant in raised areas if there is any chance of water standing around the roots.

For both drainage and acidification, add large amounts of peat moss or other organic material to the planting soil, up to three-quarters peat moss by volume for soils that tend to be heavy. Never add manure; it is alkaline. Dig a planting hole somewhat broader and deeper than the roots of the young plant. Never cramp the roots into a small hole, but spread the roots in the hole.

Set high-bush blueberry plants about 4 feet apart. Choose two varieties for cross-pollination. Since the rabbiteye plants grow much larger, you can set them up to 8 feet apart, although they can also be set closer and blended into each other.

Do not feed plants the first year. In succeeding years use cottonseed meal, ammonium

sulfate, or any product suitable for camellias, azaleas, or rhododendrons.

Blueberries require constant light moisture in the soil, and cultivating damages their shallow roots. For both these reasons, you should mulch the plants heavily. Use any organic material such as straw, leaves, peat moss, or a combination, and renew it regularly to keep it about 6 inches deep. Some materials will use nitrogen as they decay, so you will have to compensate with extra feeding.

See page 52 for instructions on pruning blueberries.

Blueberries suffer from very few difficulties, but birds will take them all unless you net the plants. Nurseries carry suitable netting.

Always taste blueberries before picking. Some look fully mature when still quite acid.

Approximately the same varieties are used throughout the country since the conditions for growing them are so similar.

'Coville'

'Dixi'

Early Varieties

'Earliblue'   One of the best for all areas, this berry is large, light blue, and firm. The picking scar is small, so fruit keeps well and resists cracking. The plants are upright and comparatively hardy. Good for all zones. Widely available. Origin: New Jersey.

'Ivanhoe'   One of the best berries, this is large, light blue, and firm. The plant is very tender. Good for the South. Origin: New Jersey.

'Northland'   This is a hardy variety. The fruit is medium sized, round, moderately firm, and medium blue. The flavor is good. The plant is spreading but reaches only 4 feet at maturity. Good for the North and West. Locally available. Origin: Michigan.

'Woodward'   Among the earliest ripening of rabbiteye types, this is a short and spreading plant. Berries are large, light blue, mildly flavored and somewhat tart until fully ripe. Best performance is in low-elevation areas of the South. Origin: Georgia.

Midseason Varieties

'Berkeley'   This large, firm berry is pale blue and resists cracking. The bush is fairly upright and moderately hardy. Good for all zones and especially for the Pacific Northwest. Widely available. Origin: New Jersey.

'Bluecrop'   The fruit is large, light blue, and rather tart, but stores well and is good for cooking. The berries stand cold well, which makes the plant good for the shortest Michigan growing seasons. The plant is upright and medium hardy. Widely available. Origin: New Jersey.

'Blueray'   The fruit is very large, firm, and sweet. The plant is upright and spreading. Good for all zones, and especially recommended for Washington. Widely available. Origin: New Jersey.

'Croatan'   The fruit is medium sized and quick to ripen in warm weather. The plant is canker resistant. Good for the South and especially recommended for North Carolina. Locally available. Origin: North Carolina.

'Northblue'   Developed by the University of Minnesota to withstand the rigors of northern winters, this is a half-high-bush type reaching about 2 feet tall—a size that gives it complete snow cover in most winters. Large berries have outstanding flavor. It needs a pollinator such as 'Northsky'. Origin: Minnesota.

'Northsky'   Another blueberry developed for the trying northern winters, this plant is same size as 'Northblue', but its highly flavorful fruit is smaller and produced in smaller quantity. Fall foliage is deep red. Use 'Northblue' as a pollinator. Origin: Minnesota.

'Stanley'   This is a widely recommended variety. The fruit is medium sized and firm with good color and flavor. The bush is hardy, vigorous, and upright. Pruning is easy because there are few main branches. Good for the North and West. Origin: New Jersey.

'Tifblue'   This rabbiteye type is a Southern favorite because of its dependable production of large, tasty berries. Origin: Maryland.

Late Varieties

'Coville'   Origin: New Jersey. This is an inconsistent variety with large, light blue fruit that remains tart until near harvest. The plant is medium hardy. It is good for all zones and widely available. Origin: New Jersey.

'Delite'   This is the only rabbiteye variety that develops some sugar early. Picking is easier and flavor is excellent. The berries are medium large and may be reddish under the bloom. Origin: Georgia.

'Dixi'   The name is not an affectionate term for the South but Latin for "I have spoken" or, loosely, "That's my last word." It was given by the developer, F. V. Coville, on his retirement. The fruit is large, aromatic, flavorful, and good fresh. Popular in the Northwest. Origin: New Jersey.

'Southland'   The firm, light blue berries have a waxy bloom, and the skin may be tough late in the season. This is a particularly good Gulf Coast plant. Locally available. Origin: Georgia.

Currants and Gooseberries

Currants and gooseberries are among the most beautiful of the small fruits, but they are good home garden shrubs for other reasons as well.

You won't often see fresh fruit in the market, since

crops from the limited commercial plantings go to processors for commercial jellies and canned fruits. But since the plants are ornamental, easy to care for, and productive, northern gardeners can tuck a few among other shrubs for the bloom, fruit, and fall color. The crop can be used for jelly, pie, or just fresh eating for those who like a tart fruit.

We discuss only the red and white currants of the species *Ribes sativum* and the gooseberries *Ribes grossularia* and *R. hirtellum.* The black currant, *Ribes nigrum,* so aromatic and rich in vitamin C, was banned almost everywhere in the past because it is part of a disease cycle of five-needle pines. Spores of white pine blister rust from miles away spend part of their lives on the currants and then transfer to pines growing within about 300 feet.

The ban has been lifted in many states, and three rust-resistant varieties have been developed: 'Coronet', 'Crusader', and 'Consort'. 'Consort' is the best rust-resistant black currant to plant and makes an attractive hedge. The other *Ribes* species can also take part in transferring this disease, and they, too, are banned in some areas. Do not transport any currant or gooseberry from outside your region.

Fall or winter planting is a good idea, since the plants leaf

'White Imperial' currants

out early. In cold climates, plant right after the leaves drop, and the roots will be established before winter. Space the plants about 4 feet apart, or set them closer if more convenient, but expect them to grow less vigorously. Both do poorly in hot summer areas but may survive if planted against a north-facing wall. In most areas plant in the open, but be sure soil moisture is constant. Set the plants a little deeper than they grew in the nursery.

Currant Varieties

'Jumbo'   This American variety has large, pale green, sweet fruit. The plants are upright and vigorous. Origin: Unknown.

'Perfection'   This old variety has medium-sized red fruit in loose clusters. The plant has good foliage and is upright, vigorous, and productive. Good for Washington and Oregon. Widely available. Origin: New York.

'Red Lake' currants

Gooseberries

'Red Lake'   Recommended everywhere that currants will grow, this variety yields medium to large, light red berries in long, easy-to-pick clusters. The plants are slightly spreading. They produced the highest yield in Canadian trials and also produce well in California. Widely available. Origin: Minnesota.

'Stephens No. 9'   This is a good Great Lakes variety, with fairly large, medium red berries in medium clusters. The plants are spreading and productive. Locally available. Origin: Ontario, Canada.

'White Grape'   This is a white variety that is widely sold but is perhaps surpassed in quality by 'White Imperial', a relatively rare similar variety. Origin: Europe.

'Wilder'   This very old variety from Indiana yields dark red berries that are firm but tender and very tart. Plants are large, hardy, and long-lived. Origin: Indiana.

Gooseberry Varieties

'Clark'   This fruit is large and red when ripe. The plants are usually free of mildew.

This is a good Canadian variety. Origin: Ontario, Canada.

'Fredonia'   The large fruit is dark red when ripe. Plants are productive and vigorous, with an open growth habit. Origin: New York.

'Oregon Champion'   The medium-sized fruit is green, and the plant bears prolifically. Good for the Pacific Coast and the East. Origin: Oregon.

'Pixwell'   This is a very hardy variety for the Central and Plains States. The berries hang away from the plant, making them easy to pick, and the canes have few thorns. Fresh flavor is mediocre. Widely available. Origin: North Dakota.

'Poorman'   An American variety with red fruit, the plants are spiny and spreading. Good for the Pacific Northwest and the Central States. Locally available. Origin: Utah.

'Welcome'   This American variety bears an abundance of wine-red fruit. The flavor is sweet-tart. Widely available. Origin: Minnesota.

Grapes

In the earliest periods of human history, four foods were recognizably important. In the North there were apples and honey. In the South there were olives and grapes.

Two types of grapes are

commonly grown today: the American and the European. The American grape entered our history more recently than the vine of Europe, but it has already played an important role since its roots saved the European grape from extinction during the *Phylloxera vitifoliae* plague of the last century. This plague threatened to destroy the European grapes and the only remedy was grafting these grapes to American rootstocks. More recently American grapes have entered into sturdy hybrids that carry European wine grapes far north of their original climate area.

Grapes send their roots deep where they can, and they prefer a soil that is rich in organic material. You can encourage growth by adding an organic supplement at planting time and mulching the roots afterward. The site should have good air circulation because grapes are subject to disease in stagnant air.

Grapes need to be fed only nitrogen and may not always need that. If the leaves yellow and there is little growth in the early part of the season, they definitely need feeding. If you're not sure, try a feeding to see the result. Late feeding during the ripening period can force excessive growth and spoil the fruit.

Homegrown seedless grapes will never grow as large as those you buy at the market because commercial growers apply sprays of gibberellic acid (a plant growth hormone) to increase the fruit size. The spray simply increases the cell size within each grape and does not increase flavor or sugar content. Therefore, homegrown grapes will be more flavorful and will last on the vine longer because they will not rot the way large, crowded fruit tends to.

Harvest grapes by taste and appearance. When you think the bunch looks ripe, taste a grape near the tip. If it's good, cut the bunch.

Sometimes grapes never taste sweet, no matter how long you wait. This simply means that you have planted the wrong variety for your area. Either switch to another variety or replant the one you stubbornly insist on in a hot spot against a south wall or in a west-facing corner.

If vines overproduce and have too many bunches, the grapes will never get sweet. This can be remedied in future years by more extreme pruning in the dormant season or by thinning the grape bunches to balance the leaf area with the grape berry load.

The two kinds of grapes are pruned differently. (See page 52 for instructions.)

Grapes mildew badly and need good air circulation and often treatment with a fungicide. The classic remedy is copper sulfate. A number of pests attack grapes, especially certain beetles. Birds love grapes, but you can save the fruit by placing whole bunches in paper bags.

'Buffalo'

Grapes for the Northeast and Midwest

Many of the following grapes also grow well in the Pacific Northwest. This is mentioned in the descriptions. The American grapes are listed first, with a note when they are choice juice or wine grapes. French hybrids are listed second.

American Varieties

'Buffalo' This grape ripens in midseason. It has fairly large clusters of reddish black berries and is a good grape for wine or juice. Cane prune this vigorous vine. Performs well in the Pacific Northwest. Origin: New York.

'Catawba' Good for wine or juice, this red grape is a popular commercial variety. It requires a long season to ripen and will do well in southerly areas with the longest growing seasons. Thinning will hasten development. Widely available. Origin: North Carolina.

'Cayuga White' This variety bears white grapes in tight clusters. They are of good dessert quality. Origin: New York.

'Concord' This late grape is so well known and widely planted that it hardly needs description. Often the standard of quality in judging American grapes, the dark blue slipskin berries are rich in the characteristic "foxy" flavor, which is retained after processing. Widely available. Origin: Massachusetts.

'Delaware' The clusters and berries of this major wine grape are small, good for wine and juice, and excellent for dessert eating. The vines are subject to mildew. Origin: New Jersey.

'Edelweiss' This hardy, medium-sized grape is of good dessert quality. Origin: Minnesota.

'Fredonia' This variety should be allowed to set heavily, as it sometimes has difficulty with pollination. This is the top black grape in its season. The vines are hardy. Widely available. Origin: New York.

'Himrod' This is the top white seedless grape throughout the northern states. 'Thompson' types replace it where weather is warmer. The vines are brittle and only moderately hardy. Widely available. Origin: New York.

'Interlaken Seedless' This grape ripens early and has medium-sized clusters of small, seedless berries with greenish-white skin that adheres. The flesh is crisp and sweet. The grape resembles 'Thompson Seedless' but has more interesting flavor overtones. The vine is fairly hardy and does best with cane pruning. Widely available. A good substitute for 'Thompson Seedless' in the Pacific Northwest. Origin: New York.

'New York Muscat' Good for wine and juice, this variety's reddish black berries in medium clusters have a muscat aroma, which is rich and fruity, not "foxy." Temperatures below –15° F can cause winter injury. Origin: New York.

'Niagara' Good for wine and juice and more productive than Concord, this is the most widely planted white grape. It is vigorous and moderately hardy. Widely available. Origin: New York.

'Ontario' These white berries form fairly loose clusters. The vines are vigorous, productive, and moderately hardy and prefer quite heavy soils. Cane pruning is best. Also grown in the Pacific Northwest. Origin: Ontario, Canada.

'Schuyler' This grape resembles European grapes in flavor. It is soft and juicy with a tough skin. The vines are fairly hardy and disease resistant. Also a good choice in the Northwest. Origin: New York.

'Seneca' The small to medium berries resemble European grapes, with tender golden skin and sweet, aromatic flavor. The vine is hardy and takes cane pruning, although one parent is a European type. Good in the Pacific Northwest. Origin: New York.

'Swenson Red' This hardy red variety has good flavor and medium to large berries. Origin: Minnesota.

'Veesport' Borne in medium clusters, these black grapes are good for wine and juice and acceptable for fresh eating. The vine is vigorous. Origin: Ontario, Canada.

French Hybrids

Spur prune these vines. They are hybrids of European and less well known American grapes (not the 'Concord' type). All are primarily for wine or juice but are also good eaten fresh.

'Aurore' ('Seibel 5279')

This very early white grape is soft, with a pleasant flavor. It is a dependable producer and a vigorous grower, better in sandy than in heavy soils. Choose it if early ripening is needed. Widely available. Origin: France.

'Baco 1' ('Baco Noir')

This midseason variety produces small clusters of small black grapes. It is extremely vigorous and productive, but it tends to bud out early and is subject to frost injury. This is not a cold hardy variety. Widely available. Origin: France.

'Fredonia'

'New York Muscat'

'Interlaken'

'Seneca'

Grapes for the West

The West is grape country wherever you go, and yet many gardeners are disappointed in the fruit they harvest from their vines. The problem is usually a poor choice of varieties. More than any other fruit, grapes require the right climate and amount of heat to produce well. Too many gardeners buy vines because they like the fruit in the market or because they know a famous name.

In general western grape climates are divided into three groups. The first includes all of the West except California and the southwestern desert. Gardeners in these cool regions should choose an American grape of the "foxy"-flavored species, *Vitis labrusca.* Some of the best choices for the Pacific Northwest are indicated in the descriptions of American grapes. 'Concord', often sold by nurseries in the cool regions, is not successful in western Washington and Oregon. It requires more heat.

In California the cooler coastal areas and coastal valleys are suited to American grapes and selected European varieties with a low-heat requirement. 'Concord' does well, but the popular 'Thompson Seedless' will almost always disappoint the home gardener. 'Perlette' is similar, but it was developed for the low heat of this climate. The inland Northwest and parts of Utah, Montana, Colorado, and Idaho can also use 'Concord' and 'Niagara' from the coastal California list.

In the hot inner valleys of the California coast range, there are major commercial vineyards growing all the renowned European wine grapes. The Napa-Sonoma wine region is well known, but there are also many wine grapes grown in newer plantings in southern Santa Clara County, San Benito County near Salinas, and north of Santa Barbara.

'Tokay'

'Baco 1'

The hot Central Valley climate is perfect for the European table grapes that you see on your grocer's counters. 'Thompson', 'Ribier', and 'Emperor' all do well.

The low and high deserts are not good grape country. The earliest maturing European varieties stand the best chance of producing a crop.

Table Grapes

'Cardinal' These large, dark red berries ripen early and have firm, greenish flesh. The medium-sized clusters are extremely abundant. Use this one to cover an arbor or summerhouse. Spur pruning is best. Performs in both coastal valley and central valley climates. Origin: California.

'Concord' This grape, described earlier, does not like high California heat or the coolest Northwest summers, but does well elsewhere. Cane prune. Origin: Massachusetts.

'Delight' This grape ripens early, yielding well-filled clusters of large, greenish yellow berries with firm flesh and a distinct muscat flavor. Spur

pruning is best. Prefers coastal valley climate; locally available. Origin: California.

'Emperor' This late-ripening, large red grape has flesh so firm it seems to crunch. It is adapted to the hottest part of the San Joaquin Valley. The berries are firm and will store longer than other varieties. Spur prune. Origin: Unknown.

'Flame Seedless' This light red table grape is popular for its crisp texture, sweet flavor, and absence of seeds. Elongated, loose, medium-sized clusters ripen early along with 'Cardinal'. Prefers plenty of heat during the ripening period; the best color develops where nights are cool. Use either spur or cane pruning. Origin: California.

'Muscat of Alexandria' These late midseason, large green berries are splotched with amber and grow in loose clusters. They are not pretty but have an unparalleled musky, rich flavor. They lose flavor if held too long so are best eaten fresh from the home garden. These grapes

can also be dried as seeded raisins. Spur pruning is best. This variety requires the moderately high heat of the San Joaquin Valley or other inland valleys but not the desert. Muscats are often used to make sweet dessert wine. Unfortified muscat wine is a treat with desserts or fruit. Origin: North Africa.

'Niabell' Performing well both in coastal valleys and hot interior regions, this midseason variety produces well-filled clusters of large, black berries that are good fresh or as juice. Vines are vigorous, resist powdery mildew, and can be pruned to long canes. Cane pruning is best. Origin: California.

'Niagara' This white variety, described on opposite page, ripens mid- to late midseason. The best crop comes in coastal regions. Cane pruning is best. Origin: New York.

'Pierce' This is the hot-summer 'Concord'. Grow it in the warmer regions of central California where you want a black slipskin. It is very vigorous. Cane prune. Locally available. Origin: New York.

'Ribier' This is a beautiful, early midseason dessert grape with large, jet-black berries. It does best in hot interior valleys. The fruit tends to soften quickly in storage and lose its mild flavor. The vines are overproductive. Use short spur pruning and thin the flowers. Origin: France.

'Thompson Seedless' Ripening in early midseason, this is the top commercial seedless green grape. The clusters are well filled with rather long, mild flavored, fruit. These are excellent fresh if clusters are thinned. They are also used for raisins. Grow only in hot climates. (Try 'Perlette' or 'Delight' instead if in doubt.) Cane pruning is required. Origin: Asia Minor.

'Tokay' This late midseason variety bears large clusters of large, very firm, red grapes that are attractive but have little flavor. It does well in the Lodi area, and the cooler valley climates. Use 'Emperor' in hotter climates. Spur pruning is best. Locally available. Origin: Algeria.

'Chardonnay'

'Pinot Noir'

'Zinfandel'

Wine Grapes

The list includes three each of the best-known red and white grapes. They change character over short distances, so unless you know that they do well near you, don't count on getting the best quality.

'Cabernet Sauvignon' This is the great European black grape used to make the red Bordeaux wines of France. Cane prune for best results. Origin: France.

'Chardonnay' This popular white grape is used to make the famous French white Burgundy. It is a vigorous grower and moderate producer. The clusters of berries are small. It is best in cool coastal areas and should be cane pruned. Origin: France.

'Chenin Blanc' The vines on this white grape variety are vigorous and productive. It yields medium-sized berries and clusters. The coastal valleys and the San Joaquin Valley have the best climates for this variety. It should be cane pruned. Origin: France.

'French Colombard' This productive white variety yields a grape that is high in acid. It is adapted to coastal valleys and the Central Valley of California and bears medium-sized berries and clusters. It can be cane or spur pruned. Origin: France.

'Pinot Noir' This small black grape is used to make the French Burgundy wines. Cane prune. Origin: France.

'Zinfandel' This is a California specialty for both red and white wines. You can probably grow this better than any other, as it seems to make drinkable wine in a variety of climates. Origin: Unknown.

Grapes for the Southeast

Two quite different types of grapes are widely grown in the Southeast. Both are American in origin: the bunch grape and the muscadine grape. The bunch grape is typified by 'Concord', which was described earlier. Although this type prefers a cool climate, varieties are available for most regions. The real southern grape is, of course, the muscadine, with its smaller clusters of berries and liking for Cotton Belt weather.

Muscadine Varieties

Many muscadines are sterile and need a pollinator. The varieties described below as "perfect" will pollinate themselves and any other variety.

'Hunt' This dull black fruit ripens evenly. The quality is excellent, very good for wine and juice. The vine is vigorous and productive. This variety is unanimously recommended for home and commercial planting by the Muscadine Grape Committee. Origin: Georgia.

'Jumbo' This is a very large black muscadine of good quality. It ripens over several weeks, so it is excellent for fresh home use. The vines are disease resistant. Origin: southern United States.

'Magoon' Perfect. Reddish purple berries are medium sized and have a sprightly, aromatic flavor. The vine is productive and vigorous. Origin: Mississippi.

'Scuppernong' Most people call any similar grape a scuppernong, but this is the real variety. The fruit color varies from greenish to reddish bronze, depending on sun. It is late ripening, sweet, and juicy with aromatic flavor. Good for eating fresh or for wine. Origin: North Carolina.

'Southland' Perfect. This very large grape is purple and dull skinned with good flavor and high sugar content. The vine is moderately vigorous and productive. Good for the central and southern Gulf Coast states. Origin: Mississippi.

'Thomas' This standard grape has reddish black, small to medium berries that are very sweet and excellent for fresh juice. Locally available. Origin: southern United States.

'Topsail' Clusters of three to five berries have green fruit splotched with bronze. This sweetest of all muscadines is very good for fresh use. It is a poor producer. Vines are not very hardy but are disease resistant. Origin: North Carolina.

'Yuga' These reddish bronze berries are sweet and of excellent quality but ripen late and irregularly. They are fine for home gardens. Origin: Georgia.

Raspberries

Raspberries are the hardiest of the cane berries, and perhaps the most worthwhile home garden crop for several reasons. Prices for the market fruit are high because care and labor are expensive, and

market raspberries are subject to a long enough holding and handling period that fruit loses its finest flavor and may be bruised. Home garden fruit can be eaten at its peak.

The thing that makes a raspberry a raspberry is the fact that it pulls free of its core when you pick it. Other bramble fruits take the core with them.

The red raspberry is the most popular, but raspberries come in a variety of colors and plant forms—red, purple, yellow, and black fruits, with the red and yellow fruits growing either one or two crops on stiff canes and the purple and black fruits growing one crop on trailing canes. Because they are trailing, purples and blacks require trellising.

One-crop (single crop) raspberries produce fruit on canes that grew the previous year. Two-crop (everbearing) raspberries produce some fruit at the top of current-season canes in fall, and then produce a second crop on the rest of the cane the following year.

Raspberries are extremely hardy, so no special protection is needed except in the coldest mountain and plains climates. Where winter temperatures stay extremely low for long periods, and winds add to the chill, you should protect your plants in the following manner: Lay canes of the current season along the row or trellis, pinning portions that arch

upward. Be careful not to snap them. Where mice are not likely to be a problem, cover the canes with straw or sawdust to a depth of several inches, and then cover the mulch with poultry netting to hold it in place. If winter mouse damage is probable, bury the canes under 2 inches of earth.

In spring uncover the canes before they begin to leaf out, just as the buds swell. If the buds break while still covered, they will be extremely tender to even light frost.

Unfortunately for southern gardeners, raspberries do poorly in much of the South. They need cold winters and a long, cool spring. Everbearing plants don't like high heat.

California and Arizona gardeners are similarly unfortunate. Raspberries do not like spring and summer heat. Only the red varieties will grow and they are recommended only for coastal or mountain regions. The prime berry country on the Pacific Coast is western Washington around Puget Sound and the Willamette Valley of Oregon.

Raspberries are subject to all the same troubles as dewberries, but in the cold climates where raspberries grow best you'll have less trouble. Any verticillium in the soil rules them out entirely, however. Because black raspberries are susceptible to virus diseases carried by red raspberries, they should be planted at least 700 feet from any reds. Virus-free stock may spare you this trouble.

If you want to enlarge a planting, it is important to know the difference between black and red raspberries. Blacks and purples arch their canes to the ground and root at the tips to form new plants. If you want more plants, leave a few canes unpruned and in late summer pin the tip to the

'Latham'

ground. Throw on a little soil if you like. Then dig and separate the new plant in spring.

Red raspberries send up root suckers. You can dig and replant them just before growth begins. Take a piece of root and cut back the top.

Red and Yellow One-Crop Varieties

In these varieties, all fruit is borne on laterals that sprout from the year-old canes. There is one crop per season, either in late spring or early summer.

'Amber' This is a yellow berry that is an excellent dessert fruit. Good for the North. Origin: New York.

'Boyne' This berry excels where winters are cold and summers no more than warm. The red fruit has a strong, sweet-tart flavor. The moderately vigorous plant is subject to anthracnose. Origin: Manitoba, Canada.

'Canby' These large, firm, midseason berries are good for freezing. The plants are semithornless and do best in light soils in the West and Northwest. Origin: Oregon.

'Cuthbert' Once the leading commercial raspberry, and still unexcelled for dessert, canning, or freezing, this variety is difficult to pick, but this is not a big problem in the home garden. Good for the West. Locally available. Origin: New York.

'Fairview' These berries are large to fairly large and light red. The tall, branched canes are moderately hardy. Especially suited to western Washington and generally good for the West. Origin: Oregon.

'Hilton' This berry is the largest of all the reds, and of excellent quality. The plants are vigorous, productive, and hardy. Good for the North. Origin: New York.

'Latham' This early midseason variety is the standard eastern red raspberry. The berry is large, firm, and attractive, with a tart flavor. The plants are somewhat resistant to viral diseases. Good for the South and West. Widely available. Origin: Minnesota.

'Fallgold'

'Heritage'

'Royalty'

'Meeker' This Pacific Northwest favorite bears firm, sweet, bright red berries. The strong plants are botrytis resistant. Origin: Washington.

'Pocahontas' This recent introduction has large, firm, medium red berries with a tart flavor. The plant is winter hardy and productive. Good for the South. Origin: southern United States.

'Puyallup' These late-ripening large berries are somewhat soft. The plant does best in light soils in the Northwest, and is generally good for the West. Locally available. Origin: Washington.

'Sumner' This medium to large berry is firm and sweet, with intense flavor. Some strains crumble badly. The plants do well in heavy soil and are recommended for western Washington or along the coast to Monterey, California. Locally available. Origin: Washington.

'Sunrise' This early variety offers firm, fine-textured fruit of good quality. The plant is hardy and very tolerant of anthracnose, leaf spot, and cane blight. Good for the South. Origin: Maryland.

'Taylor' This variety offers mid- to late season crops of attractive, firm, red berries of excellent quality. The plants are vigorous and hardy. Good for the North. Locally available. Origin: New York.

'Willamette' The berries ripen in midseason and are large, round, firm, and good for freezing or canning. This is a vigorous, widely planted commercial variety. Good for the West. Origin: Oregon.

Red and Yellow Two-Crop Varieties

Two-crop raspberries produce a crop in fall at the end of new canes and another crop in early summer of the following year. In California the second crop may not survive the heat. In the Northwest these varieties may produce some fruit throughout the summer.

'Amity' A bicoastal raspberry developed and popular in the Northwest, it is proving itself also in the Northeast. The dark red, highly flavored fruit is good fresh or canned. Ripens just ahead of 'Heritage'. Origin: Oregon.

'Cherokee' The berries are large and firm, and the plant is winter hardy and productive. Good for the South, particularly the piedmont area of Virginia. Origin: Arkansas.

'Durham' These berries have very good flavor. The plants are very hardy and productive, bearing a second crop early. Good for the North. Origin: New Hampshire.

'Fallgold' The fruit is a tawny golden color with very sweet flavor; except for color this variety is similar to 'Heritage', although its performance is poorer in warmer climates. Widely available. Origin: New Hampshire.

'Fallred' The berries are of fair quality but are often crumbly. The plants are nearly thornless. The first crop appears in spring. Good for the South and North. Widely available. Origin: New Hampshire.

'Heritage' The medium-sized, firm fruit ripens in July and September. The vigorous, stiff-caned plants need little support. You can mow all the canes in late winter to get a single August crop and save pruning. Good anywhere. Origin: New York.

'Indian Summer' The fruit is large and of good quality. The first crop is light, the fall crop very late and moderately abundant. Good for all zones. Origin: New York.

'September' These medium to large berries are of good quality. The plant is vigorous and hardy—one of the best in the coldest regions. Good for the North and South. Origin: New York.

'Southland' Recommended for farther south than any other, this berry was developed in North Carolina but is not recommended for the coastal plain. It has large fruit of fair quality. Good for the South. Origin: North Carolina.

Botanically, the strawberry "berry" is a receptacle. The "seeds" are the true fruit.

Strawberries

If you have grown strawberries for any length of time, you know that flavor and yield are not predictable but vary from year to year depending on spring growing conditions. Strawberries are also very

regional in their adaptation and the best variety in one state may be only fair in another. A good nursery can be a big help, since the staff keeps abreast of developments in plant breeding and offers plants that should succeed. Your county agricultural agent can help, too, especially if you've had trouble in the past. Two types of strawberries are available: standard and everbearing. The traditional "everbearing" types actually bear two crops per season, one in summer and one in fall. Recently strawberries have been developed that are truly everbearing, producing fruit spring, summer, and fall. These are listed in catalogs as "everbearers" or "dayneutrals" (since they fruit regardless of day length). These are very good for decorative hanging baskets, since they even fruit on unrooted runners.

To encourage vigorous growth of regular varieties, remove blossoms that appear the year the plants are set out. The year that everbearing kinds are planted, remove all blossoms until the middle of July. The later blossoms will produce a late summer and fall crop.

Purple Varieties

These raspberry plants are tall and stiff and bear a single crop on year-old canes.

'Amethyst' This early berry is of high quality. Good for the North. Origin: Iowa.

'Brandywine' This hybrid berry is of mixed ancestry; growth and fruit are closest to the purple types. Vigorous with 10-foot canes, it produces good crops of tart, red-purple berries especially fine for jam. Origin: New York.

'Clyde' This early berry is large, firm, dark purple, and of excellent quality. The plant is vigorous. Good for the North. Origin: New York.

'Royalty' This very vigorous purple-red hybrid similar to 'Brandywine' produces fruit that is large, sweet, good for eating fresh and for cooking and preserving. The plant is immune to the raspberry aphid, which carries a debilitating virus disease. Origin: New York.

'Sodus' This midseason berry is large, firm, and of good quality but tart. The plants are productive. Good for the North. Origin: New York.

Black (Blackcap) Varieties

Gardeners in the South and West should be aware that black raspberries are least able to tolerate mild climates. They need cold and do poorly in western Washington, although they are planted in the Willamette Valley and elsewhere in Oregon. They bear a single crop on year-old canes.

'Allen' This variety produces large, attractive berries on a vigorous and productive plant. Good for the North. Origin: New York.

'Black Hawk' This late variety bears large berries of good flavor and yield. Good for the North. Widely available. Origin: Iowa.

'Bristol' These attractive, glossy black berries are large, firm, and of good quality. They must be fully ripe or you can't pick them. Good for the South and West. Widely available. Origin: New York.

'Cumberland' This favored variety has large, firm berries of fine flavor. The plants are vigorous and productive. Good for the South and North. Widely available. Origin: Maryland.

'Logan' ('New Logan') This variety produces heavy crops of large, glossy, good-quality berries. The plants hold up in drought and tolerate mosaic and other raspberry diseases. Good for the South and North. Origin: Illinois.

'Manteo' The fruit resembles 'Cumberland', but the plant survives farther south than any other. Good for the South. Locally available. Origin: North Carolina.

'Munger' The medium-sized fruit is of good quality. The plants are especially recommended for western Oregon. They are worth trying in western Washington, but may succumb to disease. Good for the West. Origin: Oregon.

Plant strawberries in soil with good drainage, and mound the planting site if you're not sure about the drainage. The new leaf bud in the center of each plant should sit exactly level with the soil surface. Never plant strawberries deep.

Gardeners who grow strawberries in containers in a disease-free soil mix don't have to worry about verticillium wilt and red stele (root rot). Both are caused by a soilborne fungus. Whether growing strawberries in containers or in garden soil, always ask for plants that are certified as disease free.

Winter protection is needed where alternate freezing and thawing of the soil may cause the plants to heave and break the roots. Low temperatures also injure the crowns of the plants. In the fall, after the soil has frozen to a depth of 1 inch, place a straw mulch 3 or 4 inches deep over the plants. Remove most of the mulch in spring when the centers of a few plants show a yellow-green color. You can leave an inch of loose straw, even add some fresh straw between rows. The plants will come up through it, and it will help retain moisture in the soil and keep mud off the berries.

In northern areas, and as far south as North Carolina, strawberries should be set out in early spring. In the warmer regions of North Carolina, plants can be set out in fall or winter as well, and you can expect a light crop from these plants about five months later.

In northern Cotton Belt climates, set out plants in September for the highest yield of spring berries. Waiting until later will diminish the crop.

In the warmest Gulf climates and into Florida, you must order cold-stored plants, or plants from the north, for planting from February to March. You can also obtain a quick crop by planting northern plants in early November for winter fruiting, but the crop will be smaller. The runners from February plantings can be transplanted in May and August to increase the size of your planting.

Western planting seasons impose unique restrictions on strawberry growth. In the coldest areas of California and from the Great Basin to Colorado, plant as early in spring as possible since there is good moisture in the soil to start the plants. If soils are usually too wet for planting, protect a mounded bed with plastic to keep the soil friable.

In the Northwest, especially the coolest areas of western Washington, plant early in fall so plants can become established before real cold sets in, or else wait until early spring. Watch for washouts from heavy fall rain. Weed carefully in spring so weeds don't compete.

In milder-winter areas of California, use chilled plants (stored at 34° F for a short period) and set them out in October and early November. The low desert is a chancy area for strawberries, but October planting may give results.

Anywhere in California, the berries will do better with a plastic mulch, which increases winter soil temperature and keeps the berries off the soil. Irrigate by furrows for raised beds or by drip irrigation.

'Sequoia'

Varieties for the South

'Albritton' This late berry is large and uniform in size and is excellent fresh and for freezing. It develops a rich flavor in North Carolina. Origin: North Carolina.

'Blakemore' These early berries are small and firm, with a high acid and pectin content. They have only fair flavor but are excellent for preserves. The plants are vigorous, with good runner production and high resistance to virus diseases and verticillium wilt. They are adapted to a wide range of soil types from Virginia to Georgia and westward to Oklahoma and southern Missouri. Origin: Maryland.

'Cardinal' These large, firm, dark red berries are sweet and good for fresh eating and processing. The heavy midseason crop comes on plants resistant to leaf spot, leaf scorch, and powdery mildew. Widely available. Origin: Unknown.

'Daybreak' These medium red berries are large and very attractive with good flavor

and preserving quality. The plants are very productive. Locally available. Origin: Louisiana.

'Dixieland' This early berry is deep red, firm, acid in flavor, and excellent for freezing. Plants are sturdy and vigorous. Origin: Maryland.

'Earlibelle' This widely adapted early variety produces large, firm fruit that is good for canning and freezing. The plants are medium sized, with good runner production and resistance to leaf spot and leaf scorch. Origin: North Carolina.

'Florida Ninety' These berries are very large, with very good flavor and quality. The plant is a heavy producer of fruit and runners. Origin: Florida.

'Guardian' These large, deep red, midseason berries are firm, uniform in size, and attractive. They have good dessert quality and freeze well. The plants are vigorous and productive and resist many diseases. Origin: Maryland.

'Surecrop'

'Olympus'

'Headliner' These midseason berries are of good quality. The plants are vigorous and productive, make runners freely, and resist leaf spot. Locally available. Origin: Louisiana.

'Marlate' This very large, attractive fruit is good fresh and freezes well. The plant is extremely hardy and is, therefore, a productive and dependable late variety. Origin: Maryland.

'Pocahontas' This berry is good fresh, frozen, or in preserves. The plants are vigorous and resist leaf scorch. They are adapted from southern New England to Norfolk, Virginia. Origin: Maryland.

'Redchief' The fruit is medium to large and of uniform deep red color with a firm, glossy surface. The plant is extremely productive and resistant to red stele (root rot). Origin: Maryland.

'Sunrise' These berries are medium sized, symmetrical, and firm and have very good flavor. The flesh is too pale for freezing. A vigorous grower, the plant resists red stele, leaf scorch, and mildew. Origin: Maryland.

'Surecrop' This early berry is large, round, glossy, firm, and of good dessert quality. The large plants should be spaced 6 to 9 inches apart for top production. Resists red stele, verticillium wilt, leaf spot, leaf scorch, and drought. Good in all zones. Widely available. Origin: Maryland.

'Suwannee' This is a medium to large, early, tender berry of very good quality either fresh or frozen. It is a poor shipper but excellent for the home garden. Locally available. Origin: Maryland.

'Tennessee Beauty' This late berry is medium sized, attractive, glossy red, and firm and has good flavor. It is good for freezing. The plants are productive of both fruit and runners. They resist leaf spot, leaf scorch, and virus diseases. Origin: Tennessee.

Varieties for the Northeast and Midwest

'Ardmore' These large, late midseason berries are yellowish red outside and lighter inside and have good flavor. The plants are productive in heavy silt loam. Origin: Missouri.

'Canoga' This late-ripening, heavy-bearing cultivar produces large, sweet fruit that lasts well because of firm flesh and tough skin. Origin: New York.

'Catskill' These large midseason berries are of good dessert quality and are excellent for freezing. The fruit is not firm enough for shipping, but the plant is a productive home garden variety. It can be grown over a wide range of soil types from New England and New Jersey to southern Minnesota. Widely available. Origin: New York.

'Cyclone' This variety yields large, flavorful berries that are good for freezing. The plant is hardy, resists foliage diseases, and is well adapted to the North Central States. Widely available. Origin: Iowa.

'Dunlap' This early to midseason fruit is medium sized, with dark crimson skin and deep red flesh. It does not ship well but is a good home garden fruit. Plants are hardy and adapted to a wide range of soil types in northern Illinois, Iowa, Wisconsin, Minnesota, North Dakota, South Dakota, and Nebraska. Origin: Illinois.

'Earlidawn' If you want the first strawberrries in your neighborhood, this is the cultivar to plant. The berries are medium to large on heavy-bearing plants. Origin: Maryland.

'Fletcher' These berries are medium sized, with a medium red, glossy, tender skin and excellent flavor. They are very good for freezing. The plants are well adapted to New York and New England. Origin: New York.

'Holiday' The first fruit ripens at midseason, then ripening continues into summer with full-sized, large, bright red fruit. The plant is a heavy producer. Origin: New York.

'Midway'

'Sparkle'

'Honeoye' This conical, bright red fruit boasts exceptional flavor among the early midseason cultivars. The yield is high, and the fruit is highly resistant to berry rot. Origin: New York.

'Howard 17' ('Premier') These are early, medium-sized berries of good quality. The plants are productive and resistant to leaf and virus diseases. Locally available in the Northeast. Origin: Massachusetts.

'Midland' This very early variety bears large, glossy berries with deep red flesh. They are good to excellent fresh and also freeze well. The plant does best when grown in the hill system. It is adapted from southern New England to Virginia and west to Iowa and Kansas. Origin: Maryland.

'Midway' These large berries are of good to very good dessert quality and are also good for freezing. Plants are susceptible to leaf spot, leaf scorch, and verticillium wilt. They are widely planted in Michigan. Widely available. Origin: Maryland.

'Raritan' These midseason berries are large, firm, and flavorful. The plants are medium sized. Origin: New Jersey.

'Redstar' These late berries are large and of good to very good dessert quality. The plants resist virus diseases, leaf spot, and leaf scorch. They are grown from southern New England south to Maryland and west to Missouri and Iowa. Origin: Maryland.

'Robinson' These exceptionally large, "picture-perfect" strawberries are amply produced on vigorous, easy-growing plants. The prolonged fruiting period begins in midseason. Origin: Michigan.

'Sparkle' This is a productive midseason variety with bright red, attractive berries that are fairly soft and have good flavor. The berry size is good in early pickings, but small in later ones. Widely available. Origin: New Jersey.

'Trumpeter' These medium-sized late berries are soft and glossy and have very good flavor. This is a hardy and productive home garden variety for the upper Mississippi valley and the Plains States. Origin: Minnesota.

Everbearing Varieties for the Northeast and Midwest

'Gem' ('Superfection', 'Brilliant', 'Gem Everbearing', and 'Giant Gem') This variety yields small, glossy red, tart fruit of good dessert quality. Widely available. Origin: Michigan.

'Geneva' The large, vigorous plants fruit well in June and throughout the summer and early autumn. The berries are soft and highly flavored. Origin: New York.

'Ogallala' Berries are dark red, soft, and medium sized and have a tart flavor that makes them good for freezing. Plants are vigorous and hardy. Widely available. Origin: Nebraska.

'Ozark Beauty' An everbearing variety for the cooler climate zones, this plant produces poorly in mild climates. The berries are bright red inside and out, are large, sweet, and of good flavor. Only the mother plants produce in any one season, yielding crops in the summer and fall. Runner plants produce the following season. Widely available. Origin: Arkansas.

'Tribute' This day-neutral strawberry with tart fruit is resistant to disease. Origin: Maryland.

'Tristar' This day-neutral variety is resistant to diseases. The fruit is sweet and highly recommended. Origin: Maryland.

Netting is the best way to protect ripe berries from birds.

Varieties for the West

The western strawberry-growing regions are divided into three areas: the Rockies and the Great Basin, western Washington and Oregon, and California.

The Rockies and the Great Basin

Recommended for these areas are the following varieties described earlier: 'Cyclone', 'Dunlap', 'Gem', 'Ogallala', 'Ozark Beauty', 'Sparkle', and 'Trumpeter'.

Western Washington and Oregon

'Hood' These midseason berries are large, conical, bright red, and glossy. They are held high in upright clusters and are good fresh or in preserves. Plants are resistant to mildew but susceptible to red stele. Origin: Oregon.

'Northwest' This late midseason variety produces large fruit at first, then smaller fruit later in the season. The berry has crimson skin and red flesh and is firm, well flavored, and good fresh, in preserves, or for freezing. The plants are very productive and so resistant to virus diseases that they can be planted where virus has killed other varieties. Origin: Washington.

'Olympus' The late midseason fruit is held well up on arching stems. The berries are medium to large, bright red throughout, tender, and firm. The plants are vigorous but produce few runners. They resist red stele and virus diseases but are somewhat susceptible to botrytis infection. Origin: Washington.

'Puget Beauty' The large, glossy, very attractive fruit has light crimson skin. The flesh is highly flavored, excellent fresh, and good for freezing and preserves. The plants are large and upright with moderate runner production. They resist mildew but are somewhat susceptible to red stele. Locally available. Origin: Washington.

'Quinault' An everbearer with a moderately early crop, heavier in July through September, the fruit is large and soft with good color. The plant produces good runners. Origin: Washington.

'Rainier' These late midseason berries are large, firm, and of good quality. The plants are vigorous with large leaf blades but moderate runner production. Origin: Washington.

'Shuksan' This midseason variety bears large firm berries that are bright red, glossy, and broadly wedge shaped. The fruit is good for freezing, and the plants are vigorous. Origin: Washington.

California

'Douglas' This large, uniform, midseason fruit is light red and firm. The plant is very vigorous and produces early berries when planted in October. Good in Southern California. Origin: California.

'Pajaro' The large, uniform, early fruit has red skin and firm flesh. This berry does well along the central coast for spring and summer berries. It does well in the Central Valley for early season berries. Locally available. Origin: California.

'Sequoia' This is an early variety that may even bear in December. The exceptionally large fruit is dark red and tender with soft flesh of excellent flavor. Harvest frequently for best quality. The plant is erect and vigorous, with many runners. It is recommended for home gardens on the central and south coast. Plant in October and November. Widely available. Origin: California.

'Shasta' This large midseason berry is bright red and glossy with firm red flesh and is good for freezing or preserves. The plants are fairly vigorous with a moderate number of runners. They have some resistance to mildew and virus diseases. Locally available. Origin: California.

'Tioga' This early berry is medium red and glossy, with firm flesh that is fine for preserves or freezing. The plant is vigorous, moderately resistant to virus, and fairly tolerant of salinity, but highly susceptible to verticillium wilt. It is good for late-summer planting. Origin: California.

'Tufts' This midseason berry is red, extremely firm, and very large. This is a good variety for Southern California. Origin: California.

Everbearing Varieties for California

These are heavy berry producers and do not form many runners. Plant them anytime, and they will produce medium-sized berries in just 90 days. Try 'Aptos', 'Brighton', 'Fern', and 'Hecker'.

A nut is the "pit" of a fruit.

Almond tree

NUTS

Nut trees—with the exception of the filbert, or hazelnut—grow into extremely large trees. They make excellent shade trees and are beautiful when grown to full maturity, but they are not suitable for small yards. Like any fruit tree, they are subject to a fair amount of disease and insect attack, yet their size makes adequate spraying impossible for the average homeowner. Because of this nut trees should be viewed as large shade trees that often reward you with nuts, but not always. You should not count on a large harvest every year.

All nut trees need plenty of water because they are so large and deep rooted. Where summer rainfall is adequate, little irrigation is needed. In warm dry climates where summer rainfall is uncommon, nut trees need occasional deep watering. Run the hose at a trickle for 24 hours on each quarter of the root system.

That is, water for 4 days, moving the hose to the next quadrant of the root circumference each day. This may be necessary every two weeks, monthly, or only once during the summer, depending on the weather.

Almonds

The almond's kinship to the peach is evident at all times of the year—from bare tree to peachlike blossoms and leaves, and even to the, green, inedible fruit surrounding the edible kernel that is the almond.

Almond trees produce best in California and Arizona where spring frosts and summer rain are rare. But the regions in which they bear well constitute only a fraction of their growing range. Almonds typically flower very early and they are highly vulnerable to frost. If the flowers escape

frost damage and set fruit, the developing fruit may be ruined by late frosts. The best almond climates are areas where the last frost date is not likely to damage blossoms and where summers are long, warm, and dry so the fruit can develop and ripen well.

Gardeners in less-than-ideal climates still have a chance at success with almonds. Where untimely frosts pose potential problems, select only late-flowering cultivars. If possible plant on a north slope where the lower light delays bloom and cold air drains away. Cool, moist summer regions usually don't provide enough heat for ripening fruit well, but some success is possible if you give trees the sunniest (usually western) exposure.

Since almonds are graft-compatible with peaches, nectarines, and plums, some home growers hedge their bets by grafting one or more of these fruits to the almond, assuring some sort of crop.

A mature almond tree is 20 to 30 feet tall and dome shaped with a spread roughly equal to the height. Newly planted trees will start to bear in about 4 years, and the productive lifespan is about 50 years. Most cultivars need another almond as a pollinator (exceptions are noted as self-fertile). If you don't have enough room for two trees, dig an extra large planting hole and plant two almonds close together.

See page 46 for pruning and training instructions.

Almonds grow best in a deep soil (6 feet or more) but are not fussy about type as long as it drains well and is not saline. Almonds are more drought tolerant than most other fruit trees, though drought reduces quantity and quality of the crop. Where summer rainfall is light or lacking, the best watering regime is a thorough, deep irrigation whenever the soil dries to a depth of 3 to 5 inches.

'Hall' almond

Chinese chestnut

Shot hole and *Rhizopus* fungus can ruin crops of almonds if rains occur in springtime or summer. Navel orange worms make wormy kernels if almonds are not sprayed at hull crack (split-hull) stage or harvested promptly or if old nuts are left on the tree where worms can overwinter. Mites may attack foliage. Birds are a big problem: They will flock to a heavily fruiting tree.

Varieties

'All-in-One' This variety flowers between mid- and late season and is self-fruitful. The tree is small and bears heavy crops of soft-shelled 'Nonpareil'-quality nuts. Use as pollinator for 'Nonpareil' and 'Texas'. Origin: California.

'Garden Prince' The flowers bloom in midseason and the tree is self-fruitful. This is a genetic dwarf that grows only 10 to 12 feet tall at maturity. It is small enough to be used as a container tree. The medium-sized nuts are sweet and soft shelled. Origin: California.

'Hall' ('Hall's Hardy') This late-flowering almond bears heavy crops. The medium-

sized, sweet-flavored nuts have hard shells. The tree is partially self-fruitful but bears heavier crops with pollinator 'Texas'. Origin: Kansas.

'Ne Plus Ultra' ('Neplus') This variety flowers very early and is not a good choice for late-frost regions. Nuts are large, long, and narrow in soft shells. A good pollinator for this one is 'Nonpareil'. Origin: California.

'Nonpareil' A midseason bloomer, this is the standard commercial almond. It is adapted to nearly all almond-growing areas but experiences bud failure in the hottest regions. The nut is large and thin-shelled with excellent flavor. Pollinators are 'All-in-One', 'Hall', 'Ne Plus Ultra', or 'Texas'. Origin: California.

'Texas' ('Mission') The flowers form late, and the tree bears heavily. The nuts are hard shelled and small, with a sweet almond flavor. Pollinators are 'All-in-One', 'Hall', or 'Nonpareil'. Origin: Texas.

Chestnuts

Of all the nut trees, the chestnut must be the most romantic: No poet has immortalized the spreading walnut tree, no lyricist captured the mood of filberts roasting by an open fire. But in the past 80 years

the chestnut romance has been a tragic one, for the American chestnut (*Castanea dentata*)—once an important forest and timber tree and the source of small but flavorful nuts—has been brought to the edge of extinction in its native range by an exotic bark disease. It survives in its old territory as stump sprouts, many of which reach bearing age before being hit again by the blight.

Experimentation continues in the quest for a blight-resistant seedling and the creation of hybrids between the American species and European and Asian chestnuts that will capture the American nut quality.

Currently the American chestnut can be grown safely only in blight-free regions west of the Rocky Mountains.

The chestnuts generally available for home planting include the European chestnut (also called Spanish or Italian), the Chinese chestnut, and hybrids between European and Chinese that involve the American species. The Japanese chestnut, *C. crenata*, produces a large nut but the quality is inferior to that of the others.

Chinese chestnut, *C. mollissima*, will grow in regions that produce good peaches; the tree is hardy to about –15° F. It is slightly susceptible to chestnut blight, but pruning out infected branches usually controls the problem. The mature size is 60 feet high with a spread of 40 feet. Compared to the European chestnut, the nuts are smaller—although nut size varies—and it is drier and not as highly flavored.

European chestnut (*C. sativa*) can reach 100 feet tall by 100 feet wide, though 40 to 60 feet in both directions is more typical in gardens. Its successful range is the same as that for its Chinese counterpart except that blight susceptibility rules it out of eastern gardens. Nuts are larger and more flavorful than are those of the Chinese species. European chestnuts are the chestnuts usually sold in markets.

Hybrid chestnuts are becoming more widely available as nut growers strive for the perfect combination of a large, flavorful nut on a blight-resistant tree. Most hybrids are simply sold as seedlings, usually with ancestry indicated.

Chestnuts grow rapidly, starting to bear three to five years after planting. Despite their fast growth they are long-lived trees. All chestnuts are big trees, definitely a consideration where space is limited. And for a nut crop you need another tree as a pollinator. The trees are an ongoing source of litter—from fallen catkins in spring (bearing pollen that many people find foul smelling), burrs at harvest time, and fallen autumn foliage that is prickly and tough.

The best soil is one that is deep, well-drained, not alkaline, and reasonably fertile. Young trees achieve best growth with regular watering; mature trees will need some supplementary watering only where summers are hot and dry. If the soil is good, the trees usually need no fertilizer.

See page 47 for pruning and training instructions.

Chestnut burrs split open in early fall, releasing one to three nuts. Gather up the harvest daily, dry nuts for a day or two in the sun, then store dry at a cool but not freezing temperature.

Chinese chestnut

Chestnut blight, as mentioned above, determines which chestnuts will grow where. In blight-infested regions only the Chinese chestnut and possibly a few hybrids are safe to plant. Wherever oak-root fungus is present in the West, plant the resistant European chestnut, American chestnut, or hybrids between the two.

Varieties

'Colossal'   One of a few named hybrids, this is sold in the West. In ancestry it is at least part European. The nuts are large, flavorful, and peel easily. Origin: California.

'Silver Leaf'   This hybrid is so named because the leaf undersides turn silver at the time of nut fall. Smaller than 'Colossal' but with similar sweet flavor. Origin: Unknown.

Filberts (Hazelnuts)

Filberts and hazelnuts are one and the same. Species hail from Europe, North America, and eastern and western Asia. One European species, *Corylus avellana*, is called European filbert, and both

North American species, *C. americana* and *C. cornuta*, are often referred to as American hazelnuts. But there is no strict naming convention.

The standard nut-producing filberts are inclined to grow into large, suckering shrubs that form patches or thickets in time. They are usually trained into small trees reaching 15 to 25 feet tall. This adaptability to training and relatively small size make the filbert the most versatile landscape subject of all the nut trees. As a small tree it will easily fit into small gardens with room left over for other plants; treated as a shrub it can be used singly or as a dual-purpose hedge. It is beautiful in either form.

Its rounded, ruffled leaves are attractive in their green spring and summer phase, and in fall they turn a pleasing clear to rusty yellow. During winter the long male catkins decorate bare branches. Filberts usually start to bear at 4 years and have an average 50-year productive lifespan.

The European species, *C. avellana* and *C. maxima*, have produced the principal cultivars that furnish commercial nuts. These same cultivars are top choices for the home gardener in regions where filbert blight—an incurable bark fungus—is not present. In blight-infested regions, look for hybrid filberts: crosses between European filberts and resistant American species. All European and hybrid filberts need a second filbert as a pollinator.

Immature filberts

Filbert trained as a shrub

Climate is a major consideration in choosing which filberts to grow. East of the Rocky Mountains, European filberts are best where winter lows range between –10° F and +10° F; in the West the range is –10° F to +20° F. The European-American hybrids extend the growing territory to regions that dip to –20° F. The female filbert blossoms appear in mid- to late winter or earliest spring, and they will be ruined if the temperature drops below 15° F.

Filberts do best when planted in a deep, well-drained, fertile, slightly acid soil. But as long as drainage is good, they can grow in soils that are shallow, sandy, or of low fertility. Plant in full sun for best nut production. Where summer rainfall is regular, filberts need little or no supplementary watering. But water regularly during dry spells and in dry-summer areas.

As long as leaf color remains dark green and nut production is good, plants need no fertilizer. If color and growth diminish, give plants nitrogen at the start of the growing period; check with your county or state agricultural extension for recommended amounts.

See page 48 for pruning and training instructions.

Early autumn harvest involves picking nuts from the ground after they have been released from their enclosing husks. If you want to be certain to beat hungry wildlife to the crop, pick nuts just as soon as you can twist them in their husks. Though nuts are not fully colored at that time, they are completely ripe and will color up after picking. Place nuts in the sun to dry for several days before storing in cool, but not freezing, temperatures. Heavy crops tend to come in alternate years.

The most serious—and geographically limiting—disease is eastern filbert blight, a bark fungus that attacks European filberts in particular. There is no cure; the best way to cope with the disease is to avoid planting susceptible cultivars. The eastern American *C. americana* is a carrier for the disease but is unaffected by it. The only fairly safe areas for European cultivars east of the Rocky Mountains are where no American hazelnuts grow. The blight also has spread to Washington State, and strict quarantines are in place to prevent its spread to other filbert-growing regions in the West.

A bacterial filbert blight common in the Northwest can be controlled by pruning out infected wood (sterilizing pruning shears after each cut) during the dormant season and by spraying just in advance of fall rains. The filbertworm can attack developing nuts, burrowing inside and devouring the kernels.

European Varieties

European filberts have the largest and most flavorful nuts, and are usually the best choice if blight is not present.

'Barcelona' This is the standard commercial cultivar and is also sold for home planting. The nuts are round. Both 'DuChilly' and 'Royal' will pollinate 'Barcelona'. Origin: southern Europe.

'Butler' This is a heavy cropper that will pollinate either 'Barcelona' or 'Ennis'. Origin: Oregon.

'Daviana' This light producer of thin-shelled nuts is grown mainly to pollinate 'Barcelona', 'DuChilly', and 'Royal'. Origin: England.

'DuChilly' This produces a flavorful, elongated nut that must be hand picked since most nuts do not fall freely from the husk. Origin: England.

'Ennis' This newer cultivar has the high quality of 'Barcelona' but a heavier production of larger nuts. Origin: Oregon.

'Royal' This produces a larger nut than 'Barcelona' and ripens three to four weeks ahead of all other cultivars. Origin: Oregon.

Hybrid filberts have generally smaller nuts than the European types, and the flavor is less distinctive, but plants are hardier (as mentioned above) and have some resistance to eastern filbert blight. Standard cultivars are 'Bixby', 'Buchanan', 'Potomac', 'Reed', 'Rush', and 'Winkler'.

Pecans

At heights of 70 feet or more, with a spread nearly as great, pecans are imposing shade trees. For best nut production, most pecan varieties need another tree nearby as a pollinator, so commitment to a

pecan crop requires a considerable investment in space.

One word describes the best pecan climate: hot. A tree needs a long summer with hot days and nights in order to produce fully ripe nuts. Outside of native pecan territory in the southern and south central United States, the southwest desert regions (with irrigation) offer congenial conditions. In the upper South and central Midwest, choose from among the short-season cultivars referred to below.

Pecan cultivars fall into two broad groups. The larger and more widely planted of the two is the group often called "papershell" pecans. They are reliably hardy where winter temperatures descend no lower than 0° F, and for nut production they need a growing season of 270 to 290 days. The smaller second group includes the northern or hardy pecans, which grow in regions where winter low temperatures range from -10° to +10° F; these varieties ripen where the growing season is as short as 170 to 190 days.

Papershell pecans are divided into two categories based on resistance to pecan scab disease, which is prevalent where summer weather is hot and humid. The eastern varieties are disease resistant and will grow throughout papershell territory; susceptible western cultivars are limited to desert and dry southwest areas.

'Wichita' pecan

Pecans need deep, well-drained soil that is slightly acid (pH 6 to 7 is best). They will not tolerate saline soils, a limitation in some otherwise acceptable desert regions. Within their native range, pecans get plenty of rainfall during the growing season. In dry-summer regions, or where rainfall is skimpy, give trees a deep soaking at least every 14 days so that nuts will fill out well.

See page 49 for pruning and training instructions.

Trees will start to bear at 5 to 8 years after planting and have a productive life extending at least 70 years beyond that.

Where soil is above pH 7 (neutral), zinc deficiency may show in a condition called pecan rosette—clusters of stunted leaves at branch ends. Contact your county or state agricultural agent for the best corrective measures in your area. Harvest time runs from late summer well into fall, depending on the variety. Most nuts do not fall free from the tree but have to be knocked free with a long pole.

Pecan scab is the most serious disease, especially in humid-summer regions. Proper cultivar selection will lessen the problem, though even eastern papershells are not totally immune. Aphids may appear throughout pecan territory. In the South and Southeast, the season's first generation of an insect known as the pecan nut casebearer may damage new shoots; later the second generation may infest the developing nuts. The pecan weevil is the last pest to appear, attacking nearly mature and mature nuts.

Eastern Papershell Varieties

'Desirable'   This is a heavy cropper with brittle wood. Pollinate with 'Cheyenne', 'Stuart', or 'Western Schley'. Origin: Mississippi.

'Mahan'   This produces a very large nut. Pollinators are 'Cheyenne' or 'Western Schley'. Origin: Mississippi.

'Stuart'   The nuts are large, and the tree is partially self-fruitful but bears better crops if pollinated by 'Desirable'. Origin: Mississippi.

Southeastern Papershell Varieties

The United States Pecan Field Station in Texas has developed a number of varieties, all bearing Indian-tribe names, that will succeed in the humid southeastern pecan territory. The following are widely available.

'Cherokee'   This bears a medium-sized nut. Pollinators are 'Mohawk' or 'Wichita'. Origin: Texas.

'Cheyenne'   This medium-sized nut variety can be pollinated by 'Mohawk', 'Sioux', or 'Wichita'. Origin: Texas.

'Kiowa'   This bears a large nut. Pollinators are 'Cherokee' or 'Cheyenne'. Origin: Texas.

'Mohawk'   The tree bears very large nuts and is partially self-fruitful, but will produce better if it is pollinated by 'Cheyenne' or 'Western Schley'. Origin: Texas.

'Sioux'   This is a smaller-than-average tree. Pollinators are 'Cheyenne' or 'Western Schley'. Origin: Texas.

'Eureka' walnut

'Hartley' walnut

Western Papershell Varieties

'Western Schley' ('Western')
This is a heavy producer of elongated nuts. It has a wide soil adaptability and is less affected by zinc deficiency than other cultivars. Pollinators are 'Cheyenne', 'Mohawk', or 'Wichita'. Origin: Texas.

'Wichita' This produces highly flavored, medium-sized nuts. Weak crotches and brittle wood leave it vulnerable to wind damage, and its blossoms are sensitive to late frosts. The best pollinators are 'Cherokee', 'Cheyenne', or 'Western Schley'. Origin: Texas.

Northern (Hardy) Varieties

'Major' This cultivar is the standard pecan in northern gardens. The nut is medium to small, and cracks easily. It needs pollination from a late pollen-shedding cultivar such as 'Colby' or 'Greenriver'.

Other widely available cultivars are 'Colby', 'Fritz', 'Greenriver', and 'Peruque'. 'Major' and 'Peruque' will pollinate the other cultivars. Origin: Kentucky.

Walnuts

Walnuts are both domestic and foreign. Most familiar in the marketplace—and certainly most important to the home gardener—is the so-called English or Persian walnut (*Juglans regia*), which hails

from southeastern Europe and southwestern Asia. The nuts are large, flavorful, and enclosed in relatively thin shells.

The black walnut (*J. nigra*) from the eastern United States is famous for its fine-tasting nut encased in a shell of rocklike hardness. Another eastern tree with a similar reputation is the butternut (*J. cinerea*). The western states have one important native, *J. hindsii*, the California black walnut; it is valued more as a rootstock on which to graft English walnut cultivars than for its nuts.

English walnuts are fast-growing, heavy-textured trees. The limb structure is thick and sturdy, and mature height may reach 60 feet with a spread to match. These walnuts will grow over quite a climatic range, but the key to a successful crop is proper cultivar selection.

Where winter lows are normally -20° to -30° F, plant only those designated as Carpathian walnuts. These are frequently seedling-grown trees (rather than individual varieties), the original stock of which stems from the Carpathian Mountains of eastern Europe. If late spring frosts are a feature of your area, choose cultivars that leaf out and shed their pollen late. And if your winters are fairly mild, select one of the cultivars that needs little winter chill. Where summers are hot and humid, the pecan will be a better nut tree to plant because it is not nearly as disease prone as the walnut under those conditions.

Some walnut varieties are self-fruitful, and some need pollinators—an important distinction if you have room for only one tree. If you are allergy prone, take note that walnut pollen is a well-known allergen.

Good soil for English walnuts is fairly deep and definitely well-drained. The trees need regular deep watering for production of top-quality nuts (trees are actually somewhat drought tolerant) but cannot tolerate moist soil continually at the trunk base. Countless old orchard trees have succumbed to crown rot when the orchard has been converted to a subdivision and the trees subjected to lawn watering.

If a walnut and garden must coexist, it's better to have the tree at the garden margin where routine watering won't reach the trunk. The best watering method for walnuts is basin irrigation beneath the tree's canopy. Form an inner earth ring about a foot out from the trunk so the trunk will remain dry during waterings.

Black walnut

Young trees may begin bearing at 5 years and have a life expectancy of around 100 years. Planted in good soil, young trees should need no fertilizer. Established bearing trees may be helped by an annual fertilizer application just before they break dormancy.

Harvest time begins when husks start to split and release the enclosed nuts; this usually occurs in late summer or early autumn. At that time knock all the nuts out of the trees (if you wait for natural fall, squirrels may beat you to much of the crop), remove the husks, spread the nuts in a single layer, and dry for several days.

Codling moth can damage the nut crop in all walnut-growing areas; the walnut husk fly also zeros in on the developing nuts but is more prevalent in the west. Aphids and spider mites are two pests that favor walnut leaves, as does the fungus anthracnose; oystershell scale may affect twigs and small branches.

English Walnut Varieties for Western States

'Carmelo' This variety leafs out late and bears extremely large nuts. Origin: California.

'Chandler' The leaves fill out midseason to late season. The tree bears heavily. Yields increase with 'Hartley' or 'Franquette' as the pollinator. Origin: California.

'Chico' This fairly small tree leafs out early. Origin: California.

'Concord' This variety leafs out in midseason. Along with 'Placentia', this is a good bet for the mildest-winter regions. Origin: California.

'Eureka' This variety leafs out in midseason. Both the tree and the nut are large. The tree is slower to bear than most. Use 'Chico' as a pollinator. Origin: California.

'Franquette' This variety leafs out late, which makes it well-adapted to the Northwest. The large tree is slower than average to bear and produces light crops. Use 'Chandler' or 'Hartley' as a pollinator. Origin: California.

'Hartley' This variety bears good quality nuts at an early age. The leaves fill out in midseason. Origin: California.

'Howard' This variety leafs out in midseason. The tree is small but a heavy producer. Origin: California.

'Payne' This variety leafs out early and bears heavily at an early age. Use 'Chico' or 'Eureka' as the pollinator. Origin: California.

'Placentia' Early to leaf out, this large, early-bearing tree needs very little winter chill. Good for the mild winters of Southern California. Origin: California.

English Walnut Varieties for Midwestern and Eastern States

'Adams', 'Broadview', 'Colby', and 'Mesa' are hardy named cultivars of Carpathian stock that do well in the coldest regions where walnuts can be grown.

Among standard (non-Carpathian) English walnuts, the following cultivars have proven their worth.

'Hansen' This small tree leafs out in midseason and produces thin-shelled nuts. Widely adapted. Origin: Ohio.

'McKinster' Fairly late to leaf out, this variety is a favorite in Ohio and Michigan. Origin: Ohio.

'Metcalfe' Late to leaf out, this is a productive English walnut in New York. Origin: New York.

'Somers' This tree leafs out in midseason and ripens a crop early. Origin: Michigan.

Black Walnut Varieties

'Stabler' This variety produces a nut that is fairly easy to crack for a black walnut. Origin: Maryland.

'Thomas' This is the most popular and widely available walnut. The large nuts have excellent flavor. Origin: Pennsylvania.

Catalog Sources For Fruits, Berries, and Nuts

The best place to buy fruit and nut trees is your local nursery, where you can see the plants you are buying, and don't have to pay shipping charges. But if your local nursery can't get the varieties you want, try some of these fruit and nut nurseries. In addition to the specialists listed here, many general nurseries carry fruit and nut trees.

Adams County Nursery
Box 108
Aspers, PA 17304
(717) 677-8105
Fruit specialists. Wholesale and retail.

Ahrens Strawberry Nursery
RR 1
Huntingsburg, IN 47542
(812) 683-3055
Strawberries plus other berries, grapes, and dwarf fruit trees.

W. F. Allen Co.
Box 1577
Salisbury, MD 21801
Strawberry specialists. Catalog and planting guide lists over 30 varieties. Wholesale and retail.

Bear Creek Nursery
Box 411
Northport, WA 99157
Large selection of antique and hardy apples and nuts.

Bountiful Ridge Nurseries, Inc.
Box 250
Princess Anne, MD 21853
(800) 638-9356
Fruit and nut specialists.

Brittingham Plant Farms
Box 2538
Salisbury, MD 21801
Catalog includes virus-free strawberries, also raspberries, blackberries, blueberries, and grapes.

Bunting's Nurseries
Box 306
Selbyville, DE 19975
(302) 436-8231
Strawberry specialists.

C & O Nursery
Box 116
1700 North Wenatchee Avenue
Wenatchee, WA 98801
(509) 662-7164
Fruit specialists, exclusive patented varieties. Catalog includes ornamental and shade trees.

Chestnut Hill Nursery, Inc.
Route 3, Box 267
Alachua, FL 32615
(904) 462-2820
Hybrid and Chinese chestnuts, Oriental persimmons.

The Clyde Nursery
Highway U.S. 20
Clyde, OH 43410
(419) 547-9249
Fruits and berries.

Columbia Basin Nursery
Box 458
Quincy, WA 98848
(509) 787-4411
Seedling rootstock, dwarfing apple rootstock, dwarf and standard budded fruit trees.

Cumberland Valley Nurseries, Inc.
Box 471
McMinnville, TN 37110
(800) 492-0022
Specializes in plums, peaches, and nectarines.

Emlong Nurseries
2671 W. Marquette Woods Road
Stevensville, MI 49127
(616) 429-4341
Dwarf and standard fruit trees, nut trees, and berries.

Dean Foster Nurseries
Box 127
Hardford, MI 49257
(616) 621-2419
Specializes in strawberries, also lists flowers, vegetables, dwarf fruit, and berries.

Fowler Nurseries, Inc.
525 Fowler Road
Newcastle, CA 95658
(916) 645-8191
Fruit and nut trees. Price list of over 200 fruit and nut varieties sent on request.

Louis Gerardi Nursery
1700 East Highway 50
O'Fallon, IL 62269
(618) 632-4456
Nuts and persimmons.

Hartmann's Plantation, Inc.
Box E
Grand Junction, MI 49056
(616) 253-4281
Blueberry—both northern and southern cultivars—specialists.

Hilltop Orchards & Nurseries, Inc.
Route 2
Hartford, MI 49057
(616) 621-3135
Fruit-tree specialists for commercial orchardists. Free handbook and catalog.

Ison's Nursery & Vineyard
Brooks, GA 30205
(404) 599-6870
Specializes in grapes.

Jersey Chestnut Farm
58 Van Duyne Avenue
Wayne, NY 07470
Chestnuts and persimmons.

Lawson's Nursery
Route 1, Box 294
Ball Ground, GA 30107
(404) 893-2141
Specializes in old-fashioned and unusual fruit trees. Lists over 100 varieties of antique apples.

Lennilea Farm Nursery
RD 1, Box 683
Alburtis, PA 18011
(215) 845-2027
Hardy nut tree specialists.

Henry Leuthart Nurseries, Inc.
Box 666
Montauk Highway
East Moriches, NY 11940
(516) 878-1387
Fruit catalog and guide book on dwarf and espaliered fruit trees.

Living Tree Center
Box 797
Bolinas, CA 94914
(415) 868-1786
Heirloom apple specialists; also carries pears, apricots, and persimmons.

Makielski Berry Farm & Nursery
7130 Platt Road
Ypsilanti, MI 48197
Raspberry specialists; also carries currants, gooseberries, blueberries, blackberries, and strawberries.

Mayo Nurseries
8393 Klippel Road
Lyons, NY 14489
(315) 946-6001
Fruit specialists. Catalog includes many varieties of dwarf and semidwarf apples. Wholesale and retail.

J. E. Miller Nurseries
5060 West Lake Road
Canandaigua, NY 14424
(800) 828-9630
Fruit specialists.

New York State Fruit Testing Cooperative Association
Box 462
Geneva, NY 14456
(315) 787-2205
Fruit catalog; there is a membership fee, which is refunded on the first order.

Owen's Vineyard and Nursery
Georgia Highway 85
Gay, GA 30218
(404) 538-6983
Specializes in muscadine grapes. Guidelines for growing and training are included. Southern rabbiteye blueberries are available.

Patrick's Nursery
Box 130
Ty Ty, GA 31795
(912) 382-1122
Specializes in low-chill fruit and nut trees, berries, and grapes.

Plumtree Nursery
Box 90203
Indianapolis, IN 46290
Gooseberries, currants, and grapes.

Preservation Apple Tree Company
Box 607
Mt. Gretna, PA 17064
(717) 964-2229
Wholesale and retail apples

Raintree Nursery
391 Butts Road
Morton, WA 98356
(206) 496-5410
Fruit and nut cultivars primarily for the Northwest.

Rayner Brothers
Box 1717
Salisbury, MD 21801
(301) 742-1594
Specializes in strawberries. Also carries blackberries, raspberries, blueberries, dwarf fruit trees, and nuts.

St. Lawrence Nurseries
RD 2
Potsdam, NY 13676
(315) 265-6739
Specializes in fruit and nut trees for the North; long list of apples, also crabapples, pears, plums, and grapes.

Southmeadow Fruit Gardens
15310 Red Arrow Highway
Lakeside, MI 49116
(616) 469-2865
Probably the largest collection of fruit varieties—old, new, and rare—in the United States. Illustrated catalog is priced at $8 and worth it. A condensed catalog is free.

Stanek's Garden Center
E. 2929 Twenty-seventh Avenue
Spokane, WA 99203
(509) 924-9234
Catalog of fruit, berries, flowers, and ornamentals.

Stark Brothers Nursery
Highway 54 West
Louisiana, MO 63353
(800) 325-4180
One of the biggest and oldest fruit and nut tree specialists.

Van Well Nursery
Box 1339
Wenatchee, WA 98801
(509) 663-8189
Fruits and berries.

Waynesboro Nurseries
Box 987
Waynesboro, VA 22980
(703) 942-4141
Fruit, nuts, and ornamental plants.

Weeks Berry Nursery
6494 Windsor Island Road North
Salem, OR 97303
(503) 393-8112
Small-fruit specialists.

Wiley's Nut Nursery
1116 Hickory Lane
Mansfield, OH 44905
(419) 589-5239
Hardy nut trees (chestnuts, walnuts, filberts, pecans), and persimmons.

Leslie H. Wilmoth Nursery
Route 2, Box 469
Elizabethtown, KY 47201
(502) 369-7493
Long list of hardy nut tree cultivars for the North and Midwest.

The Annual Editions Series

ANNUAL EDITIONS, including GLOBAL STUDIES, consist of over 70 volumes designed to provide the reader with convenient, low-cost access to a wide range of current, carefully selected articles from some of the most important magazines, newspapers, and journals published today. ANNUAL EDITIONS are updated on an annual basis through a continuous monitoring of over 300 periodical sources. All ANNUAL EDITIONS have a number of features that are designed to make them particularly useful, including topic guides, annotated tables of contents, unit overviews, and indexes. For the teacher using ANNUAL EDITIONS in the classroom, an Instructor's Resource Guide with test questions is available for each volume. GLOBAL STUDIES titles provide comprehensive background information and selected world press articles on the regions and countries of the world.

VOLUMES AVAILABLE

ANNUAL EDITIONS
Abnormal Psychology
Accounting
Adolescent Psychology
Aging
American Foreign Policy
American Government
American History, Pre-Civil War
American History, Post-Civil War
American Public Policy
Anthropology
Archaeology
Astronomy
Biopsychology
Business Ethics
Child Growth and Development
Comparative Politics
Computers in Education
Computers in Society
Criminal Justice
Criminology
Developing World
Deviant Behavior
Drugs, Society, and Behavior
Dying, Death, and Bereavement
Early Childhood Education
Economics

Educating Exceptional Children
Education
Educational Psychology
Environment
Geography
Geology
Global Issues
Health
Human Development
Human Resources
Human Sexuality
India and South Asia
International Business
Latin America
Macroeconomics
Management
Marketing
Marriage and Family
Mass Media
Microeconomics
Multicultural Education
Nutrition
Personal Growth and Behavior
Physical Anthropology
Psychology
Public Administration
Race and Ethnic Relations

Social Problems
Social Psychology
Sociology
State and Local Government
Teaching English as a Second
    Language
Urban Society
Violence and Terrorism
Western Civilization,
    Pre-Reformation
Western Civilization,
    Post-Reformation
Women's Health
World History, Pre-Modern
World History, Modern
World Politics

GLOBAL STUDIES
Africa
China
India and South Asia
Japan and the Pacific Rim
Latin America
Middle East
Russia, the Eurasian Republics,
    and Central/Eastern Europe
Western Europe

Cataloging in Publication Data
Main entry under title: Annual editions: Physical anthropology. 1997/98.
    1. Physical anthropology—Periodicals. I. Angeloni, Elvio, *comp.*
II. Title: Physical anthropology.
ISBN 0–697–37341–X   573'.05

Sixth Edition

Cover: Thin-boned gracile Australopithecine skull (about 2.5 million years old) found in Sterkfontein, South Africa. Photo by Jay Kelley/Anthro-Photo.

Printed in the United States of America

Printed on Recycled Paper

PHYSICAL ANTHROPOLOGY 97/98

Sixth Edition

Editor

Elvio Angeloni
Pasadena City College

Elvio Angeloni received his B.A. from UCLA in 1963, his M.A. in anthropology from UCLA in 1965, and his M.A. in communication arts from Loyola Marymount University in 1976. He has produced several films, including *Little Warrior,* winner of the Cinemedia VI Best Bicentennial Theme, and *Broken Bottles,* shown on PBS. He most recently served as an academic adviser on the instructional television series, *Faces of Culture.*

A Library of Information from the Public Press
Dushkin/McGraw-Hill
Sluice Dock, Guilford, Connecticut 06437

Visit us on the Internet—http://www.dushkin.com/

UNIT 6

Late Hominid Evolution

Seven articles examine
archaeological evidence of
human evolution.

UNIT 7

Living with the Past

Six articles discuss evolutional theory and how genetic heritage impacts on our present and our future.

The concepts in bold italics are developed in the article. For further expansion please refer to the Topic Guide and the Index.

Topic Guide

This topic guide suggests how the selections in this book relate to topics of traditional concern to students and professionals involved with the study of physical anthropology. It is useful for locating articles that relate to each other for reading and research. The guide is arranged alphabetically according to topic. Articles may, of course, treat topics that do not appear in the topic guide. In turn, entries in the topic guide do not necessarily constitute a comprehensive listing of all the contents of each selection.

TOPIC AREA	TREATED IN	TOPIC AREA	TREATED IN
Aggression	8. Machiavellian Monkeys 9. What Are Friends For? 13. These Are Real Swinging Primates 18. Apes of Wrath 19. Dim Forest, Bright Chimps 20. To Catch a Colobus 22. Ape Cultures and Missing Links 37. Eugenics Revisited	Disease	4. Curse and Blessing of the Ghetto 5. Future of AIDS 7. Racial Odyssey 37. Eugenics Revisited 39. Saltshaker's Curse 40. Dr. Darwin
Anatomy	10. Gut Thinking 15. Sex and the Female Agenda 17. What's Love Got to Do with It? 21. Ape at the Brink 22. Ape Cultures and Missing Links 25. Sunset on the Savanna 29. *Erectus* Rising 30. First Europeans 31. Did Neandertals Lose an Evolutionary "Arms" Race? 35. Neanderthal Peace 36. No Bone Unturned	DNA (Deoxyribonucleic Acid)	5. Future of AIDS 6. Black, White, Other 37. Eugenics Revisited 38. DNA Wars
		Dominance Hierarchy	9. What Are Friends For? 13. These Are Real Swinging Primates 17. What's Love Got to Do with It? 18. Apes of Wrath
		Fluorine Testing	23. Dawson's Dawn Man
		Forensic Anthropology	36. No Bone Unturned 38. DNA Wars
Archeology	29. *Erectus* Rising 30. First Europeans 31. Did Neandertals Lose an Evolutionary "Arms" Race? 32. Dawn of Creativity 33. Old Masters 34. Dating Game 35. Neanderthal Peace 36. No Bone Unturned	Genes	2. Evolution's New Heretics 4. Curse and Blessing of the Ghetto 5. Future of AIDS 6. Black, White, Other 37. Eugenics Revisited 38. DNA Wars 39. Saltshaker's Curse 40. Dr. Darwin 41. Future Evolution of *Homo Sapiens*
Australopithecines	10. Gut Thinking 23. Dawson's Dawn Man 24. Case of the Missing Link 25. Sunset on the Savanna 26. East Side Story 27. Asian Hominids *and* Kenya Tools	Genetic Drift	1. Growth of Evolutionary Science 4. Curse and Blessing of the Ghetto
		Genetic Testing	4. Curse and Blessing of the Ghetto 37. Eugenics Revisited 38. DNA Wars
Biostratigraphy	26. East Side Story	*Homo erectus*	23. Dawson's Dawn Man 24. Case of the Missing Link 29. *Erectus* Rising 30. First Europeans
Bipedalism	17. What's Love Got to Do with It? 25. Sunset on the Savanna 28. Scavenger Hunt		
		Homo sapiens	31. Did Neandertals Lose an Evolutionary "Arms" Race? 32. Dawn of Creativity 34. Dating Game 35. Neanderthal Peace 41. Future Evolution of *Homo Sapiens*
Blood Groups	6. Black, White, Other 7. Racial Odyssey 38. DNA Wars		
Brain Size	10. Gut Thinking 41. Future Evolution of *Homo Sapiens*	Hunting and Gathering	16. Why Women Change 19. Dim Forest, Bright Chimps 20. To Catch a Colobus 28. Scavenger Hunt 31. Did Neandertals Lose an Evolutionary "Arms" Race? 32. Dawn of Creativity 33. Old Masters
Catastrophism	1. Growth of Evolutionary Science		
Chain of Being	1. Growth of Evolutionary Science		
Cro-Magnons	23. Dawson's Dawn Man 24. Case of the Missing Link 30. First Europeans 31. Did Neandertals Lose an Evolutionary "Arms" Race? 32. Dawn of Creativity 33. Old Masters 34. Dating Game 35. Neanderthal Peace	Hunting Hypothesis	9. What Are Friends For? 28. Scavenger Hunt
		Locomotion	13. These Are Real Swinging Primates 17. What's Love Got to Do with It? 25. Sunset on the Savanna

TOPIC AREA	TREATED IN	TOPIC AREA	TREATED IN
Mutation	1. Growth of Evolutionary Science 5. Future of AIDS 41. Future Evolution of *Homo Sapiens*	**Reproductive Strategy**	2. Evolution's New Heretics 5. Future of AIDS 9. What Are Friends For? 10. Gut Thinking 13. These Are Real Swinging Primates 14. Natural-Born Mothers 15. Sex and the Female Agenda 17. What's Love Got to Do with It? 18. Apes of Wrath 20. To Catch a Colobus
Natural Selection	1. Growth of Evolutionary Science 2. Evolution's New Heretics 3. Keeping Up Down House 5. Future of AIDS 6. Black, White, Other 7. Racial Odyssey 10. Gut Thinking 14. Natural-Born Mothers 16. Why Women Change 39. Saltshaker's Curse 40. Dr. Darwin 41. Future Evolution of *Homo Sapiens*	**Sexuality**	13. These Are Real Swinging Primates 15. Sex and the Female Agenda 16. Why Women Change 17. What's Love Got to Do with It? 18. Apes of Wrath 20. To Catch a Colobus 22. Ape Cultures and Missing Links
Neanderthals	23. Dawson's Dawn Man 24. Case of the Missing Link 30. First Europeans 31. Did Neandertals Lose an Evolutionary "Arms" Race? 32. Dawn of Creativity 34. Dating Game 35. Neanderthal Peace	**Social Relationships**	2. Evolution's New Heretics 9. What Are Friends For? 12. Dian Fossey and Digit 13. These Are Real Swinging Primates 15. Sex and the Female Agenda 16. Why Women Change 17. What's Love Got to Do with It? 18. Apes of Wrath 19. Dim Forest, Bright Chimps 20. To Catch a Colobus 22. Ape Cultures and Missing Links 33. Old Masters
Paleoanthropology	22. Ape Cultures and Missing Links 23. Dawson's Dawn Man 24. Case of the Missing Link 25. Sunset on the Savanna 26. East Side Story 27. Asian Hominids *and* Kenya Tools 29. *Erectus* Rising 30. First Europeans 31. Did Neandertals Lose an Evolutionary "Arms" Race? 32. Dawn of Creativity 34. Dating Game 35. Neanderthal Peace	**Species**	1. Growth of Evolutionary Science 2. Evolution's New Heretics 35. Neanderthal Peace
		Taxonomy	1. Growth of Evolutionary Science 6. Black, White, Other 7. Racial Odyssey 29. *Erectus* Rising 30. First Europeans
Piltdown Hoax	23. Dawson's Dawn Man 24. Case of the Missing Link	**Technology**	5. Future of AIDS 21. Ape at the Brink 29. *Erectus* Rising 31. Did Neandertals Lose an Evolutionary "Arms" Race? 32. Dawn of Creativity 33. Old Masters 34. Dating Game 35. Neanderthal Peace
Primates	8. Machiavellian Monkeys 9. What Are Friends For? 10. Gut Thinking 11. Mind of the Chimpanzee 12. Dian Fossey and Digit 13. These Are Real Swinging Primates 15. Sex and the Female Agenda 17. What's Love Got to Do with It? 18. Apes of Wrath 19. Dim Forest, Bright Chimps 20. To Catch a Colobus 21. Ape at the Brink 22. Ape Cultures and Missing Links	**Theology**	1. Growth of Evolutionary Science
		Uniformitarianism	1. Growth of Evolutionary Science
Race	6. Black, White, Other 7. Racial Odyssey	**Viruses**	5. Future of AIDS

Natural Selection

As the twentieth century draws to a close and we reflect upon where science has taken us over the past 100 years, it should come as no surprise that the field of genetics has swept us along a path of insight into the human condition as well as heightened controversy as to how to handle this potentially dangerous knowledge of ourselves.

Certainly, Gregor Mendel in the late nineteenth century could not have anticipated that his study of pea plants would ultimately lead to the better understanding of over 3,000 genetically caused diseases, such as sickle-cell anemia, Huntington's chorea, and Tay-Sachs. Nor could he have foreseen the present-day controversies over such matters as surrogate motherhood, cloning, and genetic engineering.

The significance of Mendel's work, of course, was his discovery that hereditary traits are conferred by particular units that we now call "genes," a then-revolutionary notion that has been followed by a better understanding of how and why such units change. It is knowledge of the process of "mutation," or alteration of the chemical structure of the gene, that is now providing us with the potential to control the genetic fate of individuals.

The other side of the evolutionary coin, as discussed in the unit's first and third articles, "The Growth of Evolutionary Science" and "Keeping Up Down House," is natural selection, a concept provided by Charles Darwin and Alfred Wallace. Natural selection refers to the "weeding out" of unfavorable mutations and the perpetuation of favorable ones. The pace and manner in which such forces become evident in the fossil record has been the subject of a great deal of discussion among scientists (see Roger Lewin's essay "Evolution's New Heretics") and, indeed, of some controversy among nonscientists.

It seems that as we gain a better understanding of both of these processes, mutation and natural selection, and grasp their relevance to human beings (as described by Jonathan Marks in his article, "Black, White, Other," and by Boyce Rensberger in "Racial Odyssey"), we draw nearer to that time when we may even control the evolutionary direction of our species. Knowledge itself, of course, is neutral—its potential for good or ill being determined by those who happen to be in a position to use it. Consider the possibility of eliminating some of the harmful hereditary traits discussed in "Curse and Blessing of the Ghetto" by Jared Diamond. While it is true that many deleterious genes do get weeded out of the population by means of natural selection, there are other harmful ones, Diamond points out, that may actually have a good side to them and will therefore be perpetuated. It may be, for example, that some men are dying from a genetically caused overabundance of iron in their blood systems in a trade-off that allows some women to absorb sufficient amounts of the element to guarantee their own survival. The question of whether or not we should eliminate such a gene would seem to depend on which sex we decide should reap the benefit.

The issue of just what is a beneficial application of scientific knowledge is a matter for debate. Who will have the final word as to how these technological breakthroughs will be employed in the future? Even with the best of intentions, how can we be certain of the long-range consequences of our actions in such a complicated field? Note, for example, the sweeping effects of ecological change upon the viruses of the world, which in turn seem to be paving the way for new waves of human epidemics. Generally speaking, there is an element of purpose and design in our machinations. Yet, even with this clearly in mind, the whole process seems to be escalating out of human control. As Geoffrey Cowley points out in "The Future of AIDS," it seems that the whole world has become an experimental laboratory in which we know not what we do until we have already done it.

As we read the essays in this unit and contemplate the significance of genetic diseases for human evolution, we can hope that a better understanding of congenital diseases will lead to a reduction of human suffering. At the same time, we must remain aware that, rather than reduce the misery that exists in the world, someone, at some time, may actually use the same knowledge to increase it.

Looking Ahead: Challenge Questions

In nature, how is it that design can occur without a designer, orderliness without purpose?

Charles Darwin

What is "natural selection"? Does it operate upon groups within a species or solely upon individuals?

How and why might the ABO blood group be related to epidemic diseases?

Discuss whether or not people should be told that they are going to die of a disease from which they are presently suffering and for which there is no cure.

How is it possible to test for deleterious genes?

Why is Tay-Sachs disease so common among Eastern European Jews?

How do ecological changes cause new viruses to emerge?

What do you predict for the future of the AIDS epidemic?

Discuss whether or not the human species can be subdivided into racial categories. How and why did the concept of race develop?

The Growth of Evolutionary Science

Douglas J. Futuyma

Today, the theory of evolution is an accepted fact for everyone but a fundamentalist minority, whose objections are based not on reasoning but on doctrinaire adherence to religious principles.
—James D. Watson, 1965*

In 1615, Galileo was summoned before the Inquisition in Rome. The guardians of the faith had found that his "proposition that the sun is the center [of the solar system] and does not revolve about the earth is foolish, absurd, false in theology, and heretical, because expressly contrary to Holy Scripture." In the next century, John Wesley declared that "before the sin of Adam there were no agitations within the bowels of the earth, no violent convulsions, no concussions of the earth, no earthquakes, but all was unmoved as the pillars of heaven." Until the seventeenth century, fossils were interpreted as "stones of a peculiar sort, hidden by the Author of Nature for his own pleasure." Later they were seen as remnants of the Biblical deluge. In the middle of the eighteenth century, the great French naturalist Buffon speculated on the possibility of cosmic and organic evolution and was forced by the clergy to recant: "I abandon everything in my book respecting the formation of the earth, and generally all of which may be contrary to the narrative of Moses." For had not St. Augustine written, "Nothing is to be accepted save on the authority of Scripture, since greater is that authority than all the powers of the human mind"?

When Darwin published *The Origin of Species,* it was predictably met by a chorus of theological protest. Darwin's theory, said Bishop Wilberforce, "contradicts the revealed relations of creation to its Creator." "If the Darwinian theory is true," wrote another clergyman, "Genesis is a lie, the whole framework of the book of life falls to pieces, and the revelation of God to man, as we Christians know it, is a delusion and a snare." When *The Descent of Man* appeared, Pope Pius IX was moved to write that Darwinism is "a system which is so repugnant at once to history, to the tradition of all peoples, to exact science, to observed facts, and even to Reason herself, [that it] would seem to need no refutation, did not alienation from God and the leaning toward materialism, due to depravity, eagerly seek a support in all this tissue of fables."[1] Twentieth-century creationism continues this battle of medieval theology against science.

One of the most pervasive concepts in medieval and post-medieval thought was the "great chain of being," or *scala naturae.*[2] Minerals, plants, and animals, according to his concept, formed a gradation, from the lowliest and most material to the most complex and spiritual, ending in man, who links the animal series to the world of intelligence and spirit. This "scale of nature" was the manifestation of God's infinite benevolence. In his goodness, he had conferred existence on all beings of which he could conceive, and so created a complete chain of being, in which there were no gaps. All his creatures must have been created at once, and none could ever cease to exist, for then the perfection of his divine plan would have been violated. Alexander Pope expressed the concept best:

Vast chain of being! which from God
  began,
Natures aethereal, human, angel, man,
Beast, bird, fish, insect, what no eye
  can see,
No glass can reach; from Infinite to
  thee,
From thee to nothing.—On superior
  pow'rs
Were we to press, inferior might on
  ours;
Or in the full creation leave a void,
Where, one step broken, the great
  scale's destroy'd;
From Nature's chain whatever link you
  strike,
Tenth, or ten thousandth, breaks the
  chain alike.

Coexisting with this notion that all of which God could conceive existed so as to complete his creation was the idea that all things existed for man. As the philosopher Francis Bacon put it, "Man, if we look to final causes, may be regarded as the centre of the world . . . for the whole world works together in the service of man . . . all things seem to be going about man's business and not their own."

"Final causes" was another funda-

*James D. Watson, a molecular biologist, shared the Nobel Prize for his work in discovering the structure of DNA.

mental concept of medieval and post-medieval thought. Aristotle had distinguished final causes from efficient causes, and the Western world saw no reason to doubt the reality of both. The "efficient cause" of an event is the mechanism responsible for its occurrence: the cause of a ball's movement on a pool table, for example, is the impact of the cue or another ball. The "final cause," however, is the goal, or purpose for its occurrence: the pool ball moves because I wish it to go into the corner pocket. In post-medieval thought there was a final cause—a purpose—for everything; but purpose implies intention, or foreknowledge, by an intellect. Thus the existence of the world, and of all the creatures in it, had a purpose; and that purpose was God's design. This was self-evident, since it was possible to look about the world and see the palpable evidence of God's design everywhere. The heavenly bodies moved in harmonious orbits, evincing the intelligence and harmony of the divine mind; the adaptations of animals and plants to their habitats likewise reflected the divine intelligence, which had fitted all creatures perfectly for their roles in the harmonious economy of nature.

Before the rise of science, then, the causes of events were sought not in natural mechanisms but in the purposes they were meant to serve, and order in nature was evidence of divine intelligence. Since St. Ambrose had declared that "Moses opened his mouth and poured forth what God had said to him," the Bible was seen as the literal word of God, and according to St. Thomas Aquinas, "Nothing was made by God, after the six days of creation, absolutely new." Taking Genesis literally, Archbishop Ussher was able to calculate that the earth was created in 4004 B.C. The earth and the heavens were immutable, changeless. As John Ray put it in 1701 in *The Wisdom of God Manifested in the Works of the Creation,* all living and nonliving things were "created by God at first, and by Him conserved to this Day in the same State and Condition in which they were first made."[3]

The evolutionary challenge to this view began in astronomy. Tycho Brahe found that the heavens were not immutable when a new star appeared in the constellation Cassiopeia in 1572. Copernicus displaced the earth from the center of the universe, and Galileo found that the perfect heavenly bodies weren't so perfect: the sun had spots that changed from time to time, and the moon had craters that strongly implied alterations of its surface. Galileo, and after him Buffon, Kant, and many others, concluded that change was natural to all things.

A flood of mechanistic thinking ensued. Descartes, Kant, and Buffon concluded that the causes of natural phenomena should be sought in natural laws. By 1755, Kant was arguing that the laws of matter in motion discovered by Newton and other physicists were sufficient to explain natural order. Gravitation, for example, could aggregate chaotically dispersed matter into stars and planets. These would join with one another until the only ones left were those that cycled in orbits far enough from each other to resist gravitational collapse. Thus order might arise from natural processes rather than from the direct intervention of a supernatural mind. The "argument from design"—the claim that natural order is evidence of a designer—had been directly challenged. So had the universal belief in final causes. If the arrangement of the planets could arise merely by the laws of Newtonian physics, if the planets could be born, as Buffon suggested, by a collision between a comet and the sun, then they did not exist for any purpose. They merely came into being through impersonal physical forces.

From the mutability of the heavens, it was a short step to the mutability of the earth, for which the evidence was far more direct. Earthquakes and volcanoes showed how unstable terra firma really is. Sedimentary rocks showed that materials eroded from mountains could be compacted over the ages. Fossils of marine shells on mountaintops proved that the land must once have been under the sea. As early as 1718, the Abbé Moro and the French academician Bernard de Fontenelle

had concluded that the Biblical deluge could not explain the fossilized oyster beds and tropical plants that were found in France. And what of the great, unbroken chain of being if the rocks were full of extinct species?

To explain the facts of geology, some authors—the "catastrophists"—supposed that the earth had gone through a series of great floods and other catastrophes that successively extinguished different groups of animals. Only this, they felt, could account for the discovery that higher and lower geological strata had different fossils. Buffon, however, held that to explain nature we should look to the natural causes we see operating around us: the gradual action of erosion and the slow buildup of land during volcanic eruptions. Buffon thus proposed what came to be the foundation of geology, and indeed of all science, the principle of uniformitarianism, which holds that the same causes that operate now have always operated. By 1795, the Scottish geologist James Hutton had suggested that "in examining things present we have data from which to reason with regard to what has been." His conclusion was that since "rest exists not anywhere," and the forces that change the face of the earth move with ponderous slowness, the mountains and canyons of the world must have come into existence over countless aeons.

If the entire nonliving world was in constant turmoil, could it not be that living things themselves changed? Buffon came close to saying so. He realized that the earth had seen the extinction of countless species, and supposed that those that perished had been the weaker ones. He recognized that domestication and the forces of the environment could modify the variability of many species. And he even mused, in 1766, that species might have developed from common ancestors:

If it were admitted that the ass is of the family of the horse, and different from the horse only because it has varied from the original form, one could equally well say that the ape is of the family of man, that he is a degenerate man, that man and ape have a common origin; that, in fact, all the families among plants as well as animals have come from a single stock, and that all animals are descended from a single animal, from which have sprung in the course of time, as a result of process or of degeneration, all the other races of animals. For if it were once shown that we are justified in establishing these families; if it were granted among animals and plants there has been (I do not say several species) but even a single one, which has been produced in the course of direct descent from another species . . . then there would no longer be any limit to the power of nature, and we should not be wrong in supposing that, with sufficient time, she has been able from a single being to derive all the other organized beings.[4]

This, however, was too heretical a thought; and in any case, Buffon thought the weight of evidence was against common descent. No new species had been observed to arise within recorded history, Buffon wrote; the sterility of hybrids between species appeared an impossible barrier to such a conclusion; and if species had emerged gradually, there should have been innumerable intermediate variations between the horse and ass, or any other species. So Buffon concluded: "But this [idea of a common ancestor] is by no means a proper representation of nature. We are assured by the authority of revelation that all animals have participated equally in the grace of direct Creation and that the first pair of every species issued fully formed from the hands of the Creator."

Buffon's friend and protégé, Jean Baptiste de Monet, the Chevalier de Lamarck, was the first scientist to take the big step. It is not clear what led Lamarck to his uncompromising belief in evolution; perhaps it was his studies of fossil molluscs, which he came to believe were the ancestors of similar species living today. Whatever the explanation, from 1800 on he developed the notion that fossils were not evidence of extinct species but of ones that had gradually been transformed into living species. To be sure, he wrote, "an enormous time and wide variation in successive conditions must doubtless have been required to enable nature to bring the organization of animals to that degree of complexity and development in which we see it at its perfection"; but "time has no limits and can be drawn upon to any extent."

Lamarck believed that various lineages of animals and plants arose by a continual process of spontaneous generation from inanimate matter, and were transformed from very simple to more complex forms by an innate natural tendency toward complexity caused by "powers conferred by the supreme author of all things." Various specialized adaptations of species are consequences of the fact that animals must always change in response to the needs imposed on them by a continually changing environment. When the needs of a species change, so does its behavior. The animal then uses certain organs more frequently than before, and these organs, in turn, become more highly developed by such use, or else "by virtue of the operations of their own inner senses." The classic example of Lamarckism is the giraffe: by straining upward for foliage, it was thought, the animal had acquired a longer neck, which was then inherited by its offspring.

In the nineteenth century it was widely believed that "acquired" characteristics—alterations brought about by use or disuse, or by the direct influence of the environment—could be inherited. Thus it was perfectly reasonable for Lamarck to base his theory of evolutionary change partly on this idea. Indeed, Darwin also allowed for this possibility, and the inheritance of acquired characteristics was not finally prove impossible until the 1890s.

Lamarck's ideas had a wide influence; but in the end did not convince many scientists of the reality of evolution. In France, Georges Cuvier, the foremost paleontologist and anatomist of his time, was an influential opponent of evolution. He rejected Lamarck's notion of the spontaneous generation of life, found it inconceivable that changes in behavior could produce the exquisite adaptations that almost every species shows, and emphasized that in both the fossil record and among living animals there were numerous "gaps" rather than intermediate forms between species. In England, the philosophy of "natural theology" held sway in science, and the best-known naturalists

continued to believe firmly that the features of animals and plants were evidence of God's design. These devout Christians included the foremost geologist of the day, Charles Lyell, whose *Principles of Geology* established uniformitarianism once and for all as a guiding principle. But Lyell was such a thorough uniformitarian that he believed in a steady-state world, a world that was always in balance between forces such as erosion and mountain building, and so was forever the same. There was no room for evolution, with its concept of steady change, in Lyell's world view, though he nonetheless had an enormous impact on evolutionary thought, through his influence on Charles Darwin.

Darwin (1809–1882) himself, unquestionably one of the greatest scientists of all time, came only slowly to an evolutionary position. The son of a successful physician, he showed little interest in the life of the mind in his early years. After unsuccessfully studying medicine at Edinburgh, he was sent to Cambridge to prepare for the ministry, but he had only a half-hearted interest in his studies and spent most of his time hunting, collecting beetles, and becoming an accomplished amateur naturalist. Though he received his B.A. in 1831, his future was quite uncertain until, in December of that year, he was enlisted as a naturalist aboard *H.M.S. Beagle,* with his father's very reluctant agreement. For five years (from December 27, 1831, to October 2, 1836) the *Beagle* carried him about the world, chiefly along the coast of South America, which it was the *Beagle*'s mission to survey. For five years Darwin collected geological and biological specimens, made geological observations, absorbed Lyell's *Principles of Geology,* took voluminous notes, and speculated about everything from geology to anthropology. He sent such massive collections of specimens back to England that by the time he returned he had already gained a substantial reputation as a naturalist.

Shortly after his return, Darwin married and settled into an estate at Down where he remained, hardly traveling even to London, for the rest of his life. Despite continual ill health, he pursued an extraordinary range of biological studies: classifying barnacles, breeding pigeons, experimenting with plant growth, and much more. He wrote no fewer than sixteen books and many papers, read voraciously, corresponded extensively with everyone, from pigeon breeders to the most eminent scientists, whose ideas or information might bear on his theories, and kept detailed notes on an amazing variety of subjects. Few people have written authoritatively on so many different topics: his books include not only *The Voyage of the Beagle, The Origin of Species,* and *The Descent of Man,* but also *The Structure and Distribution of Coral Reefs* (containing a novel theory of the formation of coral atolls which is still regarded as correct), *A Monograph on the Sub-class Cirripedia* (the definitive study of barnacle classification), *The Various Contrivances by Which Orchids are Fertilised by Insects, The Variation of Animals and Plants Under Domestication* (an exhaustive summary of information on variation, so crucial to his evolutionary theory), *The Effects of Cross and Self Fertilisation in the Vegetable Kingdom* (an analysis of sexual reproduction and the sterility of hybrids between species), *The Expression of the Emotions in Man and Animals* (on the evolution of human behavior from animal behavior), and *The Formation of Vegetable Mould Through the Action of Worms.* There is every reason to believe that almost all these books bear, in one way or another, on the principles and ideas that were inherent in Darwin's theory of evolution. The worm book, for example, is devoted to showing how great the impact of a seemingly trivial process like worm burrowing may be on ecology and geology if it persists for a long time. The idea of such cumulative slight effects is, of course, inherent in Darwin's view of evolution: successive slight modifications of a species, if continued long enough, can transform it radically.

When Darwin embarked on his voyage, he was a devout Christian who did not doubt the literal truth of the Bible, and did not believe in evolution any more than did Lyell and the other English scientists he had met or whose books he had read. By the time he returned to England in 1836 he had made numerous observations that would later convince him of evolution. It seems likely, however, that the idea itself did not occur to him until the spring of 1837, when the ornithologist John Gould, who was working on some of Darwin's collections, pointed out to him that each of the Galápagos Islands, off the coast of Ecuador, had a different kind of mockingbird. It was quite unclear whether they were different varieties of the same species, or different species. From this, Darwin quickly realized that species are not the discrete, clear-cut entities everyone seemed to imagine. The possibility of transformation entered his mind, and it applied to more than the mockingbirds: "When comparing . . . the birds from the separate islands of the Galápagos archipelago, both with one another and with those from the American mainland, I was much struck how entirely vague and arbitrary is the distinction between species and varieties."

In July 1837 he began his first notebook on the "Transmutation of Species." He later said that the Galápagos species and the similarity between South American fossils and living species were at the origin of all his views.

During the voyage of the *Beagle* I had been deeply impressed by discovering in the Pampean formation great fossil animals covered with armour like that on the existing armadillos; secondly, by the manner in which closely allied animals replace one another in proceeding southward over the continent; and thirdly, by the South American character of most of the productions of the Galápagos archipelago, and more especially by the manner in which they differ slightly on each island of the group; none of these islands appearing to be very ancient in a geological sense. It was evident that such facts as these, as well as many others, could be explained on the supposition that species gradually become modified; and the subject has haunted me.

The first great step in Darwin's thought was the realization that evolution had occurred. The second was his brilliant insight into the possible cause of evolutionary change. Lamarck's theory of "felt needs" had not been convincing. A better one was required. It came on September 18, 1838, when after grappling with the problem for fifteen months, "I happened to read for amusement Malthus on Population, and being well prepared to appreciate the struggle for existence which everywhere goes on from long-continued observation of the habits of animals and plants, it at once struck me that under these circumstances favorable variations would tend to be preserved, and unfavorable ones to be destroyed. The result of this would be the formation of new species. Here, then, I had at last got a theory by which to work."

Malthus, an economist, had developed the pessimistic thesis that the exponential growth of human populations must inevitably lead to famine, unless it were checked by war, disease, or "moral restraint." This emphasis on exponential population growth was apparently the catalyst for Darwin, who then realized that since most natural populations of animals and plants remain fairly stable in numbers, many more individuals are born than survive. Because individuals vary in their characteristics, the struggle to survive must favor some variant individuals over others. These survivors would then pass on their characteristics to future generations. Repetition of this process generation after generation would gradually transform the species.

Darwin clearly knew that he could not afford to publish a rash speculation on so important a subject without developing the best possible case. The world of science was not hospitable to speculation, and besides, Darwin was dealing with a highly volatile issue. Not only was he affirming that evolution had occurred, he was proposing a purely material explanation for it, one that demolished the argument from design in a single thrust. Instead of publishing his theory, he patiently amassed a mountain of evidence, and finally, in 1844, collected his thoughts in an essay on natural selection. But he still didn't publish. Not until 1856, almost twenty years after he became an evolutionist,

ORIGINAL
ANCESTOR

Figure 1. *Some species of Galápagos finches. Several of the most different species are represented here; intermediate species also exist. Clockwise from lower left are a male ground-finch (the plumage of the female resembles that of the tree-finches); the vegetarian tree-finch; the insectivorous tree-finch; the warbler-finch; and the woodpecker-finch, which uses a cactus spine to extricate insects from crevices. The slight differences among these species, and among species in other groups of Galápagos animals such as giant tortoises, were one of the observations that led Darwin to formulate his hypothesis of evolution. (From D. Lack, Darwin's Finches [Oxford: Oxford University Press, 1944].)*

page "abstract" that was published on November 24, 1859, under the title *The Origin of Species by Means of Natural Selection; or, the Preservation of Favored Races in the Struggle for Life.* Because it was an abstract, he had to leave out many of the detailed observations and references to the literature that he had amassed, but these were later provided in his other books, many of which are voluminous expansions on the contents of *The Origin of Species.*

The first five chapters of the *Origin* lay out the theory that Darwin had conceived. He shows that both domesticated and wild species are variable, that much of that variation is hereditary, and that breeders, by conscious selection of desirable varieties, can develop breeds of pigeons, dogs, and other forms that are more different from each other than species or even families of wild animals and plants are from each other. The differences between related species then are no more than an exaggerated form of the kinds of variations one can find in a single species; indeed, it is often extremely difficult to tell if natural populations are distinct species or merely well-marked varieties.

Darwin then shows that in nature there is competition, predation, and a struggle for life.

Owing to this struggle, variations, however slight and from whatever cause proceeding, if they be in any degree profitable to the individuals of a species, in their infinitely complex relations to other organic beings and to their physical conditions of life, will tend to the preservation of such individuals, and will generally be inherited by the offspring. The offspring, also, will thus have a better chance of surviving, for, of the many individuals of any species which are periodically born, but a small number can survive. I have called this principle, by which each slight variation, if useful, is preserved, by the term natural selection, in order to mark its relation to man's power of selection.

Darwin goes on to give examples of how even slight variations promote survival, and argues that when populations are exposed to different conditions, different variations will be favored, so that the descendants of a species become diversified in structure, and each ancestral species can give rise to sev-

did he begin what he planned to be a massive work on the subject, tentatively titled *Natural Selection.*

Then, in June 1858, the unthinkable happened. Alfred Russel Wallace (1823–1913), a young naturalist who had traveled in the Amazon Basin and in the Malay Archipelago, had also become interested in evolution. Like Darwin, he was struck by the fact that "the most closely allied species are found in the same locality or in closely adjoining localities and . . . therefore the natural sequence of the species by affinity is also geographical." In the throes of a

malarial fever in Malaya, Wallace conceived of the same idea of natural selection as Darwin had, and sent Darwin a manuscript "On the Tendency of Varieties to Depart Indefinitely from the Original Type." Darwin's friends Charles Lyell and Joseph Hooker, a botanist, rushed in to help Darwin establish the priority of his ideas, and on July 1, 1858, they presented to the Linnean Society of London both Wallace's paper and extracts from Darwin's 1844 essay. Darwin abandoned his big book on natural selection and condensed the argument into a 490-

eral new ones. Although "it is probable that each form remains for long periods unaltered," successive evolutionary modifications will ultimately alter the different species so greatly that they will be classified as different genera, families, or orders.

Competition between species will impel them to become more different, for "the more diversified the descendants from any one species become in structure, constitution and habits, by so much will they be better enabled to seize on many and widely diversified places in the polity of nature, and so be enabled to increase in numbers." Thus different adaptations arise, and "the ultimate result is that each creature tends to become more and more improved in relation to its conditions. This improvement inevitably leads to the greater advancement of the organization of the greater number of living beings throughout the world." But lowly organisms continue to persist, for "natural selection, or the survival of the fittest, does not necessarily include progressive development—it only takes advantage of such variations as arise and are beneficial to each creature under its complex relations of life." Probably no organism has reached a peak of perfection, and many lowly forms of life continue to exist, for "in some cases variations or individual differences of a favorable nature may never have arisen for natural selection to act on or accumulate. In no case, probably, has time sufficed for the utmost possible amount of development. In some few cases there has been what we must call retrogression of organization. But the main cause lies in the fact that under very simple conditions of life a high organization would be of no service. . . ."

In the rest of *The Origin of Species*, Darwin considers all the objections that might be raised against his theory; discusses the evolution of a great array of phenomena—hybrid sterility, the slave-making instinct of ants, the similarity of vertebrate embryos; and presents an enormous body of evidence for evolution. He draws his evidence from comparative anatomy, embryology, behavior, geographic variation,

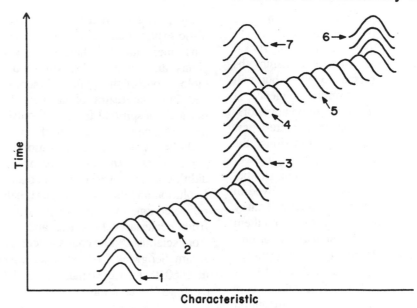

Figure 2. *Processes of evolutionary change. A characteristic that is variable (1) often shows a bell-shaped distribution—individuals vary on either side of the average. Evolutionary change (2) consists of a shift in successive generations, after which the characteristic may reach a new equilibrium (3). When the species splits into two different species (4), one of the species may undergo further evolutionary change (5) and reach a new equilibrium (6). The other may remain unchanged (7) or not. Each population usually remains variable throughout this process, but the average is shifted, ordinarily by natural selection.*

the geographic distribution of species, the study of rudimentary organs, atavistic variations ("throwbacks"), and the geological record to show how all of biology provides testimony that species have descended with modification from common ancestors.

Darwin's triumph was in synthesizing ideas and information in ways that no one had quite imagined before. From Lyell and the geologists he learned uniformitarianism: the cause of past events must be found in natural forces that operate today; and these, in the vastness of time, can accomplish great change. From Malthus and the nineteenth-century economists he learned of competition and the struggle for existence. From his work on barnacles, his travels, and his knowledge of domesticated varieties he learned that species do not have immutable essences but are variable in all their properties and blend into one another gradually. From his familiarity with the works of Whewell, Herschel, and other philosophers of science he developed a powerful method of pursuing science, the "hypothetico-deductive" method, which consists of formulating

a hypothesis or speculation, deducing the logical predictions that must follow from the hypothesis, and then testing the hypothesis by seeing whether or not the predictions are verified. This was by no means the prevalent philosophy of science in Darwin's time.[5]

Darwin brought biology out of the Middle Ages. For divine design and unknowable supernatural forces he substituted natural material causes that could be studied by the methods of science. Instead of catastrophes unknown to physical science he invoked forces that could be studied in anyone's laboratory or garden. He replaced a young, static world by one in which there had been constant change for countless aeons. He established that life had a history, and this proved the essential view that differentiated evolutionary thought from all that had gone before.

For the British naturalist John Ray, writing in 1701, organisms had no history—they were the same at that moment, and lived in the same places, doing the same things, as when they were first created. For Darwin, organisms spoke of historical change. If there has indeed been such a history, then fossils in the oldest rocks must differ from those in younger rocks: trilobites, dinosaurs, and mammoths will not be mixed together but will appear in some temporal sequence. If species come from common ancestors, they will have the same characteristics, modified for different functions: the same bones used by bats for flying will be used by horses for running. If species come from ancestors that lived in different environments, they will carry the evidence of their history with them in the form of similar patterns of embryonic development and in vestigial, rudimentary organs that no longer serve any function. If species have a history, their geographical distribution will reflect it: oceanic islands won't have elephants because they wouldn't have been able to get there.

Once the earth and its living inhabitants are seen as the products of historical change, the theological philosophy embodied in the great chain of being ceases to make sense; the plenitude, or fullness, of the world becomes not an eternal manifestation of God's bountiful creativity but an illusion. For most of earth's history, most of the present species have not existed; and many of those that did exist do so no longer. But the scientific challenge to medieval philosophy goes even deeper. If evolution has occurred, and if it has proceeded from the natural causes that Darwin envisioned, then the adaptations of organisms to their environment, the intricate construction of the bird's wing and the orchid's flower, are evidence not of divine design but of the struggle for existence. Moreover, and this may be the deepest implication of all, Darwin brought to biology, as his predecessors had brought to astronomy and geology, the sufficiency of efficient causes. No longer was there any reason to look for final causes or goals. To the questions "What purpose does this species serve? Why did God make tapeworms?" the answer is "To no purpose." Tapeworms were not put here to serve a purpose, nor were planets, nor plants, nor people. They came into existence not by design but by the action of impersonal natural laws.

By providing materialistic, mechanistic explanations, instead of miraculous ones, for the characteristics of plants and animals, Darwin brought biology out of the realm of theology and into the realm of science. For miraculous spiritual forces fall outside the province of science; all of science is the study of material causation.

Of course, *The Origin of Species* didn't convince everyone immediately. Evolution and its material cause, natural selection, evoked strong protests from ecclesiastical circles, and even from scientists.[6] The eminent geologist Adam Sedgwick, for example, wrote in 1860 that species must come into existence by creation,

a power I cannot imitate or comprehend; but in which I can believe, by a legitimate conclusion of sound reason drawn from the laws and harmonies of Nature. For I can see in all around me a design and purpose, and a mutual adaptation of parts which I *can* comprehend, and which prove that there is exterior to, and above, the mere phenomena of Nature a great prescient and designing cause. . . . The pretended physical philosophy of modern days strips man of all his moral attributes, or holds them of no account in the estimate of his origin and place in the created world. A cold atheistical materialism is the tendency of the so-called material philosophy of the present day.

Among the more scientific objections were those posed by the French paleontologist François Pictet, and they were echoed by many others. Since Darwin supposes that species change gradually over the course of thousands of generations, then, asked Pictet, "Why don't we find these gradations in the fossil record . . . and why, instead of collecting thousands of identical individuals, do we not find more intermediary forms? . . . How is it that the most ancient fossil beds are rich in a variety of diverse forms of life, instead of the few early types Darwin's theory leads us to expect? How is it that no species has been seen to evolve during human history, and that the 4000 years which separates us from the mummies of Egypt have been insufficient to modify the crocodile and the ibis?" Pictet protested that, although slight variations might in time alter a species slightly, "all known facts demonstrate . . . that the pro-

longed influence of modifying causes has an action which is constantly restrained within sufficiently confined limits."

The anatomist Richard Owen likewise denied "that . . . variability is progressive and unlimited, so as, in the course of generations, to change the species, the genus, the order, or the class." The paleontologist Louis Agassiz insisted that organisms fall into discrete groups, based on uniquely different created plans, between which no intermediates could exist. He chose the birds as a group that showed the sharpest of boundaries. Only a few years later, in 1868, the fossil *Archaeopteryx,* an exquisite intermediate between birds and reptiles, demolished Agassiz's argument, and he had no more to say on the unique character of the birds.

Within twelve years of *The Origin of Species,* the evidence for evolution had been so thoroughly accepted that the philosopher and mathematician Chauncey Wright could point out that among the students of science, "orthodoxy has been won over to the doctrine of evolution." However, Wright continued, "While the general doctrine of evolution has thus been successfully redeemed from theological condemnation, this is not yet true of the subordinate hypothesis of Natural Selection."

Natural selection turned out to be an extraordinarily difficult concept for people to grasp. St. George Mivart, a Catholic scholar and scientist, was not unusual in equating natural selection with chance. "The theory of Natural Selection may (though it need not) be taken in such a way as to lead man to regard the present organic world as formed, so to speak, *accidentally,* beautiful and wonderful as is the confessedly haphazard result." Many like him simply refused to understand that natural selection is the antithesis of chance and consequently could not see how selection might cause adaptation or any kind of progressive evolutionary change. Even in the 1940s there were those, especially among paleontologists, who felt that the progressive evolution of groups like the horses, as revealed by the fossil record, must have had some unknown cause other than natural selection. Paradoxically, then, Darwin had convinced the scientific world of evolution where his predecessors had failed; but he had not convinced all biologists of his truly original theory, the theory of natural selection.

Natural selection fell into particular disrepute in the early part of the twentieth century because of the rise of genetics—which, as it happened, eventually became the foundation of the modern theory of evolution. Darwin's supposition that variation was unlimited, and so in time could give rise to strikingly different organisms, was not entirely convincing because he had no good idea of where variation came from. In 1865, the Austrian monk Gregor Mendel discovered, from his crosses of pea plants, that discretely different characteristics such as wrinkled versus smooth seeds were inherited from generation to generation without being altered, as if they were caused by particles that passed from parent to offspring. Mendel's work was ignored for thirty-five years, until, in 1900, three biologists discovered his paper and realized that it held the key to the mystery of heredity. One of the three, Hugo de Vries, set about to explore the problem as Mendel had, and in the course of his studies of evening primroses observed strikingly different variations arise, *de novo*. The new forms were so different that de Vries believed they represented new species, which had arisen in a single step by alteration or, as he called it, mutation, of the hereditary material.

In the next few decades, geneticists working with a great variety of organisms observed many other drastic changes arise by mutation: fruit flies (*Drosophila*), for example, with white instead of red eyes or curled instead of straight wings. These laboratory geneticists, especially Thomas Hunt Morgan, an outstanding geneticist at Columbia University, asserted that evolution must proceed by major mutational steps, and that mutation, not natural selection, was the cause of evolution. In their eyes, Darwin's theory was dead on two counts: evolution was not

gradual, and it was not caused by natural selection. Meanwhile, naturalists, taxonomists, and breeders of domesticated plants and animals continued to believe in Darwinism, because they saw that populations and species differed quantitatively and gradually rather than in big jumps, that most variation was continuous (like height in humans) rather than discrete, and that domesticated species could be altered by artificial selection from continuous variation.

The bitter conflict between the Mendelian geneticists and the Darwinians was resolved in the 1930s in a "New Synthesis" that brought the opposing views into a "neo-Darwinian" theory of evolution.[7] Slight variations in height, wing length, and other characteristics proved, under careful genetic analysis, to be inherited as particles, in the same way as the discrete variations studied by the Mendelians. Thus a large animal simply has inherited more particles, or genes, for large size than a smaller member of the species has. The Mendelians were simply studying particularly well marked variations, while the naturalists were studying more subtle ones. Variations could be very slight, or fairly pronounced, or very substantial, but all were inherited in the same manner. All these variations, it was shown, arose by a process of mutation of the genes.

Three mathematical theoreticians, Ronald Fisher and J. B. S. Haldane in England and Sewall Wright in the United States, proved that a newly mutated gene would not automatically form a new species. Nor would it automatically replace the preexisting form of the gene, and so transform the species. Replacement of one gene by a mutant form of the gene, they said, could happen in two ways. The mutation could enable its possessors to survive or reproduce more effectively than the old form; if so, it would increase by natural selection, just as Darwin had said. The new characteristic that evolved in this way would ordinarily be considered an improved adaptation.

Sewall Wright pointed out, however, that not all genetic changes in

species need be adaptive. A new mutation might be no better or worse than the preexisting gene—it might simply be "neutral." In small populations such a mutation could replace the previous gene purely by chance—a process he called random genetic drift. The idea, put crudely, is this. Suppose there is a small population of land snails in a cow pasture, and that 5 percent of them are brown and the rest are yellow. Purely by chance, a greater percentage of yellow snails than of brown ones get crushed by cows' hooves in one generation. The snails breed, and there will now be a slightly greater percentage of yellow snails in the next generation than there had been. But in the next generation, the yellow ones may suffer more trampling, purely by chance. The proportion of yellow offspring will then be lower again. These random events cause fluctuations in the percentage of the two types. Wright proved mathematically that eventually, if no other factors intervene, these fluctuations will bring the population either to 100 percent yellow or 100 percent brown, purely by chance. The population will have evolved, then, but not by natural selection; and there is no improvement of adaptation.

During the period of the New Synthesis, though, genetic drift was emphasized less than natural selection, for which abundant evidence was discovered. Sergei Chetverikov in Russia, and later Theodosius Dobzhansky working in the United States, showed that wild populations of fruit flies contained an immense amount of genetic variation, including the same kinds of mutations that the geneticists had found arising in their laboratories. Dobzhansky and other workers went on to show that these variations affected survival and reproduction: that natural selection was a reality. They showed, moreover, that the genetic differences among related species were indeed compounded of the same kinds of slight genetic variations that they found within species. Thus the taxonomists and the geneticists converged onto a neo-Darwinian theory of evolution: evolution is due not to mutation *or* natural selection, but to both. Random mutations provide abundant genetic variation; natural selection, the antithesis of randomness, sorts out the useful from the deleterious, and transforms the species.

In the following two decades, the paleontologist George Gaylord Simpson showed that this theory was completely adequate to explain the fossil record, and the ornithologists Bernhard Rensch and Ernst Mayr, the botanist G. Ledyard Stebbins, and many other taxonomists showed that the similarities and differences among living species could be fully explained by neo-Darwinism. They also clarified the meaning of "species." Organisms belong to different species if they do not interbreed when the opportunity presents itself, thus remaining genetically distinct. An ancestral species splits into two descendant species when different populations of the ancestor, living in different geographic regions, become so genetically different from each other that they will not or cannot interbreed when they have the chance to do so. As a result, evolution can happen without the formation of new species: a single species can be genetically transformed without splitting into several descendants. Conversely, new species can be formed without much genetic change. If one population becomes different from the rest of its species in, for example, its mating behavior, it will not interbreed with the other populations. Thus it has become a new species, even though it may be identical to its "sister species" in every respect except its behavior. Such a new species is free to follow a new path of genetic change, since it does not become homogenized with its sister species by interbreeding. With time, therefore, it can diverge and develop different adaptations.

The conflict between the geneticists and the Darwinians that was resolved in the New Synthesis was the last ma-

jor conflict in evolutionary science. Since that time, an enormous amount of research has confirmed most of the major conclusions of neo-Darwinism. We now know that populations contain very extensive genetic variation that continually arises by mutation of pre-existing genes. We also know what genes are and how they become mutated. Many instances of the reality of natural selection in wild populations have been documented, and there is extensive evidence that many species form by the divergence of different populations of an ancestral species.

The major questions in evolutionary biology now tend to be of the form, "All right, factors x and y both operate in evolution, but how important is x compared to y?" For example, studies of biochemical genetic variation have raised the possibility that nonadaptive, random change (genetic drift) may be the major reason for many biochemical differences among species. How important, then, is genetic drift compared to natural selection? Another major question has to do with rates of evolution: Do species usually diverge very slowly, as Darwin thought, or does evolution consist mostly of rapid spurts, interspersed with long periods of constancy? Still another question is raised by mutations, which range all the way from gross changes of the kind Morgan studied to very slight alterations. Does evolution consist entirely of the substitution of mutations that have very slight effects, or are major mutations sometimes important too? Partisans on each side of all these questions argue vigorously for their interpretation of the evidence, but they don't doubt that the major factors of evolution are known. They simply emphasize one factor or another. Minor battles of precisely this kind go on continually in every field of science; without them there would be very little advancement in our knowledge.

Within a decade or two of *The Ori-*

gin of Species, the belief that living organisms had evolved over the ages was firmly entrenched in biology. As of 1982, the historical existence of evolution is viewed as fact by almost all biologists. To explain how the fact of evolution has been brought about, a theory of evolutionary mechanisms—mutation, natural selection, genetic drift, and isolation—has been developed.[8] But exactly what is the evidence for the fact of evolution?

NOTES

1. Andrew Dickson White, *A History of the Warfare of Science with Theology in Christendom* vol. I (London: Macmillan, 1896; reprint ed., New York: Dover, 1960).

2. A. O. Lovejoy, *The Great Chain of Being* (Cambridge, Mass.: Harvard University Press, 1936).

3. Much of this history is provided by J. C. Greene, *The Death of Adam: Evolution and its Impact on Western Thought* (Ames: Iowa State University Press, 1959).

4. A detailed history of this and other developments in evolutionary biology is given by Ernst Mayr, *The Growth of Biological Thought: Diversity, Evolution, Inheritance* (Cambridge, Mass.: Harvard University Press, 1982).

5. See D. L. Hull, *Darwin and His Critics* (Cambridge, Mass.: Harvard University Press, 1973).

6. *Ibid.*

7. E. Mayr and W. B. Provine, *The Evolutionary Synthesis* (Cambridge, Mass.: Harvard University Press, 1980).

8. Our modern understanding of the mechanisms of evolution is described in many books. Elementary textbooks include G. L. Stebbins, *Processes of Organic Evolution,* (Englewood Cliffs, N.J.: Prentice-Hall, 1971), and J. Maynard Smith, *The Theory of Evolution* (New York: Penguin Books, 1975). More advanced textbooks include Th. Dobzhansky, F. J. Ayala, G. L. Stebbins, and J. W. Valentine, *Evolution* (San Francisco: Freeman, 1977), and D. J. Futuyma, *Evolutionary Biology* (Sunderland, Mass.: Sinauer, 1979). Unreferenced facts and theories described in the text are familiar enough to most evolutionary biologists that they will be found in most or all of the references cited above.

Evolution's New Heretics

A growing number of evolutionary biologists think that the interests of groups sometimes supersede those of individuals

Roger Lewin

A native of England, Roger Lewin earned a Ph.D. in biochemistry from Liverpool University. He now lives in Cambridge, Massachusetts, where he is a freelance writer (specializing in evolution and ecology) and an associate at Harvard's Peabody Museum. Lewin is the author of several books, including In the Age of Mankind: A Smithsonian Book of Human Evolution *(Smithsonian Institution Press, 1989) and* The Sixth Extinction, *which he coauthored with Richard Leakey and which was published by Doubleday.*

Like an old-time preacher, David Sloan Wilson has the appearance of a man with a mission. An evolutionary biologist at the State University of New York at Binghamton, Wilson is given to marching up and down, flailing his arms, and proclaiming passionately, even in informal conversation. His message is as clear as it is bold: A whole generation of evolutionary biologists has been misled into believing that natural selection grinds inexorably at the level of individual interests, and only at that level. Instead, Wilson argues, biologists must recognize that groups of organisms have evolutionary interests, too, and that natural selection sometimes operates at this "higher" level. "Group selection" means that, occasionally, individuals

within a group—for instance, an ant colony, a baboon troop, a nomadic band of human hunter-gatherers, or even a human population united by a common culture—may sacrifice their own reproductive future if, by doing so, the group benefits. This benefit comes through increased fitness, that is, through contributing more offspring to the next generation than do other, competing groups. Similarly, individuals may cooperate if the common end enhances the group's fitness.

For more than twenty years, Wilson has been working unceasingly—initially very much as a loner, but now with a growing band of supporters—advocating a theory that has been viewed by some as nothing less than heretical. Although much of his writing is couched in the arcane language of mathematical models, Wilson is concerned with a form of behavior that is very basic and, intuitively anyway, easily understood: altruism.

Humans may pride themselves on being genuinely altruistic, selflessly helping others, whether it is by dying for one's country or giving a couple of dollars to a homeless person on the street. But, modern evolutionary biologists ask, can animals other than humans be described as sometimes acting altruistically? Is the honeybee that dies in the act of stinging an intruder to the hive

being altruistic? And what of a lioness that suckles the young of others in the pride as well as her own? Humans think of altruism as doing good for its own sake, but most of us would deny such motives to other animals. Two decades ago, Harvard biologist Edward O. Wilson proclaimed in his important and controversial book *Sociobiology: The New Synthesis* that altruism is "the central theoretical problem" of evolutionary biology in a social context.

Darwin was aware of apparently altruistic behaviors in nature. His theory of evolution by natural selection is principally about the survival of individuals in their "struggle for existence," in which they are always seeking ways of promoting their own reproductive success. Nevertheless, he recognized that individuals might sometimes act selflessly if, as a result, the success of the group is promoted instead. In *The Descent of Man*, Darwin used this line of argument to explain the evolution of morality. If this sounds uncannily like David Sloan Wilson's position, it is. So why has Wilson been called a heretic for championing something that Darwin expressed a century ago? How is it, as Wilson recently noted, that "the rejection of group selection was treated as a scientific advance comparable to the rejection of Lamarckism, and like Lamarckism, its memory was kept alive as an example of how not to think?"

During the century following the publication of *The Descent of Man*, Darwin's clear vision of group selection was superseded by a fuzzy view of life as a harmonious enterprise, with individuals acting toward a collective good. Most scientists, for instance, saw territoriality as individuals acting to control the density of the population for the good of all. Similarly, dominance hierarchies—the pecking orders so common among social animals—were seen as a means of reducing wasteful conflicts within the group.

This naïve version of group selection culminated in the 1962 publication of V. C. Wynne-Edwards's classic book, *Animal Dispersion in Relation to Social Behaviour*. Wynne-Edwards proposed that attributes of *all* social groups were subject to the forces of natural selection. His book was the target of immediate and blistering attack, for by then a new generation of evolutionary biologists had arisen. These scientists, building on a foundation of mathematical population genetics that increasingly emphasized the importance of individual selection, demolished the credibility of group selection at that time.

Prominent among the opponents was George C. Williams, of the State University of New York at Stony Brook, who took on group selection in his 1966 book, *Adaptation and Natural Selection*. Williams argued that, although a theoretical possibility, selection at the level of the group was an insignificant evolutionary force compared with individual selection. One reason was that the rate of evolution is far higher at the level of the individual than at that of the group; another was that, for the most part, groups are rather fluid, with members often moving between them, thus diluting the group as an evolving entity. Moreover, a group of altruists could easily be exploited by a sneakily selfish individual bent on boosting its own reproductive output at the expense of others holding back for the sake of the group.

If group selection were important, reasoned Williams, we would expect populations of sexually reproducing species to contain more females than males. This

is because the number of reproductively active females—not the number of males—ultimately determines the number of offspring in the population; and, in evolutionary terms, the more successfully reared offspring there are in a group, the more successful that group is. Under individual selection, an even sex ratio is predicted (as a result of a balance in the struggle by individuals to maximize their own reproductive success through their offspring). And, Williams observed, an even ratio of males to females is what is most often seen in the world.

With the publication of *The Selfish Gene* in 1976, Oxford University biologist Richard Dawkins moved the focus of natural selection even further away from the group. For Dawkins, it is not just individuals but the genes within them that matter. More ammunition against group selection came with the development of kin selection and game theory, mathematical models that were designed to explain altruism and cooperative behavior.

The theory of kin selection has deep roots, going back principally to the insights of another Oxford biologist, William Hamilton, and fellow Brit John Maynard Smith, of the University of Sussex. Maynard Smith and Hamilton pointed out that when an apparently altruistic individual sacrifices a measure of its reproductive opportunity to enhance that of another, the donor—or at least some of its genes—may actually benefit if it is a relative of the recipient of the favor.

Game theory (originally a method developed by mathematicians to examine economic cooperation and conflict among humans) showed that even unrelated individuals cooperate while looking to their own interests. Robert Trivers, then at Harvard University, used this approach in developing the notion of reciprocal altruism, or "I'll scratch your back now because you scratched mine a while ago and I expect you to scratch mine again."

The era of the individual in evolutionary biology was thus firmly established and was apparently unassailable. Ethologists entered field studies confident that this theoretical perspective

would powerfully inform what they would observe, particularly among primate species. "It's true. It was powerful," says Barbara Smuts, a primatologist at the University of Michigan, who has studied chimpanzees and baboons in East Africa. "But sometimes you had to stretch it and that made me uneasy." For instance, she doubted that a short-term, narrow view of self-interests could explain some features of alliances between males and females, because even after a long period of devotion to a female, a male has no guarantee that he will enjoy her mating favors in the future. Like many others, though, Smuts says, "I just thought that when I had more data it would be okay. I never thought about group selection. . . . No one did."

Frans de Waal, of the Yerkes Primate Center in Atlanta, recalls someone once describing a primate society as being like a transparent organism. "It was a powerful metaphor," he now says, "thinking of the group as an organism. But we weren't allowed to talk like that." A generation of field researchers felt the same: most accepted that group selection was discredited, and a few nursed curiosity about it but kept quiet for fear of appearing intellectually unsound.

To qualify as a "vehicle" of selection, says evolutionary biologist David Sloan Wilson, animals living in a group must share a common evolutionary fate and be in competition with other such groups.

Meanwhile, David Sloan Wilson had been laboring at his theoretical last since the mid-1970s, fashioning a theory of group selection that he believed eschewed the naïvete of earlier models. Wilson argues that the fundamental issue is the "vehicle" of selection. An individual, for instance, consists of a population of cells, but insofar as those cells share a common fate, in an evolutionary sense, the individual—and not the cells—is properly seen as the vehicle of selection. Similarly, a long-term alliance, or friendship, between a male and female baboon can be considered a vehicle because they share a common fate in the reproductive success that stems from the alliance. The female members of a pride of lions may also be a vehicle because their fitness depends on the fitness of their group, in terms of hunting success and protection against outside attack. An entire social group of vervet monkeys, for instance, may be a vehicle in their joint foraging and defense.

Group selection works, Wilson says, when groups are competing but not when they are in isolation. Suppose an individual within an isolated group provisions the offspring of other adults in the group, thus increasing their Darwinian fitness. Even though the overall fitness of the group is enhanced, the behavior will not be selected because it is not to the advantage of the individual doing it. However, Wilson argues, if the group is in competition with other such groups, the behavior will be selected because the group as a whole has an advantage relative to other groups.

One of the most persuasive examples of group selection in nature, argues Wilson, is the evolution of virulence in parasites. It is in a parasite's Darwinian interest to reproduce as bountifully as is compatible with high transmission to other hosts. A strain that multiplies too fast may rapidly kill off its host, thereby reducing its chances of transmission to other hosts.

Imagine two mice, each infected with a different strain of a certain parasite. Multiplying rapidly, the more virulent strain will have greater reproductive success than the less virulent one. Because its host-mouse soon expires, however, transmission to other mice will be lower than that of the less virulent parasite population. Natural selection therefore favors high virulence within hosts

but lower virulence between them. "If group selection were a negligible force in disease evolution, then parasites would evolve to maximize their virulence and the notion of optimal virulence would be irrelevant," observe David Wilson and University of Wisconsin philosopher Elliot Sober. Further support for group selection comes from the group behavior of social insects and the discovery of female-biased sex ratios in a variety of organisms, including fig wasps, hummingbird flower mites, and social spiders.

The main body of evolutionary biologists, however, remains convinced that individual-level selection is the key to understanding social organization and behavior and rejects group selection as wrongheaded thinking. Dawkins is blunt in his response to Wilson's version of group selection, writing recently that he is "baffled by . . . the sheer, wanton, head-in-bag perversity of the position." Williams points out that from his calculations and observations, group selection occupies only a tiny corner of the world of evolution. "It's not a matter of logical correctness," he says. "It's a matter of importance, and Wilson greatly overstates the importance of group selection." Maynard Smith is critical of Wilson for causing "more confusion than clarity by using the term group selection in many different ways," but he also praises him for being "one of the few who has made the subject interesting again."

Meanwhile, Wilson is making converts. "My resistance crumbled immediately, as soon as I saw what he was saying," recalls Smuts, who was witness to one of Wilson's informal, passionate expositions at a 1994 gathering of the Human Behavior and Evolution Society. "The logic of the vehicle of selection, at different levels, is very persuasive." She was particularly drawn to the shifts in focus from within-group competition to within-group cooperation that might occur for an individual at different times and under different circumstances, particularly in intelligent species. "Male chimpanzees are a good example," Smuts explains. "At any time, the community might be challenged by another group, so that individual interest

in competing with other males for access to females will be temporarily suspended in favor of collective group protection."

De Waal finds himself sympathetic to group selection but is still waiting to see how it might be a more powerful explanatory perspective than individual selection. He recently published *Good Natured*, a book on the evolution of morality in human and nonhuman primates. "Human morality is a classic case of something imposed on individuals for the well-being of the community," he explains. "Individuals benefit from a strong, united community, and that is why the community is valued so highly in our moral system." This might be seen as an example of what some anthropologists call cultural group selection.

Like humans, de Waal says, chimpanzees appear to value harmony in their social group. When peace is disrupted, often by bouts between competing males, a female acting as mediator may bring the two males together—sometimes diffusing the rising tension, sometimes effecting a reconciliation after an all-out fight. For instance, she might begin grooming one of the combatants, gradually luring him closer to his adversary; once the males have been drawn together, she will use her considerable, subtle social skills to get the two males to groom each other. Such intervention is a tricky and sometimes risky business. "It is a striking example of an individual taking care of relationships in which the mediator is not herself directly involved," observes de Waal. "This is what moral systems do all the time." Nevertheless, de Waal sees such cases of rudimentary morality "as an outcome of individual-level selection because the female benefits from a harmonious social context." Can human morality also be viewed as a result of individual-level selection, although expressed more generally and with more force because of language? "I suspect there is an interplay between individual and group selection in primate social systems," de Waal speculates, "but we haven't worked out what it is yet."

Some scientists believe passionately in group selection; others view it as an example of wrong-headed thinking. Still others are curious about the idea but unconvinced. If group selection does exist, altruism may be one result, but genocide would be another, less welcome outcome.

Anthropologists are deeply split over the place of group selection in human society. Some, such as Christopher Boehm of the University of Southern California, believe it has played a large role, much more than in nonhuman societies. "The egalitarian social structure you see in a foraging society is the result of the group's preventing its leaders from becoming dominant," argues Boehm. Richard Alexander, of the University of Michigan, points out that nobody "knows if group selection has been important in determining the genetic makeup of modern humans. . . . If it has been important, it has likely involved direct intergroup competition and hostility of the sort we've seen all across history."

The foray into the human realm inevitably muddies the waters of the group selection debate because of the thick overlay of culture. When one anthropologist points out that humans readily die for their country, supposedly demonstrating altruism for the good of the group, another reminds us that it is usually the poor who fight wars for the benefit of the rich and that if the poor choose not to fight, they may be flung in jail. Moreover, recent studies reveal that a person is much more likely to, say, give money to beggars when accompanied by a friend (particularly one of the opposite sex) than when alone. The motive here seems to be to enhance the social standing of the giver.

Donald Campbell, of Lehigh University, speaks for many when he says he expects there is a biological underpinning to many aspects of human social behavior as a result of group selection. An admirer of Wilson and Sober's work, he is disappointed that their arguments don't yet illuminate the way. He would like to see group selection in the human realm made "more explicitly plausible."

Finally, the sociological dimension of the group selection debate cannot be ignored, as psychiatrist Randolph Nesse, of the University of Michigan, observes. "It's not surprising that Wilson and Sober want to see human altruism as the result of group selection," he says. "Many people do. The discovery that some altruism isn't genuinely altruistic but is instead fundamentally selfish is deeply disturbing. Some would find comfort if we were able to reconcile our moral feelings with biological reality, but unfortunately it seems we can't." Nesse argues that we have to accept this reality and not seek to change the science to suit our feelings.

Not surprisingly, Wilson rejects this line of argument. "One of the great insights that is going to come out of group selection is that morality will be justified at face value," he says. "That is, it's a system designed to benefit the common good. But, you shouldn't think I'm a hopeless romantic, because a lot of nasty things are the result of group selection as well, including the ability to inflict genocide."

Keeping Up Down House

Darwin's home, a shrine of science, is being rescued.

Richard Milner

Richard Milner is an editor at Natural History *magazine.*

Although it is only an hour's ride from London, Charles Darwin's country estate, Down House, retains the countrified isolation that the great naturalist prized a century and a half ago. ("Its chief merit," he wrote, "is its extreme rurality.") Badgers and foxes still trot across Darwin's fields, and the wild English orchids he studied continue to thrive nearby. Nestled among the gently sloping hills, or "downs," for which it is named, the three-story Georgian home resonates with echoes of Darwin's personality. Several of the rooms—including the study where he wrote the *Origin of Species* and the *Descent of Man*—contain his original furnishings, books, and papers. Among Down House's collection of memorabilia is Darwin's hand-written journal of his five-year voyage of discovery aboard HMS *Beagle*. If science has shrines, this is surely one of them.

When twenty-seven-year-old Charles Darwin returned to England in 1826, fresh from his difficult, dangerous voyage, he still believed that each species came into existence separately and independently. He did not embrace the idea of evolution until the following year and required decades of uninterrupted thought, study, and experimentation to shape his theory. He realized the need for such reflection even while in the Brazilian rain forest in 1832. "The mind is a chaos of delight," he jotted in his diary, "out of which a world of future & more quiet pleasures will arise." On his journey home, he looked forward to the years of contemplation he would need to make sense of his travels. "I am convinced it is a most ridiculous thing to go round the world," he wrote his sister in 1826, "when by staying quietly, the world will go round with you."

Two years after returning to England, Darwin married his cousin Emma Wedgwood, of the prosperous pottery clan. For the first two years of their marriage, they took an apartment in "vile, smokey London" while he arranged his vast collections of preserved birds, mammals, fossils, sea creatures, and geological specimens—many of them new to science—and built his reputation in London's scientific circles. Emma gave birth to two children in London, but when a third was on its way, the pull of the countryside proved irresistible. In 1842, the Darwins went house-hunting in Kent and fell in love with the tiny village of Down. (In the mid-nineteenth century it became "Downe," but Darwin clung to the original spelling for his home.) The young couple agreed that the scenery was "absolutely beautiful," although Emma thought the 1770s-vintage house of crumbling, whitewashed brick had a "somewhat desolate air." Darwin wrote his sister Catherine, "It is really surprising to think London is only 16 miles off." And to another correspondent he added, "I think I was never in a more perfectly quiet country."

When the Darwins first moved there, Downe still retained the charm and social divisions of a medieval village. Village folk were shopkeepers, gardeners, and carpenters, who worked for the few landowners. Women still curtsied and men doffed their hats when such gentry as the Darwins rode by—even though Charles and Emma were of modest means compared with the wealthiest families. Nevertheless, Darwin, whose father and grandfather were prosperous physicians, never had to teach or take on other work to earn a living.

Down House sits on twenty acres that include fields, a garden, and a clump of woods planted by Darwin and his gardeners. Over the years, he added to the house, stuccoed its exterior, put a dovecote out back for breeding pigeons, and added a greenhouse (which still stands), a laboratory (now in ruins), and, later, a clay tennis court for his children—one of the first in England. Charles and Emma raised seven children (three others died in childhood) in this bustling Victorian household, replete with a dozen servants, gardeners, and groundskeepers.

Soon after settling at Downe, Darwin constructed a sand-covered path, known as the sandwalk, that still winds through the shady woods and then returns toward the house along a sunny, hedge-lined field. He strolled it daily, referring to it as "my thinking path." Often he would stack a few stones at the path's entrance, and knock one away with his walking stick on completing each circuit. He could anticipate a "three-flint problem," just as Sherlock Holmes had "three-pipe problems," and then head for home when all the stones were gone.

Darwin gradually turned his comfortable home into a biological field station. He cross-pollinated orchids in his greenhouse and conducted experiments on the movements of vines. While working out his theories on how life may have colonized oceanic islands, he soaked seeds in barrels of brine for weeks, then planted them to see if they would still germinate. He bred fancy pigeons, studied the "emotional expressions" of cats and dogs, and questioned local farmers endlessly on how they produced their domestic varieties of apples, hogs, and horses. He even placed a heavy, round "wormstone" in the garden, with a calibrated brass rod to measure the rate at which it sank because of the action of worms. (The wormstone can still be seen at Down House; after Darwin's day, however, it was moved too close to a yew tree, whose expanding roots lift it up as fast as the worms can bury it.)

About 1846, Darwin decided to make himself an expert at classifying species and varieties, to gain credibility for his theorizing on "the laws of life." For eight years (1846–54) he devoted himself to a painstaking study of barnacles, the small crustaceans that encrust dock pilings and ships' hulls. As the Darwin children grew up, they rarely saw their father do anything else. One of the boys once asked a neighbor's child, "Where does your father work on *his* barnacles?"

Darwin had a custom-made rolling armchair, with a cloth-covered writing board—still in the study—on which he wrote *Origin of Species* (1859), *Descent of Man* (1871), *Expression of the Emotions in Man and Animals* (1872), and a dozen other volumes on orchids, earthworms, carnivorous plants, and domestic plants and animals. (Paleontologist Steven Jay Gould calls it Darwin's "revolutionary armchair," insisting that Darwin was no "armchair revolutionary.") Despite the mysterious, debilitating illness that plagued him most of his adult life—limiting his working hours to two or three a day—Darwin also produced a hundred scientific papers and some 11,000 letters. "My life goes on like clockwork," he wrote in 1846 to his old friend Captain Robert Fitzroy, "and I am fixed on the spot where I shall end it." When he died at the house in 1882, he was not buried in the Downe churchyard as he expected, but at Westminster Abbey, a few paces away from the tomb of Sir Isaac Newton.

On a recent visit, I followed the sandwalk (although the sand is long gone) through Darwin's woods. As I stood motionless on the path, listening to the birds, a squirrel caught my eye. I knew from reading his son Frank's account that Darwin once stood so quietly here that a few young squirrels, "ran up his back and legs while their mother barked at them in an agony from the tree." I imagined that my visitor was the umpteenth great-great grandchild of Darwin's squirrels. Back at the house, curator Solene Morris set me straight. "You saw a gray squirrel," she said, "an introduced North American species. In Darwin's day, these woods swarmed with the smaller red squirrels with the lovely ear tufts, like Beatrix Potter's 'Squirrel Nutkin.' Since the 'jolly gray's' arrival in the late nineteenth century, it has steadily displaced the native reds at the rate of six miles a year.

"Why shouldn't we expect the 'struggle for existence' here in Darwin's woods as well as everywhere else," Morris continued. "It's a telling example of what Darwin was all about. Indeed, he insisted that animals and plants don't compete or evolve only in the Galápagos Islands or in the Brazilian rain forests, but also in the 'tangled bank' of every stream that meanders through the English countryside."

Most school groups that visit the house, according to Morris, come to learn about the Victorian era. "These kids haven't a clue as to what this geezer with the long white beard did," says Morris. "Their teachers bring them here simply because he lived when he did."

Emma Darwin continued to live in Down House until her death in 1896. A decade later, the family leased the estate to a girl's school, which held classes in the house from 1906 to 1922. Then, in 1927, a wealthy surgeon bought the estate, restored it with his own money, and opened it to the public as the Darwin Museum. Later he turned it over for safekeeping to the British Association for the Advancement of Science, which in 1953 transferred it to the Royal College of Surgeons.

Sad to say, over the years the house has decayed, since funds were lacking to keep it up. The greenhouse, with its old window panes perched precariously on weathered wood frames, needs rehabilitation (although descendants of Darwin's tropical plants still grow there). Darwin's laboratory roof has fallen in. But most of all, the main house desperately requires plumbing and electrical overhauls, and extensive repairs to its leaking shingle roof, rotting pillars, and peeling paint. Many rooms need major renovations, both esthetically and structurally. Floors need to be strengthened and reinforced to accommodate the greater influx of visitors that is anticipated.

In recent years Down House faced an increasingly uncertain future until the Natural History Museum in London initiated a campaign to save it once and for all. Fund-raising began just over a year and a half ago, and already $3.75 million of the $4.5 million target has been raised to restore the house and grounds, install basic visitor facilities, and create exhibits on Darwin and evolution. English Heritage, a governmental body that looks after some of England's most famous historic sites, bought the house in May from the Royal College of Surgeons with a $1.1 million grant from the Wellcome Trust. The restoration will be paid out of a $2.7 million grant from the United Kingdom's National Lottery, and about $1 million raised by the Natural History Museum. Another $800,000 is needed by the end of 1996 to save Down House for future generations.

Among those who successfully aroused public interest have been Stephen Jay Gould, who gave a well-attended fundraising lecture at London's Natural History Museum last year, and filmmaker-naturalist Sir David Attenborough, who has publicly championed Down House with the same enthusiasm he usually reserves for wonders of nature. "It was the forty years in this one place—Kent, Down House— that were the significant years of Darwin's thinking," says Attenborough, "and that surely must make it one of the most important places in the history of science." At his beloved country home, Darwin continued his quest for discovery, observing the familiar plants and animals with new eyes. The fulfillment he found at Down House recalls these lines from T. S. Eliot's "Little Gidding":

And the end of all our exploring
Will be to arrive where we started
And know the place for the first time.

Curse and Blessing of the Ghetto

Tay-Sachs disease is a choosy killer, one that for centuries targeted Eastern European Jews above all others. By decoding its lethal logic, we can learn a lot about how genetic diseases evolve—and how they can be conquered.

Jared Diamond

Contributing editor Jared Diamond is a professor of physiology at the UCLA School of Medicine.

Marie and I hated her at first sight, even though she was trying hard to be helpful. As our obstetrician's genetics counselor, she was just doing her job, explaining to us the unpleasant results that might come out of the genetic tests we were about to have performed. As a scientist, though, I already knew all I wanted to know about Tay-Sachs disease, and I didn't need to be reminded that the baby sentenced to death by it could be my own.

Fortunately, the tests would reveal that my wife and I were not carriers of the Tay-Sachs gene, and our preparenthood fears on that matter at least could be put to rest. But at the time I didn't yet know that. As I glared angrily at that poor genetics counselor, so strong was my anxiety that now, four years later, I can still clearly remember what was going through my mind: If I were an evil deity, I thought, trying to devise exquisite tortures for babies and their parents, I would be proud to have designed Tay-Sachs disease.

Tay-Sachs is completely incurable, unpreventable, and preprogrammed in the genes. A Tay-Sachs infant usually appears normal for the first few months after birth, just long enough for the parents to grow to love him. An exaggerated "startle reaction" to sounds is the first ominous sign. At about six months the baby starts to lose control of his head and can't roll over or sit without support. Later he begins to drool, breaks out into unmotivated bouts of laughter, and suffers convulsions. Then his head grows abnormally large, and he becomes blind. Perhaps what's most frightening for the parents is that their baby loses all contact with his environment and becomes virtually a vegetable. By the child's third birthday, if he's still alive, his skin will turn yellow and his hands pudgy. Most likely he will die before he's four years old.

My wife and I were tested for the Tay-Sachs gene because at the time we rated as high-risk candidates, for two reasons. First, Marie was carrying twins, so we had double the usual chance to bear a Tay-Sachs baby. Second, both she and I are of Eastern European Jewish ancestry, the population with by far the world's highest Tay-Sachs frequency.

In peoples around the world Tay-Sachs appears once in every 400,000 births. But it appears a hundred times more frequently—about once in 3,600 births—among descendants of Eastern European Jews, people known as Ashkenazim. For descendants of most other groups of Jews—Oriental Jews, chiefly from the Middle East, or Sephardic Jews, from Spain and other Mediterranean countries—the frequency of Tay-Sachs disease is no higher than in non-Jews. Faced with such a clear correlation, one cannot help but wonder: What is it about this one group of people that produces such an extraordinarily high risk of this disease?

Finding the answer to this question concerns all of us, regardless of our ancestry. Every human population is especially susceptible to certain diseases, not only because of its life-style but also because of its genetic inheritance. For example, genes put European whites at high risk for cystic fibrosis, African blacks for sickle-cell disease, Pacific Islanders for diabetes—and Eastern European Jews for ten different diseases, including Tay-Sachs. It's not that Jews are notably susceptible to genetic diseases in general; but a combination of historical factors has led to Jews' being intensively studied, and so their susceptibilities are far better known than those of, say, Pacific Islanders.

Tay-Sachs exemplifies how we can deal with such diseases; it has been the object of the most successful screening program to date. Moreover, Tay-Sachs is helping us understand how ethnic diseases evolve. Within the past couple of years discoveries by molecular biologists have provided tantalizing clues to precisely how a deadly gene can persist and spread over the centuries. Tay-Sachs may be primarily a disease of Eastern European Jews, but through this affliction of one group of people, we gain a window on how our genes simultaneously curse and bless us all.

The disease's hyphenated name comes from the two physicians—British ophthalmologist W. Tay and New York neurologist B. Sachs—who independently first recognized the disease, in 1881 and 1887, respectively. By 1896 Sachs had seen enough cases to realize that the disease was most common among Jewish children.

Not until 1962, however, were researchers able to trace the cause of the affliction to a single biochemical abnormality: the excessive accumulation in nerve cells of a fatty substance called G_{M2} ganglioside. Normally G_{M2} ganglioside is present at only modest levels in cell membranes, because it is constantly being broken down as well as synthesized. The breakdown depends on the enzyme hexosaminidase A, which is found in the tiny structures within our cells known as lysosomes. In the unfortunate Tay-Sachs victims this enzyme is lacking, and without it the ganglioside piles up and produces all the symptoms of the disease.

We have two copies of the gene that programs our supply of hexosaminidase A, one inherited from our father, the other from our mother; each of our parents, in turn, has two copies derived from their own parents. As long as we have one good copy of the gene, we can produce enough hexosaminidase A to prevent a buildup of G_{M2} ganglioside and we won't get Tay-Sachs. This genetic disease is of the sort termed recessive rather than dominant—meaning that to get it, a child must inherit a defective gene not just from one parent but from both of them. Clearly, each parent must have had one good copy of the gene along with the defective copy—if either had had two defective genes, he or she would have died of the disease long before reaching the age of reproduction. In genetic terms the diseased child is homozygous for the defective gene and both parents are heterozygous for it.

None of this yet gives any hint as to why the Tay-Sachs gene should be most common among Eastern European Jews. To come to grips with that question, we must take a short detour into history.

From their biblical home of ancient Israel, Jews spread peacefully to other Mediterranean lands, Yemen, and India. They were also dispersed violently through conquest by Assyrians, Babylonians, and Romans. Under the Carolingian kings of the eighth and ninth centuries Jews were invited to settle in France and Germany as traders and financiers. In subsequent centuries, however, persecutions triggered by the Crusades gradually drove Jews out of Western Europe; the process culminated in their total expulsion from Spain in 1492. Those Spanish Jews—called Sephardim—fled to other lands around the Mediterranean. Jews of France and Germany—the Ashkenazim—fled east to Poland and from there to Lithuania and western Russia, where they settled mostly in towns, as businessmen engaged in whatever pursuit they were allowed.

It seems unlikely that genetic accidents would have pumped up the frequency of the same gene not once but twice in the same population.

There the Jews stayed for centuries, through periods of both tolerance and oppression. But toward the end of the nineteenth century and the beginning of the twentieth, waves of murderous anti-Semitic attacks drove millions of Jews out of Eastern Europe, with most of them heading for the United States. My mother's parents, for example, fled to New York from the Lithuanian pogroms of the 1880s, while my father's parents fled from the Ukrainian pogroms of 1903–6. The more modern history of Jewish migration is probably well known to you all: most Jews who remained in Eastern Europe were exterminated during World War II, while most the survivors immigrated to the United States and Israel. Of the 13 million Jews alive today, more than three-quarters are Ashkenazim, the descendants of the Eastern European

Jews and the people most at risk for Tay-Sachs.

Have these Jews maintained their genetic distinctness through the thousands of years of wandering? Some scholars claim that there has been so much intermarriage and conversion that Ashkenazic Jews are now just Eastern Europeans who adopted Jewish culture. However, modern genetic studies refute that speculation.

First of all, there are those ten genetic diseases that the Ashkenazim have somehow acquired, by which they differ both from other Jews and from Eastern European non-Jews. In addition, many Ashkenazic genes turn out to be ones typical of Palestinian Arabs and other peoples of the Eastern Mediterranean areas where Jews originated. (In fact, by genetic standards the current Arab-Israeli conflict is an internecine civil war.) Other Ashkenazic genes have indeed diverged from Mediterranean ones (including genes of Sephardic and Oriental Jews) and have evolved to converge on genes of Eastern European non-Jews subject to the same local forces of natural selection. But the degree to which Ashkenazim prove to differ genetically from Eastern European non-Jews implies an intermarriage rate of only about 15 percent.

Can history help explain why the Tay-Sachs gene in particular is so much more common in Ashkenazim than in their non-Jewish neighbors or in other Jews? At the risk of spoiling a mystery, I'll tell you now that the answer is yes, but to appreciate it, you'll have to understand the four possible explanations for the persistence of the Tay-Sachs gene.

First, new copies of the gene might be arising by mutation as fast as existing copies disappear with the death of Tay-Sachs children. That's the most likely explanation for the gene's persistence in most of the world, where the disease frequency is only one in 400,000 births—that frequency reflects a typical human mutation rate. But for this explanation to apply to the Ashkenazim would require a mutation rate of at least one per 3,600 births—far above the frequency observed for any human gene. Furthermore, there would be no precedent for one particular gene mutating so much more often in one human population than in others.

As a second possibility, the Ashkenazim might have acquired the Tay-Sachs gene from some other people who already had the gene at high frequency. Arthur Koestler's controversial book *The Thirteenth Tribe,* for example, popularized the view that the Ashkenazim are really not a Semitic people but are instead descended from the Khazar, a Turkic tribe whose rulers converted to Judaism in the eighth century. Could the Khazar have brought the Tay-Sachs gene to Eastern Europe? This speculation makes good romantic reading, but there is no good evidence to support it. Moreover, it fails to explain why deaths of Tay-Sachs children didn't eliminate the gene by natural selection in the past 1,200 years, nor how the Khazar acquired high frequencies of the gene in the first place.

The third hypothesis was the one preferred by a good many geneticists until recently. It invokes two genetic processes, termed the founder effect and genetic drift, that may operate in small populations. To understand these concepts, imagine that 100 couples settle in a new land and found a population that then increases. Imagine further that one parent among those original 100 couples happens to have some rare gene, one, say, that normally occurs at a frequency of one in a million. The gene's frequency in the new population will now be one in 200 as a result of the accidental presence of that rare founder.

Or suppose again that 100 couples found a population, but that one of the 100 men happens to have lots of kids by his wife or that he is exceptionally popular with other women, while the other 99 men are childless or have few kids or are simply less popular. That one man may thereby father 10 percent rather than a more representative one percent of the next generation's babies, and their genes will disproportionately reflect that man's genes. In other words, gene frequencies will have drifted between the first and second generation.

Through these two types of genetic accidents a rare gene may occur with an unusually high frequency in a small expanding population. Eventually, if the gene is harmful, natural selection will bring its frequency back to normal by killing off gene bearers. But if the resultant disease is recessive—if heterozygous individuals don't get the disease and only the rare, homozygous individuals die of it—the gene's high frequency may persist for many generations.

These accidents do in fact account for the astonishingly high Tay-Sachs gene frequency found in one group of Pennsylvania Dutch: out of the 333 people in this group, 98 proved to carry the Tay-Sachs gene. Those 333 are all descended from one couple who settled in the United States in the eighteenth century and had 13 children. Clearly, one of that founding couple must have carried the gene. A similar accident may explain why Tay-Sachs is also relatively common among French Canadians, who number 5 million today but are descended from fewer than 6,000 French immigrants who arrived in the New World between 1638 and 1759. In the two or three centuries since both these founding events, the high Tay-Sachs gene frequency among Pennsylvania Dutch and French Canadians has not yet had enough time to decline to normal levels.

The same mechanisms were once proposed to explain the high rate of Tay-Sachs disease among the Ashkenazim. Perhaps, the reasoning went, the gene just happened to be overrepresented in the founding Jewish population that settled in Germany or Eastern Europe. Perhaps the gene just happened to drift up in frequency in the Jewish populations scattered among the isolated towns of Eastern Europe.

But geneticists have long questioned whether the Ashkenazim population's history was really suitable for these genetic accidents to have been significant. Remember, the founder effect and genetic drift become significant only in small populations, and the founding populations of Ashkenazim may have been quite large. Moreover, Ashkenazic communities were considerably widespread; drift would have sent gene frequencies up in some towns but down in others. And, finally, natural selection has by now had a thousand years to restore gene frequencies to normal.

Granted, those doubts are based on historical data, which are not always as precise or reliable as one might want. But within the past several years the case against those accidental explanations for Tay-Sachs disease in the Ashkenazim has been bolstered by discoveries by molecular biologists.

Like all proteins, the enzyme absent in Tay-Sachs children is coded for by a piece of our DNA. Along that particular stretch of DNA there are thousands of different sites where a mutation could occur that would result in no enzyme and hence in the same set of symptoms. If molecular biologists had discovered that all cases of Tay-Sachs in Ashkenazim involved damage to DNA at the same site, that would have been strong evidence that in Ashkenazim the disease stems from a single mutation that has been multiplied by the founder effect or genetic drift—in other words, the high incidence of Tay-Sachs among Eastern European Jews is accidental.

In reality, though, several different mutations along this stretch of DNA have been identified in Ashkenazim, and two of them occur much more frequently than in non-Ashkenazim populations. It seems unlikely that genetic accidents would have pumped up the frequency of the same gene not once but twice in the same population.

And that's not the sole unlikely coincidence arguing against accidental explanations. Recall that Tay-Sachs is caused by the excessive accumulation of one fatty substance, G_{M2} ganglioside, from a defect in one enzyme, hexosaminidase A. But Tay-Sachs is one of ten genetic diseases characteristic of Ashkenazim. Among those other nine, two—Gaucher's disease and Niemann-Pick disease—result from the accumulation of two other fatty substances similar to G_{M2} ganglioside, as a result of defects in two other enzymes similar to hexosaminidase A. Yet our bodies contain thousands of different enzymes. It would have been an incredible roll of the genetic dice if, by nothing more than chance, Ashkenazim had independently acquired mutations in three closely related enzymes—and had acquired mutations in one of those enzymes twice.

All these facts bring us to the fourth possible explanation of why the Tay-Sachs gene is so prevalent among Ashkenazim: namely, that something about them favored accumulation of G_{M2} ganglioside and related fats.

For comparison, suppose that a friend doubles her money on one stock while you are getting wiped out with your investments. Taken alone, that could just mean she was lucky on that one occasion. But suppose that she doubles her money on each of two different stocks and at the same time rings up big profits in real estate while also making a killing in bonds. That implies more than lady luck; it suggests that something about your friend—like shrewd judgment—favors financial success.

What could be the blessings of fat accumulation in Eastern European Jews? At first this question sounds weird. After all, that fat accumulation

was noticed only because of the curses it bestows: Tay-Sachs, Gaucher's, or Niemann-Pick disease. But many of our common genetic diseases may persist because they bring both blessings and curses (see "The Cruel Logic of Our Genes," *Discover*, November 1989). They kill or impair individuals who inherit two copies of the faulty gene, but they help those who receive only one defective gene by protecting them against other diseases. The best understood example is the sickle-cell gene of African blacks, which often kills homozygotes but protects heterozygotes against malaria. Natural selection sustains such genes because more heterozygotes than normal individuals survive to pass on their genes, and those extra gene copies offset the copies lost through the deaths of homozygotes.

So let us refine our question and ask, What blessing could the Tay-Sachs gene bring to those individuals who are heterozygous for it? A clue first emerged back in 1972, with the publication of the results of a questionnaire that had asked U.S. Ashkenzaic parents of Tay-Sachs children what their own Eastern European-born parents had

> *We're not a melting pot, and we won't be for a long time. Each ethnic group has some characteristic genes of its own, a legacy of its distinct history.*

died of. Keep in mind that since these unfortunate children had to be homozygotes, with two copies of the Tay-Sachs gene, all their parents had to be heterozygotes, with one copy, and half of the parents' parents also had to be heterozygotes.

As it turned out, most of those Tay-Sachs grandparents had died of the usual causes: heart disease, stroke, cancer, and diabetes. But strikingly, only one of the 306 grandparents had died of tuberculosis, even though TB was generally one of the big killers in these grandparents' time. Indeed, among the general population of large Eastern European cities in the early twentieth century, TB caused up to 20 percent of all deaths.

This big discrepancy suggested that

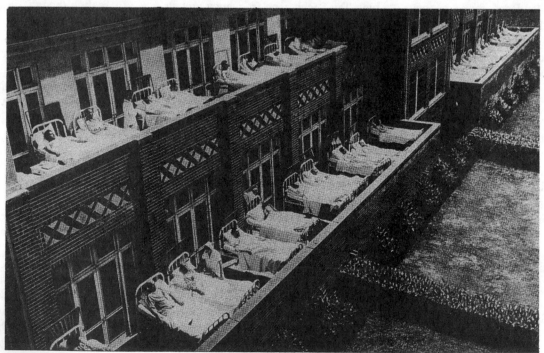

Records at a Jewish TB sanatorium in Denver indicated that among patients born in Europe between 1860 and 1910, Jews from Austria and Hungary were overrepresented. (Photo credit: AMC Cancer Research Center, Denver)

Tay-Sachs heterozygotes might somehow have been protected against TB. Interestingly, it was already well known that Ashkenazim in general had some such protection: even when Jews and non-Jews were compared within the same European city, class, and occupational group (for example, Warsaw garment workers), Jews had only half the TB death rate on non-Jews, despite their being equally susceptible to infection. Perhaps, one could reason, the Tay-Sachs gene furnished part of that well-established Jewish resistance.

A second clue to a heterozygote advantage conveyed by the Tay-Sachs gene emerged in 1983, with a fresh look at the data concerning the distributions of TB and the Tay-Sachs gene within Europe. The statistics showed that the Tay-Sachs gene was nearly three times more frequent among Jews originating from Austria, Hungary, and Czechoslovakia—areas where an amazing 9 to 10 percent of the population were heterozygotes—than among Jews from Poland, Russia, and Germany. At the same time records from an old Jewish TB sanatorium in Denver in 1904 showed that among patients born in Europe between 1860 and 1910, Jews from Austria and Hungary were overrepresented.

Initially, in putting together these two pieces of information, you might be tempted to conclude that because the highest frequency of the Tay-Sachs gene appeared in the same geographic region that produced the most cases of TB, the gene in fact offers no protection whatsoever. Indeed, this was precisely the mistaken conclusion of many researchers who had looked at these data before. But you have to pay careful attention to the numbers here: even at its highest frequency the Tay-Sachs gene was carried by far fewer people than would be infected by TB. What the statistics really indicate is that where TB is the biggest threat, natural selection produces the biggest response.

Think of it this way: You arrive at an island where you find that all the inhabitants of the north end wear suits of armor, while all the inhabitants of the south end wear only cloth shirts. You'd

be pretty safe in assuming that warfare is more prevalent in the north—and that war-related injuries account for far more deaths there than in the south. Thus, if the Tay-Sachs gene does indeed lend heterozygotes some protection against TB, you would expect to find the gene most often precisely where you find TB most often. Similarly, the sickle-cell gene reaches its highest frequencies in those parts of Africa where malaria is the biggest risk.

But you may believe there's still a hole in the argument: If Tay-Sachs heterozygotes are protected against TB, you may be asking, why is the gene common just in the Ashkenazim? Why did it not become common in the non-Jewish populations also exposed to TB in Austria, Hungary, and Czechoslovakia?

At this point we must recall the peculiar circumstances in which the Jews of Eastern Europe were forced to live. They were unique among the world's ethnic groups in having been virtually confined to towns for most of the past 2,000 years. Being forbidden to own land, Eastern European Jews were not peasant farmers living in the countryside, but businesspeople forced to live in crowded ghettos, in an environment where tuberculosis thrived.

Of course, until recent improvements in sanitation, these towns were not very healthy places for non-Jews either. Indeed, their populations couldn't sustain themselves: deaths exceeded births, and the number of dead had to be balanced by continued emigration from the countryside. For non-Jews, therefore, there was no genetically distinct urban population. For ghetto-bound Jews, however, there could be no emigration from the countryside; thus the Jewish population was under the strongest selection to evolve genetic resistance to TB.

Those are the conditions that probably led to Jewish TB resistance, whatever particular genetic factors prove to underlie it. I'd speculate that G_{M2} and related fats accumulate at slightly higher-than-normal levels in heterozygotes, although not at the lethal

levels seen in homozygotes. (The fat accumulation in heterozygotes probably takes place in the cell membrane, the cell's "armor.") I'd also speculate that the accumulation provides heterozygotes with some protection against TB, and that that's why the genes for Tay-Sachs, Gaucher's, and Niemann-Pick disease reached high frequencies in the Ashkenazim.

Having thus stated the case, let me make clear that I don't want to overstate it. The evidence is still speculative. Depending on how you do the calculation, the low frequency of TB deaths in Tay-Sachs grandparents either barely reaches or doesn't quite reach the level of proof that statisticians require to accept an effect as real rather than as one that's arisen by chance. Moreover, we have no idea of the biochemical mechanism by which fat accumulation might confer resistance against TB. For the moment, I'd say that the evidence points to some selective advantage of Tay-Sachs heterozygotes among the Ashkenazim, and that TB resistance is the only plausible hypothesis yet proposed.

For now Tay-Sachs remains a speculative model for the evolution of ethnic diseases. But it's already a proven model of what to do about them. Twenty years ago a test was developed to identify Tay-Sachs heterozygotes, based on their lower-than-normal levels of hexosaminidase A. The test is simple, cheap, and accurate: all I did was to donate a small sample of my blood, pay $35, and wait a few days to receive the results.

If that test shows that at least one member of a couple is not a Tay-Sachs heterozygotre, then any child of theirs can't be a Tay-Sachs homozygote. If both parents prove to be heterozygotes, there's a one-in-four chance of their child being a homozygote; that can then be determined by other tests performed on the mother early in pregnancy. If the results are positive, it's early enough for her to abort, should she choose to. That critical bit of knowledge has enabled parents who had gone through the agony of bearing a Tay-Sachs baby and watching him die to find the courage to try again.

The Tay-Sachs screening program launched in the United States in 1971 was targeted at the high-risk population: Ashkenazic Jewish couples of childbearing age. So successful has this approach been that the number of Tay-Sachs babies born each year in this country has declined tenfold. Today, in fact, more Tay-Sachs cases appear here in non-Jews than in Jews, because only the latter couples are routinely tested. Thus, what used to be the classic genetic disease of Jews is so no longer.

There's also a broader message to the Tay-Sachs story. We commonly refer to the United States as a melting pot, and in many ways that metaphor is apt. But in other ways we're not a melting pot, and we won't be for a long time. Each ethnic group has some characteristic genes of its own, a legacy of its distinct history. Tuberculosis and malaria are not major causes of death in the United States, but the genes that some of us evolved to protect ourselves against them are still frequent. Those genes are frequent only in certain ethnic groups, though, and they'll be slow to melt through the population.

With modern advances in molecular genetics, we can expect to see more, not less, ethnically targeted practice of medicine. Genetic screening for cystic fibrosis in European whites, for example, is one program that has been much discussed recently; when it comes, it will surely be based on the Tay-Sachs experience. Of course, what that may mean someday is more anxiety-ridden parents-to-be glowering at more dedicated genetics counselors. It will also mean fewer babies doomed to the agonies of diseases we may understand but that we'll never be able to accept.

The Future of AIDS

New research suggests HIV is not a new virus but an old one that grew deadly.
Can we turn the process around?

Geoffrey Cowley

Ten years ago, Benjamin B. got what might have been a death sentence. Hospitalized for colon surgery, the Australian retiree received a blood transfusion tainted with the AIDS virus. That he's alive at all is remarkable, but that's only half of the story. Unlike most long-term HIV survivors, he has suffered no symptoms and no loss of immune function. He's as healthy today as he was in 1983—and celebrating his 81st birthday. Benjamin B. is just one of five patients who came to the attention of Dr. Brett Tindall, an AIDS researcher at the University of New South Wales, as he was preparing a routine update on transfusion-related HIV infections last year. All five were infected by the same donor. And seven to 10 years later, none has suffered any effects.

The donor turned out to be a gay man who had contracted the virus during the late 1970s or early '80s, then given blood at least 26 times before learning he was infected. After tracking him down, Tindall learned, to his amazement, that the man was just as healthy as the people who got his blood. "We know that HIV causes AIDS," Tindall says. "We also know that a few patients remain well for long periods, but we've never known why. Is it the vitamins they take? Is it some gene they have in common? This work suggests it has more to do with the virus. I think we've found a harmless strain."

He may also have found the viral equivalent of a fossil, a clue to the origin, evolution and future of the AIDS epidemic. HIV may not be a new and inherently deadly virus, as is commonly assumed, but an old one that has recently acquired deadly tendencies. In a forthcoming book, Paul Ewald, an evolutionary biologist at Amherst College, argues that HIV may have infected people benignly for decades, even centuries, before it started causing AIDS. He traces its virulence to the social upheavals of the 1960s and '70s, which not only sped its movement through populations but rewarded it for reproducing more aggressively within the body.

The idea may sound radical, but it's not just flashy speculation. It reflects a growing awareness that parasites, like everything else in nature, evolve by natural selection, changing their character to adapt to their environments. Besides transforming our understanding of AIDS, the new view could yield bold strategies for fighting it. Viruses can evolve tens of thousands of times faster than plants or animals, and few evolve as fast as HIV. Confronted by a drug or an immune reaction, the virus readily mutates out of its range. A few researchers are now trying to exploit that very talent, using drugs to force HIV to mutate until it can no longer function. A Boston team, led by medical student Yung-Kang Chow, made headlines last month by showing that the technique works perfectly in a test tube. Human trials are now in the works, but better drug treatment isn't the only hope rising from an evolutionary outlook. If rapid spread is what turned HIV into a killer, then condoms and clean needles may ultimately do more than prevent new infections. Used widely enough, they might drive the AIDS virus toward the benign form sighted in Australia.

I. WHERE DID HIV COME FROM?

Viruses are the ultimate parasites. Unlike bacteria, which absorb nutrients, excrete waste and reproduce by dividing, they have no life of their own. They're mere shreds of genetic information, encoded in DNA or RNA, that can integrate themselves into a living cell and use its machinery to run off copies of themselves. Where the first one came from is anyone's guess, but today's viruses are, like any plant or animal, simply descendants of earlier forms.

Most scientists agree that the human immunodeficiency viruses—HIV-1 and HIV-2—are basically ape or monkey viruses. Both HIVs are genetically similar to viruses found in African primates, the so called SIVs. In fact, as the accompanying tree illustrates, the HIVs have more in common with simian viruses than they do with each other. HIV-2, found mainly in West Africa, is so similar to the SIV that infects the sooty mangabey—an ash-colored monkey from the same region—that it doesn't really qualify as a separate viral species. "When you see HIV-2," says Gerald Myers, head of the HIV database project at the Los

1. NATURAL SELECTION

Alamos National Laboratory, "you may not be looking at a human virus but at a mangabey virus in a human." HIV-1, the virus responsible for the vast majority of the world's AIDS cases, bears no great resemblance to HIV-2 or the monkey SIVs, but it's very similar to SIV cpz, a virus recovered in 1990 from a wild chimpanzee in the West African nation of Gabon.

The prevailing theory holds that humans were first infected through direct contact with primates, and that the SIVs they contracted have since diverged by varying degrees from their ancestors. It's possible, of course, that the HIVs and SIVs evolved separately, or even that humans were the original carriers. But the primates-to-people scenario has a couple of points in its favor. First, the SIVs are more varied than the HIVs, which suggests they've been evolving longer. Second, it's easier to imagine people being infected by chimps or monkeys than vice versa. Humans have hunted and handled other primates for thousands of years. Anyone who was bitten or scratched, or who cut himself butchering an animal, could have gotten infected.

Until recently, it was unclear whether people could contract SIV directly from primates, but a couple of recent accidents have settled that issue. In one case, reported last summer by the Centers for Disease Control, a lab technician at a primate-research center jabbed herself with a needle containing blood from an infected macaque. The infection didn't take—she produced antibodies to SIV only for a few months—but she was just lucky. Another lab worker, who handled monkey tissues while suffering from skin lesions, has remained SIV-positive for two years. In a recent survey of 472 blood samples drawn from primate handlers, health officials found that three of those tested positive as well. No one knows whether the people with SIV eventually develop AIDS, but the potential for cross-species transmission is now clear.

Far less clear is when the first such transmission took place. The most common view holds that since AIDS is a new epidemic, the responsible vi-

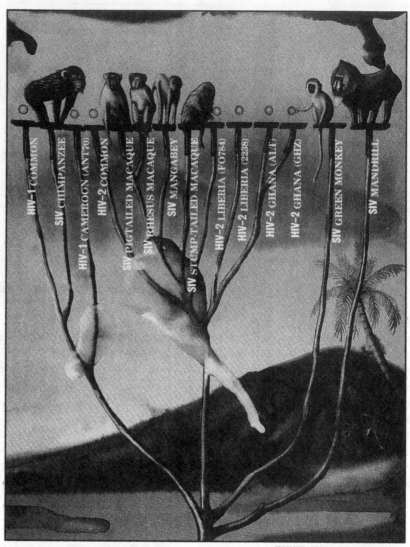

ILLUSTRATION BY ALEXIS ROCKMAN

HIV'S EXTENDED FAMILY: Today's viruses are descendants of earlier forms. This family tree shows that the human AIDS viruses, HIV–1 and HIV–2, are more closely related to viruses found in primates (the SIVs) than to each other.

ruses must have entered humans within the past few decades. That's a reasonable suspicion, but it raises a sticky question. Why, if people have been handling primates in Africa for thousands of years, did SIV take until now to jump species?

One possibility is that humans recently opened some avenue that hasn't existed in the past. Some theorists argue, for example, that AIDS was spawned by a polio-vaccination program carried out in Africa during the late 1950s. During that four-year effort, 325,000 Africans received an oral polio vaccine produced in kidney cells from African green monkeys. Even if the vaccine was contaminated with SIV—which hasn't been estab-

lished—blaming it for AIDS would be hasty. For the SIVs found in African green monkeys bear too little resemblance to HIV-1, the primary human AIDS virus, to be its likely progenitor. In order to link HIV-1 to those early lots of polio vaccine, someone would have to show that they contained a monkey virus never yet found in actual monkeys.

The alternative view—that the HIVs are old viruses—is just as hard to prove, but it requires fewer tortured assumptions. Dr. Jay Levy, an AIDS researcher at the University of California, San Francisco, puts it this way: "We know that all these other primates harbor lentiviruses [the class that includes the SIVs and HIVs]. Why should

28

humans be any exception?'' If the HIVs were on hand before the AIDS epidemic began, the key question is not where they came from but whether they always caused the disease.

II. WAS HIV ONCE LESS DEADLY?

If HIV had always caused AIDS, one would expect virus and illness to emerge together in the historical record. Antibodies to HIV have been detected in rare blood samples dating back to 1959, yet the first African AIDS cases were described in the early 1980s, when the disease started decimating the cities of Rwanda, Zaire, Zambia and Uganda. When Dr. Robert Biggar, an epidemiologist at the National Cancer Institute, pored over African hospital records looking for earlier descriptions of AIDS-like illness he didn't find any. It's possible, of course, that the disease was there all along, just too rare to be recognized as a distinct condition. But the alternative view is worth considering. There are intriguing hints that HIV hasn't always been so deadly.

Any population of living things, from fungi to rhinoceri, includes genetically varied individuals, which pass essential traits along to their progeny. As Charles Darwin discerned more than a century ago, the individuals best designed to exploit a particular environment tend to produce the greatest number of viable offspring. As generations pass, beneficial traits become more and more pervasive in the population. There's no universal recipe for reproductive success; different environments favor different traits. But by preserving some and discarding others, every environment molds the species it supports.

Viruses aren't exempt from the process. Their purpose, from a Darwinian perspective, is simply to make as many copies of themselves as they can. Other things being equal, those that replicate fastest will become the most plentiful within the host, and so stand the best chance of infecting other hosts on contact. But there's a catch. If a microbe

reproduces too aggressively inside its host, or invades too many different tissues, it may kill the host—and itself—without getting passed along at all. The most successful virus, then, is not necessarily the most or the least virulent. It's one that exploits its host most effectively.

As Ewald and others have shown, that mandate can drive different microbes to very different levels of nastiness. Because they travel via social contact between people, cold and flu viruses can't normally afford to immobilize us. To stay in business, they need hosts who are out coughing, sneezing, shaking hands and sharing pencils. But the incentives change when a parasite has other ways of getting around. Consider tuberculosis or diphtheria. Both deadly diseases are caused by bacteria that can survive for weeks or months outside the body. They can reproduce aggressively in the host, ride a cough into the external environment, then wait patiently for another host to come along. By the same token, a parasite that can travel from person to person via mosquito or some other vector has little reason to be gentle. As long as malaria sufferers can still feed hungry mosquitoes, their misery is of little consequence to the microbe. Indeed, a

host who can't wield a fly swatter may be preferable to one who can.

These patterns aren't set in stone. A shift in circumstances may push a normally mild-mannered parasite toward virulence, or vice versa. One of the most devastating plagues in human history was caused by a mere influenza virus, which swept the globe in 1918, leaving 20 million corpses in its wake. Many experts still regard the disaster as an accident, triggered by the random reshuffling of viral genes. But from an evolutionary perspective, it's no coincidence that the flu grew so deadly when it did. World War I was

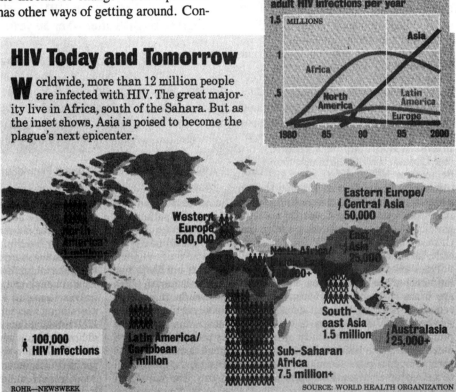

HIV Today and Tomorrow

Worldwide, more than 12 million people are infected with HIV. The great majority live in Africa, south of the Sahara. But as the inset shows, Asia is poised to become the plague's next epicenter.

Estimated/projected new adult HIV infections per year

Eastern Europe/Central Asia 50,000

Western Europe 500,000

South-east Asia 1.5 million

Australasia 25,000+

Sub-Saharan Africa 7.5 million+

Latin America/Caribbean 1 million

100,000 HIV Infections

ROHR—NEWSWEEK

SOURCE: WORLD HEALTH ORGANIZATION

raging in 1918. Great numbers of soldiers were huddled in the European trenches, where even the most ravaged host stood an excellent chance of infecting many others. For a flu virus, the incentives favoring restraint would have vanished in those circumstances. Rather than rendering the host useless, extreme virulence would simply make him more infectious.

Ewald suspects that HIV has recently undergone a similar transformation. Unlike influenza viruses, which infect cells in the respiratory tract and spread through coughs and sneezes, the HIVs insinuate themselves into white blood cells. Infected cells (or the new viruses they produce) can pass between people, but only during sex or other exchanges of body fluid. Confined to an isolated population where no carrier had numerous sex partners, a virus like HIV would gain nothing from replicating aggressively within the body; it would do best to lie low, leaving the host alive and mildly infectious for many years. But if people's sexual networks suddenly expanded, fresh hosts would become more plentiful, and infected hosts more dispensable. An HIV strain that replicated wildly might kill people in three years instead of 30, but by making them more infectious while they lasted, it would still come out ahead.

Is that what actually happened? There's no question that social changes have hastened the spread of HIV. Starting in the 1960s, war, tourism and commercial trucking forced the outside world on Africa's once isolated villages. At the same time, drought and industrialization prompted mass migrations from the countryside into newly teeming cities. Western monogamy had never been common in Africa, but as the French medical historian Mirko Grmek notes in his book "History of AIDS," urbanization shattered social structures that had long constrained sexual behavior. Prostitution exploded, and venereal disease flourished. Hypodermic needles came into wide use during the same period, creating yet another

mode of infection. Did these trends actually turn a chronic but relatively benign infection into a killer? The evidence is circumstantial, but it's hard to discount.

If Ewald is right, and HIV's deadliness is a consequence of its rapid spread, then the nastiest strains should show up in the populations where it's moving the fastest. To a surprising degree, they do. It's well known, for example, that HIV-2 is far less virulent than HIV-1. "Going on what we've seen so far, we'd have to say that HIV-1 causes AIDS in 90 percent of those infected, while HIV-2 causes AIDS in 10 percent or less," says Harvard AIDS specialist Max Essex. "Maybe everyone infected with HIV-2 will progress to AIDS after 40 or 50 years, but that's still in the realm of reduced virulence." From Ewald's perspective, it's no surprise that HIV-2, the strain found in West Africa, is the gentle one. West Africa has escaped much of the war, drought and urbanization that fueled the spread of HIV-1 in the central and eastern parts of the continent. "HIV-2 appears to be adapted for slow transmission in areas with lower sexual contact," he concludes, "and HIV-1 for more rapid transmission in areas with higher sexual contact."

The same pattern shows up in the way each virus affects different populations. HIV-2 appears particularly mild in the stable and isolated West African nation of Senegal. After following a group of Senegalese prostitutes for six years, Harvard researchers found that those testing positive for HIV-2 showed

virtually no sign of illness. In laboratory tests, researchers at the University of Alabama found that Senegalese HIV-2 didn't even kill white blood cells when allowed to infect them in a test tube. Yet HIV-2 is a killer in the more urban and less tradition-bound Ivory Coast. In a survey of hospital patients in the city of Abidjan, researchers from the U.S. Centers for Disease Control found that HIV-2 was associated with AIDS nearly as often as HIV-1.

The variations within HIV-1 are less clear-cut, but they, too, lend support to Ewald's idea. Though the evidence is mixed, there are hints that IV drug users (whose transmission rates have remained high for the past decade) may be contracting deadlier strains of HIV-1 than gay men (whose transmission rates have plummeted). In a 1990 study of infected gay men, fewer than 8 percent of those not receiving early treatment developed AIDS each year. In a more recent study of IV drug users, the proportion of untreated carriers developing AIDS each year was more than 17 percent.

Together, these disparities suggest that HIV assumes different personalities in different settings, becoming more aggressive when it's traveling rapidly through a population. But because so many factors affect the health of infected people, the strength of the connection is unclear. "This is exactly the right way to think about virulence," says virologist Stephen Morse of New York's Rockefeller University. "Virulence should be dynamic, not static. The question is, how dynamic?

A Hypothetical History of AIDS

Why did HIV suddenly emerge as a global killer? According to one theory, the virus has infected people for centuries, but recent social changes have altered its character.

BEFORE 1960
Rural Africans contracted benign ancestral forms of HIV from primates. Because the viruses spread so slowly

among people, they couldn't afford to become virulent.
1960 TO 1975
War, drought, commerce and urbanization shattered African social institutions. HIV spread rapidly, becoming more virulent as transmission accelerated.
1975 TO PRESENT
Global travel placed HIV in broader circulation. Shifting sexual mores and

modern medical practices, such as blood transfusion, made many populations susceptible.

THE FUTURE
If social changes can turn a benign virus deadly, the process should be reversible. Simply slowing transmission may help drive fast-killing strains out of circulation.

We know that a pathogen like HIV has a wide range of potentials, but we can't yet say just what pressures are needed to generate a particular outcome."

The best answers to Morse's question may come from laboratory studies. A handful of biologists are now devising test-tube experiments to see more precisely how transmission rates shape a parasite's character. Zoologist James Bull of the University of Texas at Austin has shown, for example, that a bacteriophage (a virus that infects bacteria) kills bacterial cells with great abandon when placed in a test tube and given plenty of new cells to infect. Like HIV in a large, active sexual network, it can afford to kill individual hosts without wiping itself out in the process. Yet the same virus becomes benign when confined to individual cells and their offspring (a situation perhaps akin to pre-epidemic HIV's). With a good animal model, researchers might someday manage to test Ewald's hypotheses about HIV with the same kind of precision.

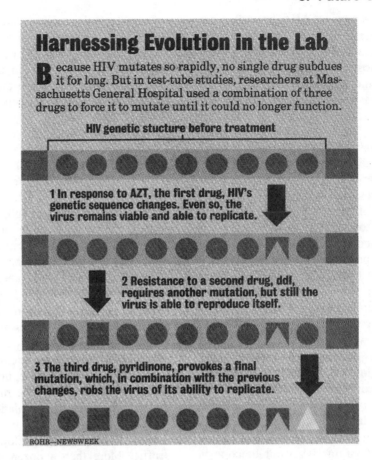

Harnessing Evolution in the Lab

Because HIV mutates so rapidly, no single drug subdues it for long. But in test-tube studies, researchers at Massachusetts General Hospital used a combination of three drugs to force it to mutate until it could no longer function.

HIV genetic stucture before treatment

1 In response to AZT, the first drug, HIV's genetic sequence changes. Even so, the virus remains viable and able to replicate.

2 Resistance to a second drug, ddI, requires another mutation, but still the virus is able to reproduce itself.

3 The third drug, pyridinone, provokes a final mutation, which, in combination with the previous changes, robs the virus of its ability to replicate.

ROHR—NEWSWEEK

III. CAN HIV BE TAMED?

Until recently, medical science seemed well on its way to controlling the microbial world. Yet after 10 years and billions of dollars in research, HIV still has scientists over a barrel. The secret of its success can be summed up in one word: mutability. Because HIV's method of replication is so error prone (its genes mutate at a million times the rate of our own), it produces extremely varied offspring, even within an individual host. Whenever a drug or immune response successfully attacks one variant, another arises to flourish in its place. Even when an AIDS drug works broadly enough to check HIV's growth, it rarely works for long. AZT, for example, can help prevent symptoms for a couple of years. But people on AZT still get AIDS, as the viral populations in their bodies evolve toward resistant forms.

There may not be a drug or vaccine on earth that could subdue such a protean parasite. But from a Darwinian perspective, killing HIV is not the only way to combat AIDS. We know the virus changes rapidly in response to outside pressures. Logic suggests that if we simply applied the right pressures—within a community, or even within a patient's body—we might begin to tame it.

It's well known that condoms and clean needles can save lives by preventing HIV infection. From an evolutionary perspective, there is every reason to think they could do more. Used widely enough, those same humble implements might push the virus toward more benevolent forms, simply by depriving virulent strains of the high transmission rates they need to survive. Gay men are already engaged in that exercise. Studies suggest that, thanks to safer sex, the rate of new infections among gays declined five to tenfold during the 1980s. There are tantalizing hints that HIV has grown less noxious in the same population over the same period. In a 1991 study, researchers at the National Institutes of Health (NIH) calculated the rates at which infected people from different risk groups were developing AIDS each year. They found that as of 1987, the rate declined sharply among gay men, suggesting the virus was taking longer to cause illness. Part of the change was due to AZT, which can delay the onset of symptoms. But when the NIH researchers corrected for AZT use, there was still a mysterious shortage of AIDS cases. From Ewald's viewpoint, the shortfall was not only unsurprising but predictable.

How far could such a trend be pushed? Would broader, better prevention efforts eventually turn today's deadliest HIV-1 into something as benign as Senegal's HIV-2? No one knows. But the prospect of domesticating the AIDS virus, even partially, should excite public-health officials. Condoms and clean needles are exceedingly cheap medicine. They can save lives even if they fail to change the course of evolution—and judging from the available evidence, they might well succeed.

In the meantime, more than 12 million people are carrying today's HIV, and those who get AIDS are still dying. Fortunately, as Yung-Kang Chow and his colleagues at Massachusetts General Hospital showed last month, there's more than one way to manipulate viral evolution. The researchers managed, in a test-tube experiment, to outsmart HIV at its own game. Their trick was to combine three drugs—AZT, ddI and pyridinone—that disarm the same part of the virus (an enzyme called reverse transcriptase).

Any of those drugs can foil HIV's efforts to colonize host cells. When HIV encounters them individually, or even in pairs, it gradually mutates into resistant forms and goes on about its business. But each mutation makes the virus slightly less efficient—and as Chow's group demonstrated, there comes a point where mutation itself hobbles the virus (see chart on previous page). By engineering an HIV mutant that contained three different mutations (one in response to each of the three drugs), the researchers ended up with a virus that was too deformed to function at all. If virgin HIV can't function in the presence of the three drugs—and if triply mutated HIV can't function at all—then the three-drug regimen should, theoretically, do wonders for patients.

It's a long way from the test tube to the clinic; many treatments have shown great promise in lab experiments, only to prove ineffective or highly toxic in people. Upcoming clinical trials will determine whether patients actually benefit from Chow's combination of drugs. The beauty of the new approach, however, is that it's not limited to any particular combination. While the Boston team experiments with drugs directed against reverse transcriptase, researchers at New York's Aaron Diamond AIDS Research Center are trying the same tack against another viral target (an enzyme called protease). "This virus has impressed us again and again with its ability to change," says Dr. David Ho, director

of the Aaron Diamond Center. "It always has a new strategy to counter our efforts. Now we're asking it to make a tradeoff. We're saying, 'Go ahead and mutate, because we think that if you mutate in the right place, you'll do less damage to the patient'."

IV. CAN THE NEXT AIDS BE AVOIDED?

The forces that brought us this plague can surely bring us others. By encroaching on rain forests and wilderness areas, humanity is placing itself in ever-closer contact with other animal species and their obscure, deadly parasites. Other activities, from irrigation to the construction of dams and cities, can create new diseases by expanding the range of the rodents or insects that carry them. Stephen Morse, the Rockefeller virologist, studies the movement of microbes among populations and species, and he worries that human activities are speeding the flow of viral traffic. More than a dozen new diseases have shown up in humans since the 1960s, nearly all of them the result of once exotic parasites exploiting new opportunities. "The primary problem," Morse concludes, "is no longer virological but social."

The Ebola virus is often cited as an example of the spooky pathogens in our future. The virus first struck in August 1976, when a trader arrived at a mission hospital in northern Zaire, fever raging and blood oozing from every orifice. Within days the man died, and nearly half of the nurses at the hospital were stricken. Thirty-nine died, and as hospital patients contracted the virus, it spread to 58 neighboring villages. Ebola fever ended up striking 1,000 people in Zaire and nearby Sudan, killing 500. Epidemiologists feared it would spread more widely, but the outbreaks subsided as quickly as they had begun. From a

Darwinian perspective, that's no great surprise. A parasite that kills that rapidly has little chance of sustaining a chain of infection unless it can survive independently of its host.

More worrisome is a virus like HTLV, a relative of HIV that infects the same class of blood cells and is riding the same waves through new populations. Though recognized only since the 1970s, the HTLVs (HTLV-1 and HTLV-2) appear to be ancient. About one in 20 HTLV-1 infections leads eventually to leukemia, lymphoma or a paralyzing neurologic disorder called TSP. The virus is less aggressive than HIV-1—it typically takes several decades to cause any illness—but its virulence seems to vary markedly from one setting to the next. In Japan, the HTLV-related cancers typically show up in 60-year-olds who were infected by their mothers in the womb. In the Caribbean, where the virus is more often transmitted through sex, the average latency is much shorter. It's not unusual for people to develop symptoms in their 40s.

HTLV may not mutate as readily as HIV, but it is subject to the same natural forces. If human activities can turn one virus into a global killer, it's only prudent to suspect they could do the same to another. "HTLV is a threat," says Ewald, "not because it has escaped from some secluded source, but because it may evolve increased virulence." HTLV-1 is only one tenth as prevalent as HIV in the United States, but it has gained a strong foothold among IV drug users, whose shared needles are a perfect breeding ground for virulent strains.

No one knows whether HTLV could cause an epidemic like AIDS. Fortunately, we don't have to wait passively to find out. We're beginning to see how our actions mold the character of our parasites. No one saw the last epidemic coming. This time, that's not an excuse.

Black, White, Other

Racial categories are cultural constructs masquerading as biology

Jonathan Marks

New York-born Jonathan Marks earned an undergraduate degree in natural science at Johns Hopkins. After getting his Ph.D. in anthropology, Marks did a post-doc in genetics at the University of California at Davis and is now an associate professor of anthropology at Yale University. He is the coauthor, with Edward Staski, of the introductory textbook Evolutionary Anthropology *(San Diego: Harcourt, Brace Jovanovich, 1992). His new book,* Human Biodiversity: Genes, Race, and History *is published (1995) by Aldine de Gruyter.*

While reading the Sunday edition of the *New York Times* one morning last February, my attention was drawn by an editorial inconsistency. The article I was reading was written by attorney Lani Guinier. (Guinier, you may remember, had been President Clinton's nominee to head the civil rights division at the Department of Justice in 1993. Her name was hastily withdrawn amid a blast of criticism over her views on political representation of minorities.) What had distracted me from the main point of the story was a photo caption that described Guinier as being "half-black." In the text of the article, Guinier had described herself simply as "black."

How can a person be black and half black at the same time? In algebraic terms, this would seem to describe a situation where $x = \frac{1}{2}x$, to which the only solution is $x = 0$.

The inconsistency in the *Times* was trivial, but revealing. It encapsulated a longstanding problem in our use of racial categories—namely, a confusion between biological and cultural heredity. When Guinier is described as "half-black," that is a statement of biological ancestry, for one of her two parents is black. And when Guinier describes herself as black, she is using a cultural category, according to which one can either be black or white, but not both.

Race—as the term is commonly used—is inherited, although not in a strictly biological fashion. It is passed down according to a system of folk heredity, an all-or-nothing system that is different from the quantifiable heredity of biology. But the incompatibility of the two notions of race is sometimes starkly evident—as when the state decides that racial differences are so important that interracial marriages must be regulated or outlawed entirely. Miscegenation laws in this country (which stayed on the books in many states through the 1960s) obliged the legal system to define who belonged in what category. The resulting formula stated that anyone with one-eighth or more black ancestry was a "negro." (A similar formula, defining Jews, was promulgated by the Germans in the Nuremberg Laws of the 1930s.)

Applying such formulas led to the biological absurdity that having one black great-grandparent was sufficient to define a person as black, but having seven white great grandparents was insufficient to define a person as white. Here, race and biology are demonstrably at odds. And the problem is not semantic but conceptual, for race is presented as a category of nature.

Human beings come in a wide variety of sizes, shapes, colors, and forms—or, because we are visually oriented primates, it certainly seems that way. We also come in larger packages called populations; and we are said to belong to even larger and more confusing units, which have long been known as races. The history of the study of human variation is to a large extent the pursuit of those human races—the attempt to identify the small number of fundamentally distinct kinds of people on earth.

This scientific goal stretches back two centuries, to Linnaeus, the father of biological systematics, who radically established *Homo sapiens* as one species within a group of animals he called Primates. Linnaeus's system of naming groups within groups logically implied further breakdown. He consequently sought to establish a number of subspecies within *Homo sapiens*. He identified five: four geographical species (from Europe, Asia, Africa, and America) and one grab-bag subspecies called *monstrosus*. This category was dropped by subsequent researchers (as was Linnaeus's use of criteria such as personality and dress to define his subspecies).

While Linnaeus was not the first to divide humans on the basis of the continents on which they lived, he had given the division a scientific stamp.

1. NATURAL SELECTION

But in attempting to determine the proper number of subspecies, the heirs of Linnaeus always seemed to find different answers, depending upon the criteria they applied. By the mid-twentieth century, scores of anthropologists—led by Harvard's Earnest Hooton—had expended enormous energy on the problem. But these scholars could not convince one another about the precise nature of the fundamental divisions of our species.

Part of the problem—as with the *Times's* identification of Lani Guinier—was that we humans have two constantly intersecting ways of thinking about the divisions among us. On the one hand, we like to think of "race"—as Linnaeus did—as an objective, biological category. In this sense, being a member of a race is supposed to be the equivalent of being a member of a species or of a phylum—except that race, on the analogy of subspecies, is an even narrower (and presumably more exclusive and precise) biological category.

The other kind of category into which we humans allocate ourselves—when we say "Serb" or "Hutu" or "Jew" or "Chicano" or "Republican" or "Red Sox fan"—is cultural. The label refers to little or nothing in the natural attributes of its members. These members may not live in the same region and may not even know many others like themselves. What they share is neither strictly nature nor strictly community. The groupings are constructions of human social history.

Membership in these *un*biological groupings may mean the difference between life and death, for they are the categories that allow us to be identified(and accepted or vilified) socially. While membership in (or allegiance to) these categories may be assigned or adopted from birth, the differentia that mark members from nonmembers are symbolic and abstract; they serve to distinguish people who cannot be readily distinguished by nature. So important are these symbolic distinctions that some of the strongest animosities are often expressed between very similar-looking peoples. Obvious examples are Bosnian Serbs and Muslims, Irish and English, Huron and Iroquois.

Obvious natural variation is rarely so important as cultural difference. One simply does not hear of a slaughter of the short people at the hands of the tall, the glabrous at the hands of the hairy, the red-haired at the hands of the brown-haired. When we do encounter genocidal violence between different looking peoples, the two groups are invariably socially or culturally distinct as well. Indeed, the tragic frequency of hatred and genocidal violence between biologically indistinguishable peoples implies that biological differences such as skin color are not motivations but, rather, excuses. They allow nature to be invoked to reinforce group identities and antagonisms that would exist without these physical distinctions. But are there any truly "racial" biological distinctions to be found in our species?

Obviously, if you compare two people from different parts of the world (or whose ancestors came from different parts of the world), they will differ physically, but one cannot therefore define three or four or five basically different kinds of people, as a biological notion of race would imply. The anatomical properties that distinguish people—such as pigmentation, eye form, body build—are not clumped in discrete groups, but distributed along geographical gradients, as are nearly all the genetically determined variants detectable in the human gene pool.

These gradients are produced by three forces. Natural selection adapts populations to local circumstances (like climate) and thereby differentiates them from other populations. Genetic drift (random fluctuations in a gene pool) also differentiates populations from one another, but in non-adaptive ways. And gene flow (via intermarriage and other child-producing unions) acts to homogenize neighboring populations.

In practice, the operations of these forces are difficult to discern. A few features, such as body build and the graduated distribution of the sickle cell anemia gene in populations from western Africa, southern Asia, and the Mediterranean can be plausibly related to the effects of selection. Others, such as the graduated distribution of a small

deletion in the mitochondrial DNA of some East Asian, Oceanic, and Native American peoples, or the degree of flatness of the face, seem unlikely to be the result of selection and are probably the results of random biohistorical factors. The cause of the distribution of most features, from nose breadth to blood group, is simply unclear.

The overall result of these forces is evident, however. As Johann Friedrich Blumenbach noted in 1775, "you see that all do so run into one another, and that one variety of mankind does so sensibly pass into the other, that you cannot mark out the limits between them." (Posturing as an heir to Linnaeus, he nonetheless attempted to do so.) But from humanity's gradations in appearance, no defined groupings resembling races readily emerge. The racial categories with which we have become so familiar are the result of our imposing arbitrary cultural boundaries in order to partition gradual biological variation.

Unlike graduated biological distinctions, culturally constructed categories are ultrasharp. One can be French or German, but not both; Tutsi or Hutu, but not both; Jew or Catholic, but not both; Bosnian Muslim or Serb, but not both; black or white, but not both. Traditionally, people of "mixed race" have been obliged to choose one and thereby identify themselves unambiguously to census takers and administrative bookkeepers—a practice that is now being widely called into question.

A scientific definition of race would require considerable homogeneity within each group, and reasonably discrete differences between groups, but three kinds of data militate against this view: First, the groups traditionally described as races are not at all homogeneous. Africans and Europeans, for instance, are each a collection of biologically diverse populations. Anthropologists of the 1920s widely recognized *three* European races: Nordic, Alpine, and Mediterranean. This implied that races could exist within races. American anthropologist Carleton Coon identified *ten* European races in 1939. With such protean use, the term race came to have little value in describing actual biological entities within *Homo sapiens*. The scholars were not only grappling with a broad north-south gradient in human appearance across Europe, they were trying to bring the data into line with their belief in profound and fundamental constitutional differences between groups of people.

But there simply isn't one European race to contrast with an African race, nor three, nor ten: the question (as scientists long posed it) fails to recognize the actual patterning of diversity in the human species. Fieldwork revealed, and genetics later quantified, the existence of far more biological diversity within any group than between groups. Fatter and thinner people exist everywhere, as do people with type O and type A blood. What generally varies from one population to the next is the *proportion* of people in these groups expressing the trait or gene. Hair color varies strikingly among Europeans and native Australians, but little among other peoples. To focus on discovering differences between presumptive races, when the vast majority of detectable variants do not help differentiate them, was thus to define a very narrow—if not largely illusory—problem in human biology. (The fact that Africans are biologically more diverse than Europeans, but have rarely been split into so many races, attests to the cultural basis of these categorizations.)

Second, differences between human groups are only evident when contrasting geographical extremes. Noting these extremes, biologists of an earlier era sought to identify representatives of "pure," primordial races presumably located in Norway, Senegal, and Thailand. At no time, however, was our species composed of a few populations within which everyone looked pretty much the same. Ever since some of our ancestors left Africa to spread out through

the Old World, we humans have always lived in the "in-between" places. And human populations have also always been in genetic contact with one another. Indeed, for tens of thousands of years, humans have had trade networks; and where goods flow, so do genes. Consequently, we have no basis for considering *extreme* human forms the most pure, or most representative, of some ancient primordial populations. Instead, they represent populations adapted to the most disparate environments.

And third, between each presumptive "major" race are unclassifiable populations and people. Some populations of India, for example, are darkly pigmented (or "black"), have Europeanlike ("Caucasoid") facial features, but inhabit the continent of Asia (which should make them "Asian"). Americans might tend to ignore these "exceptions" to the racial categories, since immigrants to the United States from West Africa, Southeast Asia, and northwest Europe far outnumber those from India. The very existence of unclassifiable peoples undermines the idea that there are just three human biological groups in the Old World. Yet acknowledging the biological distinctiveness of such groups leads to a rapid proliferation of categories. What about Australians? Polynesians? The Ainu of Japan?

Categorizing people is important to any society. It is, at some basic psychological level, probably necessary to have group identity about who and what you are, in contrast to who and what you are not. The concept of race, however, specifically involves the recruitment of biology to validate those categories of self-identity.

Mice don't have to worry about that the way humans do. Consequently, classifying them into subspecies entails less of a responsibility for a scientist than classifying humans into sub-

species does. And by the 1960s, most anthropologists realized they could not defend any classification of *Homo sapiens* into biological subspecies or races that could be considered reasonably objective. They therefore stopped doing it, and stopped identifying the endeavor as a central goal of the field. It was a biologically intractable problem—the old square-peg-in-a-round-hole enterprise; and people's lives, or welfares, could well depend on the ostensibly scientific pronouncement. Reflecting on the social history of the twentieth century, that was a burden anthropologists would no longer bear.

This conceptual divorce in anthropology—of cultural from biological phenomena was one of the most fundamental scientific revolutions of our time. And since it affected assumptions so rooted in our everyday experience, and resulted in conclusions so counterintuitive—like the idea that the earth goes around the sun, and not vice-versa—it has been widely underappreciated.

Kurt Vonnegut, in *Slaughterhouse Five*, describes what he remembered being taught about human variation: "At that time, they were teaching that there was absolutely no difference between anybody. They may be teaching that still." Of course there are biological differences between people, and between populations. The question is: How are those differences patterned? And the answer seems to be: Not racially. Populations are the only readily identifiable units of humans, and even they are fairly fluid, biologically similar to populations nearby, and biologically different from populations far away.

In other words, the message of contemporary anthropology is: You may group humans into a small number of races if you want to, but you are denied biology as a support for it.

Racial Odyssey

Boyce Rensberger

The human species comes in an artist's palette of colors: sandy yellows, reddish tans, deep browns, light tans, creamy whites, pale pinks. It is a rare person who is not curious about the skin colors, hair textures, bodily structures and facial features associated with racial background. Why do some Africans have dark brown skin, while that of most Europeans is pale pink? Why do the eyes of most "white" people and "black" people look pretty much alike but differ so from the eyes of Orientals? Did one race evolve before the others? If so, is it more primitive or more advanced as a result? Can it be possible, as modern research suggests, that there is no such thing as a pure race? These are all honest, scientifically worthy questions. And they are central to current research on the evolution of our species on the planet Earth.

Broadly speaking, research on racial differences has led most scientists to three major conclusions. The first is that there are many more differences among people than skin color, hair texture and facial features. Dozens of other variations have been found, ranging from the shapes of bones to the consistency of ear wax to subtle variations in body chemistry.

The second conclusions is that the overwhelming evolutionary success of the human species is largely due to its great genetic variability. When migrating bands of our early ancestors

reached a new environment, at least a few already had physical traits that gave them an edge in surviving there. If the coming centuries bring significant environmental changes, as many believe they will, our chances of surviving them will be immeasurably enhanced by our diversity as a species.

There is a third conclusion about race that is often misunderstood. Despite our wealth of variation and despite our constant, everyday references to race, no one has ever discovered a reliable way of distinguishing one race from another. While it is possible to classify a great many people on the basis of certain physical features, there are no known feature or groups of features that will do the job in all cases.

Skin color won't work. Yes, most Africans from south of the Sahara and their descendants around the world have skin that is darker than that of most Europeans. But there are millions of people in India, classified by some anthropologists as members of the Caucasoid, or "white," race who have darker skins than most Americans who call themselves black. And there are many Africans living in sub-Sahara Africa today whose skins are no darker than the skins of many Spaniards, Italians, Greeks or Lebanese.

What about stature as a racial trait? Because they are quite short, on the average, African Pygmies have been considered racially distinct from other dark-skinned Africans. If stature, then, is a racial criterion, would one include

in the same race the tall African Watusi and the Scandinavians of similar stature?

The little web of skin that distinguishes Oriental eyes is said to be a particular feature of the Mongoloid race. How, then, can it be argued that the American Indian, who lacks this epicanthic fold, is Mongoloid?

Even more hopeless as racial markers are hair color, eye color, hair form, the shapes of noses and lips or any of the other traits put forth as typical of one race or another.

NO NORMS

Among the tall people of the world there are many black, many white and many in between. Among black people of the world there are many with kinky hair, many with straight or wavy hair, and many in between. Among the broad-nosed, full-lipped people of the world there are many with dark skins, many with light skins and many in between.

How did our modern perceptions of race arise? One of the first to attempt a scientific classification of peoples was Carl von Linné, better known as Linnaeus. In 1735, he published a classification that remains the standard today. As Linnaeus saw it there were four races, classifiable geographically and by skin color. The names Linnaeus gave them were *Homo sapiens Africanus nigrus* (black African human being), *H. sapiens Americanus rube-*

scens (red American human being), *H. sapiens Asiaticus fuscusens* (brownish Asian human being), and *H. sapiens Europaeus albescens* (white European human being). All, Linnaeus recognized, were members of a single human species.

A species includes all individuals that are biologically capable of interbreeding and producing fertile offspring. Most matings between species are fruitless, and even when they succeed, as when a horse and a donkey interbreed and produce a mule, the progeny are sterile. When a poodle mates with a collie, however, the offspring are fertile, showing that both dogs are members of the same species.

Even though Linnaeus's system of nomenclature survives, his classifications were discarded, especially after voyages of discovery revealed that there were many more kinds of people than could be pigeonholed into four categories. All over the world there are small populations that don't fit. Among the better known are:

- The so-called Bushmen of southern Africa, who look as much Mongoloid as Negroid.
- The Negritos of the South Pacific, who do look Negroid but are very far from Africa and have no known links to that continent.
- The Ainu of Japan, a hairy aboriginal people who look more Caucasoid than anything else.
- The Lapps of Scandinavia, who look as much like Eskimos as like Europeans.
- The aborigines of Australia, who often look Negroid but many of whom have straight or wavy hair and are often blond as children.
- The Polynesians, who seem to be a blend of many races, the proportions differing from island to island.

To accommodate such diversity, many different systems of classification have been proposed. Some set up two or three dozen races. None has ever satisfied all experts.

CLASSIFICATION SYSTEM

Perhaps the most sweeping effort to impose a classification upon all the peoples of the world was made by the American anthropologist Carleton Coon. He concluded there are five basic races, two of which have major subdivisions: Caucasoids; Mongloids; full-size Australoids (Australian aborigines); dwarf Australoids (Negritos—Andaman Islanders and similar peoples); full-size Congoids (African Negroids); dwarf Congoids (African Pygmies); and Capoids (the so-called Bushmen and Hottentots).

In his 1965 classic, *The Living Races of Man,* Coon hypothesized that before A.D. 1500 there were five pure races—five centers of human population that were so isolated that there was almost no mixing.

Each of these races evolved independently, Coon believed, diverging from a pre-*Homo sapiens* stock that was essentially the same everywhere. He speculated that the common ancestor evolved into *Homo sapiens* in five separate regions at five different times, beginning about 35,000 years ago. The populations that have been *Homo sapiens* for the shortest periods of time, Coon said, are the world's "less civilized" races.

The five pure races remained distinct until A.D. 1500; then Europeans started sailing the world, leaving their genes—as sailors always have—in every port and planting distant colonies. At about the same time, thousands of Africans were captured and forcibly settled in many parts of the New World.

That meant the end of the five pure races. But Coon and other experts held that this did not necessarily rule out the idea of distinct races. In this view, there *are* such things as races; people just don't fit into them very well anymore.

The truth is that there is really no hard evidence to suggest that five or any particular number of races evolved independently. The preponderance of evidence today suggests that as traits typical of fully modern people arose in any one place, they spread quickly to all human populations. Advances in intelligence were almost certainly the fastest to spread. Most anthropologists and geneticists now believe that human beings have always been subject to migrating and mixing. In other words, there probably never were any such things as pure races.

Race mixing has not only been a fact of human history but is, in this day of unprecedented global mobility, taking place at a more rapid rate than ever. It is not farfetched to envision the day when, generations hence, the entire "complexion" of major population centers will be different. Meanwhile, we can see such changes taking place before our eyes, for they are a part of everyday reality.

HYBRID VIGOR

Oddly, those who assert scientific validity for their notions of pure and distinct races seem oblivious of a basic genetic principle that plant and animal breeders know well: too much inbreeding can lead to proliferation of inferior traits. Crossbreeding with different strains often produces superior combinations and "hybrid vigor."

The striking differences among people may very well be a result of constant genetic mixing. And as geneticists and ecologists know, in diversity lies strength and resilience.

To understand the origin and proliferation of human differences, one must first know how Darwinian evolution works.

Evolution is a two-step process. Step one is mutation: somehow a gene in the ovary or testes of an individual is altered, changing the molecular configuration that stores instructions for forming a new individual. The children who inherit that gene will be different in some way from their ancestors.

Step two is selection: for a racial difference, or any other evolutionary change to arise, it must survive and be passed through several generations. If the mutation confers some disadvantage, the individual dies, often during embryonic development. But if the change is beneficial in some way, the individual should have a better chance of thriving than relatives lacking the advantage.

DISEASE ORIGINS

The gene for sickle cell anemia, a disease found primarily among black people, appears to have evolved because its presence can render its bearer resistant to malaria. Such a trait would have obvious value in tropical Africa.

A person who has sickle cell anemia must have inherited genes for the disease from both parents. If a child inherits only one sickle cell gene, he or she will be resistant to malaria but will not have the anemia. Paradoxically, inheriting genes from both parents does not seem to affect resistance to malaria.

In the United States, where malaria is practically nonexistent, the sickle cell gene confers no survival advantage and is disappearing. Today only about 1 out of every 10 American blacks carries the gene.

Many other inherited diseases are found only in people from a particular area. Tay-Sachs disease, which often kills before the age of two, is almost entirely confined to Jews from parts of Eastern Europe and their descendants elsewhere. Paget's disease, a bone disorder, is found most often among those of English descent. Impacted wisdom teeth are a common problem among Asians and Europeans but not among Africans. Children of all races are able to digest milk because their bodies make lactase, the enzyme that breaks down lactose, or milk sugar. But the ability to digest lactose in adulthood is a racially distributed trait.

About 90 percent of Orientals and blacks lose this ability by the time they reach adulthood and become quite sick when they drink milk.

Even African and Asian herders who keep cattle or goats rarely drink fresh milk. Instead, they first treat the milk with fermentation bacteria that break down lactose, in a sense predigesting it. They can then ingest the milk in the form of yogurt or cheese without any problem.

About 90 percent of Europeans and their American descendants, on the other hand, continue to produce the enzyme throughout their lives and can drink milk with no ill effects.

NATURAL SELECTION

If a new trait is beneficial, it will bring reproductive success to its bearer. After several generations of multiplication, bearers of the new trait may begin to outnumber nonbearers. Darwin called this natural selection to distinguish it from the artificial selection exercised by animal breeders.

Skin color is the human racial trait most generally thought to confer an evolutionary advantage of this sort. It has long been obvious in the Old World that the farther south one goes, the darker the skin color. Southern Europeans are usually somewhat darker than northern Europeans. In North Africa, skin colors are darker still, and, as one travels south, coloration reaches its maximum at the Equator. The same progressions holds in Asia, with the lightest skins to the north. Again, as one moves south, skin color darkens, reaching in southern India a "blackness" equal to that of equatorial Africans.

This north-south spectrum of skin color derives from varying intensities of the same dark brown pigment called melanin. Skin cells simply have more or less melanin granules to be seen against a background that is pinkish because of the underlying blood vessels. All races can increase their melanin concentration by exposure to the sun.

What is it about northerly latitudes in the Northern Hemisphere that favors less pigmentation and about southerly latitudes that favors more? Exposure to intense sunlight is not the only reason why people living in southerly latitudes are dark. A person's susceptibility to rickets and skin cancer, his ability to withstand cold and to see in the dark may also be related to skin color.

The best-known explanation says the body can tolerate only a narrow range of intensities of sunlight. Too much causes sunburn and cancer, while too little deprives the body of vitamin D, which is synthesized in the skin under the influence of sunlight. A dark complexion protects the skin from the harmful effects of intense sunlight. Thus, albinos born in equatorial regions have a high rate of skin cancer. On the other hand, dark skin in northerly latitudes screens out sunlight needed for the synthesis of vitamin D. Thus, dark-skinned children living in northern latitudes had high rates of rickets—a bone-deforming disease caused by a lack of vitamin D—before their milk was routinely fortified. In the sunny tropics, dark skin admits enough light to produce the vitamin.

Recently, there has been some evidence that skin colors are linked to differences in the ability to avoid injury from the cold. Army researchers found that during the Korean War blacks were more susceptible to frostbite than were whites. Even among Norwegian soldiers in World War II, brunettes had a slightly higher incidence of frostbite than did blonds.

EYE PIGMENTATION

A third link between color and latitude involves the sensitivity of the eye to various wavelengths of light. It is known that dark-skinned people have more pigmentation in the iris of the eye and at the back of the eye where the image falls. It has been found that the less pigmented the eye, the more sensitive it is to colors at the red end of the spectrum. In situations illuminated with reddish light, the northern European can see more than a dark African sees.

It has been suggested that Europeans developed lighter eyes to adapt to the longer twilights of the North and their greater reliance on firelight to illuminate caves.

Although the skin cancer-vitamin D hypothesis enjoys wide acceptance, it may well be that resistance to cold, possession of good night vision and other yet unknown factors all played roles in the evolution of skin colors.

Most anthropologists agree that the original human skin color was dark brown, since it is fairly well established that human beings evolved in the tropics of Africa. This does not, however, mean that the first people were Negroids, whose descendants, as they moved north, evolved into light-skinned Caucasoids. It is more likely that the skin color of various populations changed several times from dark to light and back as people moved from one region to another.

Consider, for example, that long before modern people evolved, *Homo erectus* had spread throughout Africa, Europe and Asia. The immediate ancestor of *Homo sapiens, Homo erectus,* was living in Africa 1.5 million years ago and in Eurasia 750,000 years ago. The earliest known forms of *Homo sapiens* do not make their appearance until somewhere between 250,000 and 500,000 years ago. Although there is no evidence of the skin color of any hominid fossil, it is probable that the *Homo erectus* population in Africa had dark skin. As subgroups spread into northern latitudes, mutations that reduced pigmentation conferred survival advantages on them and lighter skins came to predominate. In other words, there were probably black *Homo erectus* peoples in Africa and white ones in Europe and Asia.

Did the black *Homo erectus* populations evolve into today's Negroids and the white ones in Europe into today's Caucasoids? By all the best evidence, nothing like this happened. More likely, wherever *Homo sapiens* arose it proved so superior to the *Homo erectus* populations that it eventually replaced them everywhere.

If the first *Homo sapiens* evolved in Africa, they were probably dark-skinned; those who migrated northward into Eurasia lost their pigmentation. But it is just as possible that the first *Homo sapiens* appeared in northern climes, descendants of white-skinned *Homo erectus.* These could have migrated southward toward Africa, evolving darker skins. All modern races, incidentally, arose long after the brain had reached its present size in all parts of the world.

North-south variations in pigmentation are quite common among mammals and birds. The tropical races tend to be darker in fur and feather, the desert races tend to be brown, and those near the Arctic Circle are lighter colored.

There are exceptions among humans. The Indians of the Americas, from the Arctic to the southern regions of South America, do not conform to the north-south scheme of coloration. Though most think of Indians as being reddish-brown, most Indians tend to be relatively light skinned, much like their presumed Mongoloid ancestors in Asia. The ruddy complexion that lives in so many stereotypes of Indians is merely what years of heavy tanning can produce in almost any light-skinned person. Anthropologists explain the color consistency as a consequence of the relatively recent entry of people into the Americas—probably between 12,000 and 35,000 years ago. Perhaps they have not yet had time to change.

Only a few external physical differences other than color appear to have adaptive significance. The strongest cases can be made for nose shape and stature.

WHAT'S IN A NOSE

People native to colder or drier climates tend to have longer, more beak-shaped noses than those living in hot and humid regions. The nose's job is to warm and humidify air before it reaches sensitive lung tissues. The colder or drier the air is, the more surface area is needed inside the nose to get it to the right temperature or humidity. Whites tend to have longer and beakier noses than blacks or Orientals. Nevertheless, there is great variation within races. Africans in the highlands of East Africa have longer noses than Africans from the hot, humid lowlands, for example.

Stature differences are reflected in the tendency for most northern peoples to have shorter arms, legs and torsos and to be stockier than people from the tropics. Again, this is an adaptation to heat or cold. One way of reducing heat loss is to have less body surface, in relation to weight or volume, from which heat can escape. To avoid overheating, the most desirable body is long limbed and lean. As a result, most Africans tend to be lankier than northern Europeans. Arctic peoples are the shortest limbed of all.

Hair forms may also have a practical role to play, but the evidence is weak. It has been suggested that the more tightly curled hair of Africans insulates the top of the head better than does straight or wavy hair. Contrary to expectation, black hair serves better in this role than white hair. Sunlight is absorbed and converted to heat at the outer surface of the hair blanket; it radiates directly into the air. White fur, common on Arctic animals that need to absorb solar heat, is actually transparent and transmits light into the hair blanket, allowing the heat to form within the insulating layer, where it is retained for warmth.

Aside from these examples, there is little evidence that any of the other visible differences among the world's people provide any advantage. Nobody knows, for example, why Orientals have epicanthic eye folds or flatter facial profiles. The thin lips of Caucasoids and most Mongoloids have no known advantages over the Negroid's full lips. Why should middle-aged and older Caucasoid men go bald so much more frequently than the men of other races? Why does the skin of Bushmen wrinkle so heavily in the middle and later years? Or why does the skin of Negroids resist wrinkling so well? Why do the Indian men in one part of South America have blue penises? Why do Hottentot women have such unusually large buttocks?

1. NATURAL SELECTION

There are possible evolutionary explanations for why such apparently useless differences arise.

One is a phenomenon known as sexual selection. Environmentally adaptive traits arise, Darwin thought, through natural selection—the environment itself chooses who will thrive or decline. In sexual selection, which Darwin also suggested, the choice belongs to the prospective mate.

In simple terms, ugly individuals will be less likely to find mates and reproduce their genes than beautiful specimens will. Take the blue penis as an example. Women might find it unusually attractive or perhaps believe it to be endowed with special powers. If so, a man born with a blue penis will find many more opportunities to reproduce his genes than his ordinary brothers.

Sexual selection can also operate when males compete for females. The moose with the larger antlers or the lion with the more imposing mane will stand a better chance of discouraging less well-endowed males and gaining access to females. It is possible that such a process operated among Caucasoid males, causing them to become markedly hairy, especially around the face.

ATTRACTIVE TRAITS

Anthropologists consider it probable that traits such as the epicanthic fold or the many regional differences in facial features were selected this way.

Yet another method by which a trait can establish itself involves accidental selection. It results from what biologists call genetic drift.

Suppose that in a small nomadic band a person is born with perfectly parallel fingerprints instead of the usual loops, whorls or arches. That person's children would inherit parallel fingerprints, but they would confer no survival advantages. But if our family decides to strike out on its own, it will become the founder of a new band consisting of its own descendants, all with parallel fingerprints.

Events such as this, geneticists and anthropologists believe, must have oc-

curred many times in the past to produce the great variety within the human species. Among the apparently neutral traits that differ among populations are:

Ear Wax
There are two types of ear wax. One is dry and crumbly and the other is wet and sticky. Both types can be found in every major population, but the frequencies differ. Among northern Chinese, for example, 98 percent have dry ear wax. Among American whites, only 16 percent have dry ear wax. Among American blacks the figure is 7 percent.

Scent Glands
As any bloodhound knows, every person has his or her own distinctive scent. People vary in the mixture of odoriferous compounds exuded through the skin—most of it coming from specialized glands called apocrine glands. Among whites, these are concentrated in the armpits and near the genitals and anus. Among blacks, they may also be found on the chest and abdomen. Orientals have hardly any apocrine glands at all. In the words of the Oxford biologist John R. Baker, "The Europids and Negrids are smelly, the Mongoloids scarcely or not at all." Smelliest of all are northern European, or so-called Nordic, whites. Body odor is rare in Japan. It was once thought to indicate a European in the ancestry and to be a disease requiring hospitalization.

Blood Groups
Some populations have a high percentage of members with a particular blood group. American Indians are overwhelmingly group O—100 percent in some regions. Group A is most common among Australian aborigines and the Indians in western Canada. Group B is frequent in northern India, other parts of Asia and western Africa.

Advocates of the pure-race theory once seized upon blood groups as possibly unique to the original pure races. The proportions of groups found today,

they thought, would indicate the degree of mixing. It was subsequently found that chimpanzees, our closest living relatives, have the same blood groups as humans.

Taste
PTC (phenylthiocarbamide) is a synthetic compound that some people can taste and other cannot. The ability to taste it has no known survival value, but it is clearly an inherited trait. The proportion of persons who can taste PTC varies in different populations: 50 to 70 percent of Australian aborigines can taste it, as can 60 to 80 percent of all Europeans. Among East Asians, the percentage is 83 to 100 percent, and among Africans, 90 to 97 percent.

Urine
Another indicator of differences in body chemistry is the excretion of a compound known as BAIB (beta-amino-isobutyric acid) in urine. Europeans seldom excrete large quantities, but high levels of excretion are common among Asians and American Indians. It had been shown that the differences are not due to diet.

No major population has remained isolated long enough to prevent any unique genes from eventually mixing with those of neighboring groups. Indeed, a map showing the distribution of so-called traits would have no sharp boundaries, except for coastlines. The intensity of a trait such as skin color, which is controlled by six pairs of genes and can therefore exist in many shades, varies gradually from one population to another. With only a few exceptions, every known genetic possibility possessed by the species can be found to some degree in every sizable population.

EVER-CHANGING SPECIES

One can establish a system of racial classification simply by listing the features of populations at any given moment. Such a concept of race is, however, inappropriate to a highly mobile and ever-changing species such as *Homo sapiens*. In the short view, races may seem distinguishable, but in biology's long haul, races come and go. New ones arise and blend into neighboring groups to create new and racially stable populations. In time, genes from these groups flow into other neighbors, continuing the production of new permutations.

Some anthropologists contend that at the moment American blacks should be considered a race distinct from African blacks. They argue that American blacks are a hybrid of African blacks and European whites. Indeed, the degree of mixture can be calculated on the basis of a blood component known as the Duffy factor.

In West Africa, where most of the New World's slaves came from, the Duffy factor is virtually absent. It is present in 43 percent of American whites. From the number of American blacks who are now "Duffy positive" it can be calculated that whites contributed 21 percent of the genes in the American black population. The figure is higher for blacks in northern and western states and lower in the South. By the same token, there are whites who have black ancestors. The number is smaller because of the tendency to identify a person as black even if only a minor fraction of his ancestors were originally from Africa.

The unwieldiness of race designations is also evident in places such as Mexico where most of the people are, in effect, hybrids of Indians (Mongoloid by some classifications) and Spaniards (Caucasoid). Many South American populations are tri-hybrids—mixtures of Mongoloid, Caucasoid and Negroid. Brazil is a country where the mixture has been around long enough to constitute a racially stable population. Thus, in one sense, new races have been created in the United States, Mexico and Brazil. But in the long run, those races will again change.

Sherwood Washburn, a noted anthropologist, questions the usefulness of racial classification: "Since races are open systems which are intergrading, the number of races will depend on the purpose of the classification. I think we should require people who propose a classification of races to state in the first place why they wish to divide the human species."

The very notion of a pure race, then, makes no sense. But, as evolutionists know full well, a rich genetic diversity within the human species most assuredly *does*.

PRIMATES

Primates are fun. They are active, intelligent, colorful, emotionally expressive, and unpredictable. In other words, observing them is like holding up an opaque mirror to ourselves. The image may not be crystal clear or, indeed, what some would consider flattering, but it is certainly familiar enough to be illuminating.

Primates are, of course, but one of many orders of mammals that adaptively radiated into the variety of ecological niches vacated at the end of the Age of Reptiles about 65 million years ago. Whereas some mammals took to the sea (cetaceans), and some took to the air (chiroptera, or bats), primates are characterized by an arboreal or forested adaptation. Whereas some mammals can be identified by their food-getting habits, such as the meat-eating carnivores, primates have a penchant for eating almost anything and are best described as omnivorous. In taking to the trees, primates did not simply develop a full-blown set of distinguishing characteristics that set them off easily from other orders of mammals, the way the rodent order can be readily identified by its gnawing set of front teeth. Rather, each primate seems to represent degrees of anatomical, biological, and behavioral characteristics on a continuum of progress with respect to the particular traits we humans happen to be interested in.

None of this is meant to imply, of course, that the living primates are our ancestors. Since the prosimians, monkeys, and apes are our contemporaries, they are no more our ancestors than we are theirs, and, as living end-products of evolution, we have all descended from a common stock in the distant past.

So, if we are interested primarily in our own evolutionary past, why study primates at all? Because, by the criteria we have set up as significant milestones in the evolution of humanity, an inherent reflection of our own bias, primates have not evolved as far as we have. They and their environments, therefore, may represent glimmerings of the evolutionary stages and ecological circumstances through which our own ancestors may have gone. What we stand to gain, for instance, is an educated guess as to how our own ancestors might have appeared and behaved as semierect creatures before becoming bipedal. It is in the spirit of this type of inquiry that Peter Radetsky, in his essay "Gut Thinking," investigates the relationship between diet, social organization, and intelligence. Aside from being a pleasure to observe, then, living primates can teach us something about our past.

Another reason for studying primates is that they allow us to test certain notions too often taken for granted. For instance, Barbara Smuts, in "What Are Friends For?" reveals that friendship bonds, as illustrated by the olive baboons of East Africa, have little if anything to do with a sexual division of labor or even sexual exclusivity between a pair-bonded male and female. Smuts challenges the traditional male-oriented idea that primate societies are dominated solely by males and for males.

This unit demonstrates that relationships between the sexes are subject to wide variation, that the kinds of answers obtained depend upon the kinds of questions asked, and that we have to be very careful in making inferences about human beings from any one particular primate study. We may, if we are not careful, draw conclusions that say more about our own skewed perspectives than about that which we claim to understand. Still another benefit of primate field research is that it provides us with perspectives that the bones and stones of the fossil hunters will never reveal: a sense of the richness and variety of social patterns that must have existed in the primate order for many tens of millions of years. (See James Shreeve's report "Machiavellian Monkeys," Jane Goodall's "The Mind of the Chimpanzee," and Sy Montgomery's essay "Dian Fossey and Digit.")

There is a sense of urgency in the study of primates as we contemplate the dreaded possibility of their imminent extinction. We have already lost 14 species of lemurs in the past thousand years (most of which were larger than the contemporary forms), and the fate of most other free-ranging primates is in the balance. It is ironic that future generations may envy us as having been among the first and last people to be able to observe our closest living relatives in their natural habitats. If for no other reason, we need to collect as much information as we can about primates now, while they are still with us.

Looking Ahead: Challenge Questions

What is the role of deception among primates, and how might it have led to greater intelligence?

Why is friendship important to olive baboons? What implications does this have for the origins of pair-bonding in hominid evolution?

How is it possible to objectively study and assess emotional and mental states of nonhuman primates?

Why is the mountain gorilla in danger of extinction?

What makes fruit-eating spider monkeys so much smarter than leaf-eating howlers? What are the implications of this for human evolution?

Machiavellian Monkeys

The sneaky skills of our primate cousins suggest that we may owe our great intelligence to an inherited need to deceive.

James Shreeve

This is a story about frauds, cheats, liars, faithless lovers, incorrigible con artists, and downright thieves. You're gonna love 'em.

Let's start with a young rascal named Paul. You'll remember his type from your days back in the playground. You're minding your own business, playing on the new swing set, when along comes Paul, such a little runt that you hardly notice him sidle up to you. All of a sudden he lets out a scream like you've run him through with a white-hot barbed harpoon or something. Of course the teacher comes running, and the next thing you know you're being whisked inside with an angry finger shaking in our face. That's the end of recess for you. But look out the window: there's Paul, having a great time on *your* swing. Cute kid.

Okay, you're a little older now and a little smarter. You've got a bag of chips stashed away in your closet, where for once your older brother won't be able to find them. You're about to open the closet door when he pokes his head in the room. Quickly you pretend to be fetching your high tops; he gives you a look but he leaves. You wait a couple of minutes, lacing up the sneakers in case he walks back in, then you dive for the chips. Before you can get the bag open, he's over your shoulder, snatching it out of your hands. "Nice try, punk," he says through a mouthful,

"but I was hiding outside your room the whole time."

This sort of trickery is such a common part of human interaction that we hardly notice how much time we spend defending ourselves against it or perpetrating it ourselves. What's so special about the fakes and cheaters here, however, is that they're not human. Paul is a young baboon, and your big brother is, well, a chimpanzee. With some admittedly deceptive alterations of scenery and props, the situations have been lifted from a recent issue of *Primate Report*. The journal is the work of Richard Byrne and Andrew Whiten, two psychologists at the University of St. Andrews in Scotland, and it is devoted to cataloging the petty betrayals of monkeys and apes as witnessed by primatologists around the world. It is a testament to the evolutionary importance of what Byrne and Whiten call Machiavellian intelligence— a facility named for the famed sixteenth-century author of *The Prince*, the ultimate how-to guide to prevailing in a complex society through the judicious application of cleverness, deceit, and political acumen.

Deception is rife in the natural world. Stick bugs mimic sticks. Harmless snakes resemble deadly poisonous ones. When threatened, blowfish puff themselves up and cats arch their backs and bristle their hair to seem bigger than they really are. All these animals could be said to practice deception because they fool other animals—usu-

ally members of other species—into thinking they are something that they patently are not. Even so, it would be overreading the situation to attribute Machiavellian cunning to a blowfish, or to accuse a stick bug of being a lying scoundrel. Their deceptions, whether in their looks or in their actions, are programmed genetic responses. Biology leaves them no choice but to dissemble: they are just being true to themselves.

The kind of deception that interests Byrne and Whiten—what they call tactical deception—is a different kettle of blowfish altogether. Here an animal has the mental flexibility to take an "honest" behavior and use it in such a way that another animal—usually a member of the deceiver's own social group—is misled, thinking that a normal, familiar state of affairs is under way, while, in fact, something quite different is happening.

Take Paul, for example. The real Paul is a young chacma baboon that caught Whiten's attention in 1983, while he and Byrne were studying foraging among the chacma in the Drakensberg Mountains of southern Africa. Whiten saw a member of Paul's group, an adult female named Mel, digging in the ground, trying to extract a nutritious plant bulb. Paul approached and looked around. There were no other baboons within sight. Suddenly he let out a yell, and within seconds his mother came running, chasing the star-

Reprinted with permission from *Discover* magazine, June 1991, pp. 69-73. © 1991 by Discover Magazine.

tled Mel over a small cliff. Paul then took the bulb for himself.

In this case the deceived party was Paul's mother, who was misled by his scream into believing that Paul was being attacked, when actually no such attack was taking place. As a result of her apparent misinterpretation Paul was left alone to eat the bulb that Mel had carefully extracted—a morsel, by the way, that he would not have had the strength to dig out on his own.

If Paul's ruse had been an isolated case, Whiten might have gone on with his foraging studies and never given it a second thought. But when he compared his field notes with Byrne's, he noticed that both their notebooks were sprinkled with similar incidents and had been so all summer long. After they returned home to Scotland, they boasted about their "dead smart" baboons to their colleagues in pubs after conferences, expecting them to be suitably impressed. Instead the other researchers countered with tales about their own shrewd vervets or Machiavellian macaques.

"That's when we realized that a whole phenomenon might be slipping through a sieve," says Whiten. Researchers had assumed that this sort of complex trickery was a product of the sophisticated human brain. After all, deceitful behavior seemed unique to humans, and the human brain is unusually large, even for primates—"three times as big as you would expect for a primate of our size," notes Whiten, if you're plotting brain size against body weight.

But if primates other than humans deceived one another on a regular basis, the two psychologists reasoned, then it raised the extremely provocative possibility that the primate brain, and ultimately the human brain, is an instrument crafted for social manipulation. Humans evolved from the same evolutionary stock as apes, and if tactical deception was an important part of the lives of our evolutionary ancestors, then the sneakiness and subterfuge that human beings are so manifestly capable of might not be simply a result of our great intelligence and oversize

brain, but a driving force behind their development.

To Byrne and Whiten these were ideas worth pursuing. They fit in with a theory put forth some years earlier

---

Suddenly Paul let out a yell, and his mother came running, chasing Mel over a small cliff.

---

by English psychologist Nicholas Humphrey. In 1976 Humphrey had eloquently suggested that the evolution of primate intelligence might have been spurred not by the challenges of environment, as was generally thought, but rather by the complex cognitive demands of living with one's own companions. Since then a number of primatologists had begun to flesh out his theory with field observations of politically astute monkeys and apes.

Deception, however, had rarely been reported. And no wonder: If chimps, baboons, and higher primates generally are skilled deceivers, how could one ever know it? The best deceptions would by their very nature go undetected by the other members of the primate group, not to mention by a human stranger. Even those ruses that an observer could see through would have to be rare, for if used too often, they would lose their effectiveness. If Paul always cried wolf, for example, his mother would soon learn to ignore his ersatz distress. So while the monkey stories swapped over beers certainly suggested that deception was widespread among higher primates, it seemed unlikely that one or even a few researchers could observe enough instances of it to scientifically quantify how much, by whom, when, and to what effect.

Byrne and Whiten's solution was to extend their pub-derived data base with a more formal survey. In 1985 they sent a questionnaire to more than 100 primatologists working both in the field and in labs, asking them to report back any incidents in which they felt their subjects had perpetrated decep-

tion on one another. The questionnaire netted a promising assortment of deceptive tactics used by a variety of monkeys and all the great apes. Only the relatively small-brained and socially simple lemur family, which includes bush babies and lorises, failed to elicit a single instance. This supported the notion that society, sneakiness, brain size, and intelligence are intimately bound up with one another. The sneakier the primate, it seemed, the bigger the brain.

Byrne and Whiten drew up a second, much more comprehensive questionnaire in 1989 and sent it to hundreds more primatologists and animal behaviorists, greatly increasing the data base. Once again, when the results were tallied, only the lemur family failed to register a single case of deception.

All the other species, however, represented a simian rogues' gallery of liars and frauds. Often deception was used to distract another animal's attention. In one cartoonish example, a young baboon, chased by some angry elders, suddenly stopped, stood on his hind legs, and stared at a spot on the horizon, as if he noticed the presence of a predator or a foreign troop of baboons. His pursuers braked to a halt and looked in the same direction, giving up the chase. Powerful field binoculars revealed that no predator or baboon troop was anywhere in sight.

Sometimes the deception was simply a matter of one animal hiding a choice bit of food from the awareness of those strong enough to take it away. One of Jane Goodall's chimps, for example, named Figan, was once given some bananas after the more dominant members of the troop had wandered off. In the excitement, he uttered some loud "food barks"; the others quickly returned and took the bananas away. The next day Figan again waited behind the others and got some bananas. This time, however, he kept silent, even though the human observers, Goodall reported, "could hear faint choking sounds in his throat."

Concealment was a common ruse in sexual situations as well. Male monkeys and chimpanzees in groups have fairly strict hierarchies that control their access to females. Animals at the top of the order intimidate those lower down, forcing them away from females. Yet one researcher reported seeing a male stump-tailed macaque of a middle rank leading a female out of sight of the more dominant males and then mating with her silently, his climax unaccompanied by the harsh, low-pitched grunts that the male stump-tailed normally makes. At one point during the tryst the female turned and stared into his face, then covered his mouth with her hand. In another case a subordinate chimpanzee, aroused by the presence of a female in estrus, covered his erect penis with his hand when a dominant male approached, thus avoiding a likely attack.

In one particularly provocative instance a female hamadryas baboon slowly shuffled toward a large rock, appearing to forage, all the time keeping an eye on the most dominant male in the group. After 20 minutes she ended up with her head and shoulders visible to the big, watchful male, but with her hands happily engaged in the elicit activity of grooming a favorite subordinate male, who was hidden from view behind the rock.

Baboons proved singularly adept at a form of deception that Byrne and Whiten call "using a social tool." Paul's scam is a perfect example: he fools his mother into acting as a lever to pry the plant bulb away from the adult female, Mel. But can it be said unequivocally that he intended to deceive her? Perhaps Paul had simply learned through trial and error that letting out a yell brought his mother running and left him with food, in which case there is no reason to endow his young baboon intellect with Machiavellian intent. How do we know that Mel didn't actually threaten Paul in some way that Byrne and Whiten, watching, could not comprehend? While we're at it, how do we know that any of the primate deceptions reported here were really deliberate, conscious acts?

"It has to be said that there is a whole school of psychology that would

deny such behavior even to humans," says Byrne. The school in question—strict behaviorism—would seek an explanation for the baboons' behavior not by trying to crawl inside their head but by carefully analyzing observable behaviors and the stimuli that might be triggering them. Byrne and Whiten's strategy against such skepticism was to be hyperskeptical themselves. They accepted that trial-and-error learning or simple conditioning, in which an animal's actions are reinforced by a reward, might account for a majority of the incidents reported to them—even when they believed that tactical deception was really taking place. But when explaining things "simply" led to a maze of extraordinary coincidences and tortuous logic, the evidence for deliberate deception seemed hard to dismiss.

Society, sneakiness, brain size, and intelligence are intimately bound up with one another.

Paul, for instance, *might* have simply learned that screaming elicits the reward of food, via his mother's intervention. But Byrne witnessed him using the same tactic several times, and in each case his mother was out of sight, able to hear his yell but not able to see what was really going on. If Paul was simply conditioned to scream, why would he do so only when his mother could not see who was—or was not—attacking her son?

Still, it is possible that she was not intentionally deceived. But in at least one other, similar case there is virtually no doubt that the mother was responding to a bogus attack, because the alleged attacker was quite able to verbalize his innocence. A five-year-old male chimp named Katabi, in the process of weaning, had discovered that the best way to get his reluctant mother to suckle him was to convince her he needed reassurance. One day Katabi approached a human observer—Japanese primatologist Toshisada Nish-

ida—and began to screech, circling around the researcher and waving an accusing hand at him. The chimp's mother and her escort immediately glared at Nishida, their hair erect. Only by slowly backing away from the screaming youngster did Nishida avoid a possible attack from the two adult chimps.

"In fact I did nothing to him," Nishida protested. It follows that the adults were indeed misled by Katabi's hysterics—unless there was some threat in Nishida unknown even to himself.

"If you try hard enough," says Byrne, "you can explain every single case without endowing the animal with the ability to deceive. But if you look at the whole body of work, there comes a point where you have to strive officiously to deny it."

The cases most resistant to such officious denials are the rarest—and the most compelling. In these interactions the primate involved not only employed tactical deception but clearly understood the concept. Such comprehension would depend upon one animal's ability to "read the mind" of another: to attribute desires, intentions, or even beliefs to the other creature that do not necessarily correspond to its own view of the world. Such mind reading was clearly evident in only 16 out of 253 cases in the 1989 survey, all of them involving great apes.

For example, consider Figan again, the young chimp who suppressed his food barks in order to keep the bananas for himself. In his case, mind reading is not evident: he might simply have learned from experience that food barks in certain contexts result in a loss of food, and thus he might not understand the nature of his own ruse, even if the other chimps are in fact deceived.

But contrast Figan with some chimps observed by Dutch primatologist Frans Plooij. One of these chimps was alone in a feeding area when a metal box containing food was opened electronically. At the same moment another chimp happened to approach. (Sound familiar? It's your older brother again.) The first chimp quickly closed the metal box (that's you hiding your chips), walked away, and sat down, looking around as if nothing had happened. The second chimp departed, but after going some distance away he hid behind a tree and peeked back at the first chimp. When the first chimp thought the coast was clear, he opened the box. The second chimp ran out, pushed the other aside, and ate the bananas.

Chimp One might be a clever rogue, but Chimp Two, who counters his deception with a ruse of his own, is the true mind reader. The success of his ploy is based on his insight that Chimp One was trying to deceive *him* and on his ability to adjust his behavior accordingly. He has in fact performed a prodigious cognitive leap—proving himself capable of projecting himself into another's mental space, and becoming what Humphrey would call a natural psychologist.

Niccolò Machiavelli might have called him good raw material. It is certainly suggestive that only the great apes—our closest relatives—seem capable of deceits based on such mind reading, and chimpanzees most of all. This does not necessarily mean that chimps are inherently more intelligent: the difference may be a matter of social organization. Orangutans live most of their lives alone, and thus they would not have much reason to develop such a complex social skill. And gorillas live in close family groups, whose members would be more familiar,

harder to fool, and more likely to punish an attempted swindle. Chimpanzees, on the other hand, spend their lives in a shifting swirl of friends and relations, where small groups constantly form and break apart and reform with new members.

"What an opportunity for lying and cheating!" muses Byrne. Many anthropologists now believe that the social life of early hominids—our first non-ape ancestors—was much like that of chimps today, with similar opportunities to hone their cognitive skills on one another. Byrne and Whiten stop just short of saying that mind reading is the key to understanding the growth of human intelligence. But it would be disingenuous to ignore the possibility. If you were an early hominid who could comprehend the subjective impressions of others and manipulate them to your own ends, you might well have a competitive advantage over those less psychosocially nimble, perhaps enjoying slightly easier access to food and to the mating opportunities that would ensure your genetic survival.

Consider too how much more important your social wits would be in a world where the targets of your deceptions were constantly trying to outsmart *you*. After millennia of intrigue and counterintrigue, a hominid species might well evolve a brain three times bigger than it "should" be—and capable of far more than deceiving other hominids. "The ability to attribute

other intentions to other people could have been an enormous building block for many human achievements, including language," says Whiten. "That this leap seems to have been taken by chimps and possibly the other great apes puts that development in human mentality quite early."

So did our intellect rise to its present height on a tide of manipulation and deceit? Some psychologists, even those who support the notion that the evolution of intelligence was socially driven, think that Byrne and Whiten's choice of the loaded adjective *Machiavellian* might be unnecessarily harsh.

"In my opinion," says Humphrey, "the word gives too much weight to the hostile use of intelligence. One of the functions of intellect in higher primates and humans is to keep the social unit together and make it able to successfully exploit the environment. A lot of intelligence could better be seen as driven by the need for cooperation and compassion." To that, Byrne and Whiten only point out that cooperation is itself an excellent Machiavellian strategy—sometimes.

The Scottish researchers are not, of course, the first to have noticed this. "It is good to appear clement, trustworthy, humane, religious, and honest, and also to be so," Machiavelli advised his aspiring Borgia prince in 1513. "But always with the mind so disposed that, when the occasion arises not to be so, you can become the opposite."

What Are Friends For?

Among East African baboons, friendship means companions, health, safety . . . and, sometimes, sex

Barbara Smuts

Virgil, a burly adult male olive baboon, closely followed Zizi, a middle-aged female easily distinguished by her grizzled coat and square muzzle. On her rump Zizi sported a bright pink swelling, indicating that she was sexually receptive and probably fertile. Virgil's extreme attentiveness to Zizi suggested to me—and all rival males in the troop—that he was her current and exclusive mate.

Zizi, however, apparently had something else in mind. She broke away from Virgil, moved rapidly through the troop, and presented her alluring sexual swelling to one male after another. Before Virgil caught up with her, she had managed to announce her receptive condition to several of his rivals. When Virgil tried to grab her, Zizi screamed and dashed into the bushes with Virgil in hot pursuit. I heard sounds of chasing and fighting coming from the thicket. Moments later Zizi emerged from the bushes with an older male named Cyclops. They remained together for several days, copulating

often. In Cyclops's presence, Zizi no longer approached or even glanced at other males.

Primatologists describe Zizi and other olive baboons (*Papio cynocephalus anubis*) as promiscuous, meaning that both males and females usually mate with several members of the opposite sex within a short period of time. Promiscuous mating behavior characterizes many of the larger, more familiar primates, including chimpanzees, rhesus macaques, and gray langurs, as well as olive, yellow, and chacma baboons, the three subspecies of savanna baboon. In colloquial usage, promiscuity often connotes wanton and random sex, and several early studies of primates supported this stereotype. However, after years of laboriously recording thousands of copulations under natural conditions, the Peeping Toms of primate fieldwork have shown that, even in promiscuous species, sexual pairings are far from random.

Some adult males, for example, typically copulate much more often than

others. Primatologists have explained these differences in terms of competition: the most dominant males monopolize females and prevent lower-ranking rivals from mating. But exceptions are frequent. Among baboons, the exceptions often involve scruffy, older males who mate in full view of younger, more dominant rivals.

A clue to the reason for these puzzling exceptions emerged when primatologists began to question an implicit assumption of the dominance hypothesis—that females were merely passive objects of male competition. But what if females were active arbiters in this system? If females preferred some males over others and were able to express these preferences, then models of mating activity based on male dominance alone would be far too simple.

Once researchers recognized the possibility of female choice, evidence for it turned up in species after species. The story of Zizi, Virgil, and Cyclops is one of hundreds of examples of female primates rejecting the sexual advances of particular males and enthusiastically cooperating with others. But what is the basis for female choice? Why might they prefer some males over others?

This question guided my research on the Eburru Cliffs troop of olive baboons, named after one of their favorite sleeping sites, a sheer rocky outcrop rising several hundred feet above the floor of the Great Rift Valley, about 100 miles northwest of Nairobi, Kenya. The 120 members of Eburru Cliffs spent their days wandering through open grassland studded with occasional acacia thorn trees. Each night they retired to one of a dozen sets of cliffs that provided protection from nocturnal predators such as leopards.

Most previous studies of baboon sexuality had focused on females who, like Zizi, were at the peak of sexual receptivity. A female baboon does not mate when she is pregnant or lactating, a period of abstinence lasting about eighteen months. The female then goes into estrus, and for about two weeks out of every thirty-five-day cycle, she mates. Toward the end of this two week period she may ovulate, but usually the female undergoes four or five estrous cycles before she conceives. During pregnancy, she once again resumes a chaste existence. As a result, the typical female baboon is sexually active for less than 10 percent of her adult life. I thought that by focusing on the other 90 percent, I might learn something new. In particular, I suspected that routine, day-to-day relationships between males and pregnant or lactating (nonestrous) females might provide clues to female mating preferences.

Nearly every day for sixteen months, I joined the Eburru Cliffs baboons at their sleeping cliffs at dawn and traveled several miles with them while they foraged for roots, seeds, grass, and occasionally, small prey items, such as baby gazelles or hares (see "Predatory Baboons of Kekopey," *Natural History*, March 1976). Like all savanna baboon troops, Eburru Cliffs functioned as a cohesive unit organized around a core of related females, all of whom were born in the troop. Unlike the females, male savanna baboons

leave their natal troop to join another where they may remain for many years, so most of the Eburru Cliffs adult males were immigrants. Since membership in the troop remained relatively constant during the period of my study, I learned to identify each individual. I relied on differences in size, posture, gait, and especially, facial features. To the practiced observer, baboons look as different from one another as human beings do.

As soon as I could recognize individuals, I noticed that particular females tended to turn up near particular males again and again. I came to think of these pairs as friends. Friendship among animals is not a well-documented phenomenon, so to convince skeptical colleagues that baboon friendship was real, I needed to develop objective criteria for distinguishing friendly pairs.

I began by investigating grooming, the amiable simian habit of picking through a companion's fur to remove dead skin and ectoparasites (see "Little Things That Tick Off Baboons," *Natural History,* February 1984). Baboons spend much more time grooming than is necessary for hygiene, and previous research had indicated that it is a good measure of social bonds.

Although eighteen adult males lived in the troop, each nonestrous female performed most of her grooming with just one, two, or occasionally, three males. For example, of Zizi's twenty-four grooming bouts with males, Cyclops accounted for thirteen, and a second male, Sherlock, accounted for all the rest. Different females tended to favor different males as grooming partners.

Another measure of social bonds was simply who was observed near whom. When foraging, traveling, or resting, each pregnant or lactating female spent a lot of time near a few males and associated with the others no more often than expected by chance. When I compared the identities of favorite grooming partners and frequent companions, they overlapped almost completely. This enabled me to develop a formal definition of friendship: any male that scored high on both

grooming and proximity measures was considered a friend.

Virtually all baboons made friends; only one female and the three males who had most recently joined the troop lacked such companions. Out of more than 600 possible adult female-adult male pairs in the troop, however, only about one in ten qualified as friends; these really were special relationships.

Several factors seemed to influence which baboons paired up. In most cases, friends were unrelated to each other, since the male had immigrated from another troop. (Four friendships, however, involved a female and an adolescent son who had not yet emigrated. Unlike other friends, these related pairs never mated.) Older females tended to be friends with older males; younger females with younger males. I witnessed occasional May-December romances, usually involving older females and young adult males. Adolescent males and females were strongly rule-bound, and with the exception of mother-son pairs, they formed friendships only with one another.

Regardless of age or dominance rank, most females had just one or two male friends. But among males, the number of female friends varied greatly from none to eight. Although high-ranking males enjoyed priority of access to food and sometimes mates, dominant males did not have more female friends than low-ranking males. Instead it was the older males who had lived in the troop for many years who had the most friends. When a male had several female friends, the females were often closely related to one another. Since female baboons spend a lot of time near their kin, it is probably easier for a male to maintain bonds with several related females at once.

When collecting data, I focused on one nonestrous female at a time and kept track of her every movement toward or away from any male; similarly, I noted every male who moved toward or away from her. Whenever the female and a male moved close enough to exchange intimacies, I wrote down exactly what happened. When foraging together, friends tended to remain a few yards apart. Males more often wandered away from females than the reverse, and females, more often than males, closed the gap. The female behaved as if she wanted to keep the male within calling distance, in case she needed his protection. The male, however, was more likely to make approaches that brought them within actual touching distance. Often, he would plunk himself down right next to his friend and ask her to groom him by holding a pose with exaggerated stillness. The female sometimes responded by grooming, but more often, she exhibited the most reliable sign of true intimacy: she ignored her friend and simply continued whatever she was doing.

In sharp contrast, when a male who was not a friend moved close to a female, she dared not ignore him. She stopped whatever she was doing and held still, often glancing surreptitiously at the intruder. If he did not move away, she sometimes lifted her tail and presented her rump. When a female is not in estrus, this is a gesture of appeasement, not sexual enticement. Immediately after this respectful acknowledgement of his presence, the female would slip away. But such tense interactions with nonfriend males were rare, because females usually moved away before the males came too close.

These observations suggest that females were afraid of most of the males in their troop, which is not surprising: male baboons are twice the size of females, and their canines are longer and sharper than those of a lion. All Eburru Cliffs males directed both mild and severe aggression toward females. Mild aggression, which usually involved threats and chases but no body contact, occurred most often during feeding competition or when the male redirected aggression toward a female after losing a fight with another male. Females and juveniles showed aggression toward other females and juveniles in similar circumstances and occasionally inflicted superficial wounds. Severe aggression by males, which involved body contact and sometimes biting, was less common and also more puzzling, since there was no apparent cause.

An explanation for at least some of these attacks emerged one day when I was watching Pegasus, a young adult male, and his friend Cicily, sitting together in the middle of a small clearing. Cicily moved to the edge of the clearing to feed, and a higher-ranking female, Zora, suddenly attacked her. Pegasus stood up and looked as if he were about to intervene when both females disappeared into the bushes. He sat back down, and I remained with him. A full ten minutes later, Zora appeared at the edge of the clearing; this was the first time she had come into view since her attack on Cicily. Pegasus instantly pounced on Zora, repeatedly grabbed her neck in his mouth and lifted her off the ground, shook her whole body, and then dropped her. Zora screamed continuously and tried to escape. Each time, Pegasus caught her and continued his brutal attack. When he finally released her five minutes later she had a deep canine gash on the palm of her hand that made her limp for several days.

This attack was similar in form and intensity to those I had seen before and labeled "unprovoked." Certainly, had I come upon the scene after Zora's aggression toward Cicily, I would not have understood why Pegasus attacked Zora. This suggested that some, perhaps many, severe attacks by males actually represented punishment for actions that had occurred some time before.

Whatever the reasons for male attacks on females, they represent a serious threat. Records of fresh injuries indicated that Eburru Cliffs adult females received canine slash wounds from males at the rate of one for every female each year, and during my study, one female died of her injuries. Males probably pose an even greater threat to infants. Although only one infant was killed during my study, observers in Botswana and Tanzania have seen recent male immigrants kill several young infants.

Protection from male aggression, and from the less injurious but more frequent aggression of other females and juveniles, seems to be one of the main advantages of friendship for a

female baboon. Seventy times I observed an adult male defend a female or her offspring against aggression by another troop member, not infrequently a high-ranking male. In all but six of these cases, the defender was a friend. Very few of these confrontations involved actual fighting; no male baboon, subordinate or dominant, is anxious to risk injury by the sharp canines of another.

Males are particularly solicitous guardians of their friends' youngest infants. If another male gets too close to an infant or if a juvenile female plays with it too roughly, the friend may intervene. Other troop members soon learn to be cautious when the mother's friend is nearby, and his presence provides the mother with a welcome respite from the annoying pokes and prods of curious females and juveniles obsessed with the new baby. Male baboons at Gombe Park in Tanzania and Amboseli Park in Kenya have also been seen rescuing infants from chimpanzees and lions. These several forms of male protection help to explain why females in Eburru Cliffs stuck closer to their friends in the first few months after giving birth than at any other time.

The male-infant relationship develops out of the male's friendship with the mother, but as the infant matures, this new bond takes on a life of its own. My co-worker Nancy Nicolson found that by about nine months of age, infants actively sought out their male friends when the mother was a few yards away, suggesting that the male may function as an alternative caregiver. This seemed to be especially true for infants undergoing unusually early or severe weaning. (Weaning is generally a gradual, prolonged process, but there is tremendous variation among mothers in the timing and intensity of weaning. See "Mother Baboons," *Natural History*, September 1980). After being rejected by the mother, the crying infant often approached the male friend and sat huddled against him until its whimpers subsided. Two of the infants in Eburru Cliffs lost their mothers when they were still quite young. In each case, their bond with the mother's friend subsequently intensified, and—perhaps as a result—both infants survived.

A close bond with a male may also improve the infant's nutrition. Larger than all other troop members, adult males monopolize the best feeding sites. In general, the personal space surrounding a feeding male is inviolate, but he usually tolerates intrusions by the infants of his female friends, giving them access to choice feeding spots.

Although infants follow their male friends around rather than the reverse, the males seem genuinely attached to their tiny companions. During feeding, the male and infant express their pleasure in each other's company by sharing spirited, antiphonal grunting duets. If the infant whimpers in distress, the male friend is likely to cease feeding, look at the infant, and grunt softly, as if in sympathy, until the whimpers cease. When the male rests, the infants of his female friends may huddle behind him, one after the other, forming a "train," or, if feeling energetic, they may use his body as a trampoline.

When I returned to Eburru Cliffs four years after my initial study ended, several of the bonds formed between males and the infants of their female friends were still intact (in other cases, either the male or the infant or both had disappeared). When these bonds involved recently matured females, their long-time male associates showed no sexual interest in them, even though the females mated with other adult males. Mothers and sons, and usually maternal siblings, show similar sexual inhibitions in baboons and many other primate species.

The development of an intimate relationship between a male and the infant of his female friend raises an obvious question: Is the male the infant's father? To answer this question definitely we would need to conduct genetic analysis, which was not possible for these baboons. Instead, I estimated paternity probabilities from observations of the temporary (a few hours or days) exclusive mating relationships, or consortships, that estrous females form with a series of different

males. These estimates were apt to be fairly accurate, since changes in the female's sexual swelling allow one to pinpoint the timing of conception to within a few days. Most females consorted with only two or three males during this period, and these males were termed likely fathers.

In about half the friendships, the male was indeed likely to be the father of his friend's most recent infant, but in the other half he was not—in fact, he had never been seen mating with the female. Interestingly, males who were friends with the mother but not likely fathers nearly always developed a relationship with her infant, while males who had mated with the female but were not her friend usually did not. Thus friendship with the mother, rather than paternity, seems to mediate the development of male-infant bonds. Recently, a similar pattern was documented for South American capuchin monkeys in a laboratory study in which paternity was determined genetically.

These results fly in the face of a prominent theory that claims males will invest in infants only when they are closely related. If males are not fostering the survival of their own genes by caring for the infant, then why do they do so? I suspected that the key was female choice. If females preferred to mate with males who had already demonstrated friendly behavior, then friendships with mothers and their infants might pay off in the future when the mothers were ready to mate again.

To find out if this was the case, I examined each male's sexual behavior with females he had befriended before they resumed estrus. In most cases, males consorted considerably more often with their friends than with other females. Baboon females typically mate with several different males, including both friends and nonfriends, but prior friendship increased a male's probability of mating with a female above what it would have been otherwise.

This increased probability seemed to reflect female preferences. Females occasionally overtly advertised their

disdain for certain males and their desire for others. Zizi's behavior, described above, is a good example. Virgil was not one of her friends, but Cyclops was. Usually, however, females expressed preferences and aversions more subtly. For example, Delphi, a petite adolescent female, found herself pursued by Hector, a middle-aged adult male. She did not run away or refuse to mate with him, but whenever he wasn't watching, she looked around for her friend Homer, an adolescent male. When she succeeded in catching Homer's eye, she narrowed her eyes and flattened her ears against her skull, the friendliest face one baboon can send another. This told Homer she would rather be with him. Females expressed satisfaction with a current consort partner by staying close to him, initiating copulations, and not making advances toward other males. Baboons are very sensitive to such cues, as indicated by an experimental study in which rival hamadryas baboons rarely challenged a male-female pair if the female strongly preferred her current partner. Similarly, in Eburru Cliffs, males were less apt to challenge consorts involving a pair that shared a long-term friendship.

Even though females usually consorted with their friends, they also mated with other males, so it is not surprising that friendships were most vulnerable during periods of sexual activity. In a few cases, the female consorted with another male more often than with her friend, but the friendship survived nevertheless. One female, however, formed a strong sexual bond with a new male. This bond persisted after conception, replacing her previous friendship. My observations suggest that adolescent and young adult females tend to have shorter, less stable friendships than do older females. Some friendships, however, last a very long time. When I returned to Eburru Cliffs six years after my study began, five couples were still together. It is possible that friendships occasionally last for life (baboons probably live twenty to thirty years in the wild), but it will require longer studies, and some very patient scientists, to find out.

By increasing both the male's chances of mating in the future and the likelihood that a female's infant will survive, friendship contributes to the reproductive success of both partners. This clarifies the evolutionary basis of friendship-forming tendencies in baboons, but what does friendship mean to a baboon? To answer this question we need to view baboons as sentient beings with feelings and goals not unlike our own in similar circumstances. Consider, for example, the friendship between Thalia and Alexander.

The affair began one evening as Alex and Thalia sat about fifteen feet apart on the sleeping cliffs. It was like watching two novices in a singles bar. Alex stared at Thalia until she turned and almost caught him looking at her. He glanced away immediately, and then she stared at him until his head began to turn toward her. She suddenly became engrossed in grooming her toes. But as soon as Alex looked away, her gaze returned to him. They went on like this for more than fifteen minutes, always with split-second timing. Finally, Alex managed to catch Thalia looking at him. He made the friendly eyes-narrowed, ears-back face and smacked his lips together rhythmically. Thalia froze, and for a second she looked into his eyes. Alex approached, and Thalia, still nervous, groomed him. Soon she calmed down, and I found them still together on the cliffs the next morning. Looking back on this event months later, I realized that it marked the beginning of their friendship. Six years later, when I returned to Eburru Cliffs, they were still friends.

If flirtation forms an integral part of baboon friendship, so does jealousy. Overt displays of jealousy, such as chasing a friend away from a potential rival, occur occasionally, but like humans, baboons often express their emotions in more subtle ways. One evening a colleague and I climbed the cliffs and settled down near Sherlock, who was friends with Cybelle, a middle-aged female still foraging on the ground below the cliffs. I observed Cybelle while my colleague watched Sherlock, and we kept up a running commentary. As long as Cybelle was feeding or interacting with females, Sherlock was relaxed, but each time she approached another male, his body would stiffen, and he would stare intently at the scene below. When Cybelle presented politely to a male who had recently tried to befriend her, Sherlock even made threatening sounds under his breath. Cybelle was not in estrus at the time, indicating that male baboon jealousy extends beyond the sexual arena to include affiliative interactions between a female friend and other males.

Because baboon friendships are embedded in a network of friendly and antagonistic relationships, they inevitably lead to repercussions extending beyond the pair. For example, Virgil once provoked his weaker rival Cyclops into a fight by first attacking Cyclops's friend Phoebe. On another occasion, Sherlock chased Circe, Hector's best friend, just after Hector had chased Antigone, Sherlock's friend.

In another incident, the prime adult male Triton challenged Cyclops's possession of meat. Cyclops grew increasingly tense and seemed about to abandon the prey to the younger male. Then Cyclops's friend Phoebe appeared with her infant Phyllis. Phyllis wandered over to Cyclops. He immediately grabbed her, held her close, and threatened Triton away from the prey. Because any challenge to Cyclops now involved a threat to Phyllis as well, Triton risked being mobbed by Phoebe and her relatives and friends. For this reason, he backed down. Males frequently use the infants of their female friends as buffers in this way. Thus, friendship involves costs as well as benefits because it makes the participants vulnerable to social manipulation or redirected aggression by others.

Finally, as with humans, friendship seems to mean something different to each baboon. Several females in Eburru Cliffs had only one friend. They were devoted companions. Louise and Pandora, for example, groomed their friend Virgil and no other male. Then there was Leda, who, with five friends, spread herself more thinly than any other female. These contrasting patterns of friendship were associated with striking personality differences. Louise and Pandora were unobtrusive females who hung around quietly with Virgil and their close relatives. Leda seemed to be everywhere at once, playing with infants, fighting with juveniles, and making friends with males. Similar differences were apparent among the males. Some devoted a great deal of time and energy to cultivating friendships with females, while others focused more on challenging other males. Although we probably will never fully understand the basis of these individual differences, they contribute immeasurably to the richness and complexity of baboon society.

Male-female friendships may be widespread among primates. They have been reported for many other groups of savanna baboons, and they also occur in rhesus and Japanese Macaques, capuchin monkeys, and perhaps in bonobos (pygmy chimpanzees). These relationships should give us pause when considering popular scenarios for the evolution of male-female relationships in humans. Most of these scenarios assume that, except for mating, males and females had little to do with one another until the development of a sexual division of labor, when, the story goes, females began to rely on males to provide meat in exchange for gathered food. This, it has been argued, set up new selection pressures favoring the development of long-term bonds between individual males and females, female sexual fidelity, and as paternity certainty increased, greater male investment in the offspring of these unions. In other words, once women began to gather and men to hunt, presto—we had the nuclear family.

This scenario may have more to do with cultural biases about women's economic dependence on men and idealized views of the nuclear family than with the actual behavior of our hominid ancestors. The nonhuman primate evidence challenges this story in at least three ways.

First, long-term bonds between the sexes can evolve in the absence of a sexual division of labor of food sharing. In our primate relatives, such relationships rest on exchanges of social, not economic, benefits.

Second, primate research shows that highly differentiated, emotionally intense male-female relationships can occur without sexual exclusivity. Ancestral men and women may have experienced intimate friendships long before they invented marriage and norms of sexual fidelity.

Third, among our closest primate relatives, males clearly provide mothers and infants with social benefits even when they are unlikely to be the fathers of those infants. In return, females provide a variety of benefits to the friendly males, including acceptance into the group and, at least in baboons, increased mating opportunities in the future. This suggests that efforts to reconstruct the evolution of hominid societies may have overemphasized what the female must supposedly do (restrict her mating to just one male) in order to obtain male parental investment.

Maybe it is time to pay more attention to what the male must do (provide benefits to females and young) in order to obtain female cooperation. Perhaps among our ancestors, as in baboons today, sex and friendship went hand in hand. As for marriage—well, that's another story.

Gut Thinking

What makes fruit-eating spider monkeys so much smarter than leaf-eating howlers? Their gourmet diet, apparently—it's gone to their heads.

Peter Radetsky

Peter Radetsky is a contributing editor of DISCOVER *and teaches science writing at the University of California at Santa Cruz. His most recent book,* Invisible Invaders: Viruses and the Scientists Who Pursue Them, *is now available in paperback. In the March* DISCOVER, *Radetsky wrote about the elusive stem cell and the controversy over patenting it.*

Life should be a breeze in the tropical forest. The weather is warm, and there's plenty of food for the asking. In theory, you need only reach out and luscious fruits and other tidbits will fall into your hands. Sadly, it's not so, says Katharine Milton—particularly if you're a monkey. Milton, a physical anthropologist at the University of California at Berkeley, has spent the last 20 years studying howler and spider monkeys in the forests of Panama. Life, she's concluded, is tough in the forest—animals need to devise all sorts of ingenious tactics just to get enough food to survive. Finding food is *so* tough, Milton thinks, that successful strategies have driven the evolution of the species. "It is the solutions to the problems of diet that have made primates primates," she says. And what pertains to forest primates pertains as well to their city-slicker cousins—us. In other words, the food we eat has made us human.

Milton began her observations in 1974 on the island of Barro Colorado in Panama. For the fledgling anthropologist—Milton was then a New York University graduate student—it was an ideal spot. The island offered a protected forest inhabited by numerous species of wild animals, including howler

monkeys—13- to 18-pound primates notorious for their terrifying, unearthly howls. It was also the site of a Smithsonian Institution research station, complete with an extensive herbarium for identifying indigenous plants.

Milton threw herself into her work. "I'd get up every morning at 4:30, go to the dining hall, stuff as much food in my face as I could, make a bunch of peanut butter sandwiches, fill my water bottle, then walk into the forest to where I'd left the monkeys the night before," she recalls, in the twang of her native Montgomery, Alabama. "I'd sit on a log in the dark, and as soon as they started to wake up at dawn, about 6, I'd start taking notes." She would then follow the monkeys as they meandered along the forest canopy some 80 feet above, note where they stopped to eat, and collect the scraps of food that dropped to the ground.

"I'd follow them until 6 P.M., when it got dark, and watch them settle down for the night. Then I'd run back with a flashlight, jumping down the trail like a little goat. I'd get to the dining hall, nobody there, but the cook would have made a plate for me and covered it with tinfoil. I'd gobble down my dinner, run to the herbarium and identify my plant scraps, take a shower, and go to bed. And the next day the same thing."

Thus passed the bulk of three years. Milton found that most of the time the howlers ate leaves and fruit in almost equal measure, but when seasonal fruits were in short supply, the animals filled up on leaves. Howler monkeys were finicky, though. They ate only tender, young leaves, and only the tips at that.

Typically, soon after awakening, a howler troop of 19 individuals would set off through the trees in single file. The group traveled with no apparent leader, but after about 45 minutes, they'd arrive at a source of food.

"They know where they're going," says Milton. "I don't know how, but they know. They appear to use a collective information pool to locate their foods. They'll just set off in a straight line right to it."

For example, one of the howlers' favorite delicacies was the leaf of the *Ceiba pentandra* tree, a 100-foot-tall monster with room-size buttresses and limbs large enough for a grown man to walk on. "It has a five-fingered leaf that's pinkish brown when it first comes out on the tree but within hours turns green," Milton says. "After that, the monkeys don't want it anymore. Somehow they know how to get there just when the emerging leaves are beginning to expand. They eat the tips, which are far more nutritious than the middles or bases."

The howlers conducted these expeditions over 75 acres, searching out as many as 25 species of plants daily. Some, like the *Ceiba pentandra* tree, were edible for only a few hours a year; others were available more often. Unerringly, the howlers tracked them down. The ranges of various howler troops overlapped, so Milton would occasionally come upon a tree filled with monkeys, with other groups in adjoining trees politely waiting their turn at the table. All of which suggested that the animals had an extraordinary collective memory, an unfailing sense of direction, refined social manners, and a built-in barometer of what foods were good for them.

This aggregate intelligence allows infant howlers to mature quickly. "After 12 to 14 months, howler mothers don't want to see their babies again," Milton says. The babies soon declare independence and rely on the group for support.

Still, despite the obvious group intelligence, the monkeys individually didn't seem particularly smart to Milton. They were relatively dull and placid—and unobservant. "I ate lunch for months in full view of dozens of howlers, and not one ever seemed to realize that I was eating, much less that what I was eating might be something they would enjoy, too," she says. "You could make noises and slurp and carry on—whatever cognitive processes are required to identify the act of eating, they don't seem to use them."

But spider monkeys did. "I saw them all the time when I was studying howlers," says Milton. "They'd go roaring by like greased lightning." Spider monkeys are the same size as howlers, and the two animals share parts of each other's ranges on Barro Colorado. But there the similarities end. Whereas howlers travel through the canopy on all fours, spiders swing along like Tarzan. Unlike the placid howlers, spiders are playful and mischievous. "They're terrible teases," says Milton. "And they're mean little devils. They remind me of people," she confides with a laugh. "Although not specifically any of my close friends."

Spider monkeys had no trouble recognizing Milton's lunch. "'*Food!*' they'd shout. '*Let's see if we can get it!*' They'd swing down toward you; they'd threaten you. They know what a banana is. They have a keen idea of what a peanut butter sandwich is. You simply cannot eat in front of them."

Intrigued, Milton decided she'd add spider monkeys to her observations. She thought it might be interesting to compare how the two species evolved from a common ancestor. But while the comparatively sedate howlers were a researcher's dream, dealing with the spider monkeys was something else again. "They were too fast for me,"

says Milton. "So I hired a young man to work with me. He would run through the forest as fast as he could, following the monkeys, and I would come behind. We communicated by calls. '*Whooooo!*' Like that. The sound really carries through the forest."

When the barnstorming spider monkeys found food, they'd finally screech to a stop, allowing Milton to catch up. "They'd just stuff themselves. Then they'd lie around and take naps."

Unlike the howlers, Milton discovered, the spider monkeys almost exclusively ate fruit, which often made up 90 percent of their diet. Even when fruit was out of season or in short supply, it constituted over half their food. But ripe fruit is even harder to find than tender leaves. To get enough, the 18 spider monkeys on the island would resort to splitting up and trying their luck on their own. "During most of the year the distribution patterns of their foods are such that if they went around in a big group, there wouldn't be enough at any one site to feed everyone," says Milton. "So they'd spend almost the whole day foraging in small subunits or by themselves. Then around twilight they'd begin to call and coalesce, and then they'd spend the night together."

As a result of this extended exploring, the spiders' territory was huge, some 750 acres, ten times that of the howler monkeys. "And that's a conservative estimate," says Milton. "Two thousand acres might be right." If the howlers displayed impressive feats of memory and direction by finding young leaves, the spider monkeys' long-distance forays after fruit were astounding. Within an enormous area they had to remember at least 100 species of fruit and where to find thousands of fruit-bearing trees. They had to remember when each fruit was ripe, how best to approach the site, and how best to return home. If a howler forgot a food source or a travel route, the others were there to take up the slack. The spiders, though, had to fend for themselves.

And they had to know how to stay in touch. Howler monkeys tended to be quiet, communicating through subtle clucks and rattles in the throat, except

at daybreak, when their eerie howls declared "This is where we are this morning." ("All the howler troops on the island participate," says Milton. "It's called the dawn chorus. The sound comes rolling by as light moves across the forest.")

Spider monkeys, on the other hand, were conspicuously noisy. They'd yelp and cry, whinnying like horses, barking like dogs—sometimes for hours at a time. "When they're cross about something," says Milton, "they'll bark incessantly, until you think they're going to fall out of the tree." And in contrast to the howlers' community messages, spider monkeys believed in individual expression. "Spider monkey vocalizations are generally individualistic. It's George giving a food call, or Mary hailing Susie, whereas a howler monkey is not saying 'Hi, Susie' but rather 'Okay, everyone, we're getting ready to move to a new food tree.' "

All that variety and independence requires lots of training. As a result, infant spider monkeys mature slowly. They are nursed and carried by their mothers for two years, and they continue to associate almost exclusively with her until they're about three or three and a half years old. Milton remembers watching female spider monkeys patiently waiting for their offspring to take off into the trees. "The mother would then slowly tag along behind," she says. "It was her way of instructing the youngster, forcing it to become independent by learning to move through the trees on its own."

Always, of course, leaving time for mischief. For example, spider monkeys loved to torment howlers. "They would steal howler babies," says Milton. "A howler mother doesn't know what to do—she's too dopey to get her baby back. Howlers would move out of the tree when they saw spiders coming. They'd sit quietly and hope the spiders didn't pick on them."

Why were the two monkeys so dissimilar? Milton wondered about the differences in their diets. Howler monkeys ate mainly leaves, sometimes exclusively leaves, a low-quality source of nutrition. Leaves are plentiful and relatively high in protein, but they're low in energy-rich carbohydrates. They also consist of some 60 percent indigestible fiber and sometimes contain toxic chemicals. How in the world did howlers get enough energy from this unpromising diet? And why did they stick to it even during seasons when there was plenty of ripe fruit in the forest?

Fruits are loaded with easily digested carbohydrates and are relatively low in fiber—they're high-quality, nutritious food. They mean instant energy. On the other hand, fruits provide little protein. So, Milton wondered, how did spider monkeys get enough protein? And why, when fruits were scarce, didn't they fill up on leaves, as howlers did? Why did they go to such extremes to find fruits?

Milton began finding some answers to these questions in 1977, when she returned to Barro Colorado after completing her doctoral thesis. She soon conducted an experiment measuring how long it took the monkeys to process their food. "I needed to look at internal features of the monkeys," she says. "I thought that perhaps the structure of their guts or efficiency of their digestion might be influencing their behavior."

She trapped howler and spider monkeys, confined them in pens, and fed them food in which she had concealed tiny plastic markers. "I used a type of thin plastic material that I cut with very fine manicure scissors into little colored plastic worms," she explains. When the monkeys excreted the remains of their food, out came the markers. Milton could therefore measure the time it took any one meal to pass through a monkey's digestive tract. The results were dramatic: howlers took 20 hours to digest their food, five times as long as spiders. "That was a *humongous* surprise," says Milton. "The difference in transit times blew me away. There had to be an explanation—they don't have a *door* in there. So I went in and looked at their guts."

When Milton came upon monkeys that had died in the forest, she took

them back to the research station, dissected them, and measured their gastrointestinal tracts. She then confirmed her figures against published material on differential gut measurements in various primates. She found that the colons of howlers were considerably wider and longer than those of spider monkeys. Food had to travel much farther and remained much longer in howler guts, and the monkeys had room for much more bulk. As a result, bacteria had a chance to ferment masses of fibrous leaves in the monkeys' colons, producing energy-rich fatty acids. Milton eventually found that howlers receive more than 30 percent of their daily energy from such fatty acids.

In contrast, spider monkey food resembled the speedy monkeys themselves, hurtling through the animals' more compact guts. Spiders were far less efficient at extracting energy from the fiber in their diet—but they didn't have to be efficient. They ate easily digestible fruits. By moving a steady stream of fruit through their gastrointestinal tracts every day, they obtained all the carbohydrates they needed and some of the protein. The rest came from supplements of young, tender leaves.

It was a striking example of evolutionary adaptation. Each monkey's physiology fit its particular diet. Spider monkeys couldn't get away with eating a howler diet of mostly leaves. With their smallish guts, they'd never keep enough bulk around long enough for fermentation to provide energy. And howlers wouldn't manage for long if they used the spider monkey tactic of eating fruit—their slow digestive tracts couldn't process nearly enough of it.

Besides, it took smarts to track down sufficient fruit, and Milton thought it unlikely that the howlers were up to the job. Nor was the howler diet of leaves up to the job of fueling the amount of brainpower necessary. The brain, a big, hungry organ, requires a disproportionate amount of energy, and leaves just don't provide enough. All of which led to the second part of the puzzle: the difference in the monkeys' mental capacities.

"I kept thinking, spider monkeys are so smart, and howler monkeys don't seem so smart. The more I thought about it, the more it seemed to make sense that if you have a high-energy diet and widely distributed foods, you're going to need a certain amount of ability to locate those foods. I became curious—I wondered how big their brains were."

Luckily for Milton, that information was available without her having to cut apart more monkeys. A scientist named Daniel Quirling had published extensive statistics about the sizes of primate brains. Spider monkey brains, he had determined, weigh twice those of howlers, 107 grams compared with 50.4. No wonder spiders are smarter.

With that, everything came together. "It was a *eureka* moment," says Milton. Here were two monkeys, the same size, living in the same forest, but so different. Compared with the howlers, spider monkeys were brighter and more lively. They matured more slowly and had more to learn; they made more ruckus, with a greater variety of vocalizations; they ate widely dispersed, high-energy foods that were harder to find—and their brains were twice as large. Why?

As far as Milton was concerned, diet was the key to these discrepancies. Eating fruits fueled the evolution of the spider monkeys' large brains. Says Milton, "It would have been a feedback process in which some slight change in the monkeys' foraging behavior conferred a benefit, which in turn permitted a modest improvement in the quality of their diet, which led to an excess of energy. Over generations, the monkeys that spent the energy on making their brain slightly bigger and more complex had an evolutionary advantage. Their improved brain allowed for more helpful changes in their behavior, and so on."

Milton realized that if such a scenario was correct, similar differences in brain size should show up in other primates with similar differences in diet—monkeys and apes that eat fruits should have larger brains than their leaf-eating counterparts. Sure enough, when Milton checked the literature, she found the pattern held true. For example, of the three great apes, lively, quick chimpanzees, our closest animal relatives, have a bigger brain for their body size than do the slower, more placid gorillas and orangutans. Chimps take some 94 percent of their diet from plants, largely in the form of ripe fruits. Gorillas and orangutans eat 99 percent plant foods, but mainly lower-quality leaves, pith, even bark. Diet had to be the key to their disparate evolution.

And what about the primate with the largest brain of all? Might large human brains also have initially been the result of a high-quality diet? Milton thinks so. "I view dietary conditions as the key pressure leading to the emergence of humans."

Her scenario goes like this: When our australopithecine ancestors emerged in Africa more than four and a half million years ago, their brains were not appreciably larger than those of today's apes, and they had massive, grinding jaws and molar teeth, suggesting that they ate mostly tough, low-quality plant material. Eventually the australopithecines were supplanted by another series of early humans with increasingly larger brains and smaller jaws and teeth, indications that their diet had become higher in quality, less fibrous and abrasive. In time these brainy ancestors refined their diet. With the introduction of meat, early humans started to eat in ways that no primates had before.

"The fossil evidence offers strong support for the view that early humans made a dietary breakthrough," Milton says. No longer were we, like spider monkeys and chimps, primarily fruit eaters. Now, drawing on the power of our large brains, we introduced tools to help us prepare food and learned to divide the responsibility for meals—a uniquely human characteristic—so that different people became experts in dif-

ferent diets. Eventually we became what Milton calls cultural omnivores. "We will eat anything, from other human beings to sea squids," she says. "But if our culture tells us not to eat something, it doesn't matter if it is the most nutritious, digestible food in the world—we won't eat it. Food for humans is more in the mind than in the item."

A case in point may involve meat itself. It is usually thought that early humans began eating meat to satisfy their need for protein, but ongoing research by Milton's former student Craig Stanford, now an anthropologist at the University of Southern California, suggests that eating meat is as much a social gesture as a dietary necessity. With Jane Goodall, Stanford studies chimpanzees at Gombe National Park in Tanzania. He has found that Gombe chimps eat about 3 percent of their diet as meat; primarily they hunt colobus monkeys.

Chimps don't routinely hunt monkeys, though. They separate into small groups to forage for fruits, and they go after a monkey only when they come upon it by chance. Even then they might not hunt—they tend to do so when a female in heat happens to be part of the foraging group. Then, once a male makes the kill, a fascinating ritual often ensues. "Immediately, within seconds, the female comes racing over with her hand out," says Stanford. "The male pulls away the carcass until the female allows him to copulate with her. Then the male shares the meat. Sometimes he will induce her to copulate, then wave the carcass in her face and pull it away until they copulate again. Then she gets some meat."

The chimps thus use meat as a commodity exchange—in this case, to elicit sexual favors. Stanford has found that foods more nutritious than meat, such as oil palm nuts, are available year-round. Such immobile foods are much easier to procure than monkeys, which

fight like the dickens and provide no more than a few ounces of meat per chimp. "It's not just a nutritional decision when they decide to hunt," Stanford says. "They have more in mind."

"Chimps appear to eat meat for social reasons more than nutritional reasons," agrees Milton. "It's kind of like a date. It's a party, a community event."

Did we humans start eating meat for similar reasons? Stanford wouldn't be surprised: "Other researchers have shown that dominant chimps withhold meat from enemies and dole it out to allies, using it in a cleverly calculated political way."

The similarity to people is reinforced by other chimp behaviors. "The whole life cycle of chimps is not that different from people's," says Stanford. Chimp babies, for example, are totally dependent on their mothers for their first four years, and they continue to hang around Mom until the age of 10 or 11. Females become sexually mature at about the age of 12; males go out into the world with the other guys at about 15. "Chimps defend their territory with lethal aggression," says Stanford. "Humans are the only other primates to do so."

Milton sees strong parallels between humans and chimps, and spider monkeys as well. Similar diet; similar aggressive, individualistic bent; similar long-term rearing of young; similar social system—and similarly large brains. All of us—spider monkeys, chimps, and humans alike—are what we eat. The behaviors and physiology that define us are the consequences of dietary-driven evolution.

"Everything comes back to diet," says Milton. "It's the pivotal feature, the kickoff. When you get right down to it, the way we behave had better translate ultimately into groceries—we're not going to be around to behave that way much longer."

The Mind of the Chimpanzee

Jane Goodall

Often I have gazed into a chimpanzee's eyes and wondered what was going on behind them. I used to look into Flo's, she so old, so wise. What did she remember of her young days? David Greybeard had the most beautiful eyes of them all, large and lustrous, set wide apart. They somehow expressed his whole personality, his serene self-assurance, his inherent dignity—and, from time to time, his utter determination to get his way. For a long time I never liked to look a chimpanzee straight in the eye— I assumed that, as is the case with most primates, this would be interpreted as a threat or at least as a breach of good manners. Not so. As long as one looks with gentleness, without arrogance, a chimpanzee will understand, and may even return the look. And then—or such is my fantasy—it is as though the eyes are windows into the mind. Only the glass is opaque so that the mystery can never be fully revealed.

I shall never forget my meeting with Lucy, an eight-year-old home-raised chimpanzee. She came and sat beside me on the sofa and, with her face very close to mine, searched in my eyes— for what? Perhaps she was looking for signs of mistrust, dislike, or fear, since many people must have been somewhat disconcerted when, for the first time, they came face to face with a grown chimpanzee. Whatever Lucy read in my eyes clearly satisfied her for she suddenly put one arm round my neck and gave me a generous and very chimp-like kiss, her mouth wide open and laid over mine. I was accepted.

For a long time after that encounter I was profoundly disturbed. I had been at Gombe for about fifteen years then and I was quite familiar with chimpanzees in the wild. But Lucy, having grown up as a human child, was like a changeling, her essential chimpanzee-ness overlaid by the various human behaviours she had acquired over the years. No longer purely chimp yet eons away from humanity, she was man-made, some other kind of being. I watched, amazed, as she opened the refrigerator and various cupboards, found bottles and a glass, then poured herself a gin and tonic. She took the drink to the TV, turned the set on, flipped from one channel to another then, as though in disgust, turned it off again. She selected a glossy magazine from the table and, still carrying her drink, settled in a comfortable chair. Occasionally, as she leafed through the magazine she identified something she saw, using the signs of ASL, the American Sign Language used by the deaf. I, of course, did not understand, but my hostess, Jane Temerlin (who was also Lucy's 'mother'), translated: 'That dog,' Lucy commented, pausing at a photo of a small white poodle. She turned the page. 'Blue,' she declared, pointing then signing as she gazed at a picture of a lady advertising some kind of soap powder and wearing a brilliant blue dress. And finally, after some vague hand movements—perhaps signed mutterings—'This Lucy's, this mine,' as she closed the magazine and laid it on her lap. She had just been taught, Jane told me, the use of the possessive pronouns during the thrice weekly ASL lessons she was receiving at the time.

The book written by Lucy's human 'father', Maury Temerlin, was entitled *Lucy, Growing Up Human*. And in fact, the chimpanzee is more like us than is any other living creature. There is close resemblance in the physiology of our two species and genetically, in the structure of the DNA, chimpanzees and humans differ by only just over one per cent. This is why medical research uses chimpanzees as experimental animals when they need substitutes for humans in the testing of some drug or vaccine. Chimpanzees can be infected with just about all known human infectious diseases including those, such as hepatitis B and AIDS, to which other non-human animals (except gorillas, orangutans and gibbons) are immune. There are equally striking similarities between humans and chimpanzees in the anatomy and wiring of the brain and nervous system, and— although many scientists have been reluctant to admit to this—in social behaviour, intellectual ability, and the emotions. The notion of an evolutionary continuity in physical structure from pre-human ape to modern man has long been morally acceptable to most scientists. That the same might hold good for mind was generally considered an absurd hypothesis—particularly by those who used, and often misused, animals in their laboratories. It is, after all, convenient to believe that the creature you are using, while it may react in disturbingly human-like ways, is, in fact, merely a mindless and, above all, unfeeling, 'dumb' animal.

When I began my study at Gombe in 1960 it was not permissible—at least not in ethological circles—to talk about an animal's mind. Only humans had minds. Nor was it quite proper to talk about animal personality. Of course everyone knew that they *did* have their own unique characters—everyone who had ever owned a dog or other pet was aware of that. But ethologists, striving to make theirs a 'hard' science, shied away from the task of trying to explain such things objectively. One respected ethologist, while acknowledging that there was 'variability between individual animals', wrote that it was best that this fact be 'swept under the carpet'. At that time ethological carpets fairly bulged with all that was hidden beneath them.

How naive I was. As I had not had an undergraduate science education I didn't realize that animals were not supposed to have personalities, or to think, or to feel emotions or pain. I had no idea that it would have been more appropriate to assign each of the chimpanzees a number rather than a name when I got to know him or her. I didn't realize that it was not scientific to discuss behaviour in terms of motivation or purpose. And no one had told me that terms such as *childhood* and *adolescence* were uniquely human phases of the life cycle, culturally determined, not to be used when referring to young chimpanzees. Not knowing, I freely made use of all those forbidden terms and concepts in my initial attempt to describe, to the best of my ability, the amazing things I had observed at Gombe.

I shall never forget the response of a group of ethologists to some remarks I made at an erudite seminar. I described how Figan, as an adolescent, had learned to stay behind in camp after senior males had left, so that we could give him a few bananas for himself. On the first occasion he had, upon seeing the fruits, uttered loud, delighted food calls: whereupon a couple of the older males had charged back, chased after Figan, and taken his bananas. And then, coming to the point of the story, I explained how, on the next occasion, Figan had actually suppressed his calls. We could hear little sounds, in his throat, but so quiet that none of the

others could have heard them. Other young chimps, to whom we tried to smuggle fruit without the knowledge of their elders, never learned such self-control. With shrieks of glee they would fall to, only to be robbed of their booty when the big males charged back. I had expected my audience to be as fascinated and impressed as I was. I had hoped for an exchange of views about the chimpanzee's undoubted intelligence. Instead there was a chill silence, after which the chairman hastily changed the subject. Needless to say, after being thus snubbed, I was very reluctant to contribute any comments, at any scientific gathering, for a very long time. Looking back, I suspect that everyone was interested, but it was, of course, not permissible to present a mere 'anecdote' as evidence for anything.

The editorial comments on the first paper I wrote for publication demanded that every *he* or *she* be replaced with *it,* and every *who* be replaced with *which.* Incensed, I, in my turn, crossed out the *its* and *whichs* and scrawled back the original pronouns. As I had no desire to carve a niche for myself in the world of science, but simply wanted to go on living among and learning about chimpanzees, the possible reaction of the editor of the learned journal did not trouble me. In fact I won that round: the paper when finally published did confer upon the chimpanzees the dignity of their appropriate genders and properly upgraded them from the status of mere 'things' to essential Beingness.

However, despite my somewhat truculent attitude, I did want to learn, and I was sensible of my incredible good fortune in being admitted to Cambridge. I wanted to get my PhD, if only for the sake of Louis Leakey and the other people who had written letters in support of my admission. And how lucky I was to have, as my supervisor, Robert Hinde. Not only because I thereby benefitted from his brilliant mind and clear thinking, but also because I doubt that I could have found a teacher more suited to my particular needs and personality. Gradually he was able to cloak me with at least some

of the trappings of a scientist. Thus although I continued to hold to most of my convictions—that animals had personalities; that they could feel happy or sad or fearful; that they could feel pain; that they could strive towards planned goals and achieve greater success if they were highly motivated—I soon realized that these personal convictions were, indeed, difficult to prove. It was best to be circumspect—at least until I had gained some credentials and credibility. And Robert gave me wonderful advice on how best to tie up some of my more rebellious ideas with scientific ribbon. 'You can't *know* that Fifi was jealous,' he admonished on one occasion. We argued a little. And then: 'Why don't you just say *If Fifi were a human child we would say she was jealous.*' I did.

It is not easy to study emotions even when the subjects are human. I know how I feel if I am sad or happy or angry, and if a friend tells me that he is feeling sad, happy or angry, I assume that his feelings are similar to mine. But of course I cannot know. As we try to come to grips with the emotions of beings progressively more different from ourselves the task, obviously, becomes increasingly difficult. If we ascribe human emotions to non-human animals we are accused of being anthropomorphic—a cardinal sin in ethology. But is it so terrible? If we test the effect of drugs on chimpanzees because they are biologically so similar to ourselves, if we accept that there are dramatic similarities in chimpanzee and human brain and nervous system, is it not logical to assume that there will be similarities also in at least the more basic feelings, emotions, moods of the two species?

In fact, all those who have worked long and closely with chimpanzees have no hesitation in asserting that chimps experience emotions similar to those which in ourselves we label pleasure, joy, sorrow, anger, boredom and so on. Some of the emotional states of the chimpanzee are so obviously similar to ours that even an inexperienced observer can understand what is going on. An infant who hurls himself screaming to the ground, face contorted, hitting out with his arms at any nearby object, banging his head, is clearly having a tantrum. Another youngster, who gambols around his mother, turning somersaults, pirouetting and, every so often, rushing up to her and tumbling into her lap, patting her or pulling her hand towards him in a request for tickling, is obviously filled with *joie de vivre*. There are few observers who would not unhesitatingly ascribe his behaviour to a happy, carefree state of well-being. And one cannot watch chimpanzee infants for long without realizing that they have the same emotional need for affection and reassurance as human children. An adult male, reclining in the shade after a good meal, reaching benignly to play with an infant or idly groom an adult female, is clearly in a good mood. When he sits with bristling hair, glaring at his subordinates and threatening them, with irritated gestures, if they come too close, he is clearly feeling cross and grumpy. We make these judgements because the similarity of so much of a chimpanzee's behaviour to our own permits us to empathize.

It is hard to empathize with emotions we have not experienced. I can imagine, to some extent, the pleasure of a female chimpanzee during the act of procreation. The feelings of her male partner are beyond my knowledge—as are those of the human male in the same context. I have spent countless hours watching mother chimpanzees interacting with their infants. But not until I had an infant of my own did I begin to understand the basic, powerful instinct of mother-love. If someone accidentally did something to frighten Grub, or threaten his well-being in any way, I felt a surge of quite irrational anger. How much more easily could I then understand the feelings of the chimpanzee mother who furiously waves her arm and barks in threat at an individual who approaches her infant too closely, or at a playmate who inadvertently hurts her child. And it was not until I knew the numbing grief that gripped me after the death of my second husband that I could even begin to appreciate the despair and sense of loss that can cause young chimps to pine

away and die when they lose their mothers.

Empathy and intuition can be of tremendous value as we attempt to understand certain complex behavioural interactions, provided that the behaviour, as it occurs, is recorded precisely and objectively. Fortunately I have seldom found it difficult to record facts in an orderly manner even during times of powerful emotional involvement. And 'knowing' intuitively how a chimpanzee is feeling—after an attack, for example—may help one to understand what happens next. We should not be afraid at least to try to make use of our close evolutionary relationship with the chimpanzees in our attempts to interpret complex behaviour.

Today, as in Darwin's time, it is once again fashionable to speak of and study the animal mind. This change came about gradually, and was, at least in part, due to the information collected during careful studies of animal societies in the field. As these observations became widely known, it was impossible to brush aside the complexities of social behaviour that were revealed in species after species. The untidy clutter under the ethological carpets was brought out and examined, piece by piece. Gradually it was realized that parsimonious explanations of apparently intelligent behaviours were often misleading. This led to a succession of experiments that, taken together, clearly prove that many intellectual abilities that had been thought unique to humans were actually present, though in a less highly developed form, in other, non-human beings. Particularly, of course, in the non-human primates and especially in chimpanzees.

When first I began to read about human evolution, I learned that one of the hallmarks of our own species was that we, and only we, were capable of making tools. *Man the Toolmaker* was an oft-cited definition—and this despite the careful and exhaustive research of Wolfgang Kohler and Robert Yerkes on the tool-using and tool-making abilities of chimpanzees. Those studies, carried out independently in the early twenties, were received with scepticism. Yet both Kohler and Yerkes were respected

scientists, and both had a profound understanding of chimpanzee behaviour. Indeed, Kohler's descriptions of the personalities and behaviour of the various individuals in his colony, published in his book *The Mentality of Apes,* remain some of the most vivid and colourful ever written. And his experiments, showing how chimpanzees could stack boxes, then climb the unstable constructions to reach fruit suspended from the ceiling, or join two short sticks to make a pole long enough to rake in fruit otherwise out of reach, have become classic, appearing in almost all textbooks dealing with intelligent behaviour in non-human animals.

By the time systematic observations of tool-using came from Gombe those pioneering studies had been largely forgotten. Moreover, it was one thing to know that humanized chimpanzees in the lab could use implements: it was quite another to find that this was a naturally occurring skill in the wild. I well remember writing to Louis about my first observations, describing how David Greybeard not only used bits of straw to fish for termites but actually stripped leaves from a stem and thus *made* a tool. And I remember too receiving the now oft-quoted telegram he sent in response to my letter: 'Now we must redefine *tool,* redefine *Man,* or accept chimpanzees as humans.'

There were, initially, a few scientists who attempted to write off the termiting observations, even suggesting that I had taught the chimps! By and large, though, people were fascinated by the information and by the subsequent observations of the other contexts in which the Gombe chimpanzees used objects as tools. And there were only a few anthropologists who objected when I suggested that the chimpanzees probably passed their tool-using traditions from one generation to the next, through observations, imitation and practice, so that each population might be expected to have its own unique tool-using culture. Which, incidentally, turns out to be quite true. And when I described how one chimpanzee, Mike, spontaneously solved a new problem by using a tool (he broke off a stick to knock a banana to the ground when he was too nervous to actually take it from my hand) I don't believe there were any raised eyebrows in the scientific community. Certainly I was not attacked viciously, as were Kohler and Yerkes, for suggesting that humans were not the only beings capable of reasoning and insight.

The mid-sixties saw the start of a project that, along with other similar research, was to teach us a great deal about the chimpanzee mind. This was Project Washoe, conceived by Trixie and Allen Gardner. They purchased an infant chimpanzee and began to teach her the signs of ASL, the American Sign Language used by the deaf. Twenty years earlier another husband and wife team, Richard and Cathy Hayes, had tried, with an almost total lack of success, to teach a young chimp, Vikki, to talk. The Hayes's undertaking taught us a lot about the chimpanzee mind, but Vikki, although she did well in IQ tests, and was clearly an intelligent youngster, could not learn human speech. The Gardners, however, achieved spectacular success with their pupil, Washoe. Not only did she learn signs easily, but she quickly began to string them together in meaningful ways. It was clear that each sign evoked, in her mind, a mental image of the object it represented. If, for example, she was asked, in sign language, to fetch an apple, she would go and locate an apple that was out of sight in another room.

Other chimps entered the project, some starting their lives in deaf signing families before joining Washoe. And finally Washoe adopted an infant, Loulis. He came from a lab where no thought of teaching signs had ever penetrated. When he was with Washoe he was given no lessons in language acquisition—not by humans, anyway. Yet by the time he was eight years old he had made fifty-eight signs in their correct contexts. How did he learn them? Mostly, it seems, by imitating the behaviour of Washoe and the other three signing chimps, Dar, Moja and Tatu. Sometimes, though, he received tuition from Washoe herself. One day, for example, she began to swagger about bipedally, hair bristling, signing *food!*

food! food! in great excitement. She had seen a human approaching with a bar of chocolate. Loulis, only eighteen months old, watched passively. Suddenly Washoe stopped her swaggering, went over to him, took his hand, and moulded the sign for *food* (fingers pointing towards mouth). Another time, in a similar context, she made the sign for *chewing gum*—but with *her* hand on *his* body. On a third occasion Washoe, apropos of nothing, picked up a small chair, took it over to Loulis, set it down in front of him, and very distinctly made the *chair* sign three times, watching him closely as she did so. The two food signs became incorporated into Loulis's vocabulary but the sign for chair did not. Obviously the priorities of a young chimp are similar to those of a human child!

When news of Washoe's accomplishments first hit the scientific community it immediately provoked a storm of bitter protest. It implied that chimpanzees were capable of mastering a human language, and this, in turn, indicated mental powers of generalization, abstraction and concept-formation as well as an ability to understand and use abstract symbols. And these intellectual skills were surely the prerogatives of *Homo sapiens*. Although there were many who were fascinated and excited by the Gardners' findings, there were many more who denounced the whole project, holding that the data was suspect, the methodology sloppy, and the conclusions not only misleading, but quite preposterous. The controversy inspired all sorts of other language projects. And, whether the investigators were sceptical to start with and hoped to disprove the Gardners' work, or whether they were attempting to demonstrate the same thing in a new way, their research provided additional information about the chimpanzee's mind.

And so, with new incentive, psychologists began to test the mental abilities of chimpanzees in a variety of different ways; again and again the results confirmed that their minds are uncannily like our own. It had long been held that only humans were capable of what is called 'cross-modal trans-

fer of information'—in other words, if you shut your eyes and someone allows you to feel a strangely shaped potato, you will subsequently be able to pick it out from other differently shaped potatoes simply by looking at them. And vice versa. It turned out that chimpanzees can 'know' with their eyes what they 'feel' with their fingers in just the same way. In fact, we now know that some other non-human primates can do the same thing. I expect all kinds of creatures have the same ability.

Then it was proved, experimentally and beyond doubt, that chimpanzees could recognize themselves in mirrors—that they had, therefore, some kind of self-concept. In fact, Washoe, some years previously, had already demonstrated the ability when she spontaneously identified herself in the mirror, staring at her image and making her name sign. But that observation was merely anecdotal. The proof came when chimpanzees who had been allowed to play with mirrors were, while anaesthetized, dabbed with spots of odourless paint in places, such as the ears or the top of the head, that they could see only in the mirror. When they woke they were not only fascinated by their spotted images, but immediately investigated, with their fingers, the dabs of paint.

The fact that chimpanzees have excellent memories surprised no one. Everyone, after all, has been brought up to believe that 'an elephant never forgets' so why should a chimpanzee be any different? The fact that Washoe spontaneously gave the name-sign of Beatrice Gardner, her surrogate mother, when she saw her after a separation of eleven years was no greater an accomplishment than the amazing memory shown by dogs who recognize their owners after separations of almost as long—and the chimpanzee has a much longer life span than a dog. Chimpanzees can plan ahead, too, at least as regards the immediate future. This, in fact, is well illustrated at Gombe, during the termiting season: often an individual prepares a tool for use on a termite mound that is several hundred yards away and absolutely out of sight.

2. PRIMATES

This is not the place to describe in detail the other cognitive abilities that have been studied in laboratory chimpanzees. Among other accomplishments chimpanzees possess pre-mathematical skills: they can, for example, readily differentiate between *more* and *less*. They can classify things into specific categories according to a given criterion—thus they have no difficulty in separating a pile of food into *fruits* and *vegetables* on one occasion, and, on another, dividing the same pile of food into *large* versus *small* items, even though this requires putting some vegetables with some fruits. Chimpanzees who have been taught a language can combine signs creatively in order to describe objects for which they have no symbol. Washoe, for example, puzzled her caretakers by asking, repeatedly, for a *rock berry*. Eventually it transpired that she was referring to Brazil nuts which she had encountered for the first time a while before. Another language-trained chimp described a cucumber as a *green banana,* and another referred to an Alka-Seltzer as a *listen drink.* They can even invent signs. Lucy, as she got older, had to be put on a leash for her outings. One day, eager to set off but having no sign for *leash,* she signalled her wishes by holding a crooked index finger to the ring on her collar. This sign became part of her vocabulary. Some chimpanzees love to draw, and especially to paint. Those who have learned sign language sometimes spontaneously label their works, 'This [is] apple'—or bird, or sweetcorn, or whatever. The fact that the paintings often look, to our eyes, remarkably unlike the objects depicted by the artists either means that the chimpanzees are poor draughtsmen or that we have much to learn regarding ape-style representational art!

People sometimes ask why chimpanzees have evolved such complex intellectual powers when their lives in the wild are so simple. The answer is, of course, that their lives in the wild are not so simple! They use—and need—all their mental skills during normal day-to-day life in their complex society. They are always having to make choices—where to go, or with whom to travel. They need highly developed social skills—particularly those males who are ambitious to attain high positions in the dominance hierarchy. Low-ranking chimpanzees must learn deception—to conceal their intentions or to do things in secret—if they are to get their way in the presence of their superiors. Indeed, the study of chimpanzees in the wild suggests that their intellectual abilities evolved, over the millennia, to help them cope with daily life. And now, the solid core of data concerning chimpanzee intellect collected so carefully in the lab setting provides a background against which to evaluate the many examples of intelligent, rational behaviour that we see in the wild.

It is easier to study intellectual prowess in the lab where, through carefully devised tests and judicious use of rewards, the chimpanzees can be encouraged to exert themselves, to stretch their minds to the limit. It is more meaningful to study the subject in the wild, but much harder. It is more meaningful because we can better understand the environmental pressures that led to the evolution of intellectual skills in chimpanzee societies. It is harder because, in the wild, almost all behaviours are confounded by countless variables; years of observing, recording and analysing take the place of contrived testing; sample size can often be counted on the fingers of one hand; the only experiments are nature's own, and only time—eventually—may replicate them.

In the wild a single observation may prove of utmost significance, providing a clue to some hitherto puzzling aspect of behaviour, a key to the understanding of, for example, a changed relationship. Obviously it is crucial to see as many incidents of this sort as possible. During the early years of my study at Gombe it became apparent that one person alone could never learn more than a fraction of what was going on in a chimpanzee community at any given time. And so, from 1964 onwards, I gradually built up a research team to help in the gathering of information about the behaviour of our closest living relatives.

Dian Fossey and Digit

Sy Montgomery

Every breath was a battle to draw the ghost of her life back into her body. At age forty-two it hurt her even to breathe.

Dian Fossey had been asthmatic as a child and a heavy smoker since her teens; X-rays of her lungs taken when she graduated from college, she remembered, looked like "a road map of Los Angeles superimposed over a road map of New York." And now, after eight years of living in the oxygen-poor heights of Central Africa's Virunga Volcanoes, breathing the cold, sodden night air, her lungs were crippled. The hike to her research camp, Karisoke, at 10,000 feet, took her graduate students less than an hour; for Dian it was a gasping two-and-a-half-hour climb. She had suffered several bouts of pneumonia. Now she thought she was coming down with it again.

Earlier in the week she had broken her ankle. She heard the bone snap when she fell into a drainage ditch near her corrugated tin cabin. She had been avoiding a charging buffalo. Two days later she was bitten by a venomous spider on the other leg. Her right knee was swollen huge and red; her left ankle was black. But she would not leave the mountain for medical treatment in the small hospital down in Ruhengeri. She had been in worse shape before. Once, broken ribs punctured a lung; another time she was bitten by a dog thought to be rabid. Only when her temperature reached 105 and her symptoms clearly matched those described in her medical book for rabies had she allowed her African staff to carry her down on a litter.

Dian was loath to leave the camp in charge of her graduate students, two of whom she had been fighting with bitterly. Kelly Steward and Sandy Harcourt, once her closest camp colleagues and confidantes, had committed the unforgivable error of falling in love with each other. Dian considered this a breach of loyalty. She yelled at them. Kelly cried and Sandy sulked.

But on this May day of 1974 Sandy felt sorry for Dian. As a gesture of conciliation, he offered to help her hobble out to visit Group 4. Splinted and steadied by a walking stick, she quickly accepted.

Group 4 was the first family of mountain gorillas Dian had contacted when she established her camp in Rwanda in September 1967. A political uprising had forced her to flee her earlier research station in Zaire. On the day she founded Karisoke—a name she coined by combining the names of the two volcanoes between which her camp nestled, Karisimbi and Visoke—poachers had led her to the group. The two Batwa tribesmen had been hunting antelope in the park—an illegal practice that had been tolerated for decades—and they offered to show her the gorillas they had encountered.

At that first contact, Dian watched the gorillas through binoculars for forty-five minutes. Across a ravine, ninety feet away, she could pick out three distinctive individuals in the fourteen-member group. There was a majestic old male, his black form silvered from shoulder to hip. This 350-pound silverback was obviously the sultan of the harem of females, the leader of the family. One old female stood out, a glare in her eyes, her lips compressed as if she had swallowed vinegar. And one youngster was "a playful little ball of disorganized black fluff . . . full of mischief and curiosity," as Dian would later describe

him in *Gorillas in the Mist*. She guessed then that he was about five years old. He tumbled about in the foliage like an animated black dustball. When the lead silverback spotted Dian behind a tree, the youngster obediently fled at his call, but Dian had the impression that the little male would rather have stayed for a longer look at the stranger. In a later contact she noticed the juvenile's swollen, extended middle finger. After many attempts at naming him, she finally called him Digit.

It was Digit, now twelve, who came over to Dian as she sat crumpled and coughing among the foliage with Sandy. Digit, a gaunt young silverback, served his family as sentry. He left the periphery of his group to knuckle over to her side. She inhaled his smell. A good smell, she noted with relief: for two years a draining wound in his neck had hunched his posture and sapped his spirit. Systemic infection had given his whole body a sour odor, not the normal, clean smell of fresh sweat. During that time Digit had become listless. Little would arouse his interest: not the sex play between the group's lead silverback and receptive females, not even visits from Dian. Digit would sit at the edge of the group for hours, probing the wound with his fingers, his eyes fixed on some distant spot as if dwelling on a sad memory.

But today Digit looked directly into Dian's eyes. He chose to remain beside her throughout the afternoon, like a quiet visitor to a shut-in, old friends with no need to talk. He turned his great domed head to her, looking at her solemnly with a brown, cognizant gaze. Normally a prolonged stare from a gorilla is a threat. But Digit's gaze bore no aggression. He seemed to say: I know. Dian would later write that she believed Digit understood she was sick. And she returned to camp that afternoon, still limping, still sick, still troubled, but whole.

"We all felt we shared something with the gorillas," one of her students would later recall of his months at Karisoke. And it is easy to feel that way after even a brief contact with these huge, solemn beings. "The face of a gorilla," wrote nature writer David Quammen after just looking at a picture of one, "offers a shock of what feels like total recognition." To be in the presence of a mountain gorilla for even one hour simply rips your soul open with awe. They are the largest of the great apes, the most hugely majestic and powerful; but it is the gaze of a gorilla that transfixes, when its eyes meet yours. The naturalist George Schaller, whose year-long study preceded Dian's, wrote that this is a look found in the eye of no other animal except, perhaps, a whale. It is not so much intelligence that strikes you, but understanding. You feel there has been an exchange.

The exchange between Digit and Dian that day was deep and long. By then Digit had known Dian for seven years. She had been a constant in his growing up from a juvenile to a young blackback and now to a silverback sentry. He had known her longer than he had known his own mother, who had died or left his group before he was five; he had known Dian longer than he had known his father, the old silverback who died of natural causes less than a year after she first observed him. When Digit was nine, his three age-mates in Group 4 departed: his half sisters were "kidnaped" by rival silverbacks, as often happens with young females. Digit then adopted Dian as his playmate, and he would often leave the rest of the group to amble to her side, eager to examine her gear, sniff her gloves and jeans, tug gently at her long brown braid.

As for Dian, her relationship with Digit was stronger than her bonds with her mother, father, or stepfather. Though she longed for a husband and babies, she never married or bore children. Her relationship with Digit endured

longer than that with any of her lovers and outlasted many of her human friendships.

In her slide lectures in the United States, Dian would refer to him as "my friend, Digit." "Friend," she admitted, was too weak a word, too casual; but she could find no other. Our words are something we share with other humans; but what Dian had with Digit was something she guarded as uniquely hers.

A mountain gorilla group is one of the most cohesive family units found among primates, a fact that impressed George Schaller. Adult orangutans live mostly alone, males and females meeting to mate. Chimpanzees' social groupings are so loosely organized, changing constantly in number and composition, that Jane Goodall couldn't make sense of them for nearly a decade. But gorillas live in tight-knit, clearly defined families. Typically a group contains a lead silverback, perhaps his adult brother, half brother, or nephew, and several adult females and their offspring.

A gorilla group travels, feeds, plays, and rests together. Seldom is an individual more than a hundred feet away from the others. The lead silverback slows his pace to that of the group's slowest, weakest member. All adults tolerate the babies and youngsters in the group, often with great tenderness. A wide-eyed baby, its fur still curly as black wool, may crawl over the great black bulk of any adult with impunity; a toddler may even step on the flat, leathery nose of a silverback. Usually the powerful male will gently set the baby aside or even dangle it playfully from one of his immense fingers.

When Dian first discovered Group 4, she would watch them through binoculars from a hidden position, for if they saw her they would flee. She loved to observe the group's three infants toddle and tumble together. If one baby found the play too rough, it would make a coughing sound, and its mother would lumber over and cradle it tenderly to her breast. Dian watched Digit and his juvenile sisters play:

wrestling, rolling, and chasing games often took them as far as fifty feet from the hulking adults. Sometimes a silverback led the youngsters in a sort of square dance. Loping from one palmlike *Senecio* tree to another, each gorilla would grab a trunk for a twirl, then spin off to embrace another trunk down the slope, until all the gorillas lay in a bouncing pileup of furry black bodies. And then the silverback would lead the youngsters up the slope again for another game.

Within a year, this cheerful silverback eventually took over leadership of the group, after the old leader died. Dian named him Uncle Bert, after her uncle Albert Chapin. With Dian's maternal aunt, Flossie (Dian named a Group 4 female after her as well), Uncle Bert had helped care for Dian after her father left the family when she was three. While Dian was in college Bert and Flossie gave her money to help with costs that her holiday, weekend, and summer jobs wouldn't cover. Naming the silverback after her uncle was the most tender tribute Dian could have offered Bert Chapin: his was the name given to the group's male magnet, its leader, protector— and the centerpiece of a family life whose tenderness and cohesion Dian, as a child, could not have imagined.

Dian was a lonely only child. Her father's drinking caused the divorce that took him out of her life; when her mother, Kitty, remarried when Dian was five, even the mention of George Fossey's name became taboo in the house. Richard Price never adopted Dian. Each night she ate supper in the kitchen with the housekeeper. Her stepfather did not allow her at the dinner table with him and her mother until she was ten. Though Dian's stepfather, a building contractor, seemed wealthy, she largely paid her own way through school. Once she worked as a machine operator in a factory.

Dian seldom spoke of her family to friends, and she carried a loathing for her childhood into her adult life. Long after Dian left the family home in California, she referred to her parents as "the Prices." She would spit on the ground whenever her stepfather's name was mentioned. When her Uncle Bert died, leaving Dian $50,000, Richard Price badgered her with cables to Rwanda, pressing her to contest the will for more money; after Dian's death he had her will overturned by a California court, claiming all her money for himself and his wife.

Her mother and stepfather tried desperately to thwart Dian's plans to go to Africa. They would not help her finance her lifelong dream to go on safari when she was twenty-eight. She borrowed against three years of her salary as an occupational therapist to go. And when she left the States three years later to begin her study of the mountain gorillas, her mother begged her not to go, and her stepfather threatened to stop her.

She chose to remain in the alpine rain forest, as alone as she had ever been. She chose to remain among the King Kong beasts whom the outside world still considered a symbol of savagery, watching their gentle, peaceful lives unfold.

Once Dian, watching Uncle Bert with his family, saw the gigantic male pluck a handful of white flowers with his huge black hands. As the young Digit ambled toward him, the silverback whisked the bouquet back and forth across the youngster's face. Digit chuckled and tumbled into Uncle Bert's lap, "much like a puppy wanting attention," Dian wrote. Digit rolled against the silverback, clutching himself in ecstasy as the big male tickled him with petals.

By the end of her first three months in Rwanda, Dian was following two gorilla groups regularly and observing another sporadically. She divided most of her time between Group 5's fifteen members, ranging on Visoke's southeastern slopes, and Group 4. Group 8, a family of nine, all adult, shared Visoke's western slopes with Group 4.

Dian still could not approach them. Gorilla families guard carefully against intrusion. Each family has at least one member who serves as sentry, typically posted at the periphery of the group to watch for danger—a rival silverback or

a human hunter. Gorilla groups seldom interact with other families, except when females transfer voluntarily out of their natal group to join the families of unrelated silverbacks or when rival silverbacks "raid" a neighboring family for females.

Adult gorillas will fight to the death defending their families. This is why poachers who may be seeking only one infant for the zoo trade must often kill all the adults in the family to capture the baby. Once Dian tracked one such poacher to his village; the man and his wives fled before her, leaving their small child behind.

At first Dian observed the animals from a distance, silently, hidden. Then slowly, over many months, she began to announce her presence. She imitated their contentment vocalizations, most often the *naoom, naoom, naoom,* a sound like belching or deeply clearing the throat. She crunched wild celery stalks. She crouched, eyes averted, scratching herself loud and long, as gorillas do. Eventually she could come close enough to them to smell the scent of their bodies and see the ridges inside the roofs of their mouths when they yawned; at times she came close enough to distinguish, without binoculars, the cuticles of their black, humanlike fingernails.

She visited them daily; she learned to tell by the contour pressed into the leaves which animal had slept in a particular night nest, made from leaves woven into a bathtub shape on the ground. She knew the sound of each individual voice belching contentment when they were feeding. But it was more than two years before she knew the touch of their skin.

Peanuts, a young adult male in Group 8, was the first mountain gorilla to touch his fingers to hers. Dian was lying on her back among the foliage, her right arm outstretched, palm up. Peanuts looked at her hand intently; then he stood, extended his hand, and touched her fingers for an instant. *National Geographic* photographer Bob Campbell snapped the shutter only a moment afterward: that the photo is blurry renders it dreamlike. The 250-pound gorilla's right hand still hangs in

midair. Dian's eyes are open but unseeing, her lips parted, her left hand brought to her mouth, as if feeling for the lingering warmth of a kiss.

Peanuts pounded his chest with excitement and ran off to rejoin his group. Dian lingered after he left; she named the spot where they touched Fasi Ya Mkoni, "the Place of the Hands." With his touch, Peanuts opened his family to her; she became a part of the families she had observed so intimately for the past two years. Soon the gorillas would come forward and welcome her into their midst.

Digit was almost always the first member of Group 4 to greet her. "I received the impression that Digit really looked forward to the daily contacts," she wrote in her book. "If I was alone, he often invited play by flopping over on his back, waving stumpy legs in the air, and looking at me smilingly as if to say, 'How can you resist me?' "

At times she would be literally blanketed with gorillas, when a family would pull close around her like a black furry quilt. In one wonderful photo, Puck, a young female of Group 5, is reclining in back of Dian and, with the back of her left hand, touching Dian's cheek—the gesture of a mother caressing the cheek of a child.

Mothers let Dian hold their infants; silverbacks would groom her, parting her long dark hair with fingers thick as bananas, yet deft as a seamstress's touch. "I can't tell you how rewarding it is to be with them," Dian told a New York crowd gathered for a slide lecture in 1982. "Their trust, the cohesiveness, the tranquility . . ." Words failed her, and her hoarse, breathy voice broke. "It is really something."

Other field workers who joined Dian at Karisoke remember similar moments. Photographer Bob Campbell recalls how Digit would try to groom his sleeves and pants and, finding nothing groomable, would pluck at the hairs of his wrist; most of the people who worked there have pictures of themselves with young gorillas on their heads or in their laps.

But with Dian it was different. Ian Redmond, who first came to Karisoke in 1976, remembers one of the first times he accompanied Dian to observe Group 4. It was a reunion: Dian hadn't been out to visit the group for a while. "The animals filed past us, and each one paused and briefly looked into my face, just briefly. And then each one looked into Dian's eyes, at very close quarters, for half a minute or so. It seemed like each one was queuing up to stare into her face and remind themselves of her place with them. It was obvious they had a much deeper and stronger relationship with Dian than with any of the other workers."

In the early days Dian had the gorillas mostly to herself. It was in 1972 that Bob Campbell filmed what is arguably one of the most moving contacts between two species on record: Digit, though still a youngster, is huge. His head is more than twice the size of Dian's, his hands big enough to cover a dinner plate. He comes to her and with those enormous black hands gently takes her notebook, then her pen, and brings them to his flat, leathery nose. He gently puts them aside in the foliage and rolls over to snooze at Dian's side.

Once Dian spotted Group 4 on the opposite side of a steep ravine but knew she was not strong enough to cross it. Uncle Bert, seeing her, led the entire group across the ravine to her. This time Digit was last in line. "Then," wrote Dian, "he finally came right to me and gently touched my hair. . . . I wish I could have given them all something in return."

At times like these, Dian wept with joy. Hers was the triumph of one who has been chosen: wild gorillas would come to her.

The great intimacy of love is onlyness, of being the loved One. It is the kind of love most valued in Western culture; people choose only one "best" friend, one husband, one wife, one God. Even our God is a jealous one, demanding "Thou shalt have no other gods before Me."

This was a love Dian sought over and over again—as the only child of parents who did not place her first, as the paramour of a succession of married lovers. The love she sought most

desperately was a jealous love, exclusive—not *agape,* the Godlike, spiritual love of all beings, not the uniform, brotherly love, *philia.* The love Dian sought was the love that singles out.

Digit singled Dian out. By the time he was nine he was more strongly attracted to her than were any of the other gorillas she knew. His only age-mates in his family, his half sisters, had left the group or been kidnaped. When Digit heard Dian belch-grunting a greeting, he would leave the company of his group to scamper to greet her. To Digit, Dian was the sibling playmate he lacked. And Dian recognized his longing as clearly as she knew her own image in a mirror.

Dian had had few playmates as a child. She had longed for a pet, but her stepfather wouldn't allow her to keep even a hamster a friend offered, because it was "dirty." He allowed her a single goldfish; she was devastated when it died and was never allowed another.

But Digit was no pet. "Dian's relationship with the gorillas is really the highest form of human-animal relationship," observed Ian Redmond. "With almost any other human-animal relationship, that involves feeding the animals or restraining the animals or putting them in an enclosure, or if you help an injured animal—you do something to the animal. Whereas Dian and the gorillas were on completely equal terms. It was nothing other than the desire to be together. And that's as pure as you can get."

When Digit was young, he and Dian played together like children. He would strut toward her, playfully whacking foliage; she would tickle him; he would chuckle and climb on her head. Digit was fascinated by any object Dian had with her: once she brought a chocolate bar to eat for lunch and accidentally dropped it into the hollow stump of a tree where she was sitting next to Digit. Half in jest, she asked him to get it back for her. "And according to script," she wrote her Louisville friend, Betty Schartzel, "Digit reached one long, hairy arm into the hole and retrieved the candy bar." But the chocolate didn't appeal to

him. "After one sniff he literally threw it back into the hole. The so-called 'wild gorillas' are really very discriminating in their tastes!"

Dian's thermoses, notebooks, gloves, and cameras were all worthy of investigation. Digit would handle these objects gently and with great concentration. Sometimes he handed them back to her. Once Dian brought Digit a hand mirror. He immediately approached it, propped up on his forearms, and sniffed the glass. Digit pursed his lips, cocked his head, and then uttered a long sigh. He reached behind the mirror in search of the body connected to the face. Finding nothing, he stared at his reflection for five minutes before moving away.

Dian took many photos of all the gorillas, but Digit was her favorite subject. When the Rwandan Office of Tourism asked Dian for a gorilla photo for a travel poster, the slide she selected was one of Digit. He is pictured holding a stick of wood he has been chewing, his shining eyes a mixture of innocence and inquiry. He looks directly into the camera, his lips parted and curved as if about to smile. "Come to meet him in Rwanda," exhorts the caption. When his poster began appearing in hotels, banks, and airports, "I could not help feeling that our privacy was on the verge of being invaded," Dian wrote.

Her relationship with Digit was one she did not intend to share. Hers was the loyalty and possessiveness of a silverback: what she felt for the gorillas, and especially Digit, was exclusive, passionate, and dangerous.

No animal, Dian believed, was truly safe in Africa. Africans see most animals as food, skins, money. "Dian had a compulsion to buy every animal she ever saw in Africa," remembers her friend Rosamond Carr, an American expatriate who lives in nearby Gisenyi, "to save it from torture." One day Dian, driving in her Combi van, saw some children on the roadside, swinging a rabbit by the ears. She took it from them, brought it back to camp, and built a spacious hutch for it. Another time it would be a chicken: visiting villagers sometimes brought one to camp, intending, of course, that it be eaten. Dian would keep it as a pet.

Dian felt compelled to protect the vulnerable, the innocent. Her first plan after high school had been to become a veterinarian; after failing chemistry and physics, she chose occupational therapy; with her degree, she worked for a decade with disabled children.

One day Dian came to the hotel in Gisenyi where Rosamond was working as a manager. Dian was holding a monkey. She had seen it at a market, packed in a carton. Rosamond remembers, "I look and see this rotten little face, this big ruff of hair, and I say, 'Dian. I'm sorry, you cannot have that monkey in this hotel!' But Dian spent the night with the animal in her room anyway.

"Luckily for me, she left the next day. I have never seen anything like the mess. There were banana peels on the ceiling, sweet potatoes on the floor; it had broken the water bottle, the glasses had been smashed and had gone down the drain of the washbasin. And with that adorable animal she starts up the mountain."

Kima, as Dian named the monkey, proved no less destructive in camp. Full grown when Dian brought her, Kima bit people, urinated on Dian's typewriter, bit the heads off all her matches, and terrorized students on their way to the latrine, leaping off the roof of Dian's cabin and biting them. Yet Dian loved her, built a hatchway allowing Kima free access to her cabin, bought her toys and dolls, and had her camp cook prepare special foods for her. Kima especially liked french fries, though she discarded the crunchy outsides and ate only the soft centers. "Everyone in camp absolutely hated that animal," Rosamond says. "But Dian loved her."

Another of Dian's rescue attempts occurred one day when she was driving down the main street of Gisenyi on a provisioning trip. Spotting a man walking a rack-ribbed dog on a leash, she slammed on the brakes. "I want to buy that dog," she announced. The Rwandan protested that it was not for sale. She got out of her Combi, lifted up the sickly animal, and drove off with it.

Rosamond learned of the incident from a friend named Rita who worked at the American embassy. For it was to Rita's home that the Rwandan man returned that afternoon to explain why the dog, which he had been taking to the vet for worming, had never made it to its destination. "Madame, a crazy woman stopped and stole your dog, and she went off with it in a gray van."

"Rita got her dog back," Rosamond continues. Dian had taken it to the hotel where she was staying overnight; when Rita tracked her down, she was feeding the dog steak in her room. "And that was typical of Dian. She had to save every animal she saw. And they loved her—every animal I ever saw her with simply loved her."

When Dian first came to Karisoke, elephants frequently visited her camp. Rosamond used to camp with Dian in those early days before the cabin was built. The elephants came so close that she remembers hearing their stomachs rumble at night. Once she asked Dian if she undressed at night. "Of course not, are you crazy?" Dian replied. "I go to bed in my blue jeans. I have to get up six times at night to see what's happening outside."

One night an elephant selected Dian's tent pole as a scratching post. Another time a wild elephant accepted a banana from Dian's hand. The tiny antelopes called duiker often wandered through camp; one became so tame it would follow Dian's laying hens around. A family of seven bushbucks adopted Karisoke as home, as did an ancient bull buffalo she named Mzee.

Dian's camp provided refuge from the poacher-infested, cattle-filled forest. For centuries the pygmylike Batwa had used these volcanic slopes as a hunting ground. And as Rwanda's human population exploded, the Virungas were the only source of bush meat left, and poaching pressure increased. Today you will find no elephants in these forests; they have all been killed by poachers seeking ivory.

If you look at the Parc National des Volcans from the air, the five volca-noes, their uppermost slopes puckered like the lips of an old woman, seem to be standing on tiptoe to withdraw from the flood of cultivation and people below. Rwanda is the most densely populated country in Africa, with more than 500 people per square mile. Almost every inch is cultivated, and more than 23,000 new families need new land each year. In rural Rwanda outside of the national parks, if you wander from a path you are more likely to step in human excrement than the scat of a wild animal. The Parc des Volcans is thoroughly ringed with *shambas,* little farm plots growing bananas, peanuts, beans, manioc, and with fields of pyrethrum, daisylike flowers cultivated as a natural insecticide for export. The red earth of the fields seems to bleed from all the human scraping.

The proud, tall Batutsi have few other areas to pasture their cattle, the pride of their existence; everywhere else are shambas. From the start Dian tried to evict the herders from the park, kidnaping their cows and sometimes even shooting them. On the Rwandan side of the mountain, cattle herds were so concentrated, she wrote, that "many areas were reduced to dustbowls." She felt guilty, but the cattle destroyed habitat for the gorillas and other wild animals the park was supposed to protect. Worse were the snares set by the Batwa. Many nights she stayed awake nursing a duiker or bushbuck whose leg had been mangled in a trap. Dian lived in fear that one of the gorillas would be next.

The Batwa do not eat gorillas; gorillas fall victim to their snares, set for antelope, by accident. But the Batwa have for centuries hunted gorillas, to use the fingers and genitals of silverbacks in magic rituals and potions. And now the hunters found a new reason to kill gorillas: they learned that Westerners would pay high prices for gorilla heads for trophies, gorilla hands for ashtrays, and gorilla youngsters for zoos.

In March 1969, only eighteen months into her study, a friend in Ruhengeri came to camp to tell Dian that a young gorilla had been captured from the southern slopes of Mount Karisimbi. All ten adults in the group had been killed so that the baby could be taken for display in the Cologne Zoo. The capture had been approved by the park conservator, who was paid handsomely for his cooperation. But something had gone wrong: the baby gorilla was dying.

Dian took the baby in, a three- to four-year-old female she named Coco. The gorilla's wrists and feet had been bound with wire to a pole when the hunters carried her away from the corpses of her family; she had spent two or three weeks in a coffinlike crate, fed only corn, bananas, and bread, before Dian came to her rescue. When Dian left the park conservator's office, she was sure the baby would die. She slept with Coco in her bed, awakening amid pools of the baby's watery feces.

A week later came another sick orphan, a four- to five-year-old female also intended for the zoo. Her family had shared Karisimbi's southern slope with Coco's; trying to defend the baby from capture, all eight members of her group had died. Dian named this baby Pucker for the huge sores that gave her face a puckered look.

It took Dian two months to nurse the babies back to health. She transformed half her cabin into a giant gorilla playpen filled with fresh foliage. She began to take them into the forest with her, encouraging them to climb trees and vines. She was making plans to release them into a wild group when the park conservator made the climb to camp. He and his porters descended with both gorillas in a box and shipped them to the zoo in West Germany. Coco and Pucker died there nine years later, within a month of one another, at an age when, in the wild, they would have been mothering youngsters the same age they had been when they were captured.

Thereafter Dian's antipoaching tactics became more elaborate. She learned from a friend in Ruhengeri that the trade in gorilla trophies was flourishing; he had counted twenty-three gorilla heads for sale in that town in one year. As loyal as a silverback, as wary as a sentry, Dian and her staff patrolled the forest for snares and destroyed the gear poachers left behind in their temporary shelters.

Yet each day dawned to the barking of poachers' dogs. A field report she submitted to the National Geographic Society in 1972 gave the results of the most recent gorilla census: though her study groups were still safe, the surrounding areas of the park's five volcanoes were literally under siege. On Mount Muhavura census workers saw convoys of smugglers leaving the park every forty-five minutes. Only thirteen gorillas were left on the slopes of Muhavura. On neighboring Mount Gahinga no gorillas were left. In the two previous years, census workers had found fresh remains of slain silverbacks. And even the slopes of Dian's beloved Karisimbi, she wrote, were covered with poachers' traps and scarred by heavily used cattle trails; "poachers and their dogs were heard throughout the region."

It was that same year, 1972, that a maturing Digit assumed the role of sentry of Group 4. In this role he usually stayed on the periphery of the group to watch for danger; he would be the first to defend his family if they were attacked. Once when Dian was walking behind her Rwandan tracker, the dark form of a gorilla burst from the bush. The male stood upright to his full height of five and a half feet; his jaw gaped open, exposing black gums and three-inch canines as he uttered two long, piercing screams at the terrified tracker. Dian stepped into view, shoving the tracker down behind her, and stared into the animal's face. They recognized each other immediately. Digit dropped to all fours and ran back to his group.

Dian wrote that Digit's new role made him more serious. No longer was he a youngster with the freedom to roll and wrestle with his playmate. But Dian was still special to him. Once when Dian went out to visit the group during a downpour, the young silverback emerged from the gloom and stood erect before his crouching human friend. He pulled up a stalk of wild celery—a favorite gorilla food that Digit had seen Dian munch on many times—peeled it with his great hands, and dropped the stalk at her feet like an offering. Then he turned and left.

As sentry, Digit sustained the wound that sapped his strength for the next two years. Dian did not observe the fight, but she concluded from tracking clues that Digit had warded off a raid by the silverback leader of Group 8, who had previously kidnaped females from Group 4. Dian cringed each time she heard him coughing and retching. Digit sat alone, hunched and indifferent. Dian worried that his growth would be retarded. In her field notes she described his mood as one of deep dejection.

This was a time when Dian was nursing wounds of her own. She had hoped that Bob Campbell, the photographer, would marry her, as Hugo van Lawick had married Jane; but Bob left Karisoke for the last time at the end of May 1972, to return to his wife in Nairobi. Then she had a long affair with a Belgian doctor, who left her to marry the woman he had been living with. Dian's health worsened. Her trips overseas for primatology conferences and lecture tours were usually paired with hospital visits to repair broken bones and heal her fragile lungs. She feared she had tuberculosis. She noted her pain in her diary telegraphically: "Very lung-sick." "Coughing up blood." "Scum in urine."

When she was in her twenties, despite her asthma, Dian seemed as strong as an Amazon. Her large-boned but lanky six-foot frame had a coltish grace; one of her suitors, another man who nearly married her, described her as "one hell of an attractive woman," with masses of long dark hair and "eyes like a Spanish dancer." But now Dian felt old and ugly and weak. She used henna on her hair to try to cover the gray. (Dian told a friend that her mother's only comment about her first appearance on a National Geographic TV special was, "Why did you dye your hair that awful orange color?") In letters to friends Dian began to sign off as "The Fossil." She referred to her house as "the Mausoleum." In a cardboard album she made for friends from construction paper and magazine cutouts, titled "The Saga of Karisoke," she pasted a picture of a mummified corpse sitting upright on a bed. She realized that many of her students disliked her. Under the picture Dian printed a caption: "Despite their protests, she stays on."

By 1976 Dian was spending less and less time in the field. Her lungs and legs had grown too weak for daily contacts; she had hairline fractures on her feet. And she was overwhelmed with paperwork. She became increasingly testy with her staff, and her students feared to knock on her door. Her students wouldn't even see her for weeks at a time, but they would hear her pounding on her battered Olivetti, a task from which she would pause only to take another drag on an Impala *filtrée* or to munch sunflower seeds. Her students were taking the field data on the gorillas by this time: when Dian went out to see the groups, she simply visited with them.

One day, she ventured out along a trail as slippery as fresh buffalo dung to find Group 4. By the time she found

them, the rain was driving. They were huddled against the downpour. She saw Digit sitting about thirty feet apart from the group. She wanted to join him but resisted; she now feared that her early contact with him had made him too human-oriented, more vulnerable to poachers. So she settled among the soaking foliage several yards from the main group. She could barely make out the humped black forms in the heavy mist.

On sunny days there is no more beautiful place on earth than the Virungas; the sunlight makes the *Senecio* trees sparkle like fireworks in midexplosion; the gnarled old *Hagenias*, trailing lacy beards of gray-green lichen and epiphytic ferns, look like friendly wizards, and the leaves of palms seem like hands upraised in praise. But rain transforms the forest into a cold, gray hell. You stare out, tunnel-visioned, from the hood of a dripping raincape, at a wet landscape cloaked as if in evil enchantment. Each drop of rain sends a splintering chill into the flesh, and your muscles clench with cold; you can cut yourself badly on the razorlike cutty grass and not

even feel it. Even the gorillas, with their thick black fur coats, look miserable and lonely in the rain.

Minutes after she arrived, Dian felt an arm around her shoulders. "I looked up into Digit's warm, gentle brown eyes," she wrote in *Gorillas in the Mist*. He gazed at her thoughtfully and patted her head, then sat by her side. As the rain faded to mist, she laid her head down in Digit's lap.

On January 1, 1978, Dian's head tracker returned to camp late in the day. He had not been able to find Group 4. But he had found blood along their trail.

Ian Redmond found Digit's body the next day. His head and hands had been hacked off. There were five spear wounds in his body.

Ian did not see Dian cry that day. She was almost supercontrolled, he remembers. No amount of keening, no incantation or prayer could release the pain of her loss. But years later she filled a page of her diary with a single word, written over and over: "Digit Digit Digit Digit . . ."

Sex and Society

Any account of hominid evolution would be remiss if it did not at least attempt to explain that most mystifying of all human experiences: our sexuality.

No other aspect of our humanity, whether it be upright posture, tool-making ability, or intelligence in general, seems to elude our intellectual grasp at least as much as it dominates our subjective consciousness. While we are a long way from reaching a consensus as to why it arose and what it is all about, there is widespread agreement that our very preoccupation with sex is in itself one of the hallmarks of being human. Even as we experience it and analyze it, we exalt it and condemn it. Beyond seemingly irrational fixations, however, there is the further tendency to project our own values upon the observations we make and the data we collect.

There are many who argue quite reasonably that the human bias has been more male- than female-oriented and that the recent "feminization" of anthropology has resulted in new kinds of research and refreshingly new theoretical perspectives. (See the articles "Sex and the Female Agenda" by Jared Diamond and "Why Women Change," also by Diamond.) Not only should we consider the source when evaluating the old theories, so goes the reasoning, but we should also welcome the source when considering the new. To take one example, traditional theory would have predicted that the reproductive competitiveness of muriqui monkeys, as described in "These Are Real Swinging Primates" by Shannon Brownlee, would be associated with greater size and aggression among males. That this is not so, that making love can be more important than making war, and that females do not necessarily

have to live in fear of competitive males, just goes to show that, even among monkeys, nothing can be taken for granted. The very idea that females are helpless in the face of male aggression is called into question by Barbara Smuts in "Apes of Wrath."

Finally, there is the question of the social significance of sexuality in humans. In "What's Love Got to Do with It?" Meredith Small shows that the chimplike bonobos of Zaire use sex to reduce tensions and cement social relations and, in so doing, have achieved a high degree of equality between the sexes. Whether or not we see parallels in the human species, says Small, depends on our willingness to interpret bonobo behavior as a "modern version of our own ancestors' sex play," and this, in turn, may depend on our prior theoretical commitments.

Looking Ahead: Challenge Questions

How can the muriqui monkeys be sexually competitive and yet gregarious and cooperative?

How does human sexuality differ from that of other creatures?

Under what circumstances do mammalian females limit their own reproductivity?

What implications does bonobo sexual behavior have for understanding human evolution?

Why do human females experience menopause?

How and why do the reproductive strategies of male and female primates differ?

How do social bonds provide females with protection against abusive males?

These Are Real Swinging Primates

There's a good evolutionary reason why the rare muriqui of Brazil should heed the dictum 'Make love, not war'

Shannon Brownlee

When I first heard of the muriqui four years ago, I knew right away that I had to see one. This is an unusual monkey, to say the least. To begin with, it's the largest primate in South America; beyond that, the males have very large testicles. We're talking gigantic, the size of billiard balls, which means that the 30-pound muriqui has *cojones* that would look more fitting on a 400-pound gorilla.

But it wasn't prurience that lured me to Brazil. My interest in the muriqui was intellectual, because more than this monkey's anatomy is extraordinary. Muriqui society is untroubled by conflict: troops have no obvious pecking order; males don't compete overtly for females; and, most un-monkeylike, these monkeys almost never fight.

The muriqui is also one of the rarest monkeys in the world. It lives in a single habitat, the Atlantic forest of southeastern Brazil. This mountainous region was once blanketed with forest from São Paulo to Salvador (*see map*), but several centuries of slash-and-burn agriculture have reduced it to fragments.

In 1969 Brazilian conservationist Alvaro Coutinho Aguirre surveyed the remaining pockets of forest and estimated that 2,000 to 3,000 muriquis survived. His data were all but ignored until Russell Mittermeier, a biologist, trained his sights on the muriquis ten years later. Known as Russel of the Apes to his colleagues, Mittermeier, an American, directs the primate program for the World Wildlife Fund. He hopscotches from forest to forest around the world looking for monkeys in trouble and setting up conservation plans for them. In 1979 he and Brazilian zoologist Celio Valle retraced Aguirre's steps and found even fewer muriquis. Today only 350 to 500 are left, scattered among four state and national parks and six other privately held plots.

In 1981 Karen Strier, then a graduate student at Harvard, approached Mittermeier for help in getting permission to observe the muriqui. He took her to a coffee plantation called Montes Claros, near the town of Caratinga, 250 miles north of Rio de Janeiro. Over the next four years she studied the social behavior of the muriqui there—and came up with a provocative theory about how the monkey's unconventional behavior, as well as its colossal testicles, evolved. She reasoned that the evolution of both could be explained, at least in part, by the muriquis' need to avoid falling out of trees.

Last June I joined Strier, now a professor at Beloit (Wis.) College, on one of her periodic journeys to Montes Claros—clear mountains, in Portuguese. We arrived there after a disagreeable overnight bus trip over bad roads. As we neared the plantation, I found it difficult to believe there was a forest—much less a monkey—within miles. Through the grimy windows of the bus I saw hillsides stripped down to russet dirt and dotted with spindly coffee plants and stucco farmhouses. There wasn't anything taller than a banana tree in sight. As the bus lurched around the last curve before our stop the forest finally appeared, an island of green amid thousands of acres of coffee trees and brown pastures.

Strier was eager to start looking for the muriquis—"There's a chance we won't see them the whole four days you're here," she said—so no sooner had we dropped our bags off at a cottage on the plantation than we set out along a dirt road into the forest. The trees closed around us—and above us, where they gracefully arched to form a vault of green filigree. Parrots screeched; leaves rustled; a large butterfly flew erratically by on transparent wings. By this time Strier had guided me onto a steep trail, along which she stopped from time to time to listen for the monkeys.

They appeared soon enough, but our first meeting was less than felicitous. After we had climbed half a mile, Strier motioned for me to stop. A muffled sound, like that of a small pig grunting contentedly, came from up ahead. We moved forward a hundred yards. Putting a finger to her lips, Strier sank to her haunches and looked up.

I did the same; twelve round black eyes stared back at me. A group of six muriquis squatted, silent, 15 feet above in the branches, watching us intently.

They began to grunt again. A sharp smell with undertones of cinnamon permeated the air. A light rain began to fall. I held out my palm to catch a drop. It was warm.

"Hey, this isn't rain!" I said.

Strier grinned and pointed to her head. "That's why I wear a hat," she said.

My enthusiasm for the muriquis waned slightly after that. We left them at dusk and retired to the cottage, where Strier described her arrival at Montes Claros four years earlier. Mittermeier acted as guide and interpreter during the first few days of her pilot study. He introduced her to the owner of the 5,000-acre plantation, Feliciano Miguel Abdala, then 73, who had preserved the 2,000-acre forest for more than 40 years. His is one of the only remaining tracts of Atlantic forest, and he agreed to let Strier use it as the site of her study. Then Mittermeier introduced her to the muriquis, assuring her they would be easy to see.

They weren't, and observing them closely is a little like stargazing on a rainy night: not only do you run the risk of getting wet, but you can also spend a lot of time looking up and never see a thing. Mittermeier was adept at spotting the monkeys in the forest, and helped Strier acquire this skill.

But brief glimpses of the monkeys weren't enough. "My strategy was to treat them like baboons, the only other species I'd ever studied," she says. "I thought I couldn't let them out of my sight." She tried to follow on the ground as they swung along in the trees. "They went berserk," she says. They threw branches, shrieked, urinated on her—or worse—and fled.

Even after the muriquis grew accustomed to her, keeping up with them wasn't easy. They travel as much as two miles a day, which is tough for someone picking her way through thick growth on the forest floor. As Strier and a Brazilian assistant learned the muriquis' habitual routes and daily patterns, they cleared trails. These helped, but the muriquis could still travel much faster than she could. "I've often thought the thing to have would be a jet

pack," Strier says. "It would revolutionize primatology. Your National Science Foundation grant would include binoculars, pencils, and a jet pack."

Observing muriquis is like stargazing on a rainy night. You may get wet, and you can spend hours looking and seeing nothing.

The monkeys move by brachiating, swinging hand over hand from branch to branch, much like a child on a jungle gym. Only one other group of monkeys brachiates; the rest clamber along branches on all fours. The muriquis' closest relatives are two other Latin American genera, the woolly monkeys and the spider monkeys— hence woolly spider monkey, its English name. But the muriqui is so unlike them that it has its own genus, *Brachyteles*, which refers to its diminutive thumb, an adaptation for swinging through the trees. Its species name is *arachnoides*, from the Greek for spider, which the muriqui resembles when its long arms, legs, and tail are outstretched.

Brachiating is a specialization that's thought to have evolved because it enables primates to range widely to feed on fruit. Curiously, though, muriquis have a stomach designed for digesting leaves. Strier found that their diet consists of a combination of the two foods. They eat mostly foliage, low-quality food for a monkey, but prefer flowers and fruits, like figs and the *caja manga*, which is similar to the mango. Year after year they return to certain trees when they bloom and bear fruit. The rest of the time the muriquis survive on leaves by passing huge quantities of them through their elongated guts, which contain special bacteria to help them digest the foliage. By the end of the day their bellies are so distended with greenery that even the males look pregnant.

We returned to the trail the next morning just after dawn. Condensation

trickled from leaves; howler monkeys roared and capuchins cooed and squeaked; a bird sang with the sweet, piercing voice of a piccolo. Then Strier had to mention snakes. "Watch out for snakes," she said blithely, scrambling on all fours up a steep bank. I followed her, treading cautiously.

The muriquis weren't where we had left them the day before. Strier led me along a ridge through a stand of bamboo, where a whisper of movement drifted up from the slope below. Maybe it was just the wind, but she thought it was the muriquis, so we sat down to wait. After a couple of hours, she confessed, "This part of research can get kind of boring."

By noon the faint noise became a distinct crashing. "That's definitely them," she said. "It's a good thing they're so noisy, or I'd never be able to find them." The monkeys, perhaps a dozen of them, swarmed uphill, breaking branches, chattering, uttering their porcine grunts as they swung along. At the crest of the ridge they paused, teetering in indecision while they peered back and forth before settling in some legume trees on the ridgetop. We crept down out of the bamboo to within a few feet of them, so close I noticed the cinnamon scent again—only this time I kept out of range.

Each monkey had its own feeding style. One hung upside down by its tail and drew the tip of a branch to its mouth; it delicately plucked the tenderest shoots with its rubbery lips. Another sat upright, grabbing leaves by the handful and stuffing its face. A female with twins—"Twins have never been seen in this species," Strier whispered as she excitedly scribbled notes—ate with one hand while hanging by the other and her tail. Her babies clung to the fur on her belly.

I had no trouble spotting the males. Their nether parts bulged unmistakably—blue-black or pink-freckled, absurd-looking monuments to monkey virility. I asked Strier what sort of obscene joke evolution was playing on the muriquis when it endowed them thus.

We were about to consider this question when a high-pitched whinnying began a few hundred yards away. Immediately a monkey just overhead pulled itself erect and let out an ear-splitting shriek, which set the entire troop to neighing like a herd of nervous horses. Then they took off down into the valley.

Strier and I had to plunge pell-mell into the underbrush or risk losing them for the rest of the day. "They're chasing the other troop," she said as we galloped downhill. A group of muriquis living on the opposite side of the forest had made a rare foray across the valley.

The monkeys we were observing swung effortlessly from tree to tree; we wrestled with thorny vines, and fell farther and farther behind. An impenetrable thicket forced us to backtrack in search of another route. By the time we caught up to the muriquis, they were lounging in a tree, chewing on unripe fruit and chuckling in a self-satisfied sort of way. The intruding troop was nowhere to be seen. "They must have scared the hell out of those other guys," said Strier, laughing.

Tolerance of another troop is odd behavior for monkeys, but not so odd as the fact that they never fight among themselves.

Such confrontations occur infrequently; muriquis ordinarily tolerate another troop's incursions. Strier thinks they challenge intruders only when there's a valuable resource to defend—like the fruit tree they were sitting in.

Tolerance of another troop is odd behavior for monkeys, but not as odd as the fact that members of a muriqui troop never fight among themselves. "They're remarkably placid," said Strier. "They wait in line to dip their hands into water collected in the bole of a tree. They have no apparent pecking order or dominance hierarchy. Males and females are equal in status, and males don't squabble over fe-

males." No other primate society is known to be so free of competition, not even that of gorillas, which have lately gained a reputation for being the gentle giants of the primate world.

Strier's portrayal of the muriqui brought to mind a bizarre episode that Katharine Milton, an anthropologist at the University of California at Berkeley, once described. While studying a troop of muriquis in another patch of the Atlantic forest, she observed a female mating with a half a dozen males in succession; that a female monkey would entertain so many suitors came as no surprise, but Milton was astonished at the sight of the males lining up behind the female "like a choo-choo train" and politely taking turns copulating. They continued in this manner for two days, stopping only to rest and eat, and never even so much as bared their teeth.

Primates aren't known for their graciousness in such matters, and I found Milton's report almost unbelievable. But Strier confirms it. She says that female muriquis come into heat about every two and a half years, after weaning their latest offspring, and repeatedly copulate during that five- to seven-day period with a number of males. Copulations, "cops" in animal-behavior lingo, last as long as 18 minutes, and average six, which for most primates (including the genus *Homo,* if Masters and Johnson are correct) would be a marathon. Yet no matter how long a male muriqui takes, he's never harassed by suitors-in-waiting.

Strier has a theory to explain the muriqui's benignity, based on a paper published in 1980 by Richard Wrangham, a primatologist at the University of Michigan. He proposed that the social behavior of primates could in large part be predicted by what the females eat.

This isn't a completely new idea. For years primatologists sought correlations between ecological conditions and social structure, but few patterns emerged—until Wrangham's ingenious insight that environment constrains the behavior of each sex differently. Specifically, food affects the sociability of females more than males.

Wrangham started with the generally accepted premise that both sexes in every species have a common aim: to leave as many offspring as possible. But each sex pursues this goal in its own way. The best strategy for a male primate is to impregnate as many females as he can. All he needs, as Wrangham points out, is plenty of sperm and plenty of females. As for the female, no matter how promiscuous she is, she can't match a male's fecundity. On average, she's able to give birth to only one offspring every two years, and her success in bearing and rearing it depends in part upon the quality of food she eats. Therefore, all other things being equal, male primates will spend their time cruising for babes, while females will look for something good to eat.

Wrangham's ingenious insight: the social behavior of primates can in large part be predicted by what the females eat.

Wrangham perceived that the distribution of food—that is, whether it's plentiful or scarce, clumped or evenly dispersed—will determine how gregarious the females of a particular species are. He looked at the behavior of 28 species and found that, in general, females forage together when food is plentiful and found in large clumps—conditions under which there's enough for all the members of the group and the clumps can be defended against outsiders. When clumps become temporarily depleted, the females supplement their diet with what Wrangham calls subsistence foods. He suggest that female savanna baboons, for example, live in groups because their favorite foods, fruits and flowers, grow in large clumps that are easy to defend. When these are exhausted they switch to seeds, insects, and grasses. The females form long-lasting relationships within their groups, and establish stable dominance hierarchies.

Chimpanzees provide an illustration of how females behave when their food isn't in clumps big enough to feed everybody. Female chimps eat flowers, shoots, leaves, and insects, but their diet is composed largely of fruits that are widely scattered and often not very plentiful. They may occasionally gather at a particularly abundant fruit tree, but when the fruit is gone they disperse to forage individually for other foods. Members of the troop are constantly meeting at fruit trees, splitting up, and gathering again.

These two types of female groups, the "bonded" savanna baboons and "fissioning" chimps, as Wrangham calls them, pose very different mating opportunities for the males of their species. As a consequence, the social behavior of the two species is different. For a male baboon, groups of females represent the perfect opportunity for him to get cops. All he has to do is exclude other males. A baboon troop includes a clan of females accompanied by a number of males, which compete fiercely for access to them. For baboons there are few advantages to fraternal cooperation, and many to competition.

Male chimpanzees fight far less over females than male baboons do, principally because there's little point—the females don't stick together. Instead, the males form strong alliances with their fellows. They roam in gangs looking for females in heat, and patrol their troop's borders against male interlopers.

Wrangham's theory made so much sense, Strier says, that it inspired researchers to go back into the field with a new perspective. She saw the muriqui as an excellent species for evaluating the model, since Wrangham had constructed it before anyone knew the first thing about this monkey. His idea would seem all the more reasonable if it could predict the muriqui's behavior.

It couldn't, at least not entirely. Strier has found that the females fit Wrangham's predictions: they stick together and eat a combination of preferred and subsistence foods, defending the preferred from other troops. But the males don't conform to the theory. "Considering that the females are foraging together, there should be relatively low pressure on the males to cooperate," she says. "It's odd: the males should compete, but they don't."

She thinks that limitations on male competition may explain muriqui behavior. First, the muriquis are too big to fight in trees. "I think these monkeys are at about the limit of size for rapid brachiation," she says. "If they were bigger, they couldn't travel rapidly through the trees. They fall a lot as

it is, and it really shakes them up. I've seen an adult fall about sixty feet, nearly to the ground, before catching hold of a branch. That means that whatever they fight about has got to be worth the risk of falling out of a tree."

Moreover, fighting may require more energy than the muriquis can afford. Milton has estimated the caloric value of the food eaten by a muriqui each day and compared it to the amount of energy she would expect a monkey of that size to need. She concluded that the muriqui had little excess energy to burn on combat.

The restriction that rapid brachiation sets on the muriqui's size discourages competition in more subtle ways, as well. Given that muriquis are polygynous, the male should be bigger than the female, as is almost invariably the case among other polygynous species—but he's not. The link between larger males and polygyny is created by sexual selection, an evolutionary force that Darwin first recognized, and which he distinguished from natural selection by the fact that it acts exclusively on one sex. Sexual selection is responsible for the manes of male lions, for instance, and for the large canines of male baboons.

In a polygynous society, the advantages to being a large male are ob-

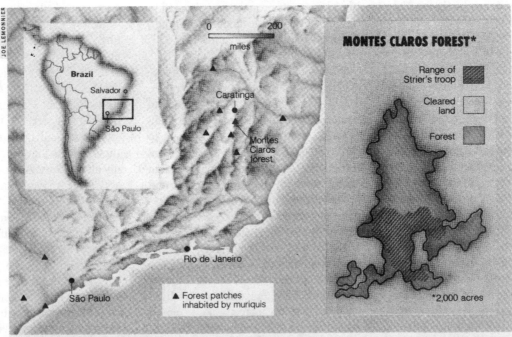

The 350 to 500 surviving muriquis live in ten patches of the Atlantic forest of southeastern Brazil.

vious: he who's biggest is most likely to win the battles over females—and pass on his genes for size. But sexual selection's push toward large males has been thwarted in the muriqui, says Strier. Any competitive benefits greater size might bring a male would be offset in part by the excessive demands on his energy and the costs of falling out of trees.

She believes that the constraints on the males' size have had a profound effect on the muriquis' social behavior. Most important, says Strier, with males and females being the same size, the females can't be dominated, which means they can pick their mates. Most female primates aren't so fortunate: if they copulate with subordinate males, they risk being attacked by dominant ones. But a female muriqui in heat can easily refuse a suitor, simply by sitting down or by moving away.

The size of the males has a profound effect on muriqui behavior. For one thing, they are simply too big to fight in trees.

Fighting not only doesn't help the male muriqui in his quest for cops; it may even harm his chances, since females can shun an aggressive male. Strier believes that females may also be responsible for the male muriquis' canine teeth not being oversized. As a rule, the male's canines are the same size as the female's only in monogamous primate species, but over the generations female muriquis may have mated more readily with males whose teeth were no bigger than their own. In sum, Strier thinks, for a male muriqui the costs of competing are far outweighted by the benefits of avoiding it.

But he has the means to vie for reproductive success and still come across as Mr. Nice Guy: his sperm. Sperm competition, as it's called, is a hot new idea in sociobiology, originally proposed to explain male bonding

in chimpanzees, and, as Milton was the first to suggest, it may explain why the muriqui has such enormous testicles.

The competition is something like a game of chance. Imagine a bucket with a hole in the bottom just big enough for a marble to pass through. People gather round, each with a handful of marbles. They drop their marbles in the bucket, mix them up, and one comes out the bottom. Whoever owns that marble is the winner.

In the sperm competition among male muriquis, the bucket is a female, the marbles are sperm, and winning means becoming a father. No male can be sure it will be his sperm that impregnates a female, since she mates with a number of his fellows. His chances are further complicated by the fact that the female muriqui, like all New World monkeys, gives no visible indication of ovulation; there may be nothing that signals the male (or the female) when during her heat that occurs. So it's to the male's advantage to continue mating as often as the female will have him.

This may sound like monkey heaven, but it puts the male on the horns of a dilemma. If he copulates as often as possible, he could run low on sperm just when the female is ovulating. On the other hand, if he refrains from copulating to save sperm, he may miss his chance at procreating altogether. Selection may have come to his aid, Strier reasons, by acting on his testicles.

Here's a plausible scenario. Suppose a male came along that could produce more sperm than the average muriqui because his testicles were bigger than average. That male would clean up in the reproductive arena. The ratio of testicle size to body weight has been correlated with high sperm count and repeated copulation over a short period in other mammals, and bigger testicles probably also increase the percentage of viable and motile sperm.

If the muriqui's testicles are anything like those of other species, then a male with extra big ones has a slight reproductive advantage. Like a player with more marbles to put in the bucket, a male that can produce more and

better sperm has a better than average chance of impregnating females and passing on this advantageous trait to his sons. Just as important, the outsized organs probably don't cost him much in metabolic energy. Thus, over generations, the muriqui's testicles have grown larger and larger.

Strier's theory has five years of data behind it, and it's the kind of theory that will stimulate researchers to reexamine their ideas about other species. Yet it isn't her only concern; she concentrates equally on the muriqui's uncertain future. On our last day in the forest we watched the monkeys cross a six-foot gap in the canopy 60 feet above us. One by one they stood poised for a moment on the end of a branch before launching themselves. Strier counted them as they appeared in silhouette against a grey sky. The total was 33, including the twins. "They're up from twenty-two in 1982," she said. "That's a very fast increase."

The muriquis at Montes Claros make up almost one-tenth of the total population of the species, and they're critical to its survival—as are all the other isolated and widely separated troops. Each group's genetic pool is limited, and eventually the troops could suffer inbreeding depression, a decline in fecundity that often appears in populations with little genetic variability.

Strier and Mittermeier predict that one day muriquis will have to be managed, the way game species are in the U.S. They may be transported between patches of forest to provide some gene flow. But that's a dangerous proposition now. There are too few muriquis to risk it, and none has ever bred or survived for long in captivity. "Before my study, conservationists would probably have moved males between forests," Strier says. "That would've been a mistake. I have tentative evidence that in a natural situation the females may be the ones that do the transferring between groups."

For now, though, she thinks the biggest concern isn't managing the monkeys but preventing their habitat from disappearing. Preserving what remains of the Atlantic forest won't be easy, and no one knows this better than Feliciano Miguel Abdala, the man responsible for there being any forest at all at Montes Claros.

Abdala has little formal education, but he's rich; he owns nine plantations besides Montes Claros. His family lives in relative splendor in Caratinga, but he likes to spend the weekdays here. His house is just beyond the edge of the forest, and sunlight filters through the bougainvillea vine entwining the front porch. Chickens can be seen through the cracks in the floorboards, scratching in the dirt under the house. Electric cords are strung crazily from the rafters, and a bare bulb dangles in the center of his office. Abdala removes his straw hat decorously and places it on a chair before sitting at his desk.

Abdala bought the 5,000 acres of Montes Claros in 1944. The region was barely settled then, and smoke still rose from the great burning heaps of slash left from clearing the forest. Abdala's land included one of the last stands of trees. I ask him why he saved it. "I am a conservationist," he says. "For a long time the local people thought I was crazy because I wouldn't cut the forest. I told them not to shoot the monkeys, and they stopped. Now all my workers are crazy, too."

I ask Abdala about his plans for his forest. He rubs his head distractedly and says, vaguely. "I hope it will continue."

Abdala believes the government should buy Montes Claros—plantation and rain forest—to create a nature reserve. He'll probably maintain the forest as long as he lives, but the land is quite valuable, and his heirs might not share his lofty sentiments.

As important as the muriquis have become to understanding social systems, and as much as U.S. conservationists may wish to see these monkeys preserved, Strier thinks that in the end it's up to the Brazilians to save them. She's expecting a three-year grant from the National Science Foundation; part of the money will go toward allowing her to observe the monkeys in other forest patches, watching for variation in their behavior as a test of her ideas. Studies like hers will be critical not only for proving theories but also for ensuring that plans for managing the muriquis will work. The rest of the money will permit her to train seven Brazilian graduate students, because she says, "the future of the muriqui lies with the Brazilians."

Natural-born Mothers

Motherhood is not as straightforward a matter as just turning on the milk.

Sarah Blaffer Hrdy

Sarah Blaffer Hrdy is an anthropologist at the University of California-Davis, a middle-aged mother of three, an ardent advocate of breast-feeding, and one of five children. As such, she has had considerable firsthand experience observing maternal fitness trade-offs. One of her various hobbies is allomaternal caretaking (aka, being a mother's helper).

On the day that the eleventh-century Italian saint Peter Damian was born, his mother was ready to call it quits. According to the saint's biography, written by his associate John of Lodi, his mother was "worn out by childbearing" and further disheartened by the reproach of an adolescent son who took her to task for bringing into the world yet another mouth to feed and one more son to add to her already existing "throng of heirs." In despair, the mother refused to nurse. The fledgling saint was on the verge of starvation when a neighbor, the kindly concubine of a local priest, intervened, reminding her that even a savage beast, a tigress or a mother lion, would suckle her own young. Could a Christian woman do less?

Five hundred years later, Italian poet Luigi Tansillo echoed similar sentiments in response to the widespread use of wet nurses—the main alternative to maternal breast-feeding in the days before the baby bottle. In a poem entitled "La Balia" (The Nurse), Tansillo wrote, "What fury, hostile to our common kind,/First led from nature's path the female mind . . . [resulting in] a babe denied its mother's breast?" In the following centuries, tens of thousands of babies in Europe were deposited in foundling homes or shipped to middlemen who contracted for a lactating woman to suckle them. In urban centers like Paris, the majority of babies were suckled by strangers. This traffic in babies led to staggering rates of infant mortality.

Reaction against wet-nursing reached a peak during the Enlightenment. In 1793, the French National Convention decreed that only mothers who nursed their own children would be eligible for state aid. The writings of Jean Jacques Rousseau inspired many reformers (although the great philosopher sent his own five children to foundling homes). Almost always, reformers invoked "natural laws," encouraging mothers to follow their instinctive urges to nurture their babies. "Look to the animals for your example," French physician and moralist Jean Emmanuel Gilibert admonished his patients. "Even though the mothers have their stomachs torn open. . . . Even though their offspring have been the cause of all their woes, their first care makes them forget all they have suffered. . . . They forget themselves, little concerned with their own happiness. . . . Woman, like all animals, is under the sway of this instinct."

For further support of their beliefs, reformers could turn to Carolus Linnaeus, the father of modern taxonomy. A physician and the father of seven children, Linnaeus was an ardent advocate of maternal breast-feeding. In 1752 he set down his views in *Nutrix noverca* (Step-Nurse), a widely read denunciation of commercial wet-nursing (which Gilibert translated from Latin into French). In the 1758 edition of his opus *Systema Naturae*, Linnaeus subsumed all warm-blooded, hairy, viviparous vertebrates into a single group—the class Mammalia—identified with milk-secreting glands of the female. (The Latin term for breasts, *mammae*, derives from the plaintive cry "mama," spontaneously uttered by young children from widely divergent linguistic groups and often conveying a single, urgent message, "suckle me.")

Linnaeus's nomenclature underscored a natural role for women based on a salient homology between women and other animals that nursed their young. Mother mammals are alchemists able to transform available fodder—grass, insects, even toxic leaves—into biological white gold. Lactation allows a mother to stockpile resources while they are available, repackage them in digestible form, and then parcel them out to growing infants at her own pace. Able to rely on its mother for food, an immature can stay safe either attached to the mother or stashed in hiding places she chooses, buffered from the vagaries and hazards of foraging in the wide world.

For most mammals, the art of reproduction is to survive poor conditions and breed again under better ones.

But motherhood is not as straightforward a matter as just turning on the milk. Mothers have to factor in recurring food shortages, predators, and social exploitation by members of their own species. Faced with poor conditions, a mother must weigh babies in hand against her own well-being, long-term survival, and—most important—the possibility of breeding again under better circumstances. Behavioral ecologists are only beginning to understand how mother mammals respond to such natural dilemmas, called fitness trade-offs.

Most mammals are iteroparous: breeding more than once, they produce offspring, either singly or in litters, over a breeding career that may last several years—twenty-five or more in the case of a woman. (A very few mammals—primarily some marsupial mice—are semelparous, breeding in one fecund burst followed by death.) In an evolutionary sense, the bottom line for iteroparous females is not the success of any particular birth but reproductive output over a lifetime. The art of iteroparity, therefore, is generally to survive poor conditions and breed again under better ones.

By drawing on help from others, however, some mothers manage to breed under circumstances that would otherwise be impossible.

Consider the case of the cotton-top tamarins. Although the birth of twins is rare among primates, the pint-sized tamarins and marmosets of South America are exceptions. Adapted

---

Hormonal Cocktails for Two

Sarah Blaffer Hrdy and C. Sue Carter

When first presented with pups, a virgin female laboratory rat generally ignores them; she may appear afraid of the tiny, squirming, naked creatures and, occasionally, may even eat them. Only after being introduced to pups many times over several days can a virgin rat be conditioned to tolerate and care for them—licking them, crouching protectively over them, retrieving them when they stray from her side. In contrast, a pregnant rat responds within minutes to pups, even prior to delivery of her own.

The idea that physiological changes might prepare the expectant mother for her new role led to a now classic experiment. In 1968 Joseph Terkel and Jay Rosenblatt, of Rutgers University, injected blood from a rat that had just given birth into a virgin female. The result was a dramatic reduction in the time it took virgins to nurture pups.

Since 1968, we have learned a great deal about what goes inside female mammals as they prepare for motherhood. During the last third of pregnancy, a cascade of endocrinological events readies and motivates mothers. Prominent in this maternal cocktail are the steroid hormones estrogen and progesterone, manufactured by the placenta and essential to maintaining pregnancy. But since the placenta is delivered along with the baby, progesterone and, a little later, estrogen levels fall around the time of birth. By themselves, these hormones cannot account for maternal responsiveness.

Enter prolactin and oxytocin, hormones essential for milk production and nursing. Prolactin is a very ancient molecule whose original function was to maintain salt and water balance in early vertebrates such as fish. Over evolutionary time, this hormone has proved very versatile and now performs diverse physiological functions in many kinds of animals. In mammals it is associated with caretaking behavior in both females and males.

But perhaps the quintessential mammal hormone is oxytocin. A muscle contractor, oxytocin (from the Greek for "swift birth") evolved in mammals and produces the uterine contractions of birth and milk ejection during lactation. Present when the mother first greets her emerging offspring, it continues to be released whenever she nurses. Oxytocin released into the brain is known to promote calming and positive social behaviors, such as pair bonding.

Studies of domestic sheep by Barry Keverne, Keith Kendrick, and their colleagues at the University of Cambridge provide the most complete picture we have of the behavioral effects of oxytocin. As a lamb moves down the birth canal, nerves stimulated during the passage trigger the release of oxytocin in the mother's nervous system. Only if oxytocin is present at birth or injected so that it reaches the brain at the same time a mother meets her newborn, will she bond with her offspring. If release of oxytocin is blocked, the ewe rejects her lamb. High levels of oxytocin also are found in mother's milk, raising the possibility that this hormone plays a role in making the mother-infant attachment mutual.

As important as these hormones can be in determining how responsive a mother will be, they do not act in a deterministic fashion. They both affect and are affected by a mother's behavior and her experience. Exposure to pups, for instance, can lead to reorganization of neural pathways in a mother rat's brain, making her respond faster to pups in the future, even with lower hormone levels. And some recent studies suggest that the hormones of breastfeeding may benefit a mother's mental health and increase her ability to deal with stress.

In many mammals, males, as well as adoptive virgin females, can be primed to exhibit parental behaviors. Prairie vole males, for instance, typically respond to a newborn pup by retrieving it and huddling over it. Geert De Vries, of the University of Massachusetts, found that such nurturing is facilitated by vasopressin, a hormone that in other contexts is associated with aggressive, territorial behavior.

for fluctuating habitats, these monkeys have the potential to breed at a staggering pace, sometimes giving birth as often as twice a year to twins whose combined weight totals up to 20 percent of the mother's. Only the help of other group members—fathers, older offspring, and transient adults—makes the mother's feat of fecundity possible. Helpers carry the offspring most of the time, except when the mother is suckling. Near weaning, helpers also provide infants with crickets and other tidbits.

Working with captive cotton-tops, Lorna Johnson, of the New England Primate Center, revealed how important helpers can be. (This species is endangered in the wild but is still well represented in research colonies.) In her analysis of breeding records over an eighteen-year period, Johnson focused on experienced parents that had already successfully reared offspring. She found that among these veterans, fully 57 percent of parents without help abandoned their young, nearly five times the rate at which parents with helpers voted "no-go."

Among marmosets and tamarins, it is usual for only one female in the group to reproduce during a breeding season. A similar situation prevails among the communally breeding dwarf mongooses of the Serengeti. Studying what keeps the other females from breeding, Purdue University biologist Scott Creel discovered that estrogen levels of these nonbreeders remained only one-third as high as in the breeding female, below that necessary for ovulation. Creel speculates that in species producing large litters, heavy young, or young designed to grow rapidly after birth, the cost of gestation and lactation is just too high for any but the most advantaged female to hazard giving birth. Often harassed and less well fed, a subordinate has such a slim chance of producing young that survive to weaning that she is better off deferring reproduction, helping instead to rear the offspring of her kin—occasionally even suckling them—and generally doing her best to be tolerated in the group and to stay alive until she can become a breeder in her own right.

While studying a closely related subspecies of dwarf mongoose, O. Anne E. Rasa, of Bonn University, learned that subordinate females have an even more pressing reason to postpone reproduction. The dominant female may destroy the pups of any rival that does breed. Earlier this year, Duke University's Leslie Digby reported that there appears to be the same pattern among wild common marmosets in Brazil. In a rare instance when a subordinate female gave birth, one of her infants was killed; the other disappeared at about the same time. For marmosets and dwarf mongooses, then, most subordinate females make the best of a grim lot by temporarily shutting down their ovaries. With luck, their time to breed does come.

Suppression of ovulation is only one of the many means for mothers to adjust the timing of their reproductive effort. In a diverse array of mammals—including bats, skunks, minks, and armadillos—ovulation occurs, but implantation of the fertilized egg in the uterine wall is delayed so as to insure birth of the offspring at the optimal season. As soon as a kangaroo mother ceases to suckle one joey, levels in her blood of the nursing hormone prolactin fall. At this signal, a tiny blastocyst (a nearly hollow globe of cells, produced by the fertilized egg, inside of which the embryo will develop) emerges from diapause (a period of developmental dormancy) and begins to grow again. In the European badger, this blastocyst-in-waiting continues to grow, but ever so slowly. Embryonic slowdown or diapause can persist for days in rodents or even months in larger mammals, until some cue signals the embryo to attach to the uterine wall and resume development. American black bears breed from May to July, but not until the female repairs to her den for winter does implantation occur, so that birth takes place to a lethargic mother in the snug safety of her winter refuge. Yet if the berry crop that year had failed, and the mother, as a result, was not in good condition, implantation might well have failed, too.

Planned parenthood primate style revolves around breast-feeding. In almost all monkeys and apes, as well as in people still living in traditional settings where infants enjoy nearly continuous contact with their mothers, babies nurse on demand. Emory University anthropologists Mel Konner and Carol Worthman report that the !Kung San of the Kalahari suckle their babies for two minutes or so as often as four times every hour, even while they sleep at night.

In eighteenth-century France, many poor mothers would give up their own babies and hire themselves out as wet nurses.

Throughout most of human evolution, mothers suckled their children on demand from infancy to the age of three or four—in some circumstances, even longer. A series of studies of hunter-gatherers from Central Africa, Botswana, and New Guinea, as well as of housewives in New England, have documented the dynamic interaction between a woman's nutritional status, her workload, and her fertility—what Harvard anthropologist Peter Ellison likes to call the ecology of the ovaries. Nipple stimulation from nearly continuous "Pleistocene style" suckling causes the pituitary to secrete more prolactin, the body's "work order" for more milk production. Through a complex and as yet poorly understood series of mediating effects involving the hypothalamus, ovulation is somehow inhibited when prolactin levels are high. The result is birth intervals as long as five years in long-suckling people like the !Kung. According to Ellison, the link between the intensity of suckling and postpartum infertility prevents a nursing mother, already energetically burdened by metabolizing for two, from being saddled with another pregnancy and the even more daunting task of metabolizing for three. (Unless, that is, she happens to be particularly well fed. Worthman and others have recently discovered that among well-nourished

Milk: It Does a Baby Good

Virginia Hayssen

A complex fluid with a long history, milk feeds us, protects us from disease, and even directs aspects of our behavior and physiology. The nutritional value of milk varies greatly from species to species. Cow's milk and human milk, for example, are both about 88 percent water and 4 percent fat, with 165 calories in an eight-ounce glass. Cow's milk, however, has 30 percent less sugar and three times more protein. For a low-fat milk, try that of black rhinos: their milk has only 0.2 percent fat. As a source of energy for their developing young, rhinos and horses use sugars instead of fats. The result is milk with only two-thirds the calories of human milk.

The cream of the crop is hooded seal milk, which is 61 percent fat, with about 1,400 calories in an eight-ounce glass. Small wonder the pups can gain forty-five pounds during their very short (four-day) nursing period. Hooded seals give birth on ice floes and must wean their pups before the ice breaks up or melts. The high-fat, low-protein milk is well suited to provide seal pups with the most important thing they need: a thick layer of blubber to insulate them against cold polar oceans.

The milk of chimpanzees living in tropical environments provides a striking contrast. There, where a mother may carry her suckling offspring with her everywhere for many months, mother's milk is very dilute, low in both fat and protein.

Mother hares, whose first priority is to provide a safe hideaway for their vulnerable babies and to keep its location a secret from predators, can only afford to let their young suckle once a day and, even then, for no more than five minutes. Not surprisingly, the milk of hares is rich in fat and protein.

Milk composition in many species changes over the course of lactation. Milk delivered in the early stages, and again as weaning approaches, often has more protein and less sugar than that produced during the interim stages. The kinds of fat in milk vary with the mother's diet. In fact, milk has different flavors depending on what mothers eat. Those flavors may later direct the food likes and dislikes of offspring.

Kangaroos are a special case in that mothers frequently suckle young of very different sizes and ages simultaneously. The youngest joey is attached to a tiny teat within the pouch and suckles constantly, while its older sibling, who may be 5,000 times larger, intermittently pokes its head into the pouch to suck on a much more elongated nipple. The milk from these adjacent teats is very different in composition.

The extent to which young mammals depend on milk for nutrition is also variable. Voles and mice rely completely upon milk for their well-being until weaning, while the young of many hoofed animals, such as deer and antelope, may begin eating grass only a few days after birth, well before weaning occurs. Mother koalas excrete a yellow-green ooze of partly digested eucalyptus leaves that their young energetically eat. The opening of the mother's pouch is directed backward, allowing the baby easy access to the nutritious slime.

In addition to nutrients, milk contains hormones and growth factors that can regulate the behavior and physiology of both mother and baby. As a mother nurses her young, subtle manipulations may be at work. Studies of rats, monkeys, and other animals have shown that nursing releases natural opiates in the mother's brain, perhaps rendering her more pliant to her baby's demands. Opiates are also present in milk, making the baby feel content, as well as well fed.

Lactation also acts as a fertility control. By suppressing ovulation, it tends to lengthen the interval between births, thus helping insure that each offspring (or litter) receives its mother's undivided attention. And experiments with rodents suggest that at least one component of milk delays puberty in female offspring by retarding ovarian development.

Another of milk's functions is immunological. Colostrum, the protein-rich fluid produced right after birth, is an important source of antibodies that confer immunity to various diseases. The protection provided by some other milk proteins, such as lysozyme and interleukin, may last throughout lactation.

The origins of milk and lactation will always remain somewhat mysterious. Without a time machine, reconstructing the early stages of any complex organ or process is difficult. As the English biologist St. George J. Mivart asked in 1871, "Is it conceivable that the young of any animal was ever saved from destruction by accidentally suckling a drop of scarcely nutritious fluid from an accidentally hypertrophied cutaneous gland of its mother?"

Nevertheless, we know that milk did evolve, and one of the proteins specific to milk—alpha-lactalbumin—may provide some clues as to how. Today, alpha-lactalbumin helps in the synthesis of lactose, but it evolved from another protein, lysozyme, common in blood and other body fluids, as well as in certain glandular secretions, including milk. Lysozyme kills bacteria and fungi, protecting animals from infection. It also protects milk from microbial attack.

Since lysozyme occurs in so many mammalian body fluids and milk, and since it gave rise to alpha-lactalbumin, this protein very likely was present in ancestral fluids that evolved into milk. The first mammals laid eggs, as the platypus and echidnas still do, and the early protomilk may have protected eggs or newly hatched young from bacterial or fungal attack. Because lysozyme is also a protein, neonates who lapped up the fluid from their mothers' bodies may have received a nutritional bonus. Eventually, the value of the fluid as a source of food and water became more important than its original, antibacterial function.

Prenatal Power Plays

David Haig

The most intimate human relationship is that between a mother and her unborn young. A fetus obtains all its nutrients and disposes of all of its wastes via its mother's blood. It shares every breath that its mother takes, every meal she eats, and draws on her fat reserves when food is scarce. What is the nature of this relationship? Do mother and fetus form one body and one flesh, a harmonious union with each attentive to the other's needs? Or is the fetus an alien intruder, a parasite that takes what it can without concern for its maternal host?

Neither the idyllic nor the parasitic vision adequately captures the complexities of pregnancy. Because they share half their genes, mother and fetus have common genetic interests, but sometimes their interests conflict because each also carries genes absent from the other. In particular, maternal and fetal genes are predicted to "disagree" over how a pregnant mother should allocate energy, time, and resources between her own needs and those of the fetus.

Mammal species vary markedly in the ability of the fetus to influence the amount of food it receives from its mother. Bush babies, for example, are small African primates with a placenta that simply absorbs uterine "milk" secreted by the glands of the mother's uterus. Other nutrients diffuse directly

from the mother's blood to fetal blood across the thin layer of maternal and placental tissues that separates the two bloodstreams. A bush baby mother is probably able to control the flow of nutrients to her fetus by contracting or relaxing the blood vessels supplying the lining of her uterus. Similar arrangements occur in a variety of animals, including pigs, cows, and whales.

By contrast, the human placenta is invasive (as are the placentas of mice, bats, sloths, and armadillos). Uterine milk is a significant source of nutrients only during the earliest stages of pregnancy. As the embryo implants within the lining of the uterus, it sends out cells that invade the blood vessels supplying the uterine lining. These invasive cells destroy the muscular wall and greatly expand the diameter of the blood vessels. The result is that the fetus has direct access to its mother's blood, and the mother, unable to constrict the vessels, cannot regulate the flow of nutrients to the placenta without starving her own tissues.

Direct access to the mother's blood also enables the placenta to release a variety of hormones into her circulatory system. These hormones probably evolved to manipulate maternal physiology for fetal benefit. For example, human placental lactogen is produced in larger quantities than any other human hormone. One of its effects is to

make maternal tissues less sensitive to the effects of insulin. If this effect went unopposed, maternal blood sugar would rise higher after meals and would remain elevated for a longer period, allowing the fetus to take a greater share of each meal. The mother is not completely powerless, however, and responds by increasing insulin production. Mothers usually maintain control of their blood sugar during pregnancy, but when they do not, gestational diabetes develops and is relieved only with the delivery of the baby and its placenta.

An appreciation of the genetic conflicts of pregnancy may help doctors understand other medical complications of pregnancy. Sometimes the placenta has inadequate access to maternal blood. One way for the placenta to compensate is to increase the flow of blood by increasing maternal blood pressure. When accompanied by excessive protein in the mother's urine, this high blood pressure can be a symptom of a life-threatening condition called preeclampsia.

Both mother and fetus, of course, share one overriding interest: the successful outcome to pregnancy. To reach that goal, the mother-child relationship appears from the very start to be marked by negotiation and compromise, although negotiations sometimes break down.

mothers, the inhibition of ovulation by suckling is less effective.)

Like delayed implantation, lactational suppression of ovulation provides made-to-order birth control. No system is foolproof, however. Saddled with an inopportune conception, a mother mammal may resort to remedies that although unmotherly to modern tastes, are nonetheless utterly natural. Possible options depend on what type of mammal she is. Untimely fetuses may be reabsorbed, spontaneously aborted, abandoned

after birth, or under some circumstances, killed and even eaten.

Golden hamsters, for instance, are highly flexible breeders adapted to the irregular rainfall and erratic food supplies in their native habitat in the arid regions of the Middle East. In addition to building a nest, licking their pups clean, protecting and suckling them—all pleasantly conventional maternal pursuits—these hamster moms may also recoup maternal resources otherwise lost in the production of pups by eating a few.

For hamsters, to quote Canadian psychologists Corinne Day and Bennett Galef, cannibalizing pups is an "organized part of normal maternal behavior which allows an individual female to adjust her litter size in accord with her capacity to rear young in the environmental conditions prevailing at the time of her parturition." Quality control can also be an issue. Among mice (but not hamsters), pups below median weight are the ones most likely to be rejected when mothers cull very large litters.

The cues mothers respond to may derive from prevailing conditions or their own internal state.

Another "rule of paw" might read: abort poor prospects sooner rather than later and, if possible, recoup resources. Recall how the mother bear's body factors in the latest update on food supplies before either canceling implantation or committing to gestation. The cues mothers respond to may derive from prevailing conditions or their own internal state. Biologist John Hoogland spent sixteen years monitoring a population of black-tailed prairie dogs in South Dakota. Mothers attempt to rear 91 percent of all litters produced; the rest of the time, they abandon their pups at birth and allow other group members to eat them, sometimes even joining in. Under closer examination, Hoogland found that the mothers that gave up on their litters weighed less. He speculated that abandonment was an adaptive response to poor body condition.

Deteriorating social conditions can also alter maternal commitment. Across a broad spectrum of animals from mice to lions, the appearance of strange males on the horizon can present a danger to unweaned infants sired by other, rival males. By killing these infants, the newcomers subvert the mother's control over the timing of her own reproduction. To minimize her loss, she breeds again sooner than she would have if she had continued to suckle her babies. Although this revised schedule of breeding is detrimental to the mother (not to mention her babies), she may ovulate again while the killer is still in the vicinity. Had the killer waited until her infants were weaned, his own window of opportunity might have long since shut, for he too is bound to be replaced by another male.

Among the strains of house mice studied by Frederick von Saal, at the University of Missouri, and by Robert Elwood, at Queen's University in Belfast, Northern Ireland, roving males that have failed to mate in the preceding seven weeks (equivalent to a three-week pregnancy, followed by four weeks of lactation) attack babies in any nest they bump into. By contrast, males that have mated during that crucial period are statistically more likely to behave "paternally," retrieving pups that have slipped out of the nest, keeping them warm, and licking them clean. Some behavioral switch accompanying ejaculation (especially if he remains near the female) transforms this potential killer into a kinder, gentler rodent. This transformation (Elwood calls it a "switch in time") saves the male from mistakenly destroying his own progeny (although, depending on circumstances, it may occasionally lead him to tolerate offspring of another male).

Male mice can also have a devastating effect on unborn young, for a pregnant mouse that encounters a strange male may reabsorb her budding embryos. This form of early abortion avoids the even greater misfortune of losing a full-term litter later on. It has become known as "the Bruce effect," after biologist Hilda Bruce, who first reported the phenomenon for laboratory mice in 1959 (at the time, its function was unclear). The Bruce effect has since been reported for deer mice, collared lemmings, and several species of voles. Elwood and others have shown that pregnant mice are especially likely to block pregnancies when confronted with males known to be infanticidal.

As bizarre as it may seem, when a mammal mother thwarts her own pregnancy, she is behaving—in strictly biological terms—just like a mother.

From the female's point of view, losing a pregnancy is scarcely an ideal strategy. Rather, her body is making the best of dismal circumstances. As bizarre as it may seem, when a mammal mother thwarts her pregnancy, she is behaving—in strictly biological terms—just like a mother. For she may soon conceive again, perhaps with a male who will stick around to help or at least keep other males away. Bruce had discovered a natural, spontaneous form of energy-conserving, early-stage abortion.

Mice are not the only animals that have to cope with infanticidal males. Among the lean and graceful langur monkeys that I studied at Mount Abu, Rajasthan, India, males pose serious threats to infants. My colleagues S. M. Mohnot and Volker Sommer, whose team has monitored the langur population at nearby Jodhpur for more than twenty years, learned that one-third of all infants born are killed by males coming into the group. Mothers initially avoid such usurpers or even fight back, but once a new male becomes ensconced in the group, he has the advantage of being able to try to kill the babies again and again, day after day.

Confronted with discouraging odds, a mother may try to deposit a nearly weaned infant with former resident males, now ousted and roving about the vicinity. This strategy rarely works. The infant will usually wend its way back to its mother, placing itself right back in harm's way. Especially if she is young with many fertile years ahead of her, a mother under persistent assault may simply stop defending her infant, leaving more intrepid kin—usually old females that have not reproduced for years—to intervene. And so it was that I once observed an aged and stiff twenty-pound female, assisted by another older female, wrest a wounded infant from the sharp-toothed jaws of a forty-pound male. The far stronger and healthier young mother watched from the sidelines. Just days before, the same young mother had made no effort to intervene when her infant fell from a jacaranda tree branch and was grabbed up by the male. Again, it was the old female who rushed to the rescue.

Liquid Assets: A Brief History of Wet-Nursing

Sarah Blaffer Hrdy

A Sumerian lullaby from the third millennium B.C. provides the first written record of wet-nursing. As the wife of Shulgi, ruler of Ur, sings her son to sleep, she promises him first a wife and then a son—complete with wet nurse. "The nursemaid joyous of heart will suckle him." Some of these nurses were from privileged backgrounds, their status elevated by contact with tiny scions.

In ancient Egypt, wet nurses were recruited from the harems of senior officials and appeared on the guest lists for royal banquets. Less fortunate wet nurses were actually, or effectively, slaves, and not all lived under the supervision of the babies' parents. The substitution by nurses of one baby for another, the source of topsy-turvy merriment in Gilbert and Sullivan, was seen as serious enough in ancient Mesopotamia to be specifically proscribed in the code of Hammurabi (1700 B.C.).

By the second century A.D., wet-nursing in Europe was an organized commercial activity. In Rome, commerce in mother's milk took place in the vegetable market around particular columns called *lactaria*, where wet nurses for hire gathered. By medieval times, wet nurses—paid, indentured, or enslaved—could be found throughout Europe.

However, the "heyday of wet nursing" (as historian George Sussman refers to it) was eighteenth-century France. It had long been a practice among the elite for infants to be nursed in their own homes or in the nearby countryside by carefully chosen nurses producing plentiful milk. Such babies tended to survive as well as if nursed by their own mothers. A growing population of urban artisans and shopkeepers, along with a rapidly increasing foundling population of abandoned babies, expanded the demand for rural wet-nursing. For various reasons, demand for affordable wet nurses far outstripped their local availability.

Parents were forced to seek wet nurses farther afield. An itinerant entrepreneur known as a *meneur* would contract with rural women and then, shortly after an infant's birth, bring a prospective wet nurse to the parents' home to pick up her charge. The *meneur*—a cartload of babies in tow—would then lead the wet nurses and their new charges back to distant rural destinations. Instances of babies being lost along the way occasionally surface in police records for Lyon and Paris.

In 1780, Lt. Gen. Charles-Pierre LeNoir, head of the Paris police (whose job it was to monitor the referral bureaus used by parents to locate wet nurses) provided a startling statistic: only 1,000 of 21,000 babies born in Paris that year were nursed by their own mothers. Infant mortality rates during this period were appalling, and there was a direct relationship between how much parents paid a wet nurse and how likely her charge was to survive.

Whether the wet nurse was adequate or not, the consequences for the mother were the same. Freed from the "drudgery"—and contraceptive effects—of nursing, mothers ovulated again, often within months. During their prime reproductive years, some women gave birth annually, with such serious health consequences as chronic anemia and prolapsed uteruses. In privileged households, the beneficiaries were often the same husbands who had insisted on using a wet nurse in the first place. Without breaking prevailing norms against sex with a lactating wife, and at no physical cost to themselves, they produced an array of legal heirs.

Down the social scale, butchers and shopkeepers faced economic ruin without the help of their wives. Hiring a wet nurse was often a financial necessity. But even with a wife's help, cou-ples could seldom afford—much less house—a choice nurse, so their infants were sent away, often to distant wet nurses. Many died. As in wealthier households, however, the production of babies was fast-paced: among the non-breast-feeding wives of butchers and silk makers in Lyon, French historian Maurice Garden documents one butcher's wife who had twenty-one children in twenty-four years.

The real losers in the system, apart from the babies, were the women whose options were truly awful to begin with. Desperate to make any kind of living, many poor rural mothers would farm out their own babies to even less fortunate women and then hire themselves out as wet nurses. Grim reminders of their plight persist in how-to manuals for selecting wet nurses. Many recommend a woman who has recently given birth but who will not be nursing any infant other than the one she is hired to care for. (In a letter from Renaissance Italy, the wife of a Florentine merchant lamented the survival of a servant girl's baby; the enterprising lady had hoped to offer the servant as a wet nurse to one of her husband's clients.)

A few destitute women managed to work the system to their advantage, after a fashion. In Russia, where scores of abandoned babies were deposited in the (usually lethal) imperial foundling hospitals in Saint Petersburg and Moscow—established by Catherine II to demonstrate how "European" Russia was—an unmarried woman might become pregnant, abandon the newborn at the hospital, and then hire herself out as a wet nurse. Like the mother of Moses, a tiny, lucky percentage (if anyone in this tragic network would be considered lucky) managed to convince or bribe employees to put them in charge of their own infants.

In the last weeks of pregnancy, langurs may respond to a usurping male by aborting rather than continuing to expend energy on a reproductive venture so unlikely to end well. Similar late-pregnancy variations on the Bruce effect have been reported for an odd assortment of large mammals, including wild horses. University of Nevada's Joel Berger, an animal behaviorist who studied wild horses in the Great Basin, watched what happens when one stallion successfully challenges another for possession of his harem. During the disruption following the changeover, 82 percent of the mares that had been impregnated in the last six months by the deposed stallion aborted their fetuses.

Infanticide, abortion, cannibalism, these are altogether natural lapses from imagined "natural laws." Why is it only in the last two decades that researchers have begun to view such behaviors as other than aberrations? Opinions, even scientific ones, are often influenced by received wisdom. As late as the 1960s, when animal behaviorists set up labs to study the maternal activities of rats, monkeys, and dogs, the categories devised to describe their behavior took for granted that mothers were instinctively nurturing. In her pioneering studies of dogs, for example, comparative psychologist Harriet Rheingold separated mothers and their pups from all other animals and then recorded behaviors that fell into her preconceived protocol of maternal activities: contact, nursing, licking, play, and so forth.

Indeed, much of the time mother mammals do carry, groom, and suckle their young. The types of maternal activities Rheingold and others investigated were those that insured that mothers passed on their genes to future generations—the primary focus of the time. Such a view of what it means to be a mother could fairly be classified as essentialist.

But the study of animal behavior has changed. With the emergence of sociobiology in the 1970s, researchers began to focus on individuals and the idiosyncratic social and environmental circumstances of each. With this new perspective, it not only became clear that one mother is not the same as another but also that not all females would be mothers. Far from essentialist or biologically determinist, most biologists today think context is critically important. Researchers like Scott Creel and Carol Worthman combined fieldwork with laboratory measures to search for the cues—inside and out—that prompt a female to opt for one reproductive strategy rather than another.

Across her life course, both a mother and her circumstances are constantly changing—as she ages, finds a new mate, loses a potential helper, stockpiles fat. In a world of leisure, plenty, and supportive social groups or in realms where offspring cost their parents little to rear, trade-offs fade from view. In contrast, overpopulation, social oppression, scarcity, bad times—none of these have ever been conducive to the development of the sort of mother characterized in Marge Piercy's poem "Magic mama" as "an aphid enrolled to sweeten the lives of others. The woman who puts down her work like knitting the moment you speak."

Real mamas must not only be magic but also multifaceted. Motherhood is more than all the licking, tending, suckling, and awe-inspiring protectiveness for which mother mammals are so justly famous. Such indeed is the art—and the tragedy—of iteroparity: offspring born at one time may be more costly to a mother or less viable than offspring born at another. Far from invalidating biological bases for maternal behavior, the extraordinary flexibility in what it means to be a mother should merely remind us that the physiological and motivational underpinnings of an archetypically pro-choice mammal are scarcely new.

Sex and the Female Agenda

Most female mammals are anything but subtle when it comes to telling males it's time for sex. Not humans. For good evolutionary reasons, women have found it's much better to keep men in the dark.

Jared Diamond

SCENE ONE: A dimly lit bedroom; a handsome man lies in bed. A beautiful young woman, in nightgown, enters. A diamond wedding ring flashes virtuously on her left hand; her right clutches a small blue strip of paper. She bends, kissing the man's ear. She: "Darling! It's time!"

Scene Two: Same bedroom, same couple, making love; details obscured by dim lighting. Camera shifts to a calendar being flipped by a graceful hand wearing the same diamond wedding ring.

Scene Three: Same couple, blissfully holding smiling baby. He: "Darling! I'm so glad Ovustick told us when it was exactly the right time!"

Last Frame: Close-up of same graceful hand, clutching same small blue strip of paper. Caption reads: OVUSTICK. HOME URINE TEST TO DETECT OVULATION.

If baboons could understand our TV ads, they'd find that one especially hilarious. Neither a male nor a female baboon needs a hormonal test kit to detect the female's ovulation, the sole time when her ovary releases an egg and she can be fertilized. Instead, the skin around the female's vagina swells and turns bright pink. She gives off a distinctive smell. And in case a dumb male still misses the point, she also crouches in front of him and presents her hindquarters. Most other female animals are similar, advertising ovulation with equally bold visual signals, odors, or behaviors.

We consider female baboons with bright pink hindquarters an oddity. In fact, though, we humans are the odd ones—our scarcely detectable ovulations make us members of a small minority in the mammalian world. Granted, quite a few other primates—the group of mammals that includes monkeys, apes, and us—also conceal their ovulations. However, even among primates, baboon-style advertisement remains the majority practice. Human males, in contrast, have no means of detecting when their partners can be fertilized; nor did the women themselves until modern, scientific times.

We're also unusual in our continuous practice of sex, which is a direct consequence of our concealed ovulations. Most other animals confine sex to a brief period of estrus around the advertised time of ovulation. At estrus, a female baboon emerges from a month of sexual abstinence to copulate up to 100 times. A female Barbary macaque does it on an average of every 17 minutes, distributing her favors at least once to every adult male in her troop. Monogamous gibbon couples go several years without sex, until the female weans her most recent infant and comes into estrus again. The gibbons relapse into abstinence as soon as the female becomes pregnant.

We humans, though, practice sex on any day of the month. Hence most human copulations involve women who are unable to conceive at that moment. Not only do we have sex at the "wrong" time of the cycle, but we continue to have sex during pregnancy and after menopause, when we know for sure that fertilization is impossible.

Human sex does seem a monumental waste of effort from a "biological" point of view. After all, most other animals are sensibly stingy of copulatory effort, and for a simple reason—sex is expensive. Just count the ways: for males, sperm production is metabolically costly, so much so that mutant worms with few sperm live longer than normal sperm producing worms. Sex takes time that could otherwise be devoted to finding food. During the sex act itself, couples locked in embrace risk being surprised and killed by a predator or enemy. Finally, fights between males competing for a female often result in serious injury to the female as well as to the males.

So why don't human females behave the way most other animals do and give dear ovulatory signals that would let us restrict sex to moments when it could do us some good?

By now, you may have decided that I'm a prime example of an ivory-tower scientist searching unnecessarily for problems to explain. I can hear several million of you protesting, "There's no problem to explain, except why Jared Diamond is such an idiot. You don't understand why we have sex all the time? Because it's fun, of course!" Unfortunately, that answer isn't enough to satisfy scientists. Humans' concealed ovulations and unceasing receptivity must have evolved for good reasons, and ones that go beyond fun. While engaged in sex, animals, too, look as if they're having fun, to judge by their intense involvement. And with respect to *Homo sapiens,* the species unique in its self-consciousness, it's especially paradoxical that a female as smart and aware as a human should be unconscious of her own ovulation, when female animals as dumb as cows are aware of it.

In speculating about the reasons for our concealed ovulations, scientists tend to focus their attention on another of our unusual features: the helpless condition of our infants, which makes lots of parental care necessary for many years. The young of most mammals start to get their own food as soon as they're weaned and become fully independent soon afterward. Hence most female mammals can and do rear their young without any assistance from the father, whom the mother never sees again after copulation. For humans, though, most food is acquired by means of complex technologies far beyond the dexterity or mental ability of a toddler. As a result, our children have to have food brought to them for over a decade after weaning, and that job is much easier for two parents than for one. Even today, it's hard for a single human mother to rear kids unassisted. It was undoubtedly much harder for our prehistoric ancestors.

Just imagine what married life would be like if women did advertise their ovulations, like female baboons with bright pink derrieres.

Consider the dilemma facing an ovulating cavewoman who has just been fertilized. In many other mammal species, the male would promptly go off in search of another ovulating fe-

male. For the cave-woman, though, that would seriously jeopardize her child's survival. She's much better off if that man sticks around. But what can she do? Her brilliant solution: remain sexually receptive all the time! Keep him satisfied by copulating whenever he wants! In that way, he'll hang around, have no need to look for new sex partners, and will even share his daily hunting bag of meat.

That in essence is the theory that was formerly popular among anthropologists—among male anthropologists, anyway. Alas for that theory, there are numerous male animals that require no such sexual bribes to induce them to remain with their mate and offspring. I already mentioned that gibbons, seeming paragons of monogamous devotion, go years without sex. Male songbirds cooperate assiduously with their mates in feeding the nestlings, although sex ceases after fertilization. Even male gorillas with a harem of several females get only a few sexual opportunities each year because their mates are usually nursing or out of estrus. Clearly, these females don't have to offer the sop of constant sex.

But there's a crucial difference between our human couples and those abstinent couples of other animal species. Gibbons, most songbirds, and gorillas live dispersed over the landscape, with each couple or harem occupying its separate territory. That means few encounters with potential extramarital sex partners. Perhaps the most distinctive feature of traditional human society is that it consists of mated couples living within large groups of other couples, with whom we have to cooperate. A father and mother must work together for years to rear their helpless children, despite being frequently tempted by other fertile adults nearby. The specter of marital disruption by extramarital sex, with its potentially disastrous consequences for parental cooperation in child-rearing, is pervasive in human societies. Somehow we evolved concealed ovulation and constant receptivity to make possible our unique combination of marriage, coparenting, and adulterous temptation. How does that combination work?

More than a dozen new theories have emerged as possible explanations. From this plethora of possibilities, two—the father-at-home theory and the many fathers theory—have survived as most plausible. Yet they are virtually opposite.

The father-at-home theory was developed by University of Michigan biologist Richard Alexander and graduate student Katharine Noonan. To understand it, imagine what married life would be like if women did advertise their ovulations, like female baboons with bright pink derrieres. A husband would infallibly recognize the day on which his wife was ovulating. On that day he would stay home and assiduously make love, in-order to fertilize her and pass on his genes. On all other days he would realize from his wife's pallid derriere that lovemaking with her was useless. He would instead wander off in search of other, unguarded pink-hued ladies so he could pass on even more of his genes. He'd feel secure in leaving his wife at home because he'd know she wasn't sexually receptive to men and couldn't be fertilized anyway.

The results of those advertised ovulations would be awful. Fathers wouldn't be at home to help rear the kids, mothers couldn't do the job unassisted, and babies would die in droves. That would be bad for both mothers and fathers because neither would succeed in propagating their genes.

Now let's picture the reverse scenario, in which a husband has no clue to his wife's fertile days. He then has to stay at home and make love with her on as many days of the month as possible if he wants to have much chance of fertilizing her. Another motive for him to stay around is to guard her against other men, since she might prove to be fertile on any day he's away. Besides, now he has less reason to wander, since he has no way of identifying when other women are fertile. The heartwarming outcome: fathers hang around and share baby care, and babies survive. That's good for both mothers and fathers, who have now succeeded in transmitting their genes. In effect, both gain: the woman, by recruiting an active co-parent; the man, because he acquires confidence that the kid he is helping to rear really carries his genes.

Naturally, infanticide horrifies us, but on reflection, one can see that the murderer gains a grisly genetic advantage.

Competing with the father-at-home theory is the many-fathers theory developed by anthropologist Sarah Hrdy of the University of California at Davis. Anthropologists have long recognized that infanticide used to be common in many human societies. Until field studies by Hrdy and others, though, zoologists had no appreciation for how often it occurs among other animals as well. Infanticide is especially likely to be committed by males against infants of females with whom they have never copulated—for example, by intruding males that have supplanted resident males and acquired their harem. The usurper "knows" that the infants killed are not his own. (Of course, animals don't carry out such subtle reasoning consciously; they evolved to behave that way instinctively.) The species in which infanticide has been documented now include our closest animal relatives, chimpanzees and gorillas, in addition to a wide range of other species from lions to African hunting dogs.

Naturally, infanticide horrifies us. But on reflection, one can see that the murderer gains a grisly genetic advantage. A female is unlikely to ovulate as long as she is nursing an infant By killing the infant, the murderous intruder terminates the mother's lactation and stimulates her to resume estrous cycles. In most cases, the murderer proceeds to fertilize the bereaved mother, who then bears an infant carrying the murderer's own genes.

Infanticide is a serious evolutionary problem for these animal mothers, who lose their genetic investment in their murdered offspring. This problem would appear to be exacerbated if the female has only a brief, conspicuously advertised estrus. A dominant male could easily monopolize her during that time. All other males would consequently know that the resulting infant was sired by their rival, and they'd have no compunctions about killing the infant.

Suppose, though, that the female has concealed ovulations and constant sexual receptivity. She can exploit these advantages to copulate with many males—even if she has to do it sneakily, when her consort isn't looking. (Hrdy, by the way, argues that the human female's capacity for repeated orgasms may have evolved to provide her with further motivation to do so.)

According to Hrdy's scenario, no male can be confident of his paternity, but many males recognize that they might have sired the mother's infant. If such a male later succeeds in driving out the mother's consort and taking her over, he avoids killing her infant because it could be his own. He might even help the infant with protection and other forms of paternal care. The mother's concealed ovulation will also serve to decrease fighting between males within her own troop, because any single copulation is unlikely to result in conception and hence is no longer worth fighting over.

In short where Alexander and Noonan view concealed ovulation as clarifying paternity and reinforcing monogamy, Hrdy sees it as confusing paternity and effectively undoing monogamy. Which theory is correct?

To find the answer, we turn to the comparative method, a technique often used by evolutionary biologists. By comparing primate species, we can learn which mating habits are shared by those species with concealed ovulation but absent from those with advertised ovulation. As we shall see, the reproductive biology of each species represents the outcome of an experiment, performed by nature, on the benefits and drawbacks of concealing ovulation.

This comparison was recently conducted by Swedish biologists Birgitta Sillén-Tullberg and Anders Møller. First they tabulated the visible signs of ovulation for 68 species of higher primates (monkeys and apes). They found that some species, including baboons and our close relatives the chimpanzees, advertise ovulation conspicuously. Others, including our close relative the gorilla, exhibit slight signs. But nearly half resemble humans in lacking visible signs. Those species include vervets, marmosets, and spider monkeys, as well as one ape, the orangutan. Thus, while concealed ovulation is still exceptional among mammals in general, it nevertheless occurs in a significant minority of higher primates.

Next, the same 68 species were categorized according to their mating system. Some, including marmosets and gibbons, turn out to be monogamous. More, such as gorillas, have harems of females controlled by a single adult male. Humans are represented in both categories, with some societies being routinely monogamous and others having female harems. But most higher primate species, including chimpanzees, have a promiscuous system in which females routinely associate and copulate with multiple males.

Sillén-Tullberg and Møller then examined whether there was any tendency for more or less conspicuous ovulations to be associated with some particular mating system. Based on a naive reading of our two competing theories, concealed ovulation should be a feature of monogamous species if the father-at-home theory is correct, but of promiscuous species if the many-fathers theory holds. In fact, almost all monogamous primate species analyzed prove to have concealed ovulation. Not a single monogamous primate species has boldly advertised ovulations, which instead are mostly confined to promiscuous species. That seems to be strong support for the father-at-home theory. But the fit of predictions to theory is only a half-fit, because the reverse correlations don't hold up. Yes, most monogamous species have concealed ovulation, yet perpetually pallid derrieres are in turn no guarantee of monogamy. Out of 32 species that hide their ovulations, 22 aren't monogamous but promiscuous or live in harems.

So regardless of what caused concealed ovulation to evolve in the first place, it can evidently be maintained under varied mating systems.

Similarly, while most species with boldly advertised ovulations are promiscuous, promiscuity doesn't require flashing a bright pink behind once a month. In fact, most promiscuous primates either have concealed ovulation or only slight signs. Harem-holding species can have any type of ovulatory signal: invisible, slightly visible, or conspicuous.

These complexities warn us that concealed ovulation will prove to serve different functions according to the particular mating system with which it coexists. To identify these changes of function, Sillén-Tullberg and Møller got the bright idea of studying the family tree of living primate species. Their underlying rationale was that some modern species that are very closely related, and thus presumably derived from a recent common ancestor, differ in mating system or in strength of ovulatory signals. This implies recent evolutionary changes, and the two researchers hoped to identify the points where those changes had taken place.

Here's an example of how the reasoning works. Comparisons of DNA show that humans, chimps, and gorillas are still about 98 percent genetically identical. Measurements of how rapidly such gene changes accumulate, plus discoveries of dated ape and protohuman fossils, show that humans, chimps, and gorillas all stem from an ancestral "missing link" that lived around 9 million years ago. Yet those three modern descendants now exhibit all three types of ovulatory signal: concealed ovulation in humans, slight signals in gorillas, bold advertisement in chimps. This means that only one of those three descendants can be like the missing link, and the other two must have evolved different signals.

A strong hint of the problem's resolution is that many living species of primitive primates—creatures like tarsiers and lemurs—have slight signs of ovulation. The simplest interpretation, then, is that the missing link inherited slight signs from a primitive ancestor,

and that gorillas in turn inherited their slight signs unchanged from the missing link. Within the last 9 million years, though, humans must have lost even those slight signs to develop our present concealed ovulation, while chimps, in contrast, went on to evolve bolder signs.

Identical reasoning can be applied to other branches of the primate family tree, to infer the ovulatory signals of other now-vanished ancestors and the subsequent changes in their descendants. As it turns out, signal switching has been rampant in primate history. There have been several independent origins of bold advertisement (including the example in chimps); many independent origins of concealed ovulation (including humans and orangutans); and several reappearances of slight signs of ovulation, either from concealed ovulation (as in some howler monkeys) or from bold advertisement (as in many macaques).

All right, so that's how we can deduce past changes in ovulatory signals. When we now turn our attention to mating systems, we can use exactly the same procedure. Again, we discover that humans and chimps evolved in opposite directions, just as they did in their ovulatory signals. Studies of living primitive primate behavior suggest that ancestral primates of 60 million years ago mated promiscuously, and that our missing link of 9 million years ago had already switched to single-male harems. Yet if we look at humans, chimps, and gorillas as they are today, we find all three types of mating system represented. Thus, while gorillas may just have retained the harems of their missing link ancestor, chimps must have reinvented promiscuity and humans invented monogamy.

Overall, it appears that monogamy has evolved independently many times in higher primates: in us, in gibbons, and in numerous groups of monkeys. Harems also seem to have evolved many times, including in the missing link. Chimps and a few monkeys apparently reinvented promiscuity, after their recent ancestors had given up promiscuity for harems.

Thus Sillén-Tullberg and Møller have reconstructed both the type of mating system and the ovulatory signal that probably coexisted in numerous primates of the remote past. Now, finally, we can put all this information together to examine what the mating system was at each of the points in our family tree when concealed ovulation evolved.

What it boils down to is that concealed ovulation has repeatedly changed and reversed its function during primate evolutionary history.

Here's what one learns. In considering those ancestral species that did have ovulatory signals and that went on to lose those signals and evolve concealed ovulation, only one was monogamous. The rest of them were promiscuous or harem-holding—one species being the human ancestor that arose from the harem-holding missing link. We thus conclude that promiscuity or harems, not monogamy, are the mating systems associated with concealed ovulation. This conclusion is as predicted by Hrdy's many-fathers theory. It doesn't agree with the father-at-home theory.

But we can also ask the reverse question: What were the ovulatory signals prevailing at each point in our family tree when monogamy evolved? We find that monogamy never evolved in species with bold advertisement of ovulation. Instead, monogamy has usually arisen in species that already had concealed ovulation, and sometimes in species that had slight ovulatory signals. This conclusion agrees with predictions of Alexander and Noonan's father-at-home theory.

How can these two apparently opposite conclusions be reconciled? Recall that Sillén-Tullberg and Møller found that almost all monogamous primates today have concealed ovulation. That result must have arisen in two steps. First, concealed ovulation arose, in a promiscuous or harem-holding species. Then, with concealed ovulation already present, the species switched to monogamy.

Perhaps, by now, you're finding our sexual history confusing. We started out with an apparently simple question deserving a simple answer: Why do we hide our ovulations and have sex on any day of the month? Instead of a simple answer, you're being told that the answer is more complex and involves two steps.

What it boils down to is that concealed ovulation has repeatedly changed and actually reversed its function during primate evolutionary history. That is, both the father-at-home and the many-fathers explanations are valid, but they operated at different times in our evolutionary history. Concealed ovulation arose at a time when our ancestors were still promiscuous or living in harems. At such times, it let the ancestral woman distribute her sexual favors to many males, none of whom could swear that he was the father of her baby but each of whom knew that he might be. As a result, none of those potentially murderous males wanted to harm the baby, and some may actually have protected or helped feed it. Once the woman had

evolved concealed ovulation for that purpose, she then used it to pick a good man, to entice or force him to stay at home with her, and to get him to provide lots of help for her baby.

On reflection, we shouldn't be surprised at this shift of function. Such shifts are very common in evolutionary biology. Natural selection doesn't proceed in a straight line toward a distant perceived goal, in the way that an engineer consciously designs a new product. Instead, some feature that serves one function in an animal begins to serve some other function as well, gets modified as a result, and may even lose the original function. The consequence is frequent reinventions of similar adaptations, and frequent losses, shifts, or even reversals of function as living things evolve.

One of the most familiar examples involves vertebrate limbs. The fins of ancestral fishes, used for swimming, evolved into the legs of ancestral reptiles, birds, and mammals, used for running or hopping on land. The front legs of certain ancestral mammals and reptile-birds then evolved into the wings of bats and modern birds respectively, to be used for flying. Bird wings and

mammal legs then evolved independently into the flippers of penguins and whales respectively, thereby reverting to a swimming function and effectively reinventing the fins of fish. At least two groups of fish descendants independently lost their limbs, to become snakes and legless lizards. In essentially the same way, features of reproductive biology—such as concealed ovulation, boldly advertised ovulation, monogamy, harems, and promiscuity—have repeatedly changed function and been transmuted, reinvented, or lost.

Think of all this the next time you are having sex for fun. Chances are it will be at a nonfertile time of the ovulatory cycle and while you're enjoying the security of a lasting monogamous relationship. At such a time, reflect on how your bliss is made paradoxically possible by precisely those features of your physiology that distinguished your remote ancestors, condemned to harems or promiscuity. Ironically, those wretched ancestors had sex only on rare days of ovulation, when they discharged the biological imperative to fertilize, robbed of leisurely pleasure by their desperate need for swift results.

WHY WOMEN CHANGE

The winners of evolution's race are those who can leave behind the most offspring to carry on their progenitors' genes. So doesn't it seem odd that human females should be hobbled in their prime by menopause?

JARED DIAMOND

Jared Diamond is a contributing editor of Discover, *a professor of physiology at the* UCLA *School of Medicine, a recipient of a MacArthur genius award, and a research associate in ornithology at the American Museum of Natural History. Expanded versions of many of his* Discover *articles appear in his book* The Third Chimpanzee: The Evolution and Future of the Human Animal, *which won Britain's 1992* COPUS *prize for best science book and the* Los Angeles Times *science book prize.*

Most wild animals remain fertile until they die. So do human males: although some may eventually become less fertile, men in general experience no shutdown of fertility, and indeed there are innumerable well-attested cases of old men, including a 94-year-old, fathering children.

But for women the situation is different. Human females undergo a steep decline in fertility from around the age of 40 and within a decade or so can no longer produce children. While some women continue to have regular menstrual cycles up to the age of 54 or 55, conception after the age of 50 was almost unknown until the recent advent of hormone therapy and artificial fertilization.

Human female menopause thus appears to be an inevitable fact of life, albeit sometimes a painful one. But to an evolutionary biologist, it is a paradoxical aberration in the animal world. The essence of natural selection is that it promotes genes for traits that increase one's number of descendants bearing those genes. How could natural selection possibly result in every female member of a species carrying genes that throttle her ability to leave more descendants? Of course, evolutionary biologists (including me) are not implying that a woman's only proper role is to stay home and care for babies and to forget about other fulfilling experiences. Instead I am using standard evolutionary reasoning to try to understand how men's and women's bodies came to be the way they are. That reasoning tends to regard menopause as among the most bizarre features of human sexuality. But it is also among the most important. Along with the big brains and upright posture that every text of human evolution emphasizes, I consider menopause to be among the biological traits essential for making us distinctively human—something qualitatively different from, and more than, an ape.

Not everyone agrees with me about the evolutionary importance of human female menopause. Many biologists see no need to discuss it farther, since they don't think it poses an unsolved problem. Their objections are of three types. First, some dismiss it as a result of a recent increase in human expected life span. That increase stems not just from public health measures developed within the last century but possibly also from the rise of agriculture 10,000 years ago, and even more likely from evolutionary changes leading to increased human survival skills within the last 40,000 years.

According to proponents of this view, menopause could not have been a frequent occurrence for most of the several million years of human evolution, because (supposedly) almost no women or men used to survive past the age of 45 or 50. Of course the female reproductive tract was programmed to shut down by age 50, since it would not have had the opportunity to operate thereafter anyway. The increase in human life span, these critics believe, has occurred much too recently in our evolutionary history for the female reproductive tract to have had time to adjust.

What this view overlooks, however, is that the human male reproductive tract and every other biological function of both women and men continue to function in most people for decades after age 50. If all other biological functions adjusted quickly to our new long life span, why was female reproduction uniquely incapable of doing so?

Furthermore, the claim that in the past few women survived until the age of menopause is based solely on paleodemography, which attempts to estimate age at time of death in ancient skeletons. Those estimates rest on unproven, implausible assumptions, such as that the recovered skeletons represent an unbiased sample of an entire ancient population, or that ancient adult skeletons' age of death can accurately be determined. While there's no question that paleodemographers can distinguish an ancient skeleton of a 10-year-old from that of a 25-year-old, they have never demonstrated that they can distinguish an ancient 40-year-old from a 55-year-old. One can hardly reason by comparison with skeletons of modern people, whose bones surely age at different rates from bones of ancients with different lifestyles, diets, and diseases.

A second objection acknowledges that human female menopause may be an ancient phenomenon but denies that it is unique to humans. Many wild animals undergo a decline in fertility with age. Some elderly individuals of many wild mammal and bird species are found to be infertile. Among animals in laboratory cages or zoos, with their lives considerably extended over expected spans in the wild by a gourmet diet, superb medical care, and protection from enemies, many elderly female rhesus monkeys and individuals of several strains of laboratory mice do become infertile. Hence some biologists object that human female menopause is merely part of a widespread phenomenon of animal menopause, not something peculiar to humans.

However, one swallow does not make a summer, nor does one sterile female constitute menopause. Establishing the existence of menopause as a biologically significant phenomenon in the wild requires far more than just coming upon the occasional sterile elderly individual in the wild or observing regular sterility in caged animals with artificially extended life spans. It requires finding a wild animal population in which a substantial proportion of females become sterile and spend a significant fraction of their life spans after the end of their fertility.

The human species does fulfill that definition, but only one wild animal species is known to do so: the short-finned pilot whale. One-quarter of all adult females killed by whalers prove to be postmenopausal, as judged by the condition of their ovaries. Female pilot whales enter menopause at the age of 30 or 40

years, have a mean survival of at least 14 years after menopause, and may live for over 60 years. Menopause as a biologically significant phenomenon is thus not strictly unique to humans, being shared at least with that one species of whale.

There is no obvious reason we had to evolve eggs that degenerate by the end of half a century. Eggs of elephants, baleen whales, and tortoises remain viable for at least 60 years.

But human female menopause remains sufficiently unusual in the animal world that its evolution requires explanation. We certainly did not inherit it from pilot whales, from whose ancestors our own ancestors parted company over 50 million years ago. In fact, we must have evolved it after we separated from the apes just 7 million to 5 million years ago, because we undergo menopause whereas chimps and gorillas appear not to (or at least not regularly).

The third and last objection acknowledges human menopause as an ancient phenomenon that is indeed unusual among animals. But these critics say that we need not seek an explanation for menopause, because the puzzle has already been solved. The solution, they say is the physiological mechanism of menopause: the senescence and exhaustion of a woman's egg supply, fixed at birth and not added to after birth. An egg is lost at each menstrual cycle. By the time a woman is 50 years old, most of that original egg supply has been depleted. The remaining eggs are half a century old and increasingly unresponsive to hormones.

But there is a fatal counterobjection to this objection. While the objection is not wrong, it is incomplete. Yes, exhaustion and aging of the egg supply are the immediate cause of human menopause, but why did natural selection program

women so that their eggs become exhausted or aged in their forties? There is no obvious reason we had to evolve eggs that degenerate by the end of half a century. Eggs of elephants, baleen whales, and tortoises remain viable for at least 60 years. A mutation only slightly altering how eggs degenerate might have sufficed for women to remain fertile until age 60 or 75.

The easy part of the menopause puzzle is identifying the physiological mechanism by which a woman's egg supply becomes depleted or impaired by the time she is around 50 years old. The challenging problem is understanding why we evolved that seemingly self-defeating detail of reproductive physiology. Apparently there was nothing physiologically inevitable about human female menopause, and there was nothing evolutionarily inevitable about it from the perspective of mammals in general. Instead the human female, but not the human male, was programmed by natural selection, at some time within the last few million years, to shut down reproduction prematurely. That premature senescence is all the more surprising because it goes against an overwhelming trend: in other respects, we humans have evolved to age more slowly, not more rapidly, than most other animals.

As a woman ages, she can do more to increase the number of people bearing her genes by devoting herself to her existing children and grandchildren than by producing yet another child.

Any theory of menopause evolution must explain how a woman's apparently counterproductive evolutionary strategy of making fewer babies could actually result in her making more. Evidently, as a woman ages, she can do more to increase the number of people bearing her genes by devoting herself to her existing children, her potential grandchildren, and her other relatives than by producing yet another child.

That evolutionary chain of reasoning rests on several cruel facts. One is that the human child depends on its parents for an extraordinarily long time, longer than in any other animal species. A baby chimpanzee, as soon as it starts to be weaned, begins gathering its own food, mostly with its own hands. (Chimpanzee use of tools, such as fishing for termites with blades of grass or cracking nuts with stones, is of great interest to human scientists but of only limited dietary significance to chimpanzees.) The baby chimpanzee also prepares its food with its own hands. But human hunter-gatherers acquire most food with tools (digging sticks, nets, spears), prepare it with other tools (knives, pounders, huskers), and then cook it in a fire made by still other tools. Furthermore, they use tools to protect themselves against dangerous predators, unlike other prey animals, which use teeth and strong muscles. Making and wielding all those tools are completely beyond the manual dexterity and mental ability of young children. Tool use and toolmaking are transmitted not just by imitation but also by language, which takes over a decade for a child to master.

As a result, human children in most societies do not become capable of economic independence until their teens or twenties. Before that, they remain dependent on their parents, especially on the mother, because mothers tend to provide more child care than do fathers. Parents not only bring food and teach tool-making but also provide protection and status within the tribe. In traditional societies, early death of either parent endangers a child's life even if the surviving parent remarries, because of possible conflicts with the stepparent's genetic interests. A young orphan who is not adopted has even worse chances of surviving.

Hence a hunter-gatherer mother who already has several children risks losing her genetic investment in them if she

does not survive until the youngest is at least a teenager. That's one cruel fact underlying human female menopause. Another is that the birth of each successive child immediately jeopardizes a mother's previous children because the mother risks dying in childbirth. In most other animal species that risk is very low. For example, in one study of 401 rhesus monkey pregnancies, only three mothers died in childbirth. For humans in traditional societies, the risk is much higher and increases with age. Even in affluent twentieth-century Western societies, the risk of dying in childbirth is seven times higher for a mother over the age of 40 than for a 20-year-old. But each new child puts the mother's life at risk not only because of the immediate risk of death in childbirth but also because of the delayed risk of death related to exhaustion by lactation, carrying a young child, and working harder to feed more mouths.

Infants of older mothers are themselves increasingly unlikely to survive or be healthy, because the risks of abortion, stillbirth, low birth weight, and genetic defects rise as the mother grows older. For instance, the risk of a fetus's carrying the genetic condition known as Down syndrome increases from one in 2,000 births for a mother under 30, one in 300 for a mother between the ages of 35 and 39, and one in 50 for a 43-year-old mother to the grim odds of one in 10 for a mother in her late forties.

Thus, as a woman gets older, she is likely to have accumulated more children, and she has been caring for them longer, so she is putting a bigger investment at risk with each successive pregnancy. But her chances of dying in or after childbirth, and the chances that the infant will die, also increase. In effect, the older mother is risking more for less potential gain. That's one set of factors that would tend to favor human female menopause and that would paradoxically result in a woman's having more surviving children by giving birth to fewer children.

But a hypothetical nonmenopausal older woman who died in childbirth, or while caring for an infant, would thereby be throwing away even more than her investment in her previous chil-

dren. That is because a woman's children eventually begin producing children of their own, and those children count as part of the woman's prior investment. Especially in traditional societies, a woman's survival is important not only to her children but also to her grandchildren.

That extended role of postmenopausal women has been explored by anthropologists Kristen Hawkes, James O'Connell, and Nicholas Blurton Jones, who studied foraging by women of different ages among the Hadza hunter-gatherers of Tanzania. The women who devoted the most time to gathering food (especially roots, honey, and fruit) were postmenopausal women. Those hardworking Hadza grandmothers put in an impressive seven hours per day, compared with a mere three hours for girls not yet pregnant and four and a half hours for women of childbearing age. As one might expect, foraging returns (measured in pounds of food gathered per hour) increased with age and experience, so that mature women achieved higher returns than teenagers. Interestingly, the grandmothers' returns were still as high as women in their prime. The combination of putting in more foraging hours and maintaining an unchanged foraging efficiency meant that the postmenopausal grandmothers brought in more food per day than women of any of the younger groups, even though their large harvests were greatly in excess of their own personal needs and they no longer had dependent young children of their own to feed.

Observations indicated that the Hadza grandmothers were sharing their excess food harvest with close relatives, such as their grandchildren and grown children. As a strategy for transforming food calories into pounds of baby, it's more efficient for an older woman to donate the calories to grandchildren and grown children than to infants of her own, because her fertility decreases with age anyway, while her children are young adults at peak fertility. Naturally, menopausal grandmothers in traditional societies contribute more to their offspring than just food. They also act as baby-sitters for grandchildren, thereby helping their adult children churn out more babies bearing Grandma's genes. And though they work hard for their grandchildren, they're less likely to die as a result of exhaustion than if they were nursing infants as well as caring for them.

Supposedly, natural selection can't weed out mutations that affect only old people, because old people are "postreproductive." But no humans, except hermits, are ever truly postreproductive.

But menopause has another virtue, one that has received little attention. That is the importance of old people to their entire tribe in preliterate societies, which means every human society in the world from the time of human origins until the rise of writing in Mesopotamia around 3300 B.C.

A common genetics argument is that natural selection cannot weed out mutations that do not damage people until they are old, because old people are supposedly "postreproductive." I believe that such statements overlook an essential fact distinguishing humans from most animal species. No humans, except hermits, are ever truly postreproductive, in the sense of being unable to aid in the survival and reproduction of other people bearing their genes. Yes, I grant that if any orangutans lived long enough in the wild to become sterile, they would count as postreproductive, since orangutans (other than mothers with one young offspring) tend to be solitary. I also grant that the contributions of very old people to modern literate societies tend to decrease with age. That new phenomenon of modern societies is at the root of the enormous problems that old

age now poses, both for the elderly themselves and for the rest of society. But we moderns get most of our information through writing, television, or radio. We find it impossible to conceive of the overwhelming importance of elderly people in preliterate societies as repositories of information and experience.

Here is an example of that role. During my field studies of bird ecology on New Guinea and adjacent southwestern Pacific islands, I live among people who traditionally were without writing, depended on stone tools, and subsisted by farming and fishing supplemented by hunting and gathering. I am constantly asking villagers to tell me the names of local birds, animals, and plants in their language, and to tell me what they know about each species. New Guineans and Pacific islanders possess an enormous fund of biological knowledge, including names for a thousand or more species, plus information about where each species occurs, its behavior, its ecology, and its usefulness to humans. All that information is important because wild plants and animals furnish much of the people's food and all their building materials, medicines, and decorations.

Again and again, when I ask about some rare bird, only the older hunters know the answer, and eventually I ask a question that stumps even them. The hunters reply, "We have to ask the old man [or the old woman]." They take me to a hut where we find an old man or woman, blind with cataracts and toothless, able to eat food only after someone else has chewed it. But that old person is the tribe's library. Because the society traditionally lacked writing, that old person knows more about the local environment than anyone else and is the sole person with accurate knowledge of events that happened long ago. Out comes the rare bird's name, and a description of it.

The accumulated experience that the elderly remember is important for the whole tribe's survival. In 1976, for instance, I visited Rennell Island, one of the Solomon Islands, lying in the southwestern Pacific's cyclone belt. When I asked about wild fruits and seeds that birds ate, my Rennellese informants named dozens of plant species by Ren-

nell language names, named for each plant species all the bird and bat species that eat its fruit, and said whether the fruit is edible for people. They ranked fruits in three categories: those that people never eat, those that people regularly eat, and those that people eat only in famine times, such as after—and here I kept hearing a Rennell term initially unfamiliar to me—the *hungi kengi*.

Those words proved to be the Rennell name for the most destructive cyclone to have hit the island in living memory—apparently around 1910, based on people's references to datable events of the European colonial administration. The hungi kengi blew down most of Rennell's forest, destroyed gardens, and drove people to the brink of starvation. Islanders survived by eating fruits of wild plant species that were normally not eaten. But doing so required detailed knowledge about which plants are poisonous, which are not poisonous, and whether and how the poison can be removed by some technique of food preparation.

When I began pestering my middle-aged Rennellese informants with questions about fruit edibility, I was brought into a hut. There, once my eyes had become accustomed to the dim light, I saw the inevitable frail old woman. She was the last living person with direct experience of which plants were found safe and nutritious to eat after the *hungi kengi*, until people's gardens began producing again. The old woman explained that she had been a child not quite of marriageable age at the time of the *hungi kengi*. Since my visit to Rennell was in 1976, and since the cyclone had struck 66 years before, the woman was probably in her early eighties. Her survival after the 1910 cyclone had depended on information remembered by aged survivors of the last big cyclone before the *hungi kengi*. Now her people's ability to survive another cyclone would depend on her own memories, which were fortunately very detailed.

Such anecdotes could be multiplied indefinitely. Traditional human societies face frequent minor risks that threaten a few individuals, and also face rare natural catastrophes or intertribal wars that threaten the lives of everybody in the society. But virtually everyone in a small traditional society is related to one another. Hence old people in a traditional society are essential to the survival not only of their children and grandchildren but also of hundreds of other people who share their genes. In preliterate societies, no one is ever postreproductive.

Any preliterate human societies that included individuals old enough to remember the last *hungi kengi* had a much better chance of surviving the next one than did societies without such old people. The old men were not at risk from childbirth or from exhausting responsibilities of lactation and child care, so they did not evolve protection by menopause. But old women who did not undergo menopause tended to be eliminated from the human gene pool because they remained exposed to the risk of childbirth and the burden of child care. At times of crises, such as a *hungi kengi,*

the prior death of such an older woman also tended to eliminate all the woman's relatives from the gene pool—a huge genetic price to pay just for the dubious privilege of continuing to produce another baby or two against lengthening odds. That's what I see as a major driving force behind the evolution of human female menopause. Similar considerations may have led to the evolution of menopause in female pilot whales. Like us, whales are long-lived, involved in complex social relationships and lifelong family ties, and capable of sophisticated communication and learning.

If one were playing God and deciding whether to make older women undergo menopause, one would do a balance sheet, adding up the benefits of menopause in one column for comparison with its costs in another column. The costs of menopause are the potential children of a woman's old age that she forgoes. The potential

benefits include avoiding the increased risk of death due to childbirth and parenting at an advanced age, and thereby gaining the benefit of improved survival for one's grandchildren, prior children, and more distant relatives. The sizes of those benefits depend on many details: for example, how large the risk of death is in and after childbirth, how much that risk increases with age, how rapidly fertility decreases with age before menopause, and how rapidly it would continue to decrease in an aging woman who did not undergo menopause. All those factors are bound to differ between societies and are not easy for anthropologists to estimate. But natural selection is a more skilled mathematician because it has had millions of years in which to do the calculation. It concluded that menopause's benefits outweigh its costs, and that women can make more by making less.

What's Love Got to Do With It?
Sex Among Our Closest Relatives Is a Rather Open Affair

Meredith F. Small

Maiko and Lana are having sex. Maiko is on top, and Lana's arms and legs are wrapped tightly around his waist. Lina, a friend of Lana's, approaches from the right and taps Maiko on the back, nudging him to finish. As he moves away, Lina enfolds Lana in her arms, and they roll over so that Lana is now on top. The two females rub their genitals together, grinning and screaming in pleasure.

This is no orgy staged for an X-rated movie. It doesn't even involve people—or rather, it involves them only as observers. Lana, Maiko, and Lina are bonobos, a rare species of chimplike ape in which frequent couplings and casual sex play characterize every social relationship—between males and females, members of the same sex, closely related animals, and total strangers. Primatologists are beginning to study the bonobos' unrestrained sexual behavior for tantalizing clues to the origins of our own sexuality.

In reconstructing how early man and woman behaved, researchers have generally looked not to bonobos but to common chimpanzees. Only about 5 million years ago human beings and chimps shared a common ancestor, and we still have much behavior in common: namely, a long period of infant dependency, a reliance on learning what to eat and how to obtain food, social bonds that persist over generations, and the need to deal as a group with many everyday conflicts. The assumption has been that chimp behavior today may be similar to the behavior of human ancestors.

Bonobo behavior, however, offers another window on the past because they, too, shared our 5-million-year-old ancestor, diverging from chimps just 2 million years ago. Bonobos have been less studied than chimps for the simple reason that they are difficult to find. They live only on a small patch of land in Zaire, in central Africa. They were first identified, on the basis of skeletal material, in the 1920s, but it wasn't until the 1970s that their behavior in the wild was studied, and then only sporadically.

Bonobos, also known as pygmy chimpanzees, are not really pygmies but welterweights. The largest males are as big as chimps, and the females of the two species are the same size. But bonobos are more delicate in build, and their arms and legs are long and slender.

On the ground, moving from fruit tree to fruit tree, bonobos often stand and walk on two legs—behavior that makes them seem more like humans than chimps. In some ways their sexual behavior seems more human as well, suggesting that in the sexual arena, at least, bonobos are the more appropriate ancestral model. Males and females frequently copulate face-to-face, which is an uncommon position in animals other than humans. Males usually mount females from behind, but females seem to prefer sex face-to-face. "Sometimes the female will let a male start to mount from behind," says Amy Parish, a graduate student at the University of California at Davis who's been watching female bonobo sexual behavior in several zoo colonies around the world. "And then she'll stop, and of course he's really excited, and then she continues face-to-face." Primatologists assume the female preference is dictated by her anatomy: her enlarged clitoris and sexual swellings are oriented far forward. Females presumably prefer face-to-face contact because it feels better.

Like humans but unlike chimps and most other animals, bonobos separate sex from reproduction. They seem to treat sex as a pleasurable activity, and they rely on it as a sort of social glue,

---

"Sex is fun. Sex makes them feel good and keeps the group together."

---

to make or break all sorts of relationships. "Ancestral humans behaved like this," proposes Frans de Waal, an ethologist at the Yerkes Regional Primate Research Center at Emory University. "Later, when we developed the family system, the use of sex for this sort of purpose became more limited, mainly occurring within families. A lot of the things we see, like pedophilia and homosexuality, may be leftovers that some now consider unacceptable in our particular society."

Depending on your morals, watching bonobo sex play may be like watching humans at their most extreme and perverse. Bonobos seem to have sex more often and in more combinations than the average person in any culture, and most of the time bonobo sex has nothing to do with making babies. Males mount females and females sometimes mount them back; females rub against other females just for fun; males stand rump to rump and press their scrotal areas together. Even juveniles participate by rubbing their genital areas against adults, although ethologists don't think that males actually insert their penises into juvenile females. Very young animals also have sex with each other: little males suck on each other's penises or French-kiss. When two animals initiate sex, others freely join in by poking their fingers and toes into the moving parts.

One thing sex does for bonobos is decrease tensions caused by potential competition, often competition for food. Japanese primatologists observing bonobos in Zaire were the first to notice that when bonobos come across a large fruiting tree or encounter piles of provisioned sugarcane, the sight of food triggers a binge of sex. The atmosphere of this sexual free-for-all is decidedly friendly, and it eventually calms the group down. "What's striking is how rapidly the sex drops off," says Nancy Thompson-Handler of the State University of New York at Stony Brook, who has observed bonobos at a site in Zaire called Lomako. "After ten minutes, sexual behavior decreases by fifty percent." Soon the group turns from sex to feeding.

But it's tension rather than food that causes the sexual excitement. "I'm sure the more food you give them, the more sex you'll get," says De Waal. "But it's not really the food, it's competition that triggers this. You can throw in a cardboard box and you'll get sexual behavior." Sex is just the way bonobos deal with competition over limited resources and with the normal tensions caused by living in a group. Anthropologist Frances White of Duke University, a bonobo observer at Lomako since 1983, puts it simply: "Sex is fun. Sex makes them feel good and therefore keeps the group together."

Sexual behavior also occurs after aggressive encounters, especially among males. After two males fight, one may reconcile with his opponent by presenting his rump and backing up against the other's testicles. He might grab the penis of the other male and stroke it. It's the male bonobo's way of shaking hands and letting everyone know that the conflict has ended amicably.

Researchers also note that female bonobo sexuality, like the sexuality of female humans, isn't locked into a monthly cycle. In most other animals, including chimps, the female's interest in sex is tied to her ovulation cycle.

"Females rule the business. It's a good species for feminists, I think."

Chimp females sport pink swellings on their hind ends for about two weeks, signaling their fertility, and they're only approachable for sex during that time. That's not the case with humans, who show no outward signs that they are ovulating, and can mate at all phases of the cycle. Female bonobos take the reverse tack, but with similar results. Their large swellings are visible for weeks before and after their fertile periods, and there is never any discernibly wrong time to mate. Like humans, they have sex whether or not they are ovulating.

What's fascinating is that female bonobos use this boundless sexuality in all their relationships. "Females rule the business—sex and food," says De Waal. "It's a good species for feminists, I think." For instance, females regularly use sex to cement relationships with other females. A genital-genital rub, better known as GG-rubbing by observers, is the most frequent behavior used by bonobo females to reinforce social ties or relieve tension. GG-rubbing takes a variety of forms. Often one female rolls on her back and extends her arms and legs. The other female mounts her and they rub their swellings right and left for several seconds, massaging their clitorises against

each other. GG-rubbing occurs in the presence of food because food causes tension and excitement, but the intimate contact has the effect of making close friends.

Sometimes females would rather GG-rub with each other than copulate with a male. Parish filmed a 15-minute scene at a bonobo colony at the San Diego Wild Animal Park in which a male, Vernon, repeatedly solicited two females, Lisa and Loretta. Again and again he arched his back and displayed his erect penis—the bonobo request for sex. The females moved away from him, tactfully turning him down until they crept behind a tree and GG-rubbed with each other.

Unlike most primate species, in which males usually take on the dangerous task of leaving home, among bonobos females are the ones who leave the group when they reach sexual maturity, around the age of eight, and work their way into unfamiliar groups. To aid in their assimilation into a new community, the female bonobos make good use of their endless sexual favors. While watching a bonobo group at a feeding tree, White saw a young female systematically have sex with each member before feeding. "An adolescent female, presumably a recent transfer female, came up to the tree, mated with all five males, went into the tree, and solicited GG-rubbing from all the females present," says White.

Once inside the new group, a female bonobo must build a sisterhood from scratch. In groups of humans or chimps, unrelated females construct friendships through the rituals of shopping together or grooming. Bonobos do it sexually. Although pleasure may be the motivation behind a female-female assignation, the function is to form an alliance.

These alliances are serious business, because they determine the pecking order at food sites. Females with powerful friends eat first, and subordinate females may not get any food at all if the resource is small. When times are rough, then, it pays to have close female friends. White describes a scene at Lomako in which an adolescent female, Blanche, benefited from her established friendship with Freda. "I was following Freda and her boyfriend, and they found a tree that they didn't expect to be there. It was a small tree, heavily in fruit with one of their favorites. Freda went straight up the tree and made a food call to Blanche. Blanche came tearing over—she was quite far away—and went tearing up the tree to join Freda, and they GG-rubbed like crazy."

Alliances also give females leverage over larger, stronger males who otherwise would push them around. Females have discovered there is strength in numbers. Unlike other species of primates, such as chimpanzees or baboons (or, all too often, humans), where tensions run high between males and females, bonobo females are not afraid of males, and the sexes mingle peacefully. "What is consistently different from chimps," says Thompson-Handler, "is the composition of parties. The vast majority are mixed, so there are males and females of all different ages."

HIDDEN HEAT

Standing upright is not a position usually—or easily—associated with sex. Among people, at least, anatomy and gravity prove to be forbidding obstacles. Yet our two-legged stance may be the key to a distinctive aspect of human sexuality: the independence of women's sexual desires from a monthly calendar.

Males in the two species most closely related to us, chimpanzees and bonobos, don't spend a lot of time worrying, "Is she interested or not?" The answer is obvious. When ovulatory hormones reach a monthly peak in female chimps and bonobos, and their eggs are primed for fertilization, their genital area swells up, and both sexes appear to have just one thing on their mind. "These animals really turn on when this happens. Everything else is dropped," says primatologist Frederick Szalay of Hunter College in New York.

Women, however, don't go into heat. And this departure from our relatives' sexual behavior has long puzzled researchers. Clear signals of fertility and the willingness to do something about it bring major evolutionary advantages: ripe eggs lead to healthier pregnancies, which leads to more of your genes in succeeding generations, which is what evolution is all about. In addition, male chimps give females that are waving these red flags of fertility first chance at high-protein food such as meat.

So why would our ancestors give this up? Szalay and graduate student Robert Costello have a simple explanation. Women gave heat up, they say, because our ancestors stood up. Fossil footprints indicate that somewhere around 3.5 million years ago hominids—non-ape primates—began walking on two legs. "In hominids, something dictated getting up. We don't know what it was," Szalay says. "But once it did, there was a problem with the signaling system." The problem was that it didn't work. Swollen genital areas that were visible when their owners were down on all fours became hidden between the legs. The mating signal was lost.

"Uprightness meant very tough times for females working with the old ovarian cycle," Szalay says. Males wouldn't notice them, and the swellings themselves, which get quite large, must have made it hard for two-legged creatures to walk around.

Those who found a way out of this quandary, Szalay suggests, were females with small swellings but with a little less hair on their rears and a little extra fat. It would have looked a bit like the time-honored mating signal. They got more attention, and produced more offspring. "You don't start a completely new trend in signaling," Szalay says. "You have a little extra fat, a little nakedness to mimic the ancestors. If there was an ever-so-little advantage because, quite simply, you look good, it would be selected for."

And if a little nakedness and a little fat worked well, Szalay speculates, then a lot of both would work even better. "Once you start a trend in sexual signaling, crazy things happen," he notes. "It's almost like: let's escalate, let's add more. That's what happens in horns with sheep. It's a particular part of the body that brings an advantage." In a few million years human ancestors were more naked than ever, with fleshy rears not found in any other primate. Since these features were permanent, unlike the monthly ups and downs of swellings, sex was free to become a part of daily life.

It's a provocative notion, say Szalay's colleagues, but like any attempt to conjure up the past from the present, there's no real proof of cause and effect. Anthropologist Helen Fisher of the American Museum of Natural History notes that Szalay is merely assuming that fleshy buttocks evolved because they were sex signals. Yet their mass really comes from muscles, which chimps don't have, that are associated with walking. And anthropologist Sarah Blaffer Hrdy of the University of California at Davis points to a more fundamental problem: our ancestors may not have had chimplike swellings that they needed to dispense with. Chimps and bonobos are only two of about 200 primate species, and the vast majority of those species don't have big swellings. Though they are our closest relatives, chimps and bonobos have been evolving during the last 5 million years just as we have, and swollen genitals may be a recent development. The current unswollen human pattern may be the ancestral one.

"Nobody really knows what happened," says Fisher. "Everybody has an idea. You pays your money and you takes your choice."

—Joshua Fischman

Female bonobos cannot be coerced into anything, including sex. Parish recounts an interaction between Lana and a male called Akili at the San Diego Wild Animal Park. "Lana had just been introduced into the group. For a long time she lay on the grass with a huge swelling. Akili would approach her with a big erection and hover over her. It would have been easy for him to do a mount. But he wouldn't. He just kept trying to catch her eye, hovering around her, and she would scoot around the ground, avoiding him. And then he'd try again. She went around full circle." Akili was big enough to force himself on her. Yet he refrained.

In another encounter, a male bonobo was carrying a large clump of branches. He moved up to a female and presented his erect penis by spreading his legs and arching his back. She rolled onto her back and they copulated. In the midst of their joint ecstasy, she reached out and grabbed a branch from the male. When he pulled back, finished and satisfied, she moved away, clutching the branch to her chest. There was no tension between them, and she essentially traded copulation

for food. But the key here is that the male allowed her to move away with the branch—it didn't occur to him to threaten her, because their status was virtually equal.

Although the results of sexual liberation are clear among bonobos, no one is sure why sex has been elevated to such a high position in this species and why it is restricted merely to reproduction among chimpanzees. "The puzzle for me," says De Waal, "is that chimps do all this bonding with kissing and embracing, with body contact. Why do bonobos do it in a sexual manner?" He speculates that the use of sex as a standard way to underscore relationships began between adult males and adult females as an extension of the mating process and later spread to all members of the group. But no one is sure exactly how this happened.

It is also unclear whether bonobo sexuality became exaggerated only after their split from the human lineage or whether the behavior they exhibit today is the modern version of our common ancestor's sex play. Anthropologist Adrienne Zihlman of the University of California at Santa Cruz, who has used the evidence of fossil

bones to argue that our earliest known non-ape ancestors, the australopithecines, had body proportions similar to those of bonobos, says, "The path of human evolution is not a straight line from either species, but what I think is important is that the bonobo information gives us more possibilities for looking at human origins."

Some anthropologists, however, are reluctant to include the details of bonobo life, such as wide-ranging sexuality and a strong sisterhood, into scenarios of human evolution. "The researchers have all these commitments to male dominance [as in chimpanzees], and yet bonobos have egalitarian relationships," says De Waal. "They also want to see humans as unique, yet bonobos fit very nicely into many of the scenarios, making humans appear less unique."

Our divergent, non-ape path has led us away from sex and toward a culture that denies the connection between sex and social cohesion. But bonobos, with their versatile sexuality, are here to remind us that our heritage may very well include a primordial urge to make love, not war.

Apes of Wrath

Barbara Smuts

Barbara Smuts is a professor of psychology and anthropology at the University of Michigan. She has been doing fieldwork in animal behavior since the early 1970s, studying baboons, chimps, and dolphins. "In my work I combine research in animal behavior with an abiding interest in feminist perspectives on science," says Smuts. She is the author of Sex and Friendship in Baboons.

Nearly 20 years ago I spent a morning dashing up and down the hills of Gombe National Park in Tanzania, trying to keep up with an energetic young female chimpanzee, the focus of my observations for the day. On her rear end she sported the small, bright pink swelling characteristic of the early stages of estrus, the period when female mammals are fertile and sexually receptive. For some hours our run through the park was conducted in quiet, but then, suddenly, a chorus of male chimpanzee pant hoots shattered the tranquility of the forest. My female rushed forward to join the males. She

greeted each of them, bowing and then turning to present her swelling for inspection. The males examined her perfunctorily and resumed grooming one another, showing no further interest.

At first I was surprised by their indifference to a potential mate. Then I realized that it would be many days before the female's swelling blossomed into the large, shiny sphere that signals ovulation. In a week or two, I thought, these same males will be vying intensely for a chance to mate with her.

The attack came without warning. One of the males charged toward us,

---

Some female primates use social bonds to escape male aggression. Can women?

---

hair on end, looking twice as large as my small female and enraged. As he rushed by he picked her up, hurled her to the ground, and pummeled her. She cringed and screamed. He ran off,

rejoining the other males seconds later as if nothing had happened. It was not so easy for the female to return to normal. She whimpered and darted nervous glances at her attacker, as if worried that he might renew his assault

In the years that followed I witnessed many similar attacks by males against females, among a variety of Old World primates, and eventually I found this sort of aggression against females so puzzling that I began to study it systematically—something that has rarely been done. My long-term research on olive baboons in Kenya showed that, on average, each pregnant or lactating female was attacked by an adult male about once a week and seriously injured about once a year. Estrous females were the target of even more aggression. The obvious question was, Why?

In the late 1970s, while I was in Africa among the baboons, feminists back in the United States were turning their attention to male violence against women. Their concern stimulated a wave of research documenting disturbingly high levels of battering, rape, sexual harassment, and murder. But although scientists investigated this kind of behavior from many perspectives, they mostly ignored the existence of similar behavior in other animals. My observations over the years have convinced me that a deeper understanding of male aggression against females in other species can help us understand its counterpart in our own.

Researchers have observed various male animals—including insects, birds, and mammals—chasing, threatening, and attacking females. Unfortunately, because scientists have rarely studied such aggression in detail, we do not know exactly how common it is. But the males of many of these species are most aggressive toward potential mates, which suggests that they sometimes use violence to gain sexual access.

Jane Goodall provides us with a compelling example of how males use violence to get sex. In her 1986 book, *The Chimpanzees of Gombe,* Goodall describes the chimpanzee dating game. In one of several scenarios, males gather around attractive estrous females and try to lure them away from other males for a one-on-one sexual expedition that may last for days or weeks. But females find some suitors more appealing than others and often resist the advances of less desirable males. Males often rely on aggression to counter female resistance. For example, Goodall describes how Evered, in "persuading" a reluctant Winkle to accompany him into the forest, attacked her six times over the course of five hours, twice severely.

Sometimes, as I saw in Gombe, a male chimpanzee even attacks an estrous female days before he tries to mate with her. Goodall thinks that a male uses such aggression to train a female to fear him so that she will be more likely to surrender to his subsequent sexual advances. Similarly, male hamadryas baboons, who form small harems by kidnapping child brides, maintain a tight rein over their females through threats and intimidation. If, when another male is nearby, a hamadryas female strays even a few feet from her mate, he shoots her a threatening stare and raises his brows. She usually responds by rushing to his side; if not, he bites the back of her neck. The neck bite is ritualized—the male does not actually sink his razor-sharp canines into her flesh—but the threat of injury is clear. By repeating this behavior hundreds of times, the male lays claim to particular females months or even years before mating with them. When a female comes into estrus, she solicits sex only from her harem master, and other males rarely challenge his sexual rights to her.

In some species, females remain in their birth communities their whole lives, joining forces with related females to defend vital food resources against other females.

These chimpanzee and hamadryas males are practicing sexual coercion: male use of force to increase the chances that a female victim will mate with him, or to decrease the chances that she will mate with someone else. But sexual coercion is much more common in some primate species than in others. Orangutans and chimpanzees are the only nonhuman primates whose males in the wild force females to copulate, while males of several other species, such as vervet monkeys and bonobos (pygmy chimpanzees), rarely if ever try to coerce females sexually. Between the two extremes lie many species, like hamadryas baboons, in which males do not force copulation but nonetheless use threats and intimidation to get sex.

These dramatic differences between species provide an opportunity to investigate which factors promote or inhibit sexual coercion. For example, we might expect to find more of it in species in which males are much larger than females—and we do. However, size differences between the sexes are far from the whole story. Chimpanzee and bonobo males both have only a slight size advantage, yet while male chimps frequently resort to force, male bonobos treat the fair sex with more respect. Clearly, then, although size matters, so do other factors. In particular, the social relationships females form with other females and with males appear to be as important.

In some species, females remain in their birth communities their whole lives, joining forces with related females to defend vital food resources against other females. In such "female bonded" species, females also form alliances against aggressive males. Vervet monkeys are one such species, and among these small and exceptionally feisty African monkeys, related females gang up against males. High-ranking females use their dense network of female alliances to rule the troop; although smaller than males, they slap persistent suitors away like annoying flies. Researchers have observed similar alliances in many other female-bonded species, including other Old World monkeys such as macaques, olive baboons, patas and rhesus monkeys; New World monkeys such as the capuchin; and prosimians such as the ring-tailed lemur.

Females in other species leave their birth communities at adolescence and spend the rest of their lives cut off from their female kin. In most such species, females do not form strong bonds with other females and rarely support one another against males. Both chimpanzees and hamadryas baboons exhibit this pattern, and, as we saw earlier, in both species females submit to sexual control by males.

This contrast between female-bonded species, in which related females gang together to thwart males, and non-female-bonded species, in which they don't, breaks down when we come to the bonobo. Female bonobos, like their close relatives the chimpanzees, leave their kin and live as adults with unrelated females. Recent field studies show that these unrelated females hang out together and engage in frequent homoerotic behavior, in which they embrace face-to-face and rapidly rub their genitals together; sex seems to cement their bonds. Examining these studies in the context of my own research has convinced me that one way females use these bonds is to form alliances against males, and that, as a consequence, male bonobos do not dominate females or attempt to coerce them sexually. How and why female bonobos, but not chimpanzees, came up with this solution to male violence remains a mystery.

Some of the factors that influence female vulnerability to male sexual coercion in different species may also help explain such variation among different groups in the same species.

Female primates also use relationships with males to help protect themselves against sexual coercion. Among olive baboons, each adult female typically forms long-lasting "friendships" with a few of the many males in her troop. When a male baboon assaults a female, another male often comes to her rescue; in my troop, nine times out of ten the protector was a friend of the female's. In return for his protection, the defender may enjoy her sexual favors the next time she comes into estrus. There is a dark side to this picture, however. Male baboons frequently threaten or attack their female friends—when, for example, one tries to form a friendship with a new male. Other males apparently recognize friendships and rarely intervene. The female, then, becomes less vulnerable to aggression from males in general, but more vulnerable to aggression from her male friends.

As a final example, consider orangutans. Because their food grows so sparsely adult females rarely travel with anyone but their dependent offspring. But orangutan females routinely fall victim to forced copulation. Female orangutans, it seems, pay a high price for their solitude.

Some of the factors that influence female vulnerability to male sexual coercion in different species may also help explain such variation among different groups in the same species. For example, in a group of chimpanzees in the Taï Forest in the Ivory Coast, females form closer bonds with one another than do females at Gombe. Taï females may consequently have more egalitarian relationships with males than their Gombe counterparts do.

Such differences between groups especially characterize humans. Among the South American Yanomamö, for instance, men frequently abduct and rape women from neighboring villages and severely beat their wives for suspected adultery. However, among the Aka people of the Central African Republic, male aggression against women has never been observed. Most human societies, of course, fall between these two extremes.

How are we to account for such variation? The same social factors that help explain how sexual coercion differs among nonhuman primates may deepen our understanding of how it varies across different groups of people. In most traditional human societies, a woman leaves her birth community when she marries and goes to live with her husband and his relatives. Without strong bonds to close female kin, she will probably be in danger of sexual coercion. The presence of close female kin, though, may protect her. For example, in a community in Belize, women live near their female relatives. A man will sometimes beat his wife if he becomes jealous or suspects her of infidelity, but when this happens, onlookers run to tell her female kin. Their arrival on the scene, combined with the presence of other glaring women, usually shames the man enough to stop his aggression.

Even in societies in which women live away from their families, kin may provide protection against abusive husbands, though how much protection varies dramatically from one society to the next. In some societies a woman's kin, including her father and brothers, consistently support her against an abusive husband, while in others they rarely help her. Why?

The key may lie in patterns of male-male relationships. Alliances between males are much more highly developed in humans than in other primates, and men frequently rely on such alliances to compete successfully against other men. They often gain more by supporting their male allies than they do by supporting female kin. In addition, men often use their alliances to defeat rivals and abduct or rape their women, as painfully illustrated by recent events in Bosnia. When women live far from close kin, among men who value their alliances with other men more than their bonds with women, they may be even more vulnerable to sexual coercion than many nonhuman primate females.

Even in societies in which women live away from their families, kin may provide protection against abusive husbands.

Like nonhuman primate females, many women form bonds with unrelated males who may protect them from other males. However, reliance on men exacts a cost—women and other primate females often must submit to control by their protectors. Such control is more elaborate in humans because allied men agree to honor one another's proprietary rights over women. In most of the world's cultures, marriage involves not only the exclusion of other men from sexual access to a man's wife—which protects the woman against rape by other men—but also entails the husband's right to complete control over his wife's sexual life, including the right to punish her for real or suspected adultery, to have sex with her whenever he wants, and even to restrict her contact with other people, especially men.

In modern industrial society, many men—perhaps most—maintain such traditional notions of marriage. At the same time, many of the traditional sources of support for women, including censure of abusive husbands by the woman's kinfolk or other community members, are eroding as more and more people end up without nearby kin or long-term neighbors. The increased vulnerability of women isolated from their birth communities, however, is not just a by-product of modern living. Historically, in highly patriarchal societies like those found in China and northern India, married women lived in households ruled by their husband's mother and male kin, and their ties with their own kin were virtually severed. In these societies, today as in the past, the husband's female kin often view the wife as a competitor for resources. Not only do they fail to support her against male coercive control, but they sometimes actively encourage it. This scenario illustrates an important point: women do not invariably support other women against men, in part because women may perceive their interests as best served through alliances with men, not with other women. When men have most of the power and control most of the resources, this looks like a realistic assessment.

Decreasing women's vulnerability to sexual coercion, then, may require fundamental changes in social alliances. Women gave voice to this essential truth with the slogan SISTERHOOD IS POWERFUL—a reference to the importance of women's ability to cooperate with unrelated women as if they were indeed sisters. However, among humans, the male-dominant social system derives support from political, economic, legal, and ideological institutions that other primates can't even dream of. Freedom from male control—including male sexual coercion—therefore requires women to form alliances with one another (and with like-minded men) on a scale beyond that shown by nonhuman primates and humans in the past. Although knowledge of other primates can provide inspiration for this task, its achievement depends on the uniquely human ability to envision a future different from anything that has gone before.

The Hominid Transition

A most intriguing and perplexing gap in the fossil record for human evolution appears in the transition from a common link between apes and humans to that which is clearly recognizable as a member of our own kind, the family *Hominidae*.

The issues involving this "black hole" cannot be resolved by simply filling it in with fossil finds. Even if we had the physical remains of the earliest hominids in front of us, which we do not have, there is no way such evidence could thoroughly answer the questions that physical anthropologists care most deeply about: How did these creatures move about and get their food? Did they cooperate and share? On what levels did they think and communicate? Did they have a sense of family, let alone a sense of self? In one way or another, all of the previous articles on primates relate to these issues, as do some of the subsequent ones on the fossil evidence. But what sets off this unit from the others is that the various authors attempt to deal with these matters head on, even in the absence of direct fossil evidence. Christophe Boesch and Hedwige Boesch-Achermann, in "Dim Forest, Bright Chimps," indicate that some aspects of "hominization" (the acquisition of such humanlike qualities as cooperative hunting and food sharing) actually may have begun in the African rain forest rather than in the dry savanna, as has usually been proposed. They base their suggestions on some remarkable first-hand observations of forest-dwelling chimpanzees.

As if to show that chimpanzee behavior may vary according to local circumstances, just as we know human behavior does, Craig Stanford, in "To Catch a Colobus,"
contrasts his observations of chimpanzee hunting in Gombe National Park with the findings of the Boesches.

Recent research, discussed in "Ape Cultures and Missing Links" by Richard Wrangham, has shown some striking resemblances between apes and humans, hinting that such qualities might have been characteristic of our common ancestor. Following this line of reasoning, teaching a bonobo how to make and use stone tools, as revealed by Sue Savage-Rumbaugh and Roger Lewin in "Ape at the Brink," allows us to make educated guesses as to the mental and physical processes of our hominid predecessors.

Taken collectively, the articles in this section show how far anthropologists are willing to go to construct theoretical formulations based upon limited data. Although making so much out of so little may be seen as a fault, and may generate irreconcilable differences among theorists, a readiness to entertain new ideas should be welcomed for what it is: a stimulus for more intensive and meticulous research.

Looking Ahead: Challenge Questions

What are the implications for human evolution of tool use, social hunting, and food-sharing among Ivory Coast chimpanzees?

What kinds of physical and mental skills are required in order to be a stone toolmaker?

How did the common ancestor of apes and humans probably get food?

What makes humans so different from the apes?

Why did our ancestors become bipedal?

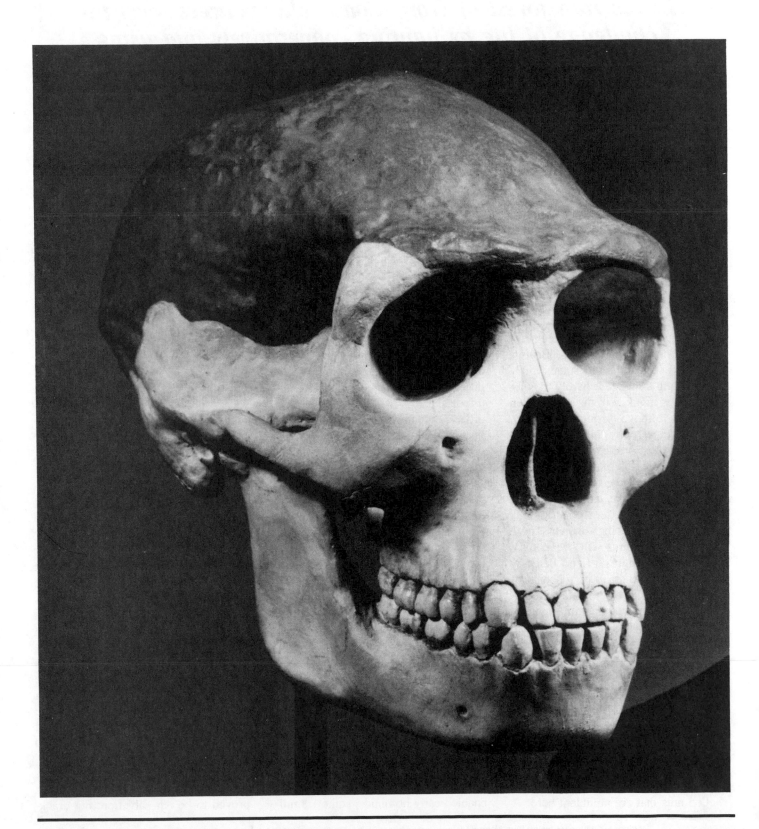

Dim Forest, Bright Chimps

In the rain forest of Ivory Coast, chimpanzees meet the challenge of life by hunting cooperatively and using crude tools

Christophe Boesch and Hedwige Boesch-Achermann

Taï National Park, Ivory Coast, December 3, 1985. Drumming, barking, and screaming, chimps rush through the undergrowth, little more than black shadows. Their goal is to join a group of other chimps noisily clustering around Brutus, the dominant male of this seventy-member chimpanzee community. For a few moments, Brutus, proud and self-confident, stands fairly still, holding a shocked, barely moving red colobus monkey in his hand. Then he begins to move through the group, followed closely by his favorite females and most of the adult males. He seems to savor this moment of uncontested superiority, the culmination of a hunt high up in the canopy. But the victory is not his alone. Cooperation is essential to capturing one of these monkeys, and Brutus will break apart and share this highly prized delicacy with most of the main participants of the hunt and with the females. Recipients of large portions will, in turn, share more or less generously with their offspring, relatives, and friends.

In 1979, we began a long-term study of the previously unknown chimpanzees of Taï National Park, 1,600 square miles of tropical rain forest in the Republic of the Ivory Coast (Côte d'Ivoire). Early on, we were most interested in the chimps' use of natural hammers—branches and stones—to crack open the five species of hard-shelled nuts that are abundant here. A

sea otter lying on its back, cracking an abalone shell with a rock, is a familiar picture, but no primate had ever before been observed in the wild using stones as hammers. East Africa's savanna chimps, studied for decades by Jane Goodall in Gombe, Tanzania, use twigs to extract ants and termites from their nests or honey from a bees' nest, but they have never been seen using hammerstones.

As our work progressed, we were surprised by the many ways in which the life of the Taï forest chimpanzees differs from that of their savanna counterparts, and as evidence accumulated, differences in how the two populations hunt proved the most intriguing. Jane Goodall had found that chimpanzees hunt monkeys, antelope, and wild pigs, findings confirmed by Japanese biologist Toshida Nishida, who conducted a long-term study 120 miles south of Gombe, in the Mahale Mountains. So we were not surprised to discover that the Taï chimps eat meat. What intrigued us was the degree to which they hunt cooperatively. In 1953 Raymond Dart proposed that group hunting and cooperation were key ingredients in the evolution of *Homo sapiens*. The argument has been modified considerably since Dart first put it forward, and group hunting has also been observed in some social carnivores (lions and African wild dogs, for instance), and even some birds of prey. Nevertheless, many anthropologists still hold that hunting cooperatively and sharing food played a central role in the drama that enabled early hominids, some 1.8 mil-

lion years ago, to develop the social systems that are so typically human.

We hoped that what we learned about the behavior of forest chimpanzees would shed new light on prevailing theories of human evolution. Before we could even begin, however, we had to habituate a community of chimps to our presence. Five long years passed before we were able to move with them on their daily trips through the forest, of which "our" group appeared to claim some twelve square miles. Chimpanzees are alert and shy animals, and the limited field of view in the rain forest—about sixty-five feet at best—made finding them more difficult. We had to rely on sound, mostly their vocalizations and drumming on trees. Males often drum regularly while moving through the forest: pant-hooting, they draw near a big buttress tree; then, at full speed they fly over the buttress, hitting it repeatedly with their hands and feet. Such drumming may resound more than half a mile in the forest. In the beginning, our ignorance about how they moved and who was drumming led to failure more often than not, but eventually we learned that the dominant males drummed during the day to let other group members know the direction of travel. On some days, however, intermittent drumming about dawn was the only signal for the whole day. If we were out of earshot at the time, we were often reduced to guessing.

During these difficult early days, one feature of the chimps' routine proved to be our salvation: nut cracking is a noisy business. So noisy, in fact, that in the early days of French colonial rule, one officer apparently even proposed the theory that some unknown tribe was forging iron in the impenetrable and dangerous jungle.

Guided by the sounds made by the chimps as they cracked open nuts, which they often did for hours at a time, we were gradually able to get within sixty feet of the animals. We still seldom saw the chimps themselves (they fled if we came too close), but even so, the evidence left after a session of nut cracking taught us a great deal about what types of nuts they were eating, what sorts of hammer and anvil tools they were using, and—thanks to the very distinctive noise a nut makes when it finally splits open—how many hits were needed to crack a nut and how many nuts could be opened per minute.

After some months, we began catching glimpses of the chimpanzees before they fled, and after a little more time, we were able to draw close enough to watch them at work. The chimps gather nuts from the ground. Some nuts are tougher to crack than others. Nuts of the *Panda oleosa* tree are the most demanding, harder than any of the foods processed by present-day hunter-gatherers and breaking open only when a force of 3,500 pounds is applied. The stone hammers used by the Taï chimps range from stones of ten ounces to granite blocks of four to forty-five pounds. Stones of any size, however, are a rarity in the forest and are seldom conveniently placed near a nut-bearing tree. By observing closely, and in some cases imitating the way the chimps handle hammerstones, we learned that they have an impressive ability to find just the right tool for the job at hand. Taï chimps could remember the positions of many of the stones scattered, often out of sight, around a panda tree. Without having to run around rechecking the stones, they would select one of appropriate size that was closest to the tree. These mental abilities in spatial representation compare with some of those of nine-year-old humans.

To extract the four kernels from inside a panda nut, a chimp must use a hammer with extreme precision. Time and time again, we have been impressed to see a chimpanzee raise a twenty-pound stone above its head, strike a nut with ten or more powerful blows, and then, using the same hammer, switch to delicate little taps from a height of only four inches. To finish the job, the chimps often break off a small piece of twig and use it to extract the last tiny fragments of kernel from the shell. Intriguingly, females crack panda nuts more often than males, a gender difference in tool use that seems to be more pronounced in the forest chimps than in their savanna counterparts.

After five years of fieldwork, we were finally able to follow the chimpanzees at close range, and gradually, we gained insights into their way of hunting. One morning, for example, we followed a group of six male chimps on a three-hour patrol that had taken them into foreign territory to the north. (Our study group is one of five chimpanzee groups more or less evenly distributed in the Taï forest.) As always during these approximately monthly incursions, which seem to be for the purpose of territorial defense, the chimps were totally silent, clearly on edge and on the lookout for trouble. Once the patrol was over, however, and they were back within their own borders, the chimps shifted their attention to hunting. They were after monkeys, the most abundant mammals in the forest. Traveling in large, multi-species groups, some of the forest's ten species of monkeys are more apt than others to wind up as a meal for the chimps. The relatively sluggish and large (almost thirty pounds) red colobus monkeys are the chimps' usual fare. (Antelope also live in the forest, but in our ten years at Taï, we have never seen a chimp catch, or even pursue, one. In contrast, Gombe chimps at times do come across fawns, and when they do, they seize the opportunity—and the fawn.)

The six males moved on silently, peering up into the vegetation and stopping from time to time to listen for the sound of monkeys. None fed or groomed; all focused on the hunt. We followed one old male, Falstaff, closely,

for he tolerates us completely and is one of the keenest and most experienced hunters. Even from the rear, Falstaff set the pace; whenever he stopped, the others paused to wait for him. After thirty minutes, we heard the unmistakable noises of monkeys jumping from branch to branch. Silently, the chimps turned in the direction of the sounds, scanning the canopy. Just then, a diana monkey spotted them and gave an alarm call. Dianas are very alert and fast; they are also about half the weight of colobus monkeys. The chimps quickly gave up and continued their search for easier, meatier prey.

Shortly after, we heard the characteristic cough of a red colobus monkey. Suddenly Rousseau and Macho, two twenty-year-olds, burst into action, running toward the cough. Falstaff seemed surprised by their precipitousness, but after a moment's hesitation, he also ran. Now the hunting barks of the chimps mixed with the sharp alarm calls of the monkeys. Hurrying behind Falstaff, we saw him climb up a conveniently situated tree. His position, combined with those of Schubert and Ulysse, two mature chimps in their prime, effectively blocked off three of the monkeys' possible escape routes. But in another tree, nowhere near any escape route and thus useless, waited the last of the hunters, Kendo, eighteen years old and the least experienced of the group. The monkeys, taking advantage of Falstaff's delay and Kendo's error, escaped.

The six males moved on and within five minutes picked up the sounds of another group of red colobus. This time, the chimps approached cautiously, nobody hurrying. They screened the canopy intently to locate the monkeys, which were still unaware of the approaching danger. Macho and Schubert chose two adjacent trees, both full of monkeys, and started climbing very quietly, taking care not to move any branches. Meanwhile, the other four chimps blocked off anticipated escape routes. When Schubert was halfway up, the monkeys finally detected the two chimps. As we watched the colobus monkeys take off in literal panic, the appropriateness of the chimpanzees' scientific name—*Pan* came to mind: with a certain stretch of the imagination, the fleeing monkeys could be shepherds and shepherdesses frightened at the sudden appearance of Pan, the wild Greek god of the woods, shepherds, and their flocks.

Taking off in the expected direction, the monkeys were trailed by Macho and Schubert. The chimps let go with loud hunting barks. Trying to escape, two colobus monkeys jumped into smaller trees lower in the canopy. With this, Rousseau and Kendo, who had been watching from the ground, sped up into the trees and tried to grab them. Only a third of the weight of the chimps, however, the monkeys managed to make it to the next tree along branches too small for their pursuers. But Falstaff had anticipated this move and was waiting for them. In the following confusion, Falstaff seized a juvenile and killed it with a bite to the neck. As the chimps met in a rush on the ground, Falstaff began to eat, sharing with Schubert and Rousseau. A juvenile colobus does not provide much meat, however, and this time, not all the chimps got a share. Frustrated individuals soon started off on another hunt, and relative calm returned fairly quickly: this sort of hunt, by a small band of chimps acting on their own at the edge of their territory, does not generate the kind of high excitement that prevails when more members of the community are involved.

So far we have observed some 200 monkey hunts and have concluded that success requires a minimum of three motivated hunters acting cooperatively. Alone or in pairs, chimps succeed less than 15 percent of the time, but when three or four act as a group, more than half the hunts result in a kill. The chimps seem well aware of the odds; 92 percent of all the hunts we observed were group affairs.

Gombe chimps also hunt red colobus monkeys, but the percentage of group hunts is much lower: only 36 percent. In addition, we learned from Jane Goodall that even when Gombe chimps do hunt in groups, their strategies are different. When Taï chimps

arrive under a group of monkeys, the hunters scatter, often silently, usually out of sight of one another but each aware of the others' positions. As the hunt progresses, they gradually close in, encircling the quarry. Such movements require that each chimp coordinate his movements with those of the other hunters, as well as with those of the prey, at all times.

Coordinated hunts account for 63 percent of all those observed at Taï but only 7 percent of those at Gombe. Jane Goodall says that in a Gombe group hunt, the chimpanzees typically travel together until they arrive at a tree with monkeys. Then, as the chimps begin climbing nearby trees, they scatter as each pursues a different target. Goodall gained the impression that Gombe chimps boost their success by hunting independently but simultaneously, thereby disorganizing their prey; our impression is that the Taï chimps owe their success to being organized themselves.

Just why the Gombe and Taï chimps have developed such different hunting strategies is difficult to explain, and we plan to spend some time at Gombe in the hope of finding out. In the meantime, the mere existence of differences is interesting enough and may perhaps force changes in our understanding of human evolution. Most currently accepted theories propose that some three million years ago, a dramatic climate change in Africa east of the Rift Valley turned dense forest into open, drier habitat. Adapting to the difficulties of life under these new conditions, our ancestors supposedly evolved into cooperative hunters and began sharing food they caught. Supporters of this idea point out that plant and animal remains indicative of dry, open environments have been found at all early hominid excavation sites in Tanzania, Kenya, South Africa, and Ethiopia. That the large majority of apes in Africa today live west of the Rift Valley appears to many anthropologists to lend further support to the idea that a change in environment caused the common ancestor of apes and humans to evolve along a different line from those remaining in the forest.

Our observations, however, suggest quite another line of thought. Life in dense, dim forest may require more sophisticated behavior than is commonly assumed: compared with their savanna relatives, Taï chimps show greater complexity in both hunting and tool use. Taï chimps use tools in nineteen different ways and have six different ways of making them, compared with sixteen uses and three methods of manufacture at Gombe.

Anthropologist colleagues of mine have told me that the discovery that some chimpanzees are accomplished users of hammerstones forces them to look with a fresh eye at stone tools turned up at excavation sites. The important role played by female Taï chimps in tool use also raises the possibility that in the course of human evolution, women may have been decisive in the development of many of the sophisticated manipulative skills characteristic of our species. Taï mothers also appear to pass on their skills by actively teaching their offspring. We have observed mothers providing their young with hammers and then stepping in to help when the inexperienced youngsters encounter difficulty. This help may include carefully showing how to position the nut or hold the hammer properly. Such behavior has never been observed at Gombe.

Similarly, food sharing, for a long time said to be unique to humans, seems more general in forest than in savanna chimpanzees. Taï chimp mothers share with their young up to 60 percent of the nuts they open, at least until the latter become sufficiently adept, generally at about six years old. They also share other foods acquired with tools, including honey, ants, and bone marrow. Gombe mothers share such foods much less often, even with their infants. Taï chimps also share meat more frequently than do their Gombe relatives, sometimes dividing a chunk up and giving portions away, sometimes simply allowing beggars to grab pieces.

Any comparison between chimpanzees and our hominid ancestors can only be suggestive, not definitive. But our studies lead us to believe that the process of hominization may have begun independently of the drying of the environment. Savanna life could even have delayed the process; many anthropologists have been struck by how slowly hominid-associated remains, such as the hand ax, changed after their first appearance in the Olduvai age.

Will we have the time to discover more about the hunting strategies or other, perhaps as yet undiscovered abilities of these forest chimpanzees? Africa's tropical rain forests, and their inhabitants, are threatened with extinction by extensive logging, largely to provide the Western world with tropical timber and such products as coffee, cocoa, and rubber. Ivory Coast has lost

90 percent of its original forest, and less than 5 percent of the remainder can be considered pristine. The climate has changed dramatically. The harmattan, a cold, dry wind from the Sahara previously unknown in the forest, has now swept through the Taï forest every year since 1986. Rainfall has diminished; all the rivulets in our study region are now dry for several months of the year.

In addition, the chimpanzee, biologically very close to humans, is in demand for research on AIDS and hepatitis vaccines. Captive-bred chimps are available, but they cost about twenty times more than wild-caught animals.

Chimps taken from the wild for these purposes are generally young, their mothers having been shot during capture. For every chimp arriving at its sad destination, nine others may well have died in the forest or on the way. Such priorities—cheap coffee and cocoa and chimpanzees—do not do the economies of Third World countries any good in the long run, and they bring suffering and death to innocent victims in the forest. Our hope is that Brutus, Falstaff, and their families will survive, and that we and others will have the opportunity to learn about them well into the future. But there is no denying that modern times work against them and us.

To Catch a Colobus

Chimpanzees in Gombe National Park band together to kill nearly
a fifth of the red colobus monkeys in their range

Craig B. Stanford

Craig B. Stanford is an assistant professor of anthropology at the University of Southern California. His first fieldwork on primates was in Peru, where he studied tamarins. For his Ph.D. at the University of California, Berkeley, Stanford traveled to India and Bangladesh to investigate ecological influences on social behavior in capped langur monkeys. Stanford hopes to expand his research to the evolution of hunting behavior in primates, including humans.

On a sunny July morning, I am sitting on the bank of Kakombe Stream in Gombe National Park, Tanzania. Forty feet above my head, scattered through large fig trees, is a group of red colobus monkeys. This is J group, whose twenty-five members I have come to know as individuals during several seasons of fieldwork. Gombe red colobus are large, long-tailed monkeys, with males sometimes weighing more than twenty pounds. Both sexes have a crown of red hair, a gray back, and buff underparts. The highlight of this particular morning has been the sighting of a new infant, born sometime in the previous two days. As the group feeds noisily on fruit and leaves overhead, I mull over the options for possible names for the infant.

While I watch the colobus monkeys, my attention is caught by the loud and excited pant-hoots of a party of chimpanzees farther down the valley. I

judge the group to be of considerable size and traveling in my direction. As the calls come closer, the colobus males begin to give high-pitched alarm calls, and mothers gather up their infants and climb higher into the tree crowns.

A moment later, a wild chorus of panthoots erupts just behind me, followed by a cacophony of colobus alarm calls, and it is obvious to both J group and to me that the chimps have arrived. The male chimps immediately climb up to the higher limbs of the tall albizia tree into which most of the colobus group have retreated. Colobus females and their offspring huddle high in the crown, while a phalanx of five adult males descends to meet the advancing ranks of four adult male chimpanzees, led by seventeen-year-old, 115-pound Frodo. Frodo is the most accomplished hunter of colobus monkeys at Gombe and the only one willing to take on several colobus males simultaneously in order to catch his prey. The other hunters keep their distance while Frodo first scans the group of monkeys, then advances upon the colobus defenders. Time and again he lunges at the colobus males, attempting to race past them and into the cluster of terrified females and infants. Each time he is driven back; at one point, the two largest males of J group leap onto Frodo's back until he retreats, screaming, a few yards away.

A brief lull in the hunt follows, during which the colobus males run to one another and embrace for reassurance, then part to renew their de-

fense. Frodo soon charges again into the midst of the colobus males, and this time manages to scatter them long enough to pluck the newborn from its mother's abdomen. In spite of fierce opposition, Frodo has caught his quarry, and he now sits calmly and eats it while the other hunters and two female chimps—their swollen pink rumps a sign that they are in estrus, a period of sexual receptivity—sit nearby begging for meat. The surviving colobus monkeys watch nervously from a few feet away. Minutes later, the mother of the dead infant attempts to approach, perhaps to try to rescue her nearly consumed offspring. She is chased, falls from the tree to the forest floor, and is pounced upon and killed by juvenile chimpanzees that have been watching the hunt from below. Seconds later, before these would-be hunters have had a chance to begin their meal, Wilkie, the chimpanzee group's dominant male, races down the tree and steals the carcass from them. He shows off his prize by charging across the forest floor, dead colobus in hand, and then, amid a frenzy of chimps eager for a morsel, he sits down to share the meat with his ally Prof and two females from the hunting party.

Until Jane Goodall observed chimpanzees eating meat in the early 1960s, they were thought to be complete vegetarians. We now know that a small but regular portion of the diet of wild chimps consists of the meat of such mammals as bush pigs, small antelopes, and a variety of monkey species.

For example, chimpanzees in the Mahale Mountains of Tanzania, the Taï forest of Ivory Coast, and in Gombe all regularly hunt red colobus monkeys. Documenting the effect of such predation on wild primate populations, however, is extremely difficult because predators—whether chimps, leopards, or eagles—are generally too shy to hunt in the presence of people. The result is that even if predation is a regular occurrence, researchers are not likely to see it, let alone study it systematically.

Gombe is one of the few primate study sites where both predators and their prey have been habituated to human observers, making it possible to witness hunts. I have spent the past four field seasons at Gombe, studying the predator-prey relationship between the 45-member Kasakela chimpanzee community and the 500 red colobus monkeys that share the same twelve square miles of Gombe National Park. Gombe's rugged terrain is composed of steep slopes of open woodland, rising above stream valleys lush with riverine forest. The chimpanzees roam across these hills in territorial communities, which divide up each day into foraging parties of from one to forty animals. So far, I have clocked in more than a thousand hours with red colobus monkeys and have regularly followed the chimps on their daily rounds, observing some 150 encounters between the monkeys and chimps and more than 75 hunts. My records, together with those of my colleagues, show that the Gombe chimps may kill more than 100 red colobus each year, or nearly one-fifth of the colobus inhabiting their range. Most of the victims are immature monkeys under two years old. Also invaluable have been the data gathered daily on the chimps for the past two decades by a team of Tanzanian research assistants.

One odd outcome of my work has been that I am in the unique position of knowing both the hunters and their victims as individuals, which makes my research intriguing but a bit heart wrenching. In October 1992, for example, a party of thirty-three chimpanzees encountered my main study group, J, in upper Kakombe valley.

The result was devastating from the monkeys' viewpoint. During the hour-long hunt, seven were killed; three were caught and torn apart right in front of me. Nearly four hours later, the hunters were still sharing and eating the meat they had caught, while I sat staring in disbelief at the remains of many of my study subjects.

Determined to learn more about the chimp-colobus relationship, however, I continued watching, that day and many others like it. I will need several more field seasons before I can measure the full impact of chimpanzee hunting on the Gombe red colobus, but several facts about hunting and its effects on the monkeys have already emerged. One major factor that determines the outcome of a hunt in Gombe is the number of male chimps involved. (Although females also hunt, the males are responsible for more than 90 percent of all colobus kills.) Red colobus males launch a courageous counterattack in response to their chimpanzee predators, but their ability to defend their group is directly proportional to the number of attackers and does not seem to be related to the number of defenders. The outcome of a hunt is thus almost always in the hands of the chimps, and in most instances, the best the monkeys can hope to do is limit the damage to a single group member rather than several. Chimpanzees have a highly fluid social grouping pattern in which males tend to travel together while females travel alone with their infants. At times, however, twenty or more male and female chimpanzees forage together. When ten or more male chimps hunt together, they are successful nine times out of ten, and the colobus have little hope of escape.

Hunting success depends on other factors as well. Unlike the shy red-tailed and blue monkeys with which they share the forest (and which are rarely hunted by the chimps), red colobus do not flee the moment they hear or see chimps approaching. Instead, the red colobus give alarm calls and adopt a vigilant wait-and-see strategy, with males positioned nearest the potential attackers. The alarm calls increase in frequency and intensity as the

chimpanzees draw closer and cease only when the chimps are sighted beneath the tree. Then, the colobus sit quietly, watching intently, and only if the chimps decide to hunt do the colobus males launch a counterattack. The monkeys' decision to stand and fight rather than flee may seem maladaptive given their low rate of successful defense. I observed, however, that when the monkeys scatter or try to flee, the chimps nearly always pursue and catch one or more of them.

Fleeing red colobus monkeys are most likely to be caught when they have been feeding on the tasty new leaves of the tallest trees, the "emergents," which rise above the canopy. When these trees are surrounded by low plant growth, they frequently become death traps because the only way colobus can escape from attacking chimps is to leap out of the tree—often into the waiting arms of more chimpanzees on the ground below.

One of my primary goals has been to learn why a party of chimps will eagerly hunt a colobus group one day while ignoring the same group under seemingly identical circumstances on another. One determinant is the number of males in the chimp party: the more males, the more likely the group will hunt. Hunts are also undertaken mainly when a mother colobus carrying a small infant is visible, probably because of the Gombe chimps' preference for baby red colobus, which make up 75 percent of all kills. The situation is quite different in the Taï forest, where half of the chimp kills are adult colobus males (*see* "Dim Forest, Bright Chimps," *Natural History,* September 1991). Christophe and Hedwige Boesch have shown that the Taï chimps hunt cooperatively, perhaps because red colobus monkeys are harder to catch in the much taller canopy of the Taï rainforest. Successful Taï chimp hunters also regularly share the spoils. In contrast, each chimp in Gombe appears to have his own hunting strategy.

The single best predictor of when Gombe chimps will hunt is the presence of one or more estrous females in the party. This finding, together with the earlier observation by Geza Teleki (formerly of George Washington University) that male hunters tend to give meat preferentially to swollen females traveling with the group, indicates that Gombe chimps sometimes hunt in order to obtain meat to offer a sexually receptive female. Since hunts also occur when no estrous females are present, this trade of sex for meat cannot be the exclusive explanation, but the implications are nonetheless intriguing. Gombe chimps use meat not only for nutrition; they also share it with their allies and withhold it from their rivals. Meat is thus a social, political, and even reproductive tool. These "selfish" goals may help explain why the Gombe chimps do not cooperate during a hunt as often as do Taï chimps.

Whatever the chimps want the monkey meat for, their predation has a severe effect on the red colobus population. Part of my work involves taking repeated censuses of the red colobus groups living in the different valleys that form the hunting range of our chimpanzees. In the core area of the range, where hunting is most intense, predation by chimps is certainly the limiting factor on colobus population growth: red colobus group size in this area is half that at the periphery of the chimps' hunting range. The number of infant and juvenile red colobus monkeys is particularly low in the core area; most of the babies there are destined to become chimpanzee food.

The proportion of the red colobus population eaten by chimps appears to fluctuate greatly from year to year, and probably from decade to decade, as the number of male hunters in the chimpanzee community changes. In the early 1980s, for instance, there were five adult and adolescent males in the Kasakela chimp community, while today there are eleven; the number of colobus kills per year has risen as the number of hunters in the community has grown.

Furthermore, a single avid hunter may have a dramatic effect. I estimate that Frodo has single-handedly killed up to 10 percent of the entire red colobus population within his hunting range. I now want to learn if chimps living in forests elsewhere in Africa are also taking a heavy toll of red colobus monkeys. If they are, then they will add support to the theory that predation is an important limiting factor on wild primate populations and may also influence some aspects of behavior. Meanwhile, I will continue to watch in awe as Frodo and his fellow hunters attack my colobus monkeys and to marvel at the courageousness of the colobus males that risk their lives to protect the other members of their group.

Ape at the Brink

Two and a half million years ago, an early human ancestor recognized that in a stone lies the possibility of a tool. Four years ago a chimp named Kanzi saw that, too.

Sue Savage-Rumbaugh and Roger Lewin

Sue Savage-Rumbaugh is a professor of biology and psychology at Georgia State University, at whose Language Research Center she conducts her research with bonobos. She has written over 100 scientific papers and two books; her volume Ape Language: From Conditioned Response to Symbol *was published in 1986. Savage-Rumbaugh's research is supported by the National Institute of Child Health and Human Development.*

Roger Lewin is the author of numerous books about human origins, including Origins, People of the Lake, *and* Origins Reconsidered, *all coauthored with Richard Leakey. In 1989 he received the inaugural Lewis Thomas Award for excellence in the communication of the life sciences, and in 1991 he was corecipient, with E. O. Wilson, of the Society for Conservation Biology's annual award for services to conservation.*

Threading my way along the sandy path toward the ocean shore, I sought out the rhythmic sound of shifting surf. The faint light of predawn arrived, and

I could see the rock coastline ahead, then the silhouette of distant mountains. I was near the small coastal village of Cascais, Portugal, attending a meeting organized by the Wenner-Gren Foundation, a group legendary in anthropological circles. Scientists invited to these meetings are kept away from the rest of the world and encouraged to examine each other's views in small, intense discussions.

Walking along the beach, I mused over the talk of the past few days. Bill Calvin, a neurobiologist at the University of Washington, had been telling us about the extraordinary accuracy and power with which humans can throw. We humans aren't the only primates with the raw ability—chimpanzees and gorillas can throw, too, as visitors to zoos sometimes discover to their chagrin. Apes do not enjoy being stared at and frequently throw things at visitors in an attempt to make them leave. But humans are far better at throwing than apes are, and the development of this skill, Bill had pointed out, was clearly important during man's evolution from an apelike ancestor. In particular, the

accurate hurling of stones became a valuable means of hunting and self-defense against predators.

Another scientist in our group, archeologist Nick Toth of Indiana University, was also interested in throwing, but for a different reason. Nick, unlike the rest of us, knew how to make the stone tools that our prehuman ancestors had utilized.

Nick was not a typical scientist. I'd recognized this right away when, in the course of discussion, he began pulling fist-size rocks out of his briefcase. He riveted the group's attention with his display and with his demonstrations of how rocks can become tools. He explained the physics of conchoidal fracture by which rocks can be made to yield good, sharp tools. Then he challenged us to accompany him to the beach to try to make the "crude" stone tools that our hominid ancestors made 2 million years ago. That afternoon I gained a newfound respect for the feats of my forebears.

It was my first attempt to emulate a Paleolithic stone knapper, and I did not find it an easy task. Neither I nor most

From *Discover* magazine, September 1994, pp. 91-96, 98. Adapted from *Kanzi: The Ape at the Brink of the Human Mind* by Sue Savage-Rumbaugh and Roger Lewin. © 1994 by Sue Savage-Rumbaugh and Roger Lewin. Reprinted by permission of John Wiley & Sons, Inc.

of the other "educated" scientists could coax even a single flake from the stones on the beach during our first half hour of trying. We even resorted to placing one stone on the ground and slamming another against it, but to no avail. Finally, instead of just watching Nick, I began to look closely at what he was doing. Why did the stones break so easily when he struck them together with such little force, while they just made a loud thud when I slammed them together as hard as I could?

I gradually recognized that Nick was not really hitting rocks together; instead he was throwing the rock in his right hand against the edge of the rock in his left hand, letting the force of the controlled throw knock off the flake. The "hammer rock" never really left his right hand, but it was nonetheless thrown, as a missile, against the "core," the rock held in place in his left hand. What had I been doing? Just slamming two rocks together as though I were clapping my hands with rocks in between.

Once I realized how Nick was actually flaking stone, I grasped the profound similarity between throwing and stone knapping. In each activity you must be able to snap the wrist rapidly forward at just the right moment during the downward motion of the forearm. This wrist-cocking action produces great force, either for achieving distance in throwing or for knocking a flake off a core. I also learned that it is important to deliver your blow to the core accurately. Several of us had bruised fingers after the afternoon's stone-knapping excursion, suggesting that, accurate though we might be as a species, as individuals we needed practice.

Bill Calvin was likely correct in his suggestion that throwing ability had been selected for in the course of human evolution. But now I saw that accurate throwers also had the potential skills for making stone tools. Could throwing as a defensive device have paved the way for the deliberate construction of stone tools?

Our conference was searching for evolutionary links between language, tools, and anatomy that could lead to the emergence of the bipedal, large-brained, technological creature that is *Homo sapiens*. The neurobiology of stone throwing and the skills of stone knapping were new to me, and as a psychobiologist, I was intrigued. Now as I walked on the beach, I attempted to integrate these ideas with my own knowledge of how apes understand language.

For the two decades I have known and studied chimpanzees, I have been attempting to discern the degree to which they can think and communicate as we do. The initial efforts of ape-language researchers, in the 1960s and early 1970s, were hurriedly greeted with acclaim. Newspapers and scientific journals declared the same message: apes can use symbols in a way that echoes the structure of human language, albeit in a modest manner. The symbols were not in the form of spoken words, of course, but were produced variously as hand gestures from American Sign Language, as col-

---

Were the earliest toolmakers doing something that was beyond the cognitive ability of apes? Or were they merely bipedal apes who were applying their apelike cognitive skills to non-apelike activities?

---

ored plastic shapes, and as arbitrary lexigrams on a computer keyboard.

But in the late 1970s and early 1980s this fascination turned to cynicism. Linguists asserted that apes were merely mimicking their caretakers and that they displayed no languagelike capacity at all. Most linguists and psychologists wanted to forget apes and move ahead with what they viewed as the "proper study of man"—generally typified by the analysis of the problem-solving strategies of freshman students.

From my earliest exposure to apes, I recognized that there would be considerable difficulty in determining whether or not they employed words with intent and meaning in the same way that we do. And so, in my research at the Language Research Center at Georgia State University, I searched for scientifically credible ways to approach the fundamental questions about apes and their intellectual and emotional capacities. By 1990, the year of the Wenner-Gren conference, I knew that at least some of this work was reaching an audience, or I would not have been invited to the conference. Perhaps there, I thought, I would have a chance to begin to tell my story—or, more accurately, Kanzi's story.

One ape out of the 11 that I have studied, a 150-pound bonobo (or pygmy chimpanzee) named Kanzi, began to learn language on his own, without drills or lessons. Kanzi, a male, was born on October 28, 1980, at the Yerkes Regional Primate Research Center's field station in Lawrenceville, Georgia. Before this time, no bonobo had been language trained. Matata, Kanzi's adoptive mother, was to be the first.

Matata proved to be a willing, though incompetent, study. She quickly understood that other chimpanzees used the keyboard to communicate and that pressing the lexigrams was what achieved this feat. However, after two years of training and 30,000 trials, she mastered only six symbols, in a limited way.

After Matata's departure, we set up the keyboard in the expectation that Kanzi would begin his language instruction—if he could learn to sit in one place long enough. Kanzi, however, had his own opinion of the keyboard, and he began at once to make it evident. Not only was he using the keyboard as a means of communicating, but he also knew what the symbols meant. For example, one of the first things he did that morning was to activate the symbol for "apple," then "chase." He then picked up an apple, looked at me, and ran away with a grin on his face. I was hesitant to believe what I knew I was seeing. Kanzi appeared to know all the things we had attempted to teach Matata, yet we had not even been attending to him. Could he simply have picked up his understanding through social exposure, as children do?

For 17 months we kept a complete record of Kanzi's utterances, either directly on the computer when he was indoors, or manually while outdoors. By the end of the period, Kanzi had a vocabulary of about 50 symbols. He was already producing combinations of words—spontaneous utterances such as "Matata group-room tickle" to ask that his mother be permitted to join in a game of tickle in the group room.

We first detected what seemed like spoken word comprehension when Kanzi was one and a half years old. We began to notice that often, when we talked about lights, Kanzi would run to the switch on the wall and flip it on and off. Kanzi seemed to be "listening in" on conversations that had nothing to do with him, in a manner that I had not experienced in other apes—even in those who had been reared in human homes. As time passed, Kanzi appeared able to understand more and more spoken words. In response, we had to do what many parents do when they don't want their children to overhear: we began to spell out some words around Kanzi. We were able to determine that Kanzi understood 150 spoken words at the end of the 17-month period.

"If an ape can begin to comprehend spoken English without being so trained," I later wrote in a scientific paper, "it would appear that the ape possessed speech and language abilities similar to our own." The dual lesson we learned from the project was that chimpanzees can acquire language skills spontaneously, through social exposure to a language-rich environment, as human children do. And, again as for humans, early exposure is critical. As Elizabeth Bates comments, "The Berlin Wall is down, and so is the wall that separates man from chimpanzee."

It was the end of a long day at the Wenner-Gren conference, and we had all eaten dinner at a restaurant in the nearby town. On our return to the hotel, I was sitting near the rear of the bus, and Nick Toth was in the very back, legs stretched out, arms folded across his chest, eyes closed, apparently asleep. Suddenly he opened an eye and beckoned me to join him. "I have something I want to ask you," he said. "Do you think Kanzi could learn to make stone tools, like early humans did?"

His question seemed to come right out of the blue. It was something I had never thought of trying. From my long experience with chimpanzees, I had gained a great respect for their abilities. But making stone tools seemed light-years beyond them. Indeed, even I could not make a worthwhile stone tool, and I'd had Nick there to teach me.

Nevertheless, I was intrigued. "What do you have in mind?" I asked. Nick sat up and quickly explained.

In 1949 the British anthropologist Kenneth P. Oakley published a classic book, *Man the Tool-Maker.* This short volume encapsulated what was widely held to set humans apart as unique: "Possession of a great capacity for conceptual thought . . . is now generally regarded by comparative psychologists as distinctive of man," he wrote. "The systematic making of tools . . . required not only for immediate use but for future use, implies a marked capacity for conceptual thought." The notion of man the tool-maker struck a receptive chord: alone among the world's species, toolmaking *Homo sapiens* fashions an elaborate culture and manufactures a powerful technology, through which the world is forever changed.

The shift to becoming a toolmaker has been seen as central to what differentiated humans from apes in an evolutionary sense. By definition, therefore, the very first members of the human family must have been toolmakers. This assumption has been challenged in the past several decades. The first members of the human family are now known to have evolved at least 5 mil-

lions years ago, perhaps as many as 8 million. And yet the first recognizable stone artifacts date only to around 2.5 million years ago. The appearance of these stone tools coincides with the first appearance of the genus *Homo,* which eventually gave rise to modern humans.

This raises an important question: Were the earliest toolmakers doing something that was beyond the cognitive ability of apes? Or were they merely bipedal apes who were applying their apelike cognitive skills to non-apelike activities?

Nick told me that he had been musing over this question for a long time, and that he had an idea in search of a collaborator. His proposal was to motivate Kanzi to make stone flakes, not to teach him with structured lessons. "We want to avoid the criticism of classical conditioning," he said. He suggested we would need a box with a transparent lid. Something enticing would be put in the box, and the lid would be secured with a length of string. Kanzi could be shown by example how to make flakes, by knocking two rocks together, but there would be no active teaching, no shaping of his hands, no breaking the task down into component parts.

Kanzi very quickly learned to discriminate between sharp flakes and dull ones, using visual inspection and his lips.

I made some suggestions about how the design of the food box, or "tool site" as we came to call it, could be improved; Nick had underestimated Kanzi's ability to tear flimsy objects apart, especially if there is food inside. Nick promised to get in touch with me after we returned to the United States. This he did within a couple of weeks, and I told him that we had made a tool site to his specifications. A week later he arrived at the Language Research Center in Atlanta with fellow archeologist Kathy Schick, their truck laden with a thousand pounds of rock.

At first we set up the tool site outside Kanzi's cage, so that Nick could show Kanzi how it was possible to gain access to the baited box. Nick struck a cobble with a hammerstone, selected a sharp flake, and then cut the string securing the lid to the box. Kanzi got the treat that was inside. Nick did this several times, after which we put the tool site inside Kanzi's enclosure. Nick knelt outside, making flakes. He handed sharp ones to me while I was inside with Kanzi, and I encouraged Kanzi to use them to cut the string. He very soon realized the utility of a sharp flake and eagerly took one from whoever was in with him. He then quickly went to the tool site to open the box. He even knocked two rocks together on several occasions, but in a rather desultory way, and without producing flakes. Nevertheless, he was clearly emulating Nick.

During that first afternoon, and throughout the project, Kanzi was never required to perform a task but was merely provided with the opportunity to participate if he wanted to. We wanted to motivate him to make and use flakes, and we hoped he would learn by example. As the days and weeks passed, he displayed a degree of persistence at the task that exceeded anything I'd seen him do.

Kanzi very quickly learned to discriminate between sharp flakes and dull ones, using visual inspection and his lips. On one occasion about three weeks into the project, one of our collaborators at Georgia State, psychologist Rose Sevcik, was striking a rock when—for the first time for her—it split, and several flakes flew off in different directions. Kanzi was watching closely and seemed to know which was the best of the flakes even before they had hit the ground. He let out a bonobo squeal of delight, rushed to pick up the sharpest flake, and was off to the tool site with it, all in one fluid motion.

Making flakes for himself, however, proved difficult. At first he was ex-

tremely tentative in the way he hit the rocks together. Almost always he used his right hand to deliver the hammer blow. He held the core in his left hand, often cradled against his chest, or sometimes braced against the floor, with his foot adding further support. Sometimes he put the core on the ground and simply struck it with the hammerstone. No one had demonstrated this "anvil" technique to him. But no matter how he held the core, he seemed unable or unwilling to deliver a powerful blow. Bonobos are three times stronger than a human of the same size, so there was no doubt that Kanzi had the muscle power to do the job. We wondered whether he was nervous about hitting his fingers; perhaps he lacked the correct wrist anatomy to produce a "snapping" action—the structure of a bonobo's arms, wrists, and hands is different from a human's (chimpanzees' wrists stiffened as they became adept knuckle walkers), and it constrains the animal's ability to deliver a sharp blow by snapping the wrist; or perhaps he was reluctant to deliver a hard blow because throughout his life we had discouraged him from slamming and breaking objects.

Then, one afternoon eight weeks into the project, I was sitting in my office when I was suddenly assailed with the sound of a BANG . . . BANG . . . BANG. I rushed to the tool-site room, and there was Kanzi, stone knapping with tremendous force. He had finally learned how to fracture rocks to make sharp flakes, albeit small ones.

During the first three months of the project Kanzi became steadily more proficient at producing flakes, in part because he seemed to have learned to aim the hammer blows at the edge of the core. But despite his willingness to deliver harder blows than he had initially, he still wasn't hitting hard enough to produce flakes bigger than about an inch long. Nevertheless, he persisted with his newfound concentration, and we in turn made the string that secured the tool site thicker and

One day Kanzi just sat there looking at me. Then at the rock. Then at me again, apparently reflecting. Suddenly he stood up and, with clear deliberation, threw the rock on the tile floor.

thicker, so that small flakes would wear out before they cut the string.

One day during the fourth month, Kanzi was having only modest success at producing flakes. He turned to me and held out the rocks, as if to say, "Here, you do it for me." He did this from time to time, and mostly I would encourage him to try some more, which is what I did that day. He just sat there looking at me, then at the rock, then at me again, apparently reflecting. Suddenly he stood up and, with clear deliberation, threw the rock on the hard tile floor. The rock shattered, producing a whole shower of flakes. Kanzi vocalized ecstatically, grabbed one of the sharpest flakes, and headed for the tool site.

There was no question that Kanzi had reasoned through the problem and had found a better solution to making flakes. No one had demonstrated the efficacy of throwing. Kanzi had just worked it out for himself. I was delighted, because it demonstrated his ingenuity in the face of a difficult problem. I quickly telephoned Nick and told him what had happened. I was so excited that I didn't give a thought to the fact that Nick might not be delighted, too. He wasn't. He was disappointed. "The Oldowan toolmakers used hard-hammer percussion, not throwing," he said.

"Oldowan" is the name applied to the earliest known stone-tool assemblages, which were found in Africa and date to 2.5 million years ago. The artifacts that make up Oldowan assemblages were produced from small cobbles, and they include about half a dozen forms of so-called core tools—

"Oldowan" is the name applied to the earliest known stone-tool assemblages, which were found in Africa and date to 2.5 million years ago.

such as hammerstones, choppers, and scrapers—and small, sharp flakes. The toolmakers were assumed to have had mental templates of these various tool types. The tools are often found in association with broken animal bones, which sometimes show signs of butchery. The clear inference is that, beginning about 2.5 million years ago, our human ancestors began exploiting their environment in a non-apelike way, by using stone tools as a means of including significant amounts of meat in their diet.

Until quite recently archeologists argued that the earliest toolmakers lived lives analogous to those of contemporary hunter-gatherers: they organized themselves into small, mobile bands, established temporary home bases, and divided the labor of hunting and gathering between male and female members of the band. This was a very humanlike way of life, albeit in primitive form, and most definitely unlike that of an ape.

In recent years, however, a reexamination of the archeological evidence has changed this picture dramatically, making it much less humanlike and more apelike. There is considerable debate over the extent to which these early members of the human family were active hunters as opposed to opportunistic scavengers. And the notion of home bases and a division of labor between the sexes has been abandoned as untenable. The earliest toolmakers are now viewed as bipedal apes who lived and foraged in social groups in a woodland-savanna environment, as baboons and chimpanzees do.

An equally important shift of perspective has taken place regarding the tool assemblages themselves. Nick Toth

began a program of experimental archeology in the 1970s, in which he became a proficient maker of Oldowan artifacts himself. "My experimental findings suggest that far too much emphasis has been put on cores at the expense of flakes," he wrote. "It seems possible that the traditional relationship might be reversed: the flakes may have been the primary tools and the cores often (although not always) simply the by-product of manufacture. . . . Thus the shape of many early cores may have been incidental to the process of manufacture and therefore indicative of neither the maker's purpose nor the artifact's function."

According to [one] new theory, the half-dozen different tool types in Oldowan assemblages were not the product of mental templates in the minds of sophisticated toolmakers.

Nick's reassessment of the Oldowan artifacts revolutionized African archeology and further changed the perception of the humanness of the earliest toolmakers. According to this new theory, the half-dozen different tool types in Oldowan assemblages were not the product of mental templates in the minds of sophisticated toolmakers. The only skill required, therefore, was that of striking flakes off a core using a hammerstone.

"If Kanzi throws the rocks, the percussion marks will be random, and we won't learn anything," Nick protested. Our different reactions reflected, I suppose, the different interests of the psychologist and the archeologist. Nick said I had to discourage Kanzi from throwing, and I pointed out that that would be difficult. "Try," said Nick. I agreed to try.

Rose Sevcik came up with the ob-

vious suggestion, which was to cover the floor with soft carpeting. The first time Kanzi went into the carpeted room, he threw the rock a few times and looked puzzled when it didn't shatter as usual. He paused a few seconds, looked around until he found a place where two pieces of carpet met, pulled back a piece, and hurled the rock. We have assembled a videotape of the tool-making project, which I show to scientific and more general audiences. Whenever the tape reaches this incident there is always a tremendous roar of approval as Kanzi—the hero—outwits the humans yet again.

By this time, spring was approaching, and we decided to take the tool site outdoors, where there was no hard floor to throw against. Forced to abandon his throwing technique, Kanzi steadily became more efficient at hardhammer percussion, delivering more forceful and more precisely aimed blows. Very consistently now, Kanzi was hitting the edge of the core and was more successful at producing flakes. The resulting cores were sometimes very simple, with just a couple of flakes removed, or, if Kanzi had persistently hammered at them, they had many small flake scars and steep, battered edges, some of which resembled eoliths, or "dawn stones," found in Europe in the decades around the turn of the century. There had been great controversy about these objects, with some arguing that they were true artifacts. They turned out to have been the product of natural forces, such as wave action or glaciation.

Just as Kanzi was becoming quite proficient at hard-hammer percussion, he foiled us yet again—which again delighted the psychologist and dismayed the archeologist. Kanzi discovered that even outside on soft ground he could exploit his throwing technique. This discovery seemed to be the result of a thoughtful analysis of the problem as well: he placed a rock carefully on the ground, stepped back, and took careful aim with the second rock, poised in his right hand. His aim was true, and the rock shattered. He continued to use this technique, and there was no way of stopping him. As far as I was concerned, we had presented Kanzi with a problem and he had figured out the best way to solve it—three times.

Kanzi had become a toolmaker. But how good a toolmaker? Could he have stood shoulder to shoulder with the makers of Oldowan tools, striking flakes off cores as effectively as they did?

Kanzi had become a toolmaker. But our question was, how good a toolmaker? Could he have stood shoulder to shoulder with the makers of Oldowan tools, striking flakes off cores as effectively as they did? Nick's experience as an Oldowan toolmaker offered us a way of addressing these questions. On that beach in Portugal, I had been impressed by how very difficult it is to produce flakes. The initial inclination of the naive stone knapper is to hit the core hard enough so that a flake will pop out of the core, as if it were being chiseled out. But as Nick demonstrated, the flakes come from the bottom of the core, not the top. The best everyday example of the principle of conchoidal fracture at work in stone toolmaking is the effect of a tiny pebble hitting a window: a cone of glass is punched out of the pane, and the exact shape of the cone is determined by the direction at which the stone hits the glass.

For effective flaking by hard-hammer percussion, three conditions have to be met. First, the core must have an acute edge (one with an angle of less than 90 degrees). Second, the core must be struck with a sharp, glancing blow, hitting about half an inch from the edge. And third, the blow must be directed through an area of high mass, such as a ridge or a bulge. With these conditions met, and starting with suitable raw material, you can form long, sharp flakes. Whatever forms are produced, they have the appearance of great simplicity. But as Nick correctly points out, "it is the process, not the product, that reveals the complexity of Oldowan toolmaking."

Nick and Kathy Schick recently drew up a list of criteria by which to assess the technological sophistication of simple tools. "It was necessary to get beyond relying on gut reaction for distinguishing between true artifacts and naturally fractured stone," explains Nick. The criteria include: flake angle, formed by its top and bottom surfaces; flake size, an indication of how efficiently flakes are being detached from the core; and the amount of step fractures and battering seen. Step fractures are unclean breaks. When a stone is hit at exactly the proper angle, a clean flake falls off, leaving the stone surface as smooth as if someone had run a knife through butter. If a stone is merely slammed into another hard surface, with little regard for the angle of the blow, it may break, but it will have a battered appearance. Well-flaked stone looks as though it has been sculpted or chiseled.

Measured against these criteria, the products of the earliest toolmakers score high. Their makers knew about angles required on the core, about sharp, glancing blows, and about seeking regions of high mass on the core. Therefore the Oldowan toolmakers displayed significant technological sophistication and perceptual skills.

What of Kanzi? His progress in hard-hammer percussion has been considerable, moving from the undirected, timid tapping together of rocks to forceful hammering. Nick describes the process of learning to make tools as being punctuational, with periods of slow change in between. "You suddenly get an insight into what is required and then slowly improve on that," he explains. Kanzi clearly had an insight into the importance of hitting the rock close to its edge; and he had important insights when he developed his throwing techniques. Despite this, however, he has not yet developed the stone-knapping skills of the Oldowan toolmakers.

Kanzi's relative low degree of technological finesse seems to imply that these early humans had indeed ceased to be apes.

Kanzi's relatively low degree of technological finesse seems to imply that these early humans had indeed ceased to be apes. It isn't yet certain, however, if Kanzi's poorer performance is the result of a cognitive or an anatomical limitation. I suspect that if Kanzi is limited in the quality of flaking through hard-hammer percussion, it is the result of biomechanical, not cognitive, constraints. Or simply lack of practice—certainly most of us working with Kanzi are unable to make stone tools ourselves. Without a good teacher and constant practice it is a very difficult skill to master. Making stone tools does not seem to be a skill that normal human beings acquire readily with little instruction, as we are asking Kanzi to do.

Nick hopes to learn whether or not Kanzi, with minimal demonstration, can acquire a skill that took, at best, many generations for our ancestors to perfect. If Kanzi does not succeed in matching the skills of the Oldowan toolmakers in the span of one research career, it would still be foolish to rule out the potential of the ape mind to do so, given a few generations of exposure to a need to use such tools.

The greatest surprise of the toolmaking project was Kanzi's development of throwing as a way of obtaining sharp edges. Not only did it reflect a problem-solving process in Kanzi's mind, but it also produced material that addresses an important archeological problem: What did early humans do *before* they made Oldowan tools? The criteria mentioned earlier to identify genuine artifacts as compared with naturally fractured stone would reject Kanzi's flakes and cores as tools. And yet they are artifacts, and they can be used as cutting tools.

Some of Kanzi's cores look rather similar to Oldowan core tools, acknowledges Nick, but most do not. "If I were surveying a Stone Age site and found some of these things, I'd definitely check them out, but I would almost certainly conclude they were naturally flaked," he says. "But after seeing these incipient flaking skills with Kanzi, we certainly have to consider it as a possible model for the earliest stone-tool making. He has taught us what we should be looking for."

Nick joked that Kanzi should be awarded an honorary doctorate, pointing out that he would need a small cap and a gown with long arms. He wasn't joking, though, when in the spring of 1991 he conferred on Kanzi the inaugural CRAFT annual award for outstanding research. Nick and Kathy are codirectors of CRAFT, the Center for Research into the Anthropological Foundations of Technology, at Indiana University. "The award is justified," says Nick, "because the work with Kanzi has given us one of our most important insights into Paleolithic technology. It has given us a view of what is possible with apes, and an insight into the cognitive background of what is necessary to go further."

FURTHER READING

Making Silent Stones Speak: Human Evolution and the Dawn of Technology. Kathy D. Schick and Nicholas Toth. Simon and Schuster, 1993. Two field researchers describe early tools and toolmakers and discuss how new technologies shape the course of human evolution.

Ape Cultures and Missing Links

Richard W. Wrangham

Richard W. Wrangham, M. A. Oxford University (New College) 1970; Ph.D. Cambridge University (St. Johns College) 1975, is professor of Anthropology at Harvard University. He has held academic appointments at the University of Michigan, King's College (Cambridge, England), Stanford University, and Bristol University, and is on the Board of Trustees at the Center for Advanced Study of Behavioral Sciences, Stanford, the Dian Fossey Gorilla Foundation, and the Jane Goodall Institute. His current research interests are the nutritional ecology of chimpanzees compared to other primates, the role of chimpanzees in the frugivore community, and functional aspects of communication. Prof. Wrangham does two to seven months of fieldwork annually in western Uganda as Director of the Kibale Chimpanzee Project. In addition to numerous articles and chapters, his books include Current Problems in Sociobiology *(1982),* Ecology and Social Evolution: Birds and Mammals *(1986),* Primate Societies *(1987), and* Chimpanzee Cultures *(1994). He is a Fellow of the American Academy of Arts and Sciences, and received the Rivers Medal from the Royal Anthropological Institute in 1993.*

. . . In recognition of Gordon Getty's extraordinary reach, I'm going to address the three questions at the heart of

This paper was the first Getty Lecture of the Leakey Foundation, presented at the American Museum of Natural History on October 21, 1994.

the Leakey Foundation's mission. I'll frame those questions in a minute. But first, ladies and gentlemen, we've all had a long day, so please fill your glasses, sit back, and relax. And incidentally, as far as I'm concerned feel free to drink your wine with your fingers, or by dipping a napkin in it, or by sucking the tablecloth, or however you choose. I say this because I want to encourage you into the spirit of our ancestors . . . but, well, I'll come back to all that in a minute.

Last Saturday, six days ago, I was in Kibale Forest in western Uganda with a party of ten chimpanzees. About eight o'clock, we met a group of sixty red colobus monkeys. The high-ranking chimps stopped and stared. The younger adult males did the same. The colobus chirped in alarm. Some chimps started climbing. Others watched from the ground. Within minutes, chimps were hunting. Two drove a party of monkeys towards a third waiting in ambush fifty feet above the ground. The colobus did their best to file away through the tree-crowns, searching for an escape among the branches. But they found their path blocked. They turned, and tried another escape. The chimpanzees kept turning them back. The hunts went on for an hour and twenty minutes. At one point, thirty monkeys were trapped on a high branch, two chimps drove them higher, till one by one they jumped. There was a chimp waiting at the landing-point. The first three just escaped. The fourth was caught. The fracas went on. There were fourteen separate hunts in an hour and a half. By the end, three colobus were dead, three chimps had

killed, and five human observers were enthralled.

How things have changed. In 1959, 100 years after *The Origin of the Species* was published, humans were the only primate known to prey on mammals. Last week's observation would have been a paper in *Science*. Today, thanks to grantees of the Leakey Foundation, it's almost routine. We know now that chimpanzees everywhere kill and eat their own prey; that to do so, they often use elaborate cooperative strategies; that the meat is held by males, who share it with friends and lovers in exchange for favors; and that they can hunt so well and so often as to kill 15–30% of their prey population per year, a higher proportion than any carnivore does.

So what does this sort of observation mean for our history? Does it suggest a cooperatively hunting, killer-ape in our past? Some people think so. But why shouldn't we focus on other apes instead? For example, think about bonobos, the sister species to the chimpanzee. Bonobos live in similar forests with similar monkeys. They like to eat meat. But they don't cooperate in hunting monkeys. They don't even kill monkeys, even though they occasionally catch them and play with them like pets! And when they do eat meat, (meat of small antelopes), it's the females, not males that hold the carcass. Should we think, because of bonobos, that our male ancestors disdained the hunt, and ceded meat to females?

I'm not going to focus on hunting this evening. I use hunting just as an example. The same issues apply to any behaviors we're interested in. Whether

Fig. 1. *Where Do We Come From? What Are We? Where Are We Going?*
Paul Gauguin (1848-1903) Museum of Fine Arts, Boston.

we're talking about hunting, or communicating, or tool-using, or anything else, we have to sort out what ape behavior today means for the human past.

In this lecture I'm going to argue that to be with chimpanzees in an African forest, is to climb into a time machine . . . that by stepping into the world of these extraordinary apes we move back six million years, to glimpse where we have come from. The glimpse isn't a perfect picture, but it's amazingly good. That's the argument.

Let's begin by looking back 25 years. In those first years of the Leakey Foundation, I couldn't have made *any* suggestion about apes as time machines without sounding very silly. At that time, with genetic and fossil data still poor, apes and humans were thought to be distantly related, not only to each other but also to their common ancestors.

Apes were certainly *fascinating* to visionaries like Louis Leakey, but then to that extraordinary man, everything was interesting. Happily, he supported Jane Goodall. And he was thrilled when she found chimpanzees modifying tools (and hunting prey), because this meant that chimps were a sort of bridge between humans and other primates. This gave flesh to the idea of evolution. But because at that time, 25 years ago, the kinship between humans and chimpanzees was thought to have ended in the distant past, maybe 15–20 million years ago, no-one was sure

what these observations meant for our history. And anyway, the idea of chimps as a bridge was undermined by an apparent gulf between apes and humans in certain critical aspects of behavior.

Certainly there were *some* similarities. Mothers were strongly attached to their infants. Many features were strikingly similar. But the parallels evaporated at a critical point: there was no evidence of serious aggression. Chimpanzees were seen to live wonderfully peaceful lives. So human society was something apart. Reviewing the chimpanzee studies of the 1960s, Robert Ardrey decisively affirmed the human-ape divide. The life of chimpanzees was an "arcadian existence of primal innocence."

This became the conventional wisdom for other apes. George Schaller and Dian Fossey found gorillas to be a gentle giant. Their new picture rightly challenged the view that gorillas were natural aggressors towards people. In so doing, it left them unconnected with modern human behavior. So the prevailing view was that "human forms of social life were largely unique to humans, created by us, subject to human manipulation according to our vision of human good."[1] Apes had nature; people had culture, and culture, it seemed, wasn't always so great.

In the first hundred years after Darwin, in every area of human thought, people were searching for new meanings

of human existence . . . and this was a common conclusion. Paul Gauguin was one of the first artists to do what the Leakey Foundation does, to search in the primitive. This painting (Figure 1) he considered his spiritual legacy. It looks to an imagined past, a primitive idyll where man and nature lived in harmony. It has on it, written in the top left-hand corner, the three questions of the Leakey Foundation, questions that go back to Thomas Carlyle's *Sartor Resartus*. On the right is a newborn child, representing "Where do we come from?" The figure plucking fruit in the centre shows our day to day existence: ("What are we?"). On the left, an old woman facing death symbolizes concern for the future: "Where are we going?"

You might think that, like ourselves today, Gauguin would have been inspired by his exploration of the human past, present and future. No; he was oppressed. Near the centre, you can see by the tree of knowledge two sinister figures. Their sombre colors show the suffering that comes from leaving nature, pain that Gauguin felt acutely. For Gauguin, human history was a story of acquired sin.

The challenge of completing the painting kept him alive during a period of depression over his daughter's death, but its conclusions left him empty. As soon as he'd finished his masterpiece, he walked out into the mountains, took a massive dose of arsenic, and lay down to wait for death. Should we feel the same depression from looking into the past? Was Gauguin right to see humans as figures of tragedy, doomed by the very abilities of brains and culture that represent the best of our achievements? No. We have new, more confident answers now, coming not from an imaginary vision of primitive Tahiti but from the real world of living primates. The story of human evolution that emerges is different from Gauguin's, still discomfiting, but much richer and more inspiring. It's a story of unfinished challenges. I'm going to address them by taking Gauguin's, and our Foundation's three questions in turn. Let's begin with the past. "Where do we come from?"

So what made a savannah-living, upright hominid out of a forest-living quadrupedal ape? And what was that ancestral species like, in how it looked and how it behaved? I claimed just now that our prehominid ancestor looked like a chimpanzee. Let me explain why I think so.

First, it's obvious that the three African apes, chimpanzees, gorillas, and bonobos, are all very similar, much more like each other than they are like any other species. The genetic evidence unambiguously supports our intuitions. Let's look at these three species.

Genetic evidence from Phil Morin, Maryellen Ruvolo and others show that West African chimps have been separate for about one-and-a-half million years from chimpanzees in East Africa. But morphologically, there's very little difference in chimpanzees across the continent. Like all the great apes, this a conservative species. Most gorillas are lowland gorillas. Recent Leakey Foundation studies are exciting because they are some of the first to watch lowland gorillas undisturbed. They are so similar to chimpanzees that people can find it hard to tell big chimpanzees and small gorillas apart.

Bonobos are the third African ape. They look so like chimpanzees that they weren't recognized to be different until 1933, when they were called "pygmy chimpanzees." But so-called pygmy chimpanzees are actually no smaller than some chimpanzees. Most people prefer to call them bonobos. They live south of the Zaïre River, where there are no gorillas or chimpanzees. Chimpanzees live north of the Zaïre River, and share much of their range with gorillas.

The evolutionary relationship among these three apes is undisputed: chimpanzees and bonobos split most recently, around 2.5 m y ago, and their common ancestor split with gorillas much earlier, about 8–10 m y ago. So where do humans fit? Probably everyone here knows of the shocking genetic evidence now showing chimpanzees to be more closely related to humans than they are to gorillas. The last four years in particular give mounting confidence to this view, as every new nuclear or mitochondrial gene is looked at, currently more than 10 genes in detail as well as from DNA hybridisation looking at the genome as a whole. This means that human ancestors are no sister group to the apes, but instead arose within the African ape tree. Our hominid ancestors apparently split from the chimp-bonobo line *after* the split from gorillas. Louis Leakey, a great iconoclast, would have loved it. Now, this surprise gives us an unexpected bonus. It implies that our ancestral 6 m y species is likely to have been very like a modern-day chimpanzee. We can see why by reconstructing our various common ancestors.

First, what was the common ancestor of chimpanzees and bonobos like? The answer depends on comparisons with gorillas, the first ape to split off, Which is more similar to gorillas? Is it chimpanzees? or bonobos? The answer is clear: In characteristics that differ between chimpanzees and bonobos, chimpanzees are consistently more like gorillas. This is true for things we can see, such as the body build, the shape of the head, or the structure of the genitals, as well as those we can't, such as chromosomes and blood groups. Chimpanzees are like small gorillas, whereas bonobos are like changed chimpanzees. So the common ancestor of chimpanzees and bonobos should have looked like a chimpanzee.

What about the common ancestor of chimpanzees and gorillas? Well, a gorilla is basically a big chimpanzee. The differences between chimpanzees and gorillas in morphology, as well as in feeding behavior, sexual anatomy, grouping patterns and social relationships, can [in large part] be explained simply by gorillas being larger. These two species are so similar that they should be in the same genus. So the common ancestor of chimpanzees and gorillas was surely an animal built on their body plan, more chimpanzee-like if it was smaller, more gorilla-like if it was larger.

Finally, the early, ape-like Australopithecine, *afarensis,* is sufficiently well-known for its body weight to be closely estimated. They were about the size of chimpanzees.

So our ape-like ancestor that gave rise to our hominid ancestors was presumably also the size of a chimpanzee, and built on the body plan of a chimpanzee. And with [*Ardipithecus*] *ramidus* suddenly presented to us, the earliest australopithecine is looking, as expected, more chimpanzee-like.

This is all very disturbing, and exciting. For years we were brought up to say that the living apes were interesting, but we mustn't think of them as our living ancestors. But now maybe one of them was!

So here's the scenario. At 8–10 m y a chimpanzee-like species gave rise to early gorillas; at 5–6 m y it calved off australopithecines; and at 2–3 m y it gave rise to bonobos; and it's still going. You *can* still argue, and some people do, that gorillas and chimpanzees are similar from parallel evolution rather than common phylogeny. If so, this argument fails. But the great thing is . . . that this question will eventually be settled when the astonishing fossil gap is filled. . . . The most reasonable view for the moment, however, is that chimpanzees are a conservative species and an amazingly good model for the ancestor of hominids. So . . . "What do we come from?" Our ancestor was likely a black-haired, knuckle-walking, large-brained, deep-voiced, heavily-built, big-mouthed, thin-enamelled fruit-eating, fission-fusion, male-bonded species living at low population density in the forests of equatorial Africa.

If we know what our ancestor looked like, naturally we get clues about how it behaved . . . that is, like modern-day chimpanzees. This helps in some ways of course. But . . . [we] can't just talk about The Chimpanzee: we have to talk about particular chimpanzee cultures, because chimps invent lots of different signals and different ways to live. For instance, look at the ways chimpanzees drink. They can put their lips to water. But they often make leaf-sponges, which they dip into water and suck. Sometimes they make drinking-brushes, dipped into narrow-holes. One population uses natural water-bottles. Another uses a pestle and mortar to smash up the juicy parts of a palm. And one, as Denise Wardill has seen this year in Burundi, uses whole leaves as bowls to scoop up water. These different drinking styles come from Guinea and Zaïre and Uganda and Tanzania and Burundi. So you make the call: what did an Ethiopian australopithecine do?

I love this list of drinking styles because it makes two other points. First, it shows how dynamic this field is. The stem-sponges were first seen less than 5 years ago, the pestle-and-mortar was reported this year, the moss-sponges and water-bowls haven't

Fig. 2. Adult female leaf-grooming.
Photo: Richard Wrangham.

yet been published. People are moving into new chimpanzee populations and seeing new traditions all the time, not just in drinking but in eating, body care, signalling, play, everything! (Figure 2) Earlier this year, Rosalind Alp found chimpanzees in Sierra Leone using leafy branches like sandals: they do this when they climb along the thorn-studded branches of capok trees, holding their leafy sandals in their hands and feet to raise their soles and palms about the spines. So did Lucy sometimes wear shoes?

The inventiveness of chimpanzees is remarkable, and sometimes one can even see it directly. Last year, I watched a lonely boy chimpanzee, eight-year-old Kakama, playing for four hours with a log. He carried it on his back, on his belly, in his groin, on his shoulders. He took it with him every time he moved. He carried it up four trees, and down again. He lay in his nest and held it above him like a mother with her baby. And he made a special nest that he didn't use himself, except to put the log in. Three months later, he did it again, watched by two of my field assistants in Kibale Forest. They recovered the log, and pinned to it a description of the behavior. Their report was headed 'Kakama's toy baby.' Imagination made wood.

As more chimpanzee populations are watched, each has its own culture. But the differences aren't understood. A tiny few can be attributed to simple ecological causes, but more appear arbitrary. The explanation of cultural differences is becoming an exciting challenge, and it involves explaining not only why traits are invented and passed on, but also why they go extinct. That's the first lesson of the drinking tools. And what it means for our big questions is both inspiring and annoying. It means that we can look to our past and see a cultural ape that could show a hundred or more inventions of tools and signs and ways to get food . . . but alas, an ape with so much invention that we can't easily predict where and what it did.

The second lesson is that some of the new observations are wonderfully suggestive about ape-hominid transi-

tions. People have argued that hominids were seed-eaters, making dramatic the hammer-tools used by chimps in West Africa. The fat-rich seeds made available by smashing nuts provide much of the calories for the Taï chimps, at some times of year. Did australopithecines harvest palm nuts along the fringes of a Pliocene swamp draining Lake Turkana? Here is an adaptation they could easily have brought with them from the forests.

Others think that hominids were root-eaters, using, like root-eating pigs, the seasonal stores of diverse savannah tubers. This is reasonable because roots could supply the fall-back food eaten when fruits were scarce. But could root-eating have started in the forest? Until the 1990s, there was no evidence of it, and it makes little sense . . . forests have few large storage organs, a tribute to their relatively even micro-climate. But now we have Annette Lanjouw's extraordinary observations of the root-eating chimpanzees of Tongo. The Tongo chimps, in eastern Zaïre, live on a lava flow. All water drains quickly: there are no streams or pools. [To get water] these chimps use their moss-sponges, up to 20 minutes a day, but it's laborious. So, when they're lucky, they have another trick. Sometimes they find a stem that excites them. Pulling the lava boulders away, they dig deep into the soil, maybe up to their shoulders, and extract a root. The prize is prized indeed. Like a prey monkey, the root is guarded by the possessor while around him his companions scream and hug and charge in joy. The root may be divided and shared. It can be carried for a kilometer or more, while it's slowly finished. What's in the root that excites them so? Its' saturated with water, according to Annette. She thinks it's a bottle.

So I like the idea of some strangely desiccated forest, on a lava flow, perhaps, or on an upland granite outcrop, leading an early Pliocene population of forest chimpanzees to become root-eaters—first for water, and only then for food . . . forest root-eating, precursor to savannah life.

And once on the savannah, can chimps help us imagine the past? In

Chambura Gorge in western Uganda, Cathy Poppenwimer's work in the last two years has uncovered a forest-based group of chimpanzees that come into the open savannahs for figs. They can nest in these isolated fig-trees. In the savannah grassland they chase the young antelope, the Uganda kob. Presumably they catch them sometimes. There are two lion dens in the gorge, but the chimps survive: leopards they chase in groups. And only two months ago, the first observations emerged of Ugandan chimpanzees using tools to fish for termites, out on the savannah rim of the gorge.

Nut-smashing, root-eating, savannah-using chimpanzees, resembling our ancestors, and capable by the way of extensive bipedalism. Using antwands, and sandals, and bowls, meat-sharing, hunting cooperatively. Strange paradox . . . a species trembling on the verge of hominization, but so conservative that it has stayed on that edge, little changed for 6 million years or more. It's hard to imagine what more one could ask for as pre-adaptations to a savannah life. But the history of chimpanzee studies shows that our imagination is limited only by what we know. We're still a long way from defining the limits of what chimpanzees do, and therefore from imagining the range of our ape ancestor's feats. We have a good answer, however, to "Where do we come from?" For one thing, we come from an ape with enough brains to invent novel cultural adaptations in every new environment.

The second of Gauguin's questions, "What are we?", goes to the heart of his anxieties. The big issue was the source of evil . . . human aggression and pain and misery. Gauguin, as we saw, thought it unnatural, the result of the loss of nature . . . a widespread romantic view, from Rousseau to Ardrey. But others saw deep roots. Dostoyevsky grappled with the question for a lifetime, and gave a stern answer in *The Brothers Karamazov:* "In every man, a demon lies hidden—the demon of rage, the demon of lustful heat at the screams of the tortured victim, the demon of lawlessness let off the chain . . ."[2] Who was right? Did humans get their

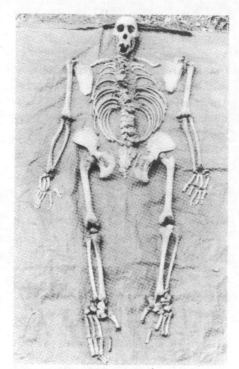

Fig. 3. Chimpanzee Ruwenzori's remains. Photo: Colin Chapman.

demons after leaving nature, or have we inherited them from our ancient forest lives?

The last two decades allow, at last, a reasonably confident comparison of human and chimpanzee behavior. The first similarities we find in the social behavior of chimpanzees and humans are those attractive ones from the era of Louis Leakey. Wherever chimpanzees are studied they form long-lasting individual social relationships, based on exchanges of gestures and favors in remarkably human-like patterns.

But the dramatic discoveries, of course, of the last two decades, have been of the violence that occasionally erupts to destroy Ardrey's "arcadian existence of primal innocence." I'm sure most people here know of the gut-wrenching episodes of male raiding that culminated in at least ten lethal attacks at Gombe and the mortal elimination of seven males that had recently set themselves up as their own independent group. Much was horrifying about that so-called warfare. It involved males that knew each other well. Associating as close companions before the split; victims were stalked and hunted like prey; the kills appeared the result of deliberate attempts to

maim often with extreme cruelty, such as the tearing of skin up an arm, or the twisting of a limb to break it.

Does this mean chimpanzees are naturally violent? Ten years ago it wasn't clear. The warfare was in Gombe, where chimpanzees were fed bananas. Maybe other populations, unaffected by human provisioning, would be found to have escaped the horror of cooperative male violence? Alas, the evidence is mounting, and it all points the same way. Here from my study site in Kibale (Figure 3), are the bones of the one of our chimpanzees killed by the neighbouring group during a period of feeding competition . . . the first [such] death known in an undisturbed population. In Mahale, border patrols, stalking, counter-chasing, and the extinction of the males of a community all suggest a comparable pattern of inter-group attacks. In Taï, reports of wounds from territorial encounters. In captivity, lethal gang attacks. In this cultural species, it may turn out that one of the least variable of all chimpanzee behaviors is the intense competition between males, the violent aggression they use against strangers, and their willingness to maim and kill those that frustrate their goals.

As the picture of chimpanzee society settles into focus, it now includes infanticide, rape, and regular battering of females by males. Some of these occur in other apes. If we leave Africa for a moment and go to Asia, we find that male orangutans rape regularly, so that perhaps half of their copulations involve force and patent resistance by the female. Orangutan rape has never been photographed in the wild, but something like it has been shown in captivity. And in the wild, adult male orangutans can't be together without violent aggression.

Back in Africa, the threat and practice of infanticide appears to lie at the heart of the mountain gorilla social system. Males fight violently for the control of groups, and can sometimes kill each other. The average female experiences infanticide at least once in her lifetime, and infanticide has been found responsible for 37% of infant deaths in these gentle giants. As in chimpanzees and orangutans, sexual coercion emerges readily in captivity. Males attack females, who copulate more willingly as a result.

There is a common theme to these relationships: male sexual aggression against females whose only defense is other males. The females of these species of ape live at risk of male brutality. The risk is not constant. For years on end a female gorilla may endure charmed days of relaxed relationships. But intermittent scenes of violence appear to pervade all their lives, so that all must be constantly on their guard.

What a change we see now from the 1968 view. Then, humans were an independent line, and the violence of our species represented novelty, perhaps arbitrary, perhaps random, perhaps a maladaptive trait, but at least without any evolutionary precedence.

Now, not merely do we see humans as descended from within the tightly related cluster of African apes, but the apes also show similar kinds of violence to ourselves. What makes this especially vivid is that these patterns of violence are generally uncommon in other primates and other animals. Deliberate raiding into neighbouring territories to ambush neighbours; sexual coercion, especially of females outside estrus . . . these are rare. The implication is that strong aspects of human violence have long evolutionary roots. "What are we?" In our aggressive urges we are not Gauguin's creatures of culture. We are apes of nature, cursed over six million years or more with a rare inheritance, a Dostoyevskyan demon.

It's a galling scenario. The implication is that for six million years or more, while we have been evolving from ape to australopithecine to human, through several foraging specialisations, while abandoning the trees and committing ourselves to earth, while brains expanded and faces shrunk and hair became short and fine, while sexuality shifted from promiscuity towards bonding, throughout all this we clung to a suite of characters so rare that it's not confirmed in any other species, and so dangerous that it threatens the survival of our species. Through all these

changes we retained intense rivalry between neighbouring groups of males, lethal coalitionary behavior, and a systematic use of violent sexual coercion. On another day, we could discuss the reasons, which look consistent and visceral: unbalanced power corrupts and pays. There's much still waiting to be explained about the conditions that lead to male-bonding. But one male-bonding is present, lethal aggression follows easily. The coincidence of demonic aggression in ourselves and our closest kin bespeaks its antiquity.

If that's what we are, "Where *are* we going?" The big issue, in taking up the third question, is whether we can go beyond our past. Gauguin eventually did. His suicide failed: (he threw up the arsenic). Eventually, he decided to send his picture to Paris, and it was his curiosity about the public response to it that dispelled his mood of morbid helplessness[3] and led to his painting a pastoral which was an optimistic counterpoint to the tragedy of his earlier fresco. The fourth ape gives us the equivalent to our Tahitian pastoral, our opportunity to be optimistic about controlling our natural demon.

Bonobos, as we saw, have apparently evolved from a chimpanzee ancestor. Yet, as we shall see, they have escaped the violence of chimpanzees. How has this happened?

Bonobos have been watched less than other apes, so generalizations are a little less secure. Still, from Kano's group in Wamba, the Stony Brook group in Lomako, and several studies in captivity, the overall pattern is clear. Bonobos have communities like chimpanzees, founded on a resident group of males and their sons. But the violence has died.

Male chimps commonly batter females. Male bonobos hardly ever attack females. And when they do, these occasional incidents suggest one main way that bonobos reduce male aggression. A female that is attacked screams, and what happens? Other females pour in on her side, and chase the offending male. Alliances among females keep males from getting out of hand. Kano saw them in Wamba. Amy Parish has been showing this very clearly in cap-

tivity, and just recently Barbara Fruth and Gottfried Hohmann have been seeing it in Lomako.

The extraordinary thing is that this doesn't happen in wild chimps. Why not? I mentioned that female chimps rarely travel together. So how can they help each other? But female bonobos are hardly ever apart: small parties are made up of *females* with the occasional male, whereas small parties of chimps are *males* with the occasional female. Do female bonobos support each other simply because they can spend time together? Yes—just like chimps in captivity. They also have to trust each other. Female bonobos invest a lot of time in developing friendly relationships with each other, using the most exotic means. If they're going to spend a lot of time together, supportive relationships are invaluable.

Bonobos have much else to recommend them, such as their famous sexual gymnastics, but I want to focus just on this use of alliances among females to deter male aggression. What does it do for the species? It means that sexual coercion doesn't pay. So males compete for mates not by being brutal, but by being socially attractive. This, I believe, lies at the heart of the bonobo changes from chimpanzees. Bonobos are neotenous, retaining a suite of juvenile characters into adulthood. They are slender, and their vocal repertoire is full of high-pitched, submissive-sounding calls. They have become sexy, friendly, mild. If only males assisted in parenting, they'd be a feminists's dream.

How did this change come about? The critical change, I believe, was the evolution of grouping patterns. Chimpanzee females travel together when fruits are abundant, but when fruits are scare they split up. That clearly suggests they travel alone to feed well. But bonobo females travel together all the time. Is there something different about the foods of bonobos?

Several years ago a number of us suggested that the key difference was that bonobos eat more piths from the forest floor. Pith-eating is a good thing if you can do it. The piths of forest herbs, like sugar-cane, provide good alternatives to fruits. And there's often a lot of it, so there's no need for foraging parties to break up if they can find a field of piths. Gorillas eat a lot of piths. It's the fields of pith that appear to allow the groups of lowland gorillas to forage together as a group.

But do bonobos eat more piths? Recently Richard Malenky and I compared pith densities and the amount of pith taken by bonobos and Lomako and chimpanzees in Zaïre. We found that bonobos passed much more pith than the chimps did over the year. They were consistently more focussed on the pith fields than chimps were. So they seem to have a good back-up food when fruits are few—and one that allows groups to stay together.

Let me, then, imagine one way that bonobos evolved from their chimp ancestors. Genetic evidence dates the split at 2–3 m y ago. We know from Liz Vrba and others that around 2.5 m y there was a major drying event. I suggest that south of the Zaïre River, gorillas and chimps, or ancestors very like them, lived together as they do now to the north of the river. Then the drying event, and what happened? Only chimps survived. As we see today, in the more seasonal areas north of the Zaïre River, gorillas give out and only chimps remain.

Then the moistness returned, and with it, the piths that gorillas like to eat, as recent studies have been finding in Gabon. But there were no gorillas. So the chimps expanded to occupy the empty niche, including gorilla-foods—in other words, the piths—alongside their previous chimp-foods—that is, the tree-fruits. And as they adapted to the new combination of gorilla foods and chimp foods, they changed. They were rarely forced to travel alone. Females lived together. They developed supportive relationships. They attacked aggressive males. Aggressive males were failures as mates. Males were juvenilized.

The details of the process can barely be guessed at the moment. Certainly, a major role was played by the prolonged sexuality of bonobos, maybe involving concealed ovulation. But the principle will surely remain, that bonobos evolved from changed circumstance; and the way it happened was for a change in the environment to allow a political change. Bonobos weren't constrained by their chimpanzee past to keep their legacy of male violence. Social strategies have different pay-offs in different contexts. They can be easily changed when the contexts change. And a remarkable feature of the alliances among bonobo females is that they are developed among strangers. In other animals, alliances are linked to kinship. In bonobos, alliances are produced from recognition of common interest, a recognition that takes brains. The development of big brains and advanced cognition has brought with it the ability to escape from the constraints of biology, even in a species with little self-consciousness. Gauguin thought us tragic: the very skills that make us human, our intellect and emotions, also bring demons. But that romantic view is wrong, almost the reverse of history as we can see it now. Our demons come from our ape past, and we need our intellect and emotions to forge the alliances that can defeat the beast. It's common sense,

Fig. 4. Reassurance behavior between adult males. Photo: Richard Wrangham.

Fig. 5. Adult male inspecting tree for fruit. Photo: Richard Wrangham.

supported by the evidence.

What does it do for us, then, to know the behavior of our closest relatives? Chimpanzees and bonobos are an extraordinary pair. One, I suggest, shows us some of the worst aspects of our past and our present; the other shows an escape from it. In thinking creatively about our future, I hope we honor our sister species, who by being different from ourselves, emphasize the unity of our humanity.

Let me return to the extraordinary achievements of the Leakey Foundation. In this talk I've referred to perhaps twenty field studies. All have input from the Leakey Foundation. I should be referring to each by name, and honoring the individual scientists that make a broad review possible. But let me honor, instead, the trustees and supporters of the Leakey Foundation, who have put their time, their money, and their spirit into helping us all.

In a mere quarter-century, this imaginative group has presided over the golden age of biological anthropology, stimulated a range of exciting discoveries, brought academics face to face

with the public that supports them, and incidentally greatly benefited primate conservation efforts.

The knowledge so gained should help us, though some fear it. For Gauguin, "primitive" was good. Those who ate from the tree of knowledge suffered. For others today, "biology" (the primitive) is fearful. Many people reject the idea that we still follow rules that we can trace to the Pliocene. So we can pretend it's not true, but much good that may do us. Denial of our demons won't make them go away. But even if we're driven to accepting the evidence of a grisly past, we're not forced into thinking it condemns us to an unchanged future. There are many challenges.

For primatologists: to understand more precisely the conditions that favor male aggression, and the conditions that suppress it. Are there populations of chimpanzees that have evolved beyond violence? Are there bonobos that are violent like chimpanzees? We can expect some such local adaptations to special conditions—can we use them to explain the ill effects of testerone poisoning in our own ape lineage?

For psychologists: what has the long legacy of aggression done to our psyche? Has it made men specially vulnerable to deindividuation—that mindless loss of self, and acceptance of gang wisdom; . . . or to dehumanization—the cruel emotional deafness to the cries of outsiders? If it prepared us for a career of obedience to authority, of heroism and impulsivity, of quick acceptance of group norms, how can our understanding of history create an enlightened world? The human future may depend on taming the demonic male. It may involve the growing political power of women. But it will not happen in quite the bonobo way. We must look to nature not to copy but to learn.

It may be tempting to condemn the aggressive apes and overly praise the

bonobo. But of course apes are not humans, even though humans may be apes. Their failure to conform to a human morality is their problem, not ours, and they deserve sympathy, respect, and admiration, not disdain. Apes provide us both with a story of our ancestry and a glimpse of a better future. So let us celebrate their lives as instructive visions of other worlds. We still know little, but every population of apes teaches something new. Of course, those populations are disappearing fast. Even the vast forests of Zaïre, uncut though they may be, are being attacked: this year, when I was in Wamba, all mammals but the tabooed bonobo were gone; and the local population had already started killing and eating the bonobos. The same problems are everywhere. We have only a few decades to recover the extraordinary evidence waiting elusively in a hundred forests still unvisited by scientists. Future generations will think ill of us for letting them slide unknown into oblivion. But if we *can* get the observers into the field, and watch the apes in nature, we will surely continue to learn and to be inspired. In that way, the next version of this lecture (in 25 years' time) won't have to be retitled *'Ape Links . . . and Missing Cultures'*!

So ladies and gentlemen, for our sakes and theirs, the apes need all the support they can get. I ask you therefore to raise your glasses, or suck your table-cloths, in celebration of all that's been done, and will be done, by Gordon Getty and the Leakey Foundation!

NOTES

1. (adapted from R. Ardrey 1967, *'The Territorial Imperative'*, p. 232, Fontana).
2. Fyodor Dostoyevsky (1880) *The Brothers Karamazov*. Translated by Constance Garnett. Random House, New York (1950). Part I, Book V, Chapter IV: "Rebellion" (pp. 282–292). p. 287
3. Thomson B. 1987 *Gauguin*. London: Thames and Hudson.

THE FOSSIL EVIDENCE

A primary focal point of this book, as well as the whole of biological anthropology, is the search for, and interpretation of, fossil evidence for hominid (meaning human or humanlike) evolution. Paleontologists are those who carry out this task by conducting the painstaking excavations and detailed analyses that serve as a basis for understanding our past. Every fragment found is cherished like a ray of light that may help to illuminate the path taken by our ancestors in the process of becoming "us." At least, that is what we would like to believe. In reality, each discovery leads to further mystery, and for every fossil-hunting paleoanthropologist who thinks his or her find supports a particular theory, there are many others anxious to express their disagreement.

How wonderful it would be, we sometimes think in moments of frustration with inconclusive data, if the fossils would just speak for themselves, and every primordial piece of humanity were to carry with it a self-evident explanation for its place in the evolutionary story. Paleoanthropology would then be more a quantitative problem of amassing enough material to reconstruct our ancestral development than a qualitative problem of interpreting what it all means. It would certainly be a simpler process, but would it be as interesting?

Most scientists tolerate, welcome, or even (dare it be said?) thrive on controversy, recognizing that diversity of opinion refreshes the mind, rouses students, and captures the imagination of the general public. After all, where would paleoanthropology be without the gadflies, the near-mythic heroes, and, lest we forget, the research funds they generate? As an example of the ongoing debates, Yves Coppens, in "East Side Story: The Origin of Humankind," puts forth the idea that by studying the stratigraphic sequences of fossil remains of species other than hominids we can tell that climatic change in East Africa had a significant impact on hominid evolutionary developments. On the other hand, in "African Origins: West Side Story," Virginia Morell challenges Coppens's claim. Even the long-held notion that bipedalism evolved in the grasslands is now being questioned as James Shreeve reveals in his report "Sunset on the Savanna."

Not all the research and theoretical speculation taking place in the field of paleoanthropology is so highly volatile. Most scientists, in fact, go about their work quietly and methodically, generating hypotheses that are much less explosive and yet have the cumulative effect of enriching our understanding of the details of human evolution. In "Scavenger Hunt," for instance, Pat Shipman tells how modern technology, in the form of the scanning electron microscope, combined with meticulous detailed analysis of cut marks on fossil animal bones, can help us better understand the locomotor and food-getting adaptations of our early hominid ancestors. In one stroke, she is able to challenge the traditional "man the hunter" theme that has pervaded most early hominid research and writing and simultaneously set forth an alternative hypothesis that will, in turn, inspire further research.

As we mull over the controversies outlined in this unit, we should not take them as reflecting an inherent weakness of the field of paleoanthropology, but rather as symbolic of its strength: the ability and willingness to scrutinize, question, and reflect (seemingly endlessly) on every bit of evidence. Even in the case of purposeful deception, as recounted in "Dawson's Dawn Man: The Hoax of Piltdown" by Kenneth L. Feder and "The Case of the Missing Link" by Robert Anderson, we should remember that it was the skepticism of scientists themselves that finally led to the revelation of fraud.

Contrary to the way the creationists would have it, an admission of doubt is not an expression of ignorance, but simply a frank recognition of the imperfect state of our knowledge. If we are to increase our understanding of ourselves, we must maintain an atmosphere of free inquiry without preconceived notions and an unquestioning commitment to a particular point of view. To paraphrase anthropologist Ashley Montagu, whereas creationism seeks certainty without proof, science seeks proof without certainty.

Looking Ahead: Challenge Questions

What effect did the Piltdown hoax have upon paleoanthropology? Who do you think perpetrated the hoax, and why?

What has climatic change in Africa had to do with hominid evolution?

Under what circumstances did bipedalism evolve?

What is the "man the hunter" hypothesis, and how might the "scavenging" theory better suit the early hominid data?

How would you draw the early hominid family tree?

UNIT 5

Dawson's Dawn Man: The Hoax at Piltdown

Kenneth L. Feder

The Piltdown Man fossil is a literal skeleton in the closet of prehistoric archaeology and human paleontology. This single specimen seemed to turn our understanding of human evolution on its head and certainly did turn the heads of not just a few of the world's most talented scientists. The story of Piltdown has been presented in detail by Ronald Millar in his 1972 book *The Piltdown Men,* by J. S. Weiner in his 1955 work *The Piltdown Forgery,* and most recently in 1986 by Charles Blinderman in *The Piltdown Inquest.* The story is useful in its telling if only to show that even scientific observers can make mistakes. This is particularly the case when trained scientists are faced with that which they are not trained to detect—intellectual criminality. But let us begin before the beginning, before the discovery of the Piltdown fossil.*

THE EVOLUTIONARY CONTEXT

We need to turn the clock back to Europe of the late nineteenth and early twentieth centuries. The concept of evolution—the notion that all animal and plant forms seen in the modern world had descended or evolved from earlier, ancestral forms—had been debated by scientists for quite some time (Greene 1959). It was not until Charles Darwin's *On the Origin of Species* was published in 1859, however, that a viable mechanism for evolution was proposed and supported with an enormous body of data. Darwin had meticulously studied his subject, collecting evidence from all over the world for more than thirty years in support of his evolutionary mechanism called *natural selection*. Darwin's arguments were so well reasoned that most scientists soon became convinced of the explanatory power of his theory. Darwin went on to apply his general theory to humanity in *The Descent of Man,* published in 1871. This book was also enormously successful, and more thinkers came to accept the notion of human evolution.

Around the same time that Darwin was theorizing about the biological origin of humanity, discoveries were being made in Europe and Asia that seemed to support the concept of human evolution from ancestral forms. In 1856, workmen building a roadway in the Neander Valley of Germany came across some remarkable bones. The head was large but oddly shaped (Figure 1). The cranium (the skull minus the mandible or jaw) was much flatter than a modern human's, the bones heavier. The face jutted out, the forehead sloped back, and massive bone ridges appeared just above the eye sockets. Around the same time, other skeletons were found in Belgium and Spain that looked very similar. The postcranial bones (all the bones below the skull) of these fossils were quite similar to those of modern humans.

There was some initial confusion about how to label these specimens. Some scientists concluded that they simply represented pathological freaks. Rudolf Virchow, the world's preeminent anatomist, explained the curious bony ridges above the eyes as the result of blows to the foreheads of the creatures (Kennedy 1975). Eventually, however, scientists realized that these creatures, then and now called *Neandertals* after their most famous find-spot, represented a primitive and ancient form of humanity.

The growing acceptance of Darwin's theory of evolution and the discovery of primitive-looking, though humanlike, fossils combined to radically shift people's opinions about human origins. In fact, the initial abhorrence many felt concerning the entire notion of human evolution from lower, more primitive forms was remarkably changed in just a few decades (Greene 1959). By the turn of the twentieth century, not only were many people comfortable with the general concept of human evolution, but there actually was also a feeling of national pride concerning the discovery of a human ancestor within one's borders.

The Germans could point to their Neandertal skeletons and claim that the first primitive human being was a German. The French could counter that their own Cro-Magnon—ancient, though not as old as the German Neandertals—was a more humanlike and advanced ancestor; therefore, the first true human was a Frenchman. Fossils had also been found in Belgium and Spain, so Belgians and Spaniards could claim for themselves a place within the story of human origin and development. Even so small a nation as Holland could lay claim to a place in human evolutionary history since a Dutchman, Eugene Dubois, in 1891 had discovered the fossilized remains of a primitive human ancestor in Java, a Dutch-owned colony in the western Pacific.

However, one great European nation did not and could not participate fully in the debate over the ultimate origins of humanity. That nation was England. Very simply, by the beginning of the second decade of the twentieth century, no fossils of human evolutionary significance had been located in England. This lack of fossils led French scientists to label English human paleontology mere "pebble-collecting" (Blinderman 1986).

The English, justifiably proud of their cultural heritage and cultural evolution, simply could point to no evidence that humanity had initially developed within their borders. The conclusion reached by most was completely unpalatable to the proud English—no one had evolved in England. The English must have originally arrived from somewhere else.

At the same time that the English were feeling like a people with no evolutionary roots of their own, many other Europeans were still uncomfortable with the fossil record as it stood in the first decade of the twentieth century. While most were happy to have human fossils in their countries, they were generally not happy with what those fossils looked like and what their appearance implied about the course of human evolution.

Java Man (now placed in the category *Homo erectus* along with Peking Man), with its small cranium—its volume was about 900 cubic centimeters (cc), compared to a modern human average of about 1,450 cc—and large eyebrow ridges seemed quite apelike (see Figure 1). Neandertal Man, with his sloping forehead and thick, heavy brow ridges appeared to many to be

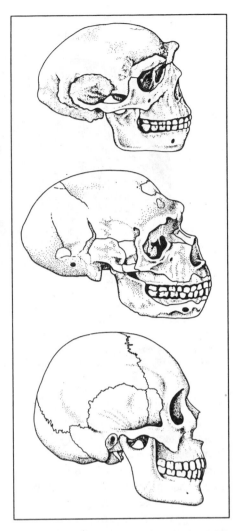

Figure 1 *Drawings showing the general differences in skull size and form between* Homo erectus *(Peking Man—500,000 years ago [top]), Neandertal Man (100,000 years ago [center]), and a modern human being [bottom]. Note the large brow ridges and forward-thrusting faces of* Homo erectus *and Neandertal, the rounded outline of the modern skull, and the absence of a chin in earlier forms. (Carolyn Whyte)*

quite ugly, stupid, and brutish. While the skulls of these fossil types were clearly not those of apes, they were equally clearly not fully human. On the other hand, the femur (thigh bone) of Java Man seemed identical to the modern form. While some emphasized what they perceived to be primitive characteristics of the postcranial skeleton of the Neandertals, this species clearly had walked on two feet; and apes do not.

All this evidence suggested that ancient human ancestors had primitive heads and, by implication, primitive

brains, seated atop rather modern-looking bodies. This further implied that it was the human body that evolved first, followed only later by the development of the brain and associated human intelligence.

Such a picture was precisely the opposite of what many people had expected and hoped for. After all, it was argued, it is intelligence that most clearly and absolutely differentiates humanity from the rest of the animal kingdom. It is in our ability to think, to communicate, and to invent that we are most distant from our animal cousins. This being the case, it was assumed that such abilities must have been evolving the longest; in other words, the human brain and the ability to think must have evolved first. Thus, the argument went, the fossil evidence for evolution should show that the brain had expanded first, followed by the modernization of the body.

Such a view is exemplified in the writings of anatomist Grafton Elliot Smith. Smith said that what most characterized human evolution must have been the "steady and uniform development of the brain along a well-defined course . . ." (as quoted in Blinderman 1986:36). Arthur Smith Woodward, ichthyologist and paleontologist at the British Museum of Natural History, later characterized the human brain as "the most complex mechanism in existence. The growth of the brain preceded the refinement of the features and of the somatic characters in general" (Dawson and Woodward 1913).

Put most simply, many researchers in evolution were looking for fossil evidence of a creature with the body of an ape and the brain of a human being. What was being discovered, however, was the reverse; both Java and Neandertal Man seemed more to represent creatures with apelike, or certainly not humanlike, brains but with humanlike bodies. Many were uncomfortable with such a picture.

A REMARKABLE DISCOVERY IN SUSSEX

Thus was the stage set for the initially rather innocuous announcement that

5. THE FOSSIL EVIDENCE

appeared in the British science journal *Nature* on December 5, 1912, concerning a fossil find in the Piltdown section of Sussex in southern England. The notice read, in part:

> Remains of a human skull and mandible, considered to belong to the early Pleistocene period, have been discovered by Mr. Charles Dawson in a gravel-deposit in the basin of the River Ouse, north of Lewes, Sussex. Much interest has been aroused in the specimen owing to the exactitude with which its geological age is said to have been fixed. . . . (p. 390)

In the December 19 issue of *Nature,* further details were provided concerning the important find:

> The fossil human skull and mandible to be described by Mr. Charles Dawson and Dr. Arthur Smith Woodward at the Geological Society as we go to press is the most important discovery of its kind hitherto made in England. The specimen was found in circumstances which seem to leave no doubt of its geological age, and the characters it shows are themselves sufficient to denote its extreme antiquity. (p. 438)

According to the story later told by those principally involved, in February 1912 Arthur Smith Woodward at the British Museum received a letter from Charles Dawson—a Sussex lawyer and amateur scientist. Woodward had previously worked with Dawson and knew him to be an extremely intelligent man with a keen interest in natural history. Dawson informed Woodward in the letter that he had come upon several fragments of a fossil human skull. The first piece had been discovered in 1908 by workers near the Barcombe Mills manor in the Piltdown region of Sussex, England. In 1911, a number of other pieces of the skull came to light in the same pit, along with a fossil hippopotamus bone and tooth.

In the letter to Woodward, Dawson expressed some excitement over the discovery and claimed to Woodward that the find was quite important and might even surpass the significance of Heidelberg Man, an important specimen found in Germany just the previous year.

Due to bad weather, Woodward was not immediately able to visit Piltdown. Dawson, undaunted, continued to work

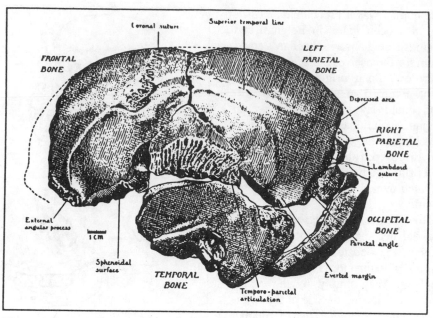

Figure 2 *This drawing with anatomical labels of the fragmentary remains of the Piltdown cranium appeared in a book written by one of the fossil's chief supporters. (From* The Evolution of Man, *by Grafton Elliot Smith, Oxford University Press)*

in the pit, finding fossil hippo and elephant teeth. Finally, in May 1912, he brought the fossil to Woodward at the museum. What Woodward saw was a skull that matched his own expectations and those of many others concerning what a human ancestor should look like. The skull, stained a dark brown from apparent age, seemed to be quite modern in many of its characteristics. The thickness of the bones of the skull, however, argued for a certain primitiveness. The association of the skull fragments with the bones of extinct animals implied that an ancient human ancestor indeed had inhabited England. By itself this was enormous news; at long last, England had a human fossil (Figure 2).

Things were to get even more exciting for English paleontologists. At the end of May 1912 Dawson, Woodward, and Pierre Teilhard de Chardin—a Jesuit priest with a great interest in geology, paleontology, and evolution whom Dawson had met in 1909—began a thorough archaeological excavation at the Piltdown site. . . . More extinct animal remains and flint tools were recovered. The apparent age of the fossils based upon comparisons to other sites indicated not only that Pilt-

down was the earliest human fossil in England, but also that, at an estimated age of 500,000 years, the Piltdown fossil represented potentially the oldest known human ancestor in the world.

Then, to add to the excitement, Dawson discovered one half of the mandible. Though two key areas—the chin, and the condyle where the jaw connects to the skull—were missing, the preserved part did not look anything like a human jaw. The upright portion or *ramus* was too wide, and the bone too thick. In fact, the jaw looked remarkably like that of an ape (Figure 3). Nonetheless, and quite significantly, the molar teeth exhibited humanlike wear. The human jaw, lacking the large canines of apes, is free to move from side to side while chewing. The molars can grind in a sideways motion in a manner impossible in monkeys or apes. The wear on human molars is, therefore, quite distinct from that of other primates. The Piltdown molars exhibited such humanlike wear in a jaw that was otherwise entirely apelike.

That the skull and the jaw had been found close together in the same geologically ancient deposit seemed to argue for the obvious conclusion that

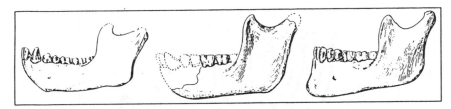

Figure 3 *Comparison of the mandibles (lower jaws) of a young chimpanzee [left], modern human [right], and Piltdown [center]. Note how much more similar the Piltdown mandible is to that of the chimp, particularly in the absence of a chin. The presence of a chin is a uniquely human trait. (From Dawson and Woodward, 1913, The Geological Society of London)*

they belonged to the same ancient creature. But what kind of creature could it have been? There were no large brow ridges like those of Java or Neandertal Man. The face was flat as in modern humans and not snoutlike as in the Neandertals. The profile of the cranium was round as it is in modern humans, not flattened as it appeared to be in the Java and Neandertal specimens (Figure 4). According to Woodward, the size of the skull indicated a cranial capacity or brain size of at least 1,100 cc (Dawson and Woodward 1913), much larger than Java and within the range of modern humanity. Anatomist Arthur Keith (1913) suggested that the capacity of the skull was actually much larger, as much as 1,500 cc, placing it close to the modern mean. But the jaw, as described above, was entirely apelike.

The conclusion drawn first by Dawson, the discoverer, and then by Woodward, the professional scientist, was that the Piltdown fossil—called *Eoanthropus dawsoni*, meaning Dawson's Dawn Man—was the single most important fossil find yet made anywhere in the world. Concerning the Piltdown discovery, the *New York Times* headline of December 19, 1912, proclaimed "Paleolithic Skull Is a Missing Link." Three days later the *Times* headline read "Darwin Theory Is Proved True."

The implications were clear. Piltdown Man, with its modern skull, primitive jaw, and great age, was the evidence many human paleontologists had been searching for: an ancient man with a large brain, a modern-looking head, and primitive characteristics below the important brain. As anatomist G. E. Smith summarized it:

The brain attained what may be termed the human rank when the jaws and face, and no doubt the body also, still retained much of the uncouthness of Man's simian ancestors. In other words, Man at first, so far as his general appearance and "build" are concerned, was merely an Ape with an overgrown brain. The importance of the Piltdown skull lies in the fact that it affords tangible confirmation of these inferences. (Smith 1927:105–6)

If Piltdown were the evolutionary "missing link" between apes and people, then neither Neandertal nor Java Man could be. Since Piltdown and Java Man lived at approximately the same time, Java might have been a more primitive offshoot of humanity that had

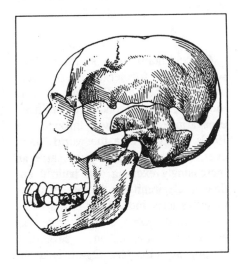

Figure 4 *Drawn reconstruction of the Piltdown skull. The portion of the skull actually recovered is shaded. As reconstructed, the cranium shows hominid (human) traits and the mandible shows pongid (ape) traits. Compare this drawing to those in Figure 1. With its humanlike head and apelike jaw, the overall appearance of the Piltdown fossil is far different from Homo erectus, Neandertal, or modern humans. (From* The Evolution of Man, *Grafton Elliot Smith, Oxford University Press)*

become extinct. Since Neandertal was much more recent than Piltdown, yet looked more primitive where it really counted (that is, the head), Neandertal must have represented some sort of primitive throwback, an evolutionary anachronism (Figure 5).

By paleontological standards the implications were breathtaking. In one sweeping blow Piltdown had presented England with its first ancestral human fossil, it had shown that human fossils found elsewhere in the world were either primitive evolutionary offshoots or later throwbacks to a more primitive type, and it had forced the rewriting of the entire story of human evolution. Needless to say, many paleontologists, especially those in England, were enthralled by the discovery in Sussex.

In March 1913, Dawson and Woodward published the first detailed account of the characteristics and evolutionary implications of the Piltdown fossil. Again and again in their discussion, they pointed out the modern characteristics of the skull and the simian appearance of the mandible. Their comments regarding the modernity of the skull and the apelike characteristics of the jaw, as you will see, turned out to be accurate in a way that few suspected at the time.

Additional discoveries were made at Piltdown. In 1913 a right canine tooth apparently belonging to the jaw was discovered by Teilhard de Chardin. It matched almost exactly the canine that had previously been proposed for the Piltdown skull in the reconstruction produced at the British Museum of Natural History. Its apelike form and wear were precisely what had been expected: "If a comparative anatomist were fitting out *Eoanthropus* with a set of canines, he could not ask for anything more suitable than the tooth in question," stated Yale University professor George Grant MacCurdy (1914: 159).

Additional artifacts, including a large bone implement, were found in 1914. Then, in 1915, Dawson wrote Woodward announcing spectacular evidence confirming the first discovery; fragments of another fossil human skull were found (possibly at a site just two miles from the first—Dawson never revealed the location). This skull, dubbed Piltdown II, looked just like the first with a rounded profile and thick cranial bones. Though no jaw was discovered, a molar recovered at the site bore the same pattern of wear as that seen in the first specimen.

Dawson died in 1916 and, for reasons not entirely clear, Woodward held back announcement of the second discovery until the following year. When the existence of a second specimen became known, many of those skeptical after the discovery of the first Piltdown fossil became supporters. One of those converted skeptics, Henry Fairfield Osborn, president of the American Museum of Natural History, suggested:

If there is a Providence hanging over the affairs of prehistoric man, it certainly manifested itself in this case, because the three minute fragments of this second Piltdown man found by Dawson are exactly those which we should have selected to confirm the comparison with the original type. (1921:581)

THE PILTDOWN ENIGMA

There certainly was no unanimity of opinion, however, concerning the significance of the Piltdown discoveries. The cranium was so humanlike and the jaw so apelike that some scientists maintained that they simply were the fossils of two different creatures; the skeptics suggested that the association of the human cranium and ape jaw was entirely coincidental. Gerrit S. Miller, Jr. (1915) of the Smithsonian Institution conducted a detailed analysis of casts of Piltdown I and concluded that the jaw was certainly that of an ape (See Figure 3). Many other scientists in the United States and Europe agreed. Anatomy professor David Waterson (1913) at the University of London, King's College, thought the mandible was that of a chimpanzee. The very well-known German scientist Franz Weidenreich concluded that Piltdown I was " . . . the artificial combination of fragments of a modern-human braincase with an orangutan-like mandible and teeth" (1943:273).

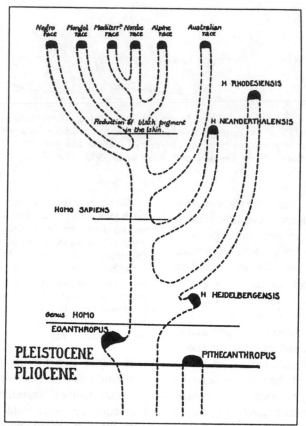

Figure 5 *Among its supporters,* Eoanthropus *(Piltdown Man) was seen as more directly ancestral to modern humanity than either* Homo erectus—*here labeled* Pithecanthropus *and depicted as an entirely separate evolutionary pathway—or Neandertal—shown here as a short-lived diversion off the main branch of human evolution. (From* The Evolution of Man, *Grafton Elliot Smith, Oxford University Press)*

Coincidentally or not, after Dawson's death no further discoveries were made in either the Piltdown I or II localities, though Woodward continued excavating at Piltdown through the 1920s. Elsewhere in the world, however, human paleontology became an increasingly exciting and fruitful endeavor. Beginning in the late 1920s as many as forty individuals of a species now called *Homo erectus* were unearthed at Zhoukoudian, a cave near Beijing in China (see Figure 1). Ironically, Davidson Black, anatomist at the Peking Union Medical College, who was instrumental in obtaining financial support for the excavation, had visited Grafton Elliot Smith's laboratory in 1914 and had become fascinated by the Piltdown find (Shapiro 1974). Further, Teilhard de Chardin participated in the excavation at the cave. The Zhoukoudian fossils were estimated to be one-half million years old. Also on Java,

another large group of fossils (close to twenty) were found at Sangiran; these were similar to those from Zhoukoudian.

Also in the 1920s, in Africa, the discovery was made of a fossil given the name *Australopithecus africanus.* It was initially estimated to be more than one million years old. In the 1930s and 1940s additional finds of this and other varieties of *Australopithecus* were made. In Europe the number of Neandertal specimens kept increasing; and even in England, in 1935, a fossil human ancestor was discovered at a place called Swanscombe.

Unfortunately for *Eoanthropus,* all of these discoveries seemed to contradict its validity. The Chinese and Sangiran *Homo erectus* evidence pointed to a fossil ancestor with a humanlike body and primitive head; these specimens were quite similar to Java Man in appearance (Java Man is also now considered to belong to the species *Homo erectus*), possessing large brow ridges, a flat skull, and a thrust-forward face while being quite modern from the neck down. Even the much older australopithecines showed clear evidence of walking on two feet; their skeletons were remarkably humanlike from the neck down, though their heads were quite apelike. Together, both of these species seemed to confirm the notion that human beings began their evolutionary history as upright apes, not as apelike people. *Eoanthropus* seemed more and more to be the evolutionary "odd man out."

How could Piltdown be explained in light of the new fossil evidence from China, Java, Europe, and Africa? Either Piltdown was a human ancestor, rendering all the manifold other discoveries members of extinct offshoots of the main line of human evolution, or else Piltdown was the remarkable coincidental find of the only known ape fossil in England within a few feet of a rather modern human skull that seemed to date back 500,000 years. Neither explanation sat well with many people.

UNMASKING THE HOAX

This sort of confusion characterized the status of Piltdown until 1949, when a new dating procedure was applied to the fossil. A measurement was made of the amount of the element fluorine in the bones. This was known to be a relative measure of the amount of time bone had been in the ground. Bones pick up fluorine in groundwater; the longer they have been buried, the more fluorine they have. Kenneth Oakley of the British Museum of Natural History conducted the test. While the fossil animal bones from the site showed varying amounts of fluorine, they exhibited as much as ten times more than did either the cranium or jaw of the fossil human. Piltdown Man, Oakley concluded, based on comparison to fluorine concentrations in bones at other sites in England, was no more than 50,000 years old (Oakley and Weiner 1955).

While this certainly cast Piltdown in a new light, the implications were just as mysterious; what was a fossil human doing with an entirely apelike jaw at a date as recent as 50,000 years ago? Then, in 1953 a more precise test was applied to larger samples of the cranium and jaw. The results were quite conclusive; the skull and jaw were of entirely different ages. The cranium possessed .10 percent fluorine, the mandible less than .03 percent (Oakley 1976). The inevitable conclusion was reached that the skull and jaw must have belonged to two different creatures.

As a result of this determination, a detailed reexamination of the fossil was conducted and the sad truth was finally revealed. The entire thing had been a hoax. The skull was that of a modern human being. Its appearance of age was due, at least in part, to its having been artificially chemically stained. The thickness of the bone may have been due to a pathological condition (Spencer 1984) or the result of a chemical treatment that had been applied, perhaps to make it appear older than it was (Montague 1960).

Those scientific supporters of *Eoanthropus* who previously had pointed out the apelike character of the jaw were more right than they could have imagined; it was, indeed, a doctored ape jaw, probably that of an orangutan. When Gerrit Miller of the Smithsonian Institution had commented on the broken condyle of the mandible by saying, "Deliberate malice could hardly have been more successful than the hazards of deposition in so breaking the fossils as to give free scope to individual judgement in fitting the parts together" (1915:1), he was using a literary device and not suggesting that anyone had purposely broken the jaw. But that is likely precisely what happened. An ape's jaw could never articulate with the base of a human skull, and so the area of connection had to be removed to give "free scope" to researchers to hypothesize how the cranium and jaw went together. Otherwise the hoax would never have succeeded. Beyond this, the molars had been filed down to artificially create the humanlike wear pattern. The canine tooth had been

stained with an artist's pigment and filed down to simulate human wear; the pulp cavity had been filled with a substance not unlike chewing gum.

It was further determined that at least one of the fragments of the Piltdown II skull was simply another piece of the first one. Oakley further concluded that all the other paleontological specimens had been planted at the site; some were probably found in England, but others had likely originated as far away as Malta and Tunisia. Some of the ostensible bone artifacts had been carved with a metal knife.

The verdict was clear; as Franz Weidenreich (1943) put it, Piltdown was like the chimera of Greek mythology—a monstrous combination of different creatures. The question of Piltdown's place in human evolution had been answered: it had no place. That left still open two important questions: who did it and why?

WHODUNNIT?

The most succinct answer that can be provided for the question "Whodunnit?" is "No one knows." It seems, however, that every writer on the subject has had a different opinion.

Each of the men who excavated at Piltdown has been accused at one time or another. . . . Charles Dawson is an obvious suspect. He is the only person who was present at every discovery. He certainly gained notoriety; even the species name is *dawsoni*. Blinderman (1986) points out, however, that much of the evidence against Dawson is circumstantial and exaggerated. Dawson did indeed stain the fossil with potassium bichromate and iron ammonium sulfate. These gave the bones a more antique appearance, but such staining was fairly common. It was felt that these chemicals helped preserve fossil bone, and Dawson was quite open about having stained the Piltdown specimens. In an unrelated attack on his character, some have even accused Dawson of plagiarism in a book he wrote on Hastings Castle (Weiner 1955), but this seems to be unfair; as Blinderman points out, the book was explicitly a compilation of previous sources and Dawson did not attempt to take credit for the work of others.

Dawson's motive might have been the fame and notoriety that accrued to this amateur scientist who could command the attention of the world's most famous scholars. But there is no direct evidence concerning Dawson's guilt, and questions remain concerning his ability to fashion the fraud. And where would Dawson have obtained the orangutan jaw?

Arthur Smith Woodward certainly possessed the opportunity and expertise to pull off the fraud. His motive might have been to prove his particular view of human evolution. That makes little sense though, since he could not have expected the kind of confirming evidence he knew his colleagues would demand. Furthermore, his behavior after Dawson's death seems to rule out Woodward as the hoaxer. His fruitlessly working the original Piltdown pit in his retirement renders this scenario nonsensical.

Even the priest Teilhard de Chardin has been accused, most recently by Harvard paleontologist and chronicler of the history of science Stephen Jay Gould (1980). The evidence marshalled against the Jesuit is entirely circumstantial, the argument strained. The mere facts that Teilhard mentioned Piltdown but little in his later writings on evolution and was confused about the precise chronology of discoveries in the pit do not add up to a convincing case.

Others have had fingers pointed at them. W. J. Sollas, a geology professor at Oxford and a strong supporter of Piltdown, has been accused from beyond the grave. In 1978, a tape-recorded statement made before his death by J. A. Douglass, who had worked in Sollas's lab for some thirty years, was made public. The only evidence provided is Douglass's testimony that on one occasion he came across a package containing the fossil-staining agent potassium bichromate in the lab—certainly not the kind of stuff to convince a jury to convict.

Even Sir Arthur Conan Doyle has come under the scrutiny of would-be Piltdown detectives. Doyle lived near Piltdown and is known to have visited the site at least once. He may have held a grudge against professional scientists who belittled his interest in and credulity concerning the paranormal. Doyle, the creator of the most logical, rational mind in literature, Sherlock Holmes, found it quite reasonable that two young English girls could take photographs of real fairies in their garden. But why would Doyle strike out at paleontologists, who had nothing to do with criticizing his acceptance of the occult? Again, there is no direct evidence to implicate Doyle in the hoax.

The most recent name added to the roster of potential Piltdown hoaxers is that of Lewis Abbott, another amateur scientist and artifact collector. Blinderman (1986) argues that Abbott is the most likely perpetrator. He had an enormous ego and felt slighted by professional scientists. He claimed to have been the one who directed Dawson to the pit at Piltdown and may even have been with Dawson when Piltdown II was discovered (Dawson said only that he had been with a friend when the bones were found). Abbott knew how to make stone tools and so was capable of forging those found at Piltdown. Again, however, the evidence, though tantalizing, includes no smoking gun.

A definitive answer to the question "whodunnit" may never be forthcoming. The lesson in Piltdown, though, is clear. Unlike the case for the Cardiff Giant where scientists were not fooled, here many were convinced by what appears to be, in hindsight, an inelegant fake. It shows quite clearly that scientists, though striving to be objective observers and explainers of the world around them, are, in the end, human. Many accepted the Piltdown evidence because they wished to—it supported a more comfortable view of human evolution. Furthermore, perhaps out of naïveté, they could not even conceive that a fellow thinker about human origins would wish to trick them; the possibility that Piltdown was a fraud probably occurred to few, if any, of them.

Nevertheless, the Piltdown story, rather than being a black mark against science, instead shows how well it ultimately works. Even before its unmasking, Piltdown had been consigned by most to a netherworld of doubt. There was simply too much evidence supporting a different human pedigree than that implied by Piltdown. Proving it a hoax was just the final nail in the coffin lid for this fallacious fossil. As a result, though we may never know the hoaxer's name, at least we know this: if the goal was to forever confuse our understanding of the human evolutionary story, the hoax ultimately was a failure.

CURRENT PERSPECTIVES HUMAN EVOLUTION

With little more than a handful of cranial fragments, human paleontologists defined an entire species, *Eoanthropus,* and recast the story of human evolution. Later, in 1922, on the basis of a single fossil tooth found in Nebraska, an ancient species of man, *Hesperopithecus,* was defined. It was presumed to be as old as any hominid species found in the Old World and convinced some that then-current evolutionary models needed to be overhauled. The tooth turned out to belong to an ancient pig. Even in the case of Peking Man, the species was defined and initially named *Sinanthropus pekinensis* on the basis of only two teeth.

Today, the situation in human paleontology is quite different. The tapestry of our human evolutionary history is no longer woven with the filaments of a small handful of gauzy threads. We can now base our evolutionary scenarios (Figure 6) on enormous quantities of data supplied by several fields of science (see Feder and Park 1989 for a detailed summary of current thinking on human evolution).

Australopithecus afarensis, for example, the oldest known hominid, dating to more than 3.5 million years ago, is represented by more than a dozen fossil individuals. The most famous specimen, known as "Lucy," is more than 40 percent complete. Its pelvis is remarkably modern and provides clear evidence of its upright, and therefore

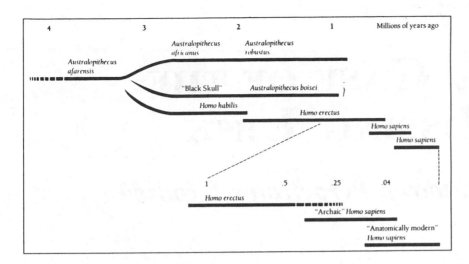

Figure 6 *Current human evolutionary chronologies are based on a large body of paleontological, archaeological, and genetic data. There is no room for—and no need for—a precociously large-brained human ancestor like* Eoanthropus *in the human pedigree. (From* Human Antiquity, *Feder and Park, Mayfield Publishing)*

humanlike, posture. Its skull, on the other hand, is quite apelike and contained a brain the size of a chimpanzee's. We even have a preserved pathway of footprints dating to the time when Lucy and her cohorts walked the earth, showing as dramatically as possible that they did so in a bipedal, humanlike fashion.

Homo erectus is known from dozens of individuals—forty from Zhoukoudian alone, nearly twenty from Java, and more than a dozen from Africa. In Kenya, the 80 percent complete skeleton of a twelve-year-old *Homo erectus* boy has been dated to more than 1.5 million years ago.

Archaic forms of *Homo sapiens,* especially the famous Neandertals, number in the hundreds. The fossil human record is rich and growing. Our evolu-

tionary scenarios are based, not on a handful of fragmentary bones, but on the remains of hundreds of individuals. Grafton Elliot Smith, Arthur Smith Woodward, and the others were quite wrong. The abundant evidence shows very clearly that human evolutionary history is characterized by the precedence of upright posture and the tardy development of the brain. It now appears that while our ancestors developed upright posture and humanlike bodies more than 3.5 million years ago, the modern human brain did not develop until as recently as 100,000 years ago.

Beyond this, human paleontologists are no longer restricted solely to the paleontological record. Exciting techniques of genetic analysis have allowed scientists to develop measures of dif-

ference between living species, including humans and our nearest extant relatives, the apes. Genetic "clocks" have been created from the results of such techniques.

For example, through DNA hybridization, scientists can quantify the difference between the genetic codes of people and chimpanzees. Here, an attempt is made to bond human and chimp DNA, much in the way the separate strands of the DNA double helix bond to produce the genetic code for a single organism. It turns out that the DNA of our two species is so similar that we can form a nearly complete bond. The opinion of most is that our two species could have split evolutionarily no more than five or six million years ago.

New dating techniques based on radioactive half-lives, biomechanical analysis of bones, scanning electron microscopy in bone and artifact examination, and many other new forms of analysis all make our evolutionary scenarios more concrete. It is to be expected that ideas will change as new data are collected and new analytical techniques are developed. Certainly our current views will be fine-tuned, and perhaps even drastic changes of opinion will take place. This is the nature of science. It is fair to suggest, however, that no longer could a handful of enigmatic bones that contradicted our mutually supportive paleontological, cultural, and genetic data bases cause us to unravel and reweave our evolutionary tapestry. Today, the discovery of a Piltdown Man likely would fool few.

*Also see next article, "The Case of the Missing Link." **ED.**

THE CASE OF THE MISSING LINK

Part I: Piltdown Perpetrator Exposed

ROBERT B. ANDERSON

ROBERT B. ANDERSON *is earth science editor at* Natural History *magazine at the American Museum of Natural History.*

THE PILTDOWN AFFAIR* *ranks as one of the most embarrassing episodes in the history of science. In the years preceding World War I, in the countryside just 30 miles south of London, at Piltdown in Sussex, someone seeded the ground with a handful of carefully doctored skull bones. Once discovered and assembled, the bones fooled Britain's leading scientists into believing they had found a fossil man, one ancient enough to be the missing link between humans and apes. The forgery successfully duped authorities for four decades and when it was discovered in 1953 Piltdown man instantly became one of the most famous hoaxes ever committed.*

Despite a plethora of circumstantial evidence pointing to a dozen different suspects, the identity of the hoaxer has remained a great mystery. The layers of intrigue surrounding it ensure that many plausible perpetrators can safely be proposed, and that, short of finding a signed confession, the mystery is unlikely ever to be solved. Yet, surprisingly, a confession in the Piltdown affair has finally come to light.

But first, the crime.

*Also see previous article, "Dawson's Dawn Man: The Hoax at Piltdown." ED.

THE HOAX BEGAN TO UNFOLD in 1908, when a laborer discovered a large skull fragment lying in a shallow gravel pit alongside the tree-lined drive leading to Barkham Manor, an estate near the village of Piltdown. He gave it to Charles Dawson, the lawyer managing the property and an accomplished amateur paleontologist. During the next four years, Dawson kept a close eye on the ditch and retrieved some stone tools and an amazing array of fossils from it: teeth and bones from extinct early Pleistocene mammals including elephants, hippopotami, and beaver. But by far the most intriguing to Dawson were the additional pieces of the human cranium.

In the early summer of 1912, the trap was set and Dawson took the bait. Stooping over the pit, he picked out a jawbone, which showed characteristics of both ape and man. Although Arthur Smith Woodward, a vertebrate paleontologist with the British Museum of Natural History, would later join Dawson in finding a few more pieces of the skull, and Pierre Teilhard de Chardin, a French Jesuit priest, would spot a strangely human canine tooth, the jaw set the deception in motion. Joined to the reconstructed cranium, it was enough to convince the fossil hunters that they had found the remains of an ancient human, one with a massive jaw defining an apelike face.

On December 18, 1912, at a meeting of the Geological Society of London, Dawson and Woodward announced their spectacular find to the world. They christened it *Eoanthropus dawsoni,* "Dawson's dawn man." Piltdown man, as the find soon came to be known, fit remarkably well with the expectations of British anthropologists. They had predicted that a missing link would be an intelligent ape. Its cranium—rivalling our own in size—seemed to confirm their belief that our large brain developed long ago, while the simian facial features disappeared relatively recently. While human intellect was no longer considered divine in origin, surely, they had argued, something so extraordinary could not have evolved overnight.

For the scientists at the British Museum, Piltdown man was a godsend. They thought the skull fragments were much older than the Neandertals and Cro-Magnons found on the continent, raising the possibility that some of the earliest humans were Englishmen. (Only the remains of Java Man, first found in 1891, were considered older.) More importantly, because of the paucity of human fossils that had been discovered so far, the bones from Sussex gave scientists something to study. In the decades that followed, anatomists debated the accuracy of various skull reconstructions, and anthropologists wrote scores of papers on Piltdown man, each offering a view of how it fit into our evolutionary history.

Only a few scientists, mostly on the other side of the Atlantic, suggested fraud. In 1914, William Gregory, a paleontologist at the American Museum of Natural History in New York, wrote presciently, "It has been suspected by some that geologically [the bones] are not that old at all; that they may even represent a deliberate hoax, a negro or Australian skull and a broken ape jaw, artificially fossilized and planted in the grave bed, to fool scientists."

Despite early suspicions, Piltdown man held its place in the human lineage until 1953, when Kenneth Oakley, a paleontologist at the British Museum, and Joseph S. Weiner and Wilfred Le Gros Clark, both anatomists at Oxford, reexamined the bones and found unmistakable signs of a forgery. The molars in the lower jaw—which came from a young female orangutan—had been filed flat to appear more human. The single canine tooth was similarly altered. The articular condyle, the hinge of the jaw, was deliberately broken off, masking the fact that it did not fit the human skull. The chin of the jaw, another distinctive feature that might have given away the jaw's simian origins, was also missing. The fragments of the cranium had come from a modern human, but one with an unusually thick skull—a supposedly primitive characteristic. Just enough of it was recovered to reconstruct the large braincase; except for a few small nose bones, the entire face and upper jaw were missing, leaving the appearance of Piltdown man open to interpretation. The bones had been soaked in a potassium bichromate solution to harden them and impart a patina of extreme age.

In retrospect, the forgery was rather crude. The molars, for example, bore file marks that were obvious under moderate magnification. Charles Blinderman, a professor of English and biology at Clark University who has authored a book on Piltdown man, observed that the whole affair was a testament "not to the skill of the hoaxer, but the the victims' credulity."

Whether or not it was the hoaxer's intention, Piltdown man effectively muddled early attempts to trace humanity's origins. The importance of new finds was clouded by the firm belief in the reality of *Eoanthropus dawsoni*. This was particularly true in the case of the Taung child, the first australopithcine from Africa, described in 1924 by Raymond Dart.

Who could have created this colossal red herring? Who had the wherewithal to alter the bones with just enough skill to fool the leading anthropologists and anatomists of the day? Who spent years setting it up?

Since 1953, researchers have written scores more papers and a half dozen books on Piltdown, each speculating on who the guilty party might have been. A dozen or so candidates have been proposed, but Charles Dawson, the discoverer of most of the bones, has been the favored culprit. After presenting his case against Dawson in his 1955 book *The Piltdown Forgery,* Weiner concludes that the circumstantial evidence was insufficient to prove beyond all reasonable doubt that Dawson was guilty. "In the circumstances, can we withhold from Dawson the one alternative possibility, remote though it seems, but which we cannot altogether dis-

prove: that he might, after all, have been implicated in a 'joke,' perhaps not even his own, which went too far?" The circumstantial evidence used to cast suspicion on others has appeared just as weak. Indeed, most investigators have conceded that a smoking gun would never be found.

INCREDIBLE AS IT MAY SEEM, however, the hoaxer did leave a confession. Although he had no intention of making them easy to find, he left behind a few crumbs for us to follow in the pages of a popular adventure novel. John Winslow, an American anthropologist, was the first to pick up the trail. In September 1983, his article "The Perpetrator of Piltdown," written with Alfred Meyer, was published in the magazine *Science 83.*

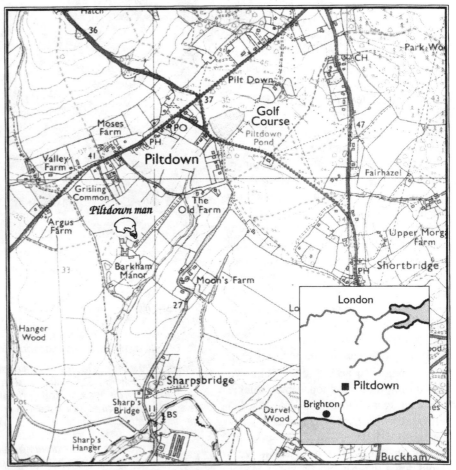

Ordnance Survey Map of Piltdown region

Computer enhancement: Paula McKenzie Nelson

5. THE FOSSIL EVIDENCE

After some two thousand hours of sleuthing, Winslow was convinced that he had found his man: Sir Arthur Conan Doyle. But Winslow prefaces his case against the famed British author with this caveat: "That Doyle has not been implicated in the hoax before now not only is a testament to the skill with which he appears to have perpetrated it, but it also explains why the case against him is circumstantial, intricate, even convoluted. For to be on Doyle's trail is, in a sense, to be on the trail of the world's greatest fictional detective himself, Sherlock Holmes." Indeed, if Doyle was adept at fabricating bizarre crimes for his stories, couldn't he have engineered the Piltdown forgery?

Winslow began to suspect Doyle after learning that he was an alumnus of Stonyhurst College—the same school attended by Charles Waterton, a notorious natural history faker. In 1825, shortly after Waterton returned from the Amazon, he wrote of a strange ape-man he had encountered in the rainforest. Having killed the beast for his collection, he claimed that, because of its weight, he could only carry out its head and shoulders. Waterton invited anyone who doubted his story to examine the creature, which he called Nondescript. But it was a fake. Waterton had taken the skin of a red howler monkey and reshaped its facial features to look humanoid.

Winslow admitted that this connection to the Piltdown hoax was slim, but he noted that Doyle had undoubtedly been exposed to many of Waterton's creations because the naturalist had donated his collections to Stonyhurst. Just as Edgar Allen Poe's amateur sleuth, C. Auguste Dupin in *The Murders in the Rue Morgue*, gave rise to Sherlock Holmes, Waterton's Nondescript may have planted the idea for Piltdown man in Doyle's mind.

When Winslow reviewed the list of people who had access to the Piltdown site during its excavation, Doyle's name surfaced again. Doyle lived some seven miles away in Crowborough and knew Dawson and Woodward. Winslow suggested that he could have easily ventured down the drive to Barkham Manor on his frequent country excursions and placed the fakes, or concealed behind

E. O. Hoppe/Lancelyn Green Collection

Lancelyn Green Collection

Top: Sir Arthur Conan Doyle, creator of Sherlock Holmes, relaxes at his country estate of Windlesham in 1912. In the same year, only eight miles away, the jaw of Piltdown man was recovered.

Above: Professor George Challenger, leader of a fictional expedition to a land inhabited by dinosaurs and primitive cave dwellers, glares from the cover of the first American serialization of The Lost World *in 1912. Doyle modelled as his character for a similar painting used in the novel's first English large paper edition. The novel was conceived shortly before the first fossils appeared at Piltdown, and was published soon afterward.*

the hedgerow that paralleled the road, he could have monitored progress on the excavation undetected. Dawson found the remains either on or near the surface of the shallow pit. The hoaxer, therefore, could have quickly deposited the bones and artifacts.

Winslow points out that Doyle, as a doctor, certainly had the knowledge necessary to fake the bones and that he was well acquainted with two people who could have been the sources for the orangutan jaw and the unusually thick cranium. Similarly, he detailed connections between Doyle and the mammal fossils and stone tools found at Piltdown—all of which he could have obtained from friends or during his travels.

But Winslow's most intriguing evidence against Doyle, he found in the author's 1912 adventure novel, *The Lost World*. In the story, four explorers, led by the ill-tempered Professor Challenger, make their way through the Amazon jungle to an isolated plateau, where dinosaurs and cave men have escaped extinction. Doyle had more than a casual interest in early human history. He had scribbled notes for *The Lost World* on the inside cover of an archeology journal and he had also written an unpublished manuscript entitled "Human Origins." As to the novel's relevance to the Piltdown case, Winslow points to a number of allusions to the crime in the book. For example, one character declares, "If you are clever and you know your business you can fake a bone as easily as you can a photograph." Even more suggestive, the plateau is populated by shaggy, red-haired ape-men, a description that allies them closely with orangutans—and the jaw of Piltdown man. Conan Doyle mentions missing links several times in the book. Winslow also makes the observation that Doyle models the plateau in the novel on the general area around Piltdown. In the story, the plateau was described "as large perhaps as Sussex, [which] has been lifted *en bloc* with all its living contents." The map of the plateau in the book seems to match the horseshoe-rimmed basin south of London known as the Weald, where Piltdown is located.

Winslow believes that the timing of Doyle's book is all important. Indeed, it shows that the Piltdown hoax was "developed hand-in-hand" with *The Lost World*. "On August 5, 1910, at a time when the Piltdown site had yielded nothing but a single skull fragment and no public announcements had yet been made," wrote Winslow, Doyle outlined his plans for *The Lost World* in a letter to his friend [Roger] Casement." According to Winslow, Doyle sent the novel off to *The Strand* magazine in December 1911. It started running in serial form in April 1912, several months before the crucial jaw was found and before there was any record of Doyle visiting the Piltdown site.

Winslow surmised that Doyle's motive for the crime was his longstanding animosity towards the scientists who attacked spiritualism, the once fashionable belief that departed souls could be contacted, especially through mediums at seances. Oddly enough, Doyle, the creator of a character who has come to symbolize rational thought, enthusiastically supported the spiritualist movement. In his later years, he devoted much time and money to the cause and spoke on both sides of the Atlantic promoting it. Doyle often hosted famous mediums and genuinely believed he and his wife had communicated with the dead.

In 1876, Edwin Lankester, an evolutionary biologist who later became director of the British Museum of Natural History, had exposed the American medium, Henry Slade, as a fake and taken him to court on charges of criminal fraud. Released on a technicality, the once popular Slade left England immediately. In the years that followed, Winslow found that Doyle often made references to the Lankester-Slade affair. In his short story "The Captain of the Pole-Star," published in 1883, Doyle stated that just because Slade had been revealed as a fraud, spiritualism as a whole should not be condemned. According to Winslow, "If science swallowed a scientific fraud like Piltdown man, then all of science, especially the destructive and arrogant evolutionists, whom Doyle called the Materialists, could be condemned."

Winslow also found that between 1906 and 1909 Lankester made a number of predictions about what future hominid discoveries would reveal. In almost every detail the Piltdown remains, the stone tools, and the fossil mammals fit the predictions.

Most Piltdown afficionados, however, did not take Winslow's accusation of Doyle seriously. In the letters column of the November issue of *Science 83*, Stephen Jay Gould wrote, "what can one say of an evidence-free argument based on speculations about motive?" Another letter read: "Although Winslow and Meyer did a fine job of research, they are totally incorrect. Once more, I have succeeded!" Signed, "Moriarty."

It's easy to agree. The allusions to Piltdown in *The Lost World* are numerous, but what do they prove? Doyle may have modeled the plateau on Sussex Weald, but what is so unusual about an author incorporating familiar surroundings into his fictional writing? And even though Winslow's suggested motive is intriguing, the problem remains: he failed to find a smoking gun.

BUT NOT EVERYONE WAS ready to give up on Doyle. Perhaps *The Lost World* contained important clues that Winslow overlooked. Richard Milner of the American Museum of Natural History focused on a cryptic diagram in Chapter 15 that he believed was the key—a puzzle that Doyle had left for someone to solve. In the story, the four explorers, having survived a number of harrowing encounters with dinosaurs and primitive men, are anxious to escape off the plateau and back to London where they can announce their discoveries to the world. Coming to their aid, a friendly native hands them a piece of bark on which a number of lines are scrawled in charcoal. The cryptic drawing turns out to be a map which shows a cave leading off the plateau. Milner found that the original diagram in *The Strand* was considerably different from those appearing in subsequent editions of the novel—probably because its accurate reproduction seemed unimportant to the story.

Milner also noticed that in the text that followed the diagram, Doyle had used a latin name, *Araucaria*, to describe the type of wood that the adventurers burned to light their way through the cave. Looking it up, he discovered that the wood came from a monkey puzzle tree. Milner remembered reading a biography of Teilhard de Chardin written by Mary and Ellen Lukas, in which Teilhard had picked the fossils "out of a gravel field near a cluster of monkey puzzle trees at Piltdown." The property was recently relandscaped, however, making it hard to check this reference.

Upon seeing the diagram and hearing of the monkey puzzle trees, I was hooked. Here was a puzzle created by Doyle more than 80 years ago. It was like being presented with a lost, unsolved Sherlock Holmes mystery. If I could find the solution, the reward would be the evidence needed to show that Doyle, in all probability, had masterminded the Piltdown hoax. Since Doyle had worked monkey puzzle trees into the narrative immediately following the puzzle, I guessed that he might have placed additional clues in this section of the text.

Some might think it unlikely that the Piltdown hoaxer would have concealed his admission of guilt in a literary work, but while researching this mystery, I came across a book that lent strong support to this possibility. In the book "Naked Is the Best Disguise," Samuel Rosenberg, a literary detective of some renown, documented the frequent personal references that Doyle often inserted into his books. Summarizing his findings, Rosenberg wrote, "As my more than 100 discoveries will testify, there was a compulsively honest second Doyle—an allegorist who left an abundance of clues which, when decoded, reveal a fascinating segment of his concealed personality."

Before you look at the solution to the puzzle in Part II, I urge you to try your hand at solving Doyle's mystery on your own. While you might want to read a copy of *The Lost World*, the puzzle and the essential section of the narrative to solve it are printed below. I also referred to a copy of the local Ordnance Survey map of the area around Piltdown (see map).

"This is a chart of the caves. What! Eighteen of them all in a row, some short, some deep, some branching, same as we saw them. It's a map, and here's a cross on it. What's the cross for? It is placed to mark one that is much deeper than the others."

THEY WERE NEATLY done in charcoal upon the white surface, and looked to me at first sight like some sort of rough musical score.

"Whatever it is, I can swear that it is of importance to us," said I. "I could read that on his face as he gave it."

"Unless we have come upon a primitive practical joker," Summerlee suggested, "which I should think would be one of the most elementary developments of man."

"It is clearly some sort of script," said Challenger.

"Looks like a guinea puzzle competition," remarked Lord John, craning his neck to have a look at it. When suddenly he stretched out his hand and seized the puzzle.

"By George!" he cried, "I believe I've got it. The boy guessed right the very first time. See here! How many marks are on that paper? Eighteen. Well, if you come to think of it there are eighteen cave openings on the hillside above us.

"He pointed up to the caves when he gave it to me," said I.

"Well, that settles it. This is a chart of the caves. What! Eighteen of them all in a row, some short, some deep, some branching, same as we saw them. It's a map, and here's a cross on it. What's the cross for? It is placed to mark one that is much deeper than the others."

"One that goes through," I cried.

"I believe our young friend has read the riddle," said Challenger. "If the cave does not go through I do not understand why this person, who has every reason to mean us well, should have drawn our attention to it. But if it does go through and comes out at the corresponding point on the other side, we should not have more than a hundred feet to descend."

"A hundred feet!" grumbled Summerlee.

"Well, our rope is still more than a hundred feet long," I cried. Surely we could get down."

"How about the Indians in the cave?" Summerlee objected.

"There are no Indians in any of the caves above our heads," said I. "They are all used as barns and storehouses. Why should we not go up now at once and spy out the land?"

There is a dry bituminous wood upon the plateau—a species of Araucaria, according to our botanist—which is always used by the Indians for torches. Each of us picked up a faggot of this, and we made our way up weed-covered steps to the particular cave which was marked in the drawing. It was, as I had said, empty, save for a great number of enormous bats, which flapped round our heads as we advanced into it. As we had no desire to draw the attention of the Indians to our proceedings, we stumbled along in the dark until we had gone round several curves and penetrated a considerable distance into the cavern. Then, at last, we lit our torches. It was a beautiful dry tunnel, with smooth grey walls covered with native symbols, a curved roof which arched over our heads, and white glistening sand beneath our feet. We hurried eagerly along it until, with a deep groan of bitter disappointment, we were brought to halt. A sheer wall of rock had appeared before us, with no chink through which a mouse could have slipped. There was no escape for us there.

We stood with bitter hearts staring at this unexpected obstacle. It was not the result of any convulsion, as in the case of the ascending tunnel. It was, and had always been, a cul-de-sac.

"Never mind, my friends," said the indomitable Challenger. "You have still my promise of a balloon."

Summerlee groaned.

"Can we be in the wrong cave?" I suggested.

"No use, young fellah," said Lord John, with his finger on our chart. "Seventeen from the right and second from the left. This is the cave sure enough."

"It was a beautiful dry tunnel, with smooth grey walls covered with native symbols, a curved roof which arched over our heads, and white glistening sand beneath our feet."

I looked at the mark to which his finger pointed, and I gave a sudden cry of joy.

"I believe I have it! Follow me! Follow me!"

I hurried back along the way we had come, my torch in my hand. "Here," said I, pointing to some matches upon the ground, "is where we lit up."

"Exactly."

"Well, it is marked as a forked cave, and in the darkness we passed the fork before the torches were lit. On the right side as we go out we should find the longer arm."

It was as I had said. We had not gone thirty yards before a great black opening loomed in the wall. We turned into it to find that we were in a much larger passage than before. Along it we hurried in breathless impatience for many hundreds of yards. Then suddenly, in the black darkness of the arch in front of us we saw a gleam of dark red light. We stared in amazement. A sheet of steady flame seemed to cross the passage and to bar our way. We hastened towards it. No sound, no heat, no movement came from it, but still the great luminous curtain glowed before us, silvering all the cave and turning the sand to powdered jewels, until as we drew closer it discovered a circular edge.

"The moon, by George!" cried Lord John. "We are through, boys! We are through!

It was indeed the full moon which shone straight down the aperture which opened upon the cliffs. It was a small rift, not larger than a window, but it was enough for all our purposes. As we craned our necks through it we could see that the descent was not a very difficult one, and that the level ground was no very great way below us. It was no wonder that from below we had not observed the place, as the cliffs curved overhead and an ascent at the spot would have seemed so impossible as to discourage close inspection. We satisfied ourselves that with the help of our rope we could find our way down, and then returned, rejoicing, to our camp to make our preparations for the next evening.

For the solution to "The Case of the Missing Link," see *Part II:* The Solution, next page.

THE CASE OF THE MISSING LINK

Part II: The Solution

ROBERT B. ANDERSON

The solution to the puzzle unfolds here in more or less the same way I discovered it. The subtle way in which the clues came together is an integral part of the proof, leaving little doubt that the Piltdown man was Sir Arthur Conan Doyle's best work of fiction. That being the case, it seems only fitting to have Sherlock Holmes solve the mystery. What follows is Dr. John Watson's account of how the great detective did it.

HOLMES REACHED DOWN into a well-worn leather brief-case and pulled out a copy of Doyle's book *The Lost World* and began: "After reading Winslow's case against Doyle, and listening to Milner's arguments that the puzzle was the key, I must admit that my mood was considerably brightened by the prospects of a real mystery to solve. As you undoubtedly noticed, Watson, the puzzle is followed by the narrator's comment: 'Whatever it is, I can swear that it is of importance to us.' To which another character replied, 'Unless we have come upon a primitive practical joker, which I should think would be one of the most elementary developments of man.' When I read those lines, I too was convinced that Doyle had committed the hoax."

"But how could you be sure?" I asked.

"Ah. That was precisely the problem, Watson. I couldn't. I began to notice all sorts of possible connections between Doyle and Piltdown. Perhaps Doyle had even left his signature on the jaw of Piltdown man by breaking off its articular condyle. But I needed something concrete—the solution to the puzzle. As you know, I could not rest until I discovered the message Doyle was hiding in those cryptic marks!"

"Indeed! Judging from your late night violin sessions, you've barely slept for a week. I thought you might go mad trying to figure this one out. I myself stared at the puzzle for hours, but could make nothing of it. Surely, it is the most difficult of riddles!"

"You think so now, but after I have laid my reasoning before you, you will wonder why you did not see it, too."

"I swear I will not."

Ignoring my pledge, Holmes began, "I read the passage immediately following the puzzle countless times. I took particular note that the adventurers in the story took the markings to be a map of the 18 caves in the hillside above them, one of which would lead them off the plateau. So, Doyle tells us the puzzle is a map! I wondered if the second line, the one marked with an "X", might lead directly to Barkham Manor and the gravel pit where Piltdown man was found."

"But Holmes, those sticks bear no resemblance to any map I've ever seen."

"Quite so," my friend agreed with a thin smile, "but as professor Stephen Jay Gould wrote in his article on the Piltdown affair, 'The mark of any good theory is that it makes coordinated sense of a string of observations otherwise independent and inexplicable.' If my map theory was correct, it would make sense of the puzzle and the accompanying narrative, which I suspected was a veiled description of the Piltdown site. The first and most important test was to compare the puzzle to an actual map. I saw that, tipped on edge, the marked line roughly matched the immediate roads around Barkham Manor."

For my benefit Holmes drew a map.

"One glance at this and I knew I was on the right track. I was bothered by this part of the line," he said, pointing to the triangular area, the Piltdown common. "It crosses the land where there is only a dashed line, representing a foot path. Trusting that this inconsistency would somehow resolve itself, I continued to search the map for other similarities. I was struck by the name of a farm located near Barkham Manor, Moon's Farm. Do you recall the narrator's description of their emergence from the cave?"

"I confess. I do not."

Without referring to the book, Holmes recited the passage word for word:

"Then suddenly, in the black darkness of the arch in front of us we saw a gleam of dark red light. We stared in

amazement. A sheet of steady flame seemed to cross the passage and to bar our way. We hastened towards it. No sound, no heat, no movement came from it, but still the great luminous curtain glowed before us, silvering all the cave and turning the sand to powdered jewels, until as we drew closer it discovered a circular edge.

'The moon, by George!' cried Lord John. 'We are through, boys! We are through!'

It was indeed the full moon which shone straight down the aperture which opened upon the cliffs."

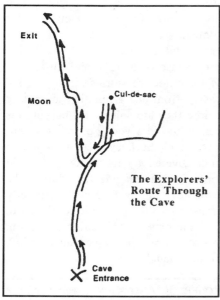

The Explorers' Route Through the Cave

Robert B. Anderson

PAUSING TO RELIGHT HIS PIPE, Holmes continued, "Moon's Farm is located exactly where it should be if my theory was correct. Moving through the cave marked with the X, the explorers find their way out by following the left branch, precisely where the farm is located. But they take the left branch only after they wander down the middle passage by mistake—the one that comes to an abrupt end. This middle tunnel corresponds to the drive leading to Barkham Manor and the pit where Piltdown Man was found. 'It was, and had always been, a cul-de-sac.' The road to Barkham Manor ended in a cul-de-sac and Doyle couldn't have chosen a better analogy for the Piltdown site. Didn't his

forgery sidetrack anthropologists down the wrong trail, leading to a dead end?"

"Yes, indeed, But what of the puzzle?" I asked excitedly.

"All of these clues supported my little theory, but I knew no one would be convinced by them alone. I was troubled by the seventeen other lines of the puzzle. I couldn't figure out what they were, or even if they were significant. Then suddenly, after studying the map blindly, the solution to the puzzle jumped out at me. In bold letters across the Piltdown common, it read 'Piltdown Golf Course.' On occasion, even a man of my abilities needs to have the answer spelled out in bold lettering."

"You've lost me, Holmes. Other than its proximity to Barkham Manor, what significance could a golf course possibly have to the Piltdown mystery?"

"Ah, you do disappoint me, Watson. How many holes are there on a course? And how many caves were there?

"Eighteen. But how does that. . . . My God! Of course! The number of lines in the puzzle!"

"Precisely. Each line in the puzzle represents one of the 18 fairways. The "x" where the explorers enter the cave corresponds to the "CH" on the map— the golf course club house. And as you may know, Doyle was an avid golfer. In his article on Piltdown, Winslow wrote that Doyle frequently vacationed in Norfolk because it:

> . . . boasted excellent golf courses, as golf was one of his favorite games. These included the Sheringham Golf Course, which abutted the East Runton deposit, a confusing collection of fossils ranging from a late Pleistocene beaver to a variety of early Pleistocene animals. The same unusual kind of mix found in the Piltdown gravel pit. It is not unlikely that the beaver fossils as well as some older fossils found at Piltdown were picked up between rounds of golf.

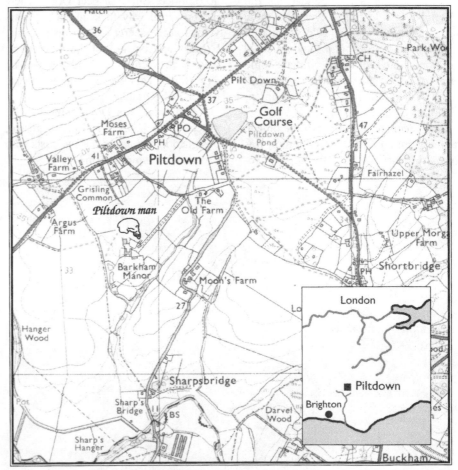

Computer enhancement: Paula McKenzie Nelson

Ordnance Survey Map of Piltdown region

5. THE FOSSIL EVIDENCE

I submit to you, Watson, that Doyle not only picked up fossils while golfing, but also deposited them while visiting the Piltdown course."

"Extraordinary, Holmes, you've done it again," I cried. "This truly is the case of the missing links."

"Yes, quite. Having solved the puzzle, I am now confident that even the most ardent skeptic will have to concede that Doyle was describing the exact location where Piltdown man would be found. The evidence is overwhelming: the mention of monkey puzzle trees which lined the drive to Barkham Manor, the fit of the roads, the location of Moon's Farm and the cul-de-sac, and finally the Piltdown golf course. And how could Doyle have known the location of the site when he wrote *The Lost World*, unless of course, he was the culprit!"

"This is incredible! After more than 80 years, the crime is solved! Doyle certainly didn't make it easy!" I exclaimed.

"On the contrary. I believe, as did Winslow, that he gave everyone more than a sporting chance. Two years after Piltdown man was found Doyle planted a curious bone tool in the pit that should have tipped off the scientists to the fact that they had fallen prey to a monumental hoax.

"In December 1914 when the object, carved from a fossil elephant femur, was formally described at a Geological Society meeting, Reginald Smith, an expert on antiquities, said that he could not imagine a use for an implement that so closely resembled a cricket bat. Smith also raised the possibility that the old bone had been 'whittled in recent times.' But rather than follow up on these suspicions, his colleagues preferred to believe that the object was a genuine Paleolithic tool, albeit one of unknown use. As Winslow noted, Doyle also excelled at cricket and had played on some of the country's top amateur teams. By placing a cricket bat 'in the hands' of Piltdown man, Doyle had given the scientists another chance to discover the hoax and its perpetrator. He must have been incredulous when, once again, the scientists failed to notice anything amiss."

"Do you think Doyle acted alone?" I asked.

"Even with all the evidence against him, one might still be tempted to think that others also had a hand in it. Indeed, we cannot say that Doyle didn't have an accomplice. But I suspect Doyle knew the only real way to keep such a marvelous crime secret was to go it alone. I think Dawson and all the others who have been wrongly accused can now be exonerated. This was a hoax, not a case of scientific fraud."

"And what was Doyle's motive?" I asked. "Was it, as Winslow suggested, designed to embarrass the scientists who had attacked spiritualism?"

"Perhaps. But given the way the crime is so intimately tied to the writing of *The Lost World*, I think we should look there for the real motive. The strange introduction to the novel seems to hold the key to the motive:

> *I have wrought my simple plan*
> *If I give one hour of joy*
> *To the boy who's half a man*
> *Or the man who's half a boy.*

These four lines, which have no apparent meaning as a preamble to the story, suggest to me that Doyle created the Piltdown forgery as a mystery for someone to solve—one that might very well outlast him. Doyle had a deep sense of history, and, in several of his detective stories, mysteries are passed from one generation to another before they are solved. By burying his confession in a young person's adventure novel, he knew that it would go undiscovered by the academics for a long time. I imagine that during the remainder of his life, he was greatly amused by his hoax as it continued to fool the experts. One thing is certain, Watson: when he passed on in 1930, he died laughing."

ROBERT B. ANDERSON *is earth science editor at* Natural History *magazine at the American Museum of Natural History.*

Sunset on the Savanna

Why do we walk? For decades anthropologists said that we became bipedal to survive on the African savanna. But a slew of new fossils have destroyed that appealing notion and left researchers groping for a new paradigm.

James Shreeve

James Shreeve is a contributing editor of Discover. *He is co-author, with anthropologist Donald Johanson, of* Lucy's Child: The Discovery of a Human Ancestor. *His latest book is* The Neandertal Enigma. *Shreeve is currently at work on his first novel.*

IT'S A WONDERFUL STORY. *Once upon a time, there was an ape who lived in the middle of a dark forest. It spent most of its days in the trees, munching languidly on fruits and berries. But then one day the ape decided to leave the forest for the savanna nearby. Or perhaps it was the savanna that moved, licking away at the edge of the forest one tree at a time until the fruits and berries all the apes had found so easily weren't so easy to find anymore.*

In either case, the venturesome ape found itself out in the open, where the air felt dry and crisp in its lungs. Life was harder on the savanna: there might be miles between one meal and another, there were seasons of drought to contend with, and large, fierce animals who didn't mind a little ape for lunch. But the ape did not run back into the forest. Instead it learned to adapt, walking from one place to another on two legs. And it learned to live by its wits. As the years passed, the ape grew smarter and smarter until it was too smart to be called an ape anymore. It lived anywhere it wanted and gradually made the whole world turn to its own purposes. Meanwhile, back in the forest, the other apes went on doing the same old thing, lazily munching on leaves and fruit. Which is why they are still just apes, even to this day.

The tale of the ape who stood up on two legs has been told many times over the past century, not in storybooks or nursery rhymes but in anthropology texts and learned scientific journals. The retellings have differed from one another in many respects: the name of the protagonist, for instance, the location in the world where his transformation took place, and the immediate cause of his metamorphosis. One part of the story, however, has remained remarkably constant: the belief that it was the shift from life in the forest to life in a more open habitat that set the ape apart by forcing it onto two legs. Bipedalism allowed hominids to see over tall savanna grass, perhaps, or escape predators, or walk more efficiently over long distances. In other scenarios, it freed the hands to make tools for hunting or gathering plants. A more recent hypothesis suggests that an erect posture exposes less skin to the sun, keeping body temperature lower in open terrain. Like the painted backdrop to a puppet theater, the savanna can accommodate any number of dramatic scenarios and possible plots.

But now that familiar stage set has come crashing down under the weight of a spectacular crop of new hominid fossils from Africa, combined with revelations about the environment of our earliest ancestors. The classic savanna hypothesis is clearly wrong, and while some still argue that the open grasslands played some role in the origins of bipedalism, a growing number of researchers are beginning to think the once unthinkable: the savanna may have had little or nothing to do with the origins of bipedalism.

"The savanna paradigm has been overthrown," says Phillip Tobias, a distinguished paleoanthropologist at the University of Witwatersrand in Johannesburg and formerly a supporter of the hypothesis. "We have to look now for some other explanation for bipedalism."

The roots of the savanna hypothesis run deep. More than 100 years ago, Charles Darwin thought that mankind's early ancestor moved from "some warm, forest-clad land" owing to "a change in its manner of procuring subsistence, or to a change in the surrounding conditions." In his view, the progenitor assumed a two-legged posture to free the hands for fashioning tools and performing other activities that in turn nourished the development of an increasingly refined intelligence. Darwin believed that this seminal event happened in Africa, where mankind's closest relatives, the African apes, still lived.

By the turn of this century, however, most anthropologists believed that the critical move to the grasslands had occurred in Asia. Though the bones of a primitive hominid—which later came to be called *Homo erectus*—had been discovered on the Southeast Asian island of Java, the change of venue had more

5. THE FOSSIL EVIDENCE

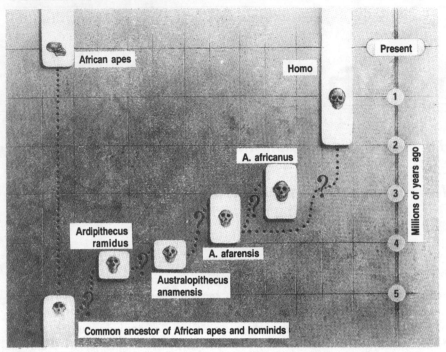

New hominid fossils have filled out our evolutionary tree, but the question of exactly how they are related to one another (and ourselves) remains unanswered.

to do with cultural values and racist reasoning than with hard evidence. Africa was the "dark continent," where progress was slowed by heat, disease, and biotic excess. In such a place, it was thought, the mind would vegetate—witness the "regressive" races that inhabited the place in modern times. The plains of central Asia, on the other hand, seemed just the sort of daunting habitat that would call out the best in an enterprising ape. In that environment, wrote the American paleontologist Henry Fairfield Osborn, "the struggle for existence was severe and evoked all the inventive and resourceful faculties of man . . . ; while the anthropoid apes were luxuriating in the forested lowlands of Asia and Europe, the Dawn Men were evolving in the invigorating atmosphere of the relatively dry uplands."

In hindsight, the contrasts made between dark and light, forest and plain, slovenly ape and resourceful man seem crudely moralistic. Higher evolution, one would think, was something reserved for the primate who had the guts and wits to go out there and grab it, as if the entrepreneurial spirit of the early twentieth century could be located in

our species' very origins. But at the time the idea was highly influential. Among those impressed was a young anatomist in South Africa named Raymond Dart. in 1925 Dart announced the discovery, near the town of Taung, of what he believed to be the skull of a juvenile "man-ape," which he called *Australopithecus africanus*. While Darwin had been correct in supposing Africa to be the home continent of our ancestors, it seemed that Osborn had been right about the creature's habitat: there are no forests around Taung, and scientists assumed there hadn't been any for millions of years. Forests might provide apes with "an easy and sluggish solution" to the problems of existence, wrote Dart, but "for the production of man a different apprenticeship was needed to sharpen the wits and quicken the higher manifestations of intellect—a more open veld country where competition was keener between swiftness and stealth, and where adroitness of thinking and movement played a preponderating role in the preservation of the species."

In the decades that followed Dart's discovery, more early hominids emerged from eastern and southern Africa, and

most researchers concluded that they made their homes in the savanna as well. The question was less *whether* the savanna played a part in the origin of bipedalism—that was obvious—than *how*. Dart originally proposed that *Australopithecus* had taken to two legs to avoid predators ("for sudden and swift bipedal movement, to elude capture"), but later he reversed this scenario and imagined his "killer ape" the eater rather than the eaten, forsaking the trees for "the more attractive fleshy foods that lay in the vast savannas of the southern plains."

Later studies suggested that the environments in eastern and southern Africa where early hominids lived were not the vast, unchanging plains Dart imagined. Instead they appeared to be variable, often characterized by seasonally semiarid terrain, a plain studded with scraggly trees and patches of denser woodland. But no matter: This "savanna mosaic" was still drier and more open than the thick forest that harbors the African apes today. It just made good sense, moreover, that our ancestors would have come down to the ground and assumed their bipedal stance in a habitat where there were not as many trees to climb around in. The satisfying darkness-into-light theme of early hominid development held up, albeit with a little less wattage.

For decades the popularity of the savanna hypothesis rested on the twin supports of its moral resonance and general plausibility: our origin should have happened this way, and it would make awfully good sense if it did. In East Africa, geography seemed to reinforce the sheer rightness of the hypothesis. Most of the earliest hominid fossils have come from the eastern branch of the great African Rift Valley; researchers believed that when these hominids were alive, the region was much like the dry open grasslands that dominate it today. The lusher, more forested western branch, meanwhile, is home to those lazy chimps and gorillas—but to no hominid fossils.

Still, despite such suggestive correspondence, until 20 years ago something was missing from the hypothesis: some hard data to link environment to human evolution. Then paleontologist Elisabeth Vrba of Yale began offering what was the strongest evidence that the drying up of African environments helped shape early human evolution. In studying the bones of antelope and other bovids from hominid sites in South Africa, Vrba noticed a dramatic change occurring between 2.5 and 2 million years ago. Many species that were adapted to wooded environments, she saw, suddenly disappeared from the fossil record, while those suited to grassy regions appeared and multiplied. This "turnover pulse" of extinctions and origins coincides with a sudden global cooling, which may have triggered the spread of savannas and the fragmentation of forests.

Other investigators, meanwhile, were documenting an earlier turnover pulse around 5 million years ago. For humans, both dates are full of significance. This earlier pulse corresponds to the date when our lineage is thought to have diverged from that of the apes and become bipedal. Vrba's second, later pulse marks the appearance of stone tools and the arrival on the scene of new hominid species, some with brains big enough to merit inclusion in the genus *Homo*. The inference was clear: our early ancestors were savanna born and savanna bred.

"All the evidence, as I see it," Vrba wrote in 1993, "indicates that the lineage of upright primates known as australopithecines, the first hominids, was one of the founding groups of the great African savanna biota."

With empirical evidence drawn from two different sources, the turnover pulse is a great improvement on the traditional savanna hypothesis (which in retrospect looks not so much like a hypothesis as a really keen idea). Best of all, Vrba's hypothesis is testable. Let's say that global climate changes did indeed create open country in East Africa, which in turn triggered a turnover of species and pushed ahead the evolution of hominids. If so, then similar turnovers in animal species should have appeared in the fossil record whenever global change occurred.

Over the last 15 years, Andrew Hill and John Kingston, both also at Yale, have been looking for signs of those dramatic shifts at some 400 sites in the Tugen Hills of Kenya. In the heart of the fossil-rich eastern branch of the Rift Valley, the Tugen hills offer a look at a succession of geologic layers from 16 million years ago to a mere 200,000 years ago—studded with fragmentary remains of ancient apes and hominids. To gauge the past climate of the Tugen Hills, the researchers have looked at the signatures of the ancient soils preserved in rock. Different plants incorporate different ratios of isotopes of carbon in

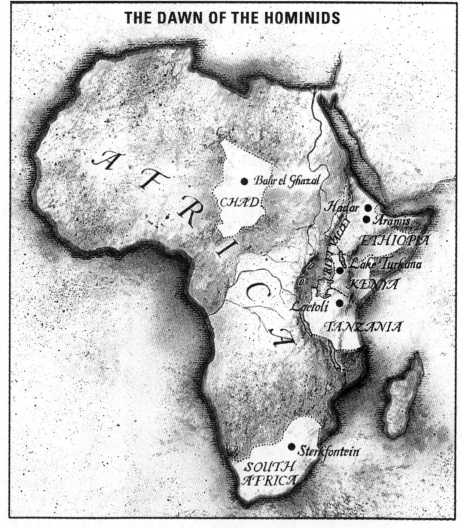

● **ARAMIS**
AGE: 4.4 million years old
SPECIES: *Ardipithecus ramidus*
BIPEDAL? Unknown
ENVIRONMENT: Dense woodlands

● **LAKE TURKANA**
AGE: 4.2 million years old
SPECIES: *Australopithecus anamensis*
BIPEDAL? Yes
ENVIRONMENT: Lakeside forests in an arid region

● **LAETOLI**
AGE: 3.5 million years old
SPECIES: *Australopithecus afarensis*
BIPEDAL? Yes
ENVIRONMENT: Primarily grasslands, with some forests

THE DAWN OF THE HOMINIDS

● **STERKFONTEIN**
AGE: 3 to 3.5 million years old
SPECIES: Probably *Australopithecus africanus*
BIPEDAL? Yes, but also still a tree climber
ENVIRONMENT: Forests

● **BAHR EL GHAZAL**
AGE: 3 to 3.5 million years old
SPECIES: Similar to *Australopithecus afarensis*
BIPEDAL? Probably
ENVIRONMENT: Gallery forests with grassy patches

● **HADAR**
AGE: 3.2 million years old
SPECIES: *Australopithecus afarensis* ("Lucy")
BIPEDAL? Yes
ENVIRONMENT: Forests and bushlands

their tissues, and when those plants die and decompose, that distinctive ratio remains in the soil. Thus grasslands and forests leave distinguishing isotopic marks. When Hill and Kingston looked at soils formed during Vrba's turnover pulses, however, they found nothing like the radical shifts to grasslands that she predicted. Instead of signs that the environment was opening up, they found that there was a little bit of grass all the time, with no dramatic changes, and no evidence that early hominids there ever encountered an open grassland.

"Elisabeth's turnover pulse hypothesis is very attractive," says Hill. "It would have been lovely if it had also been true."

Other research has also contradicted Vrba's hypothesis. Laura Bishop of Liverpool University in England, for instance, has been studying pig fossils from several East African sites. Some of those fossil animals, she has found, had limbs that were adapted not for open habitats but for heavy woods. Peter de Menocal of Columbia University's Lamont-Doherty Earth Observatory has been looking at long-term climate patterns in Africa by measuring the concentration of dust in ocean sediments. Over the past 5 million years, he has found, the African climate has cycled back and forth between dry and wet climates, but the pattern became dramatic only 2.8 million years ago, when Africa became particularly arid. Such a change could have played a role in the dawn of *Homo*, but it came over a million years too late to have had a hand in australopithecines' becoming bipedal.

WHAT MAY finally kill the savanna hypothesis—or save it—are the hominids themselves. More than anything else, walking on two feet is what makes a hominid a hominid. If those first bipedal footsteps were made on savanna, we should find fossils of the first hominids in open habitats. For almost 20 years, the earliest hominid known has been *Australopithecus afarensis*, exemplified by the 3.2-million-year-old skeleton called Lucy. Lucy had a chimp-size skull but an upright posture, which clinches the argument that hominids evolved bipedalism

before big brains. But had Lucy completely let go of the trees? It's been a matter of much debate: oddities such as her curved digits may be the anatomic underpinnings of a partially arboreal life-style or just baggage left over from her tree-climbing ancestry.

Nor did *afarensis* make clear a preference for one sort of habitat over another. Most of the fossils come from two sites: Hadar in Ethiopia, where Lucy was found, and Laetoli in Tanzania, where three hominids presumed to be *afarensis* left their footprints in a layer of newly erupted volcanic ash 3.5 million years ago. Laetoli has been considered one of the driest, barest habitats in the eastern rift and thus has given comfort to the savanna faithful. But at Hadar the *afarensis* fossils appear to have been laid down among woodlands along ancient rivers. Other ambiguous bones that may have belonged to *afarensis* and may have dated back as far as 4 million years ago have been found at nearby East African localities. Environments at these sites run the gamut from arid to lush, suggesting that Lucy and her kin may not have been confined to one particular habitat but rather lived in a broad range of them. So why give some special credit to the savanna for launching our lineage?

"I've always thought that there was scant evidence for the savanna hypothesis, based simply on the fact that hominids are extremely plastic behaviorally," says Bill Kimbel, director of science at the Institute for Human Origins in Berkeley, California.

At least *afarensis* had enough respect for conventional wisdom to stay on the right side of the Rift Valley. Or it did until last year. In November a team led by Michel Brunet of the University of Poitiers in France announced the discovery of a jawbone similar to that of *afarensis* and, at 3 to 3.5 million years old, well within the species' time range. The ecology where Brunet's hominid lived also has a familiar ring: "a vegetational mosaic of gallery forest and wooded savanna with open grassy patches." But in one important respect the fossil is out of left field— Brunet found it in Chad, in north central Africa, more than 1,500 miles from

the eastern Rift Valley, where all the other *afarensis* specimens have been found. This hominid's home is even farther west, in fact, than the dark, humid forests of the western rift, home to the great apes—clearly on the wrong side of the tracks.

In a normal year the Chad fossil would have been the biggest hominid news. But 1995 was anything but a normal year. In August, Meave Leakey of the National Museums of Kenya and her colleagues made public the discovery of a new hominid species, called *Australopithecus anamensis*, even older than *afarensis*. (The fossils were found at two sites near Lake Turkana in Kenya and derive their name from the Turkana word for "lake.") The previous spring, Tim White of the University of California at Berkeley and his associates had named a whole new genus, *Ardipithecus ramidus*, that was older still, represented by fossils found in Aramis, Ethiopia, over the previous three years. The character of *anamensis* and *ramidus* could well have decided the fate of the savanna hypothesis. If they were bipeds living in relatively open territory, they could breathe new life into a hypothesis that's struggling to survive. But if either showed that our ancestors were upright *before* leaving the forest, the idea that has dominated paleoanthropology for a century would be reduced to little more than a historical artifact.

The best hope for the savanna hypothesis rests with Leakey's new species. In its head and neck, *anamensis* shares a number of features with fossil apes, but Leakey's team also found a shinbone that is quite humanlike and emphatically bipedal—in spite of the deep antiquity of *anamensis*: the older of the two sites has been dated to 4.2 million years ago. And the region surrounding the sites was a dry, relatively open bushland.

This is good news for those, such as Peter Wheeler of Liverpool John Moores University in England, whose theories of bipedalism depend on an initial movement out of the closed-canopy forest. Wheeler maintains that standing upright exposes less body surface to the sun, making it possible for proto-hominids to keep cool enough out of the shade to exploit savanna resources. "It's pretty clear that by three and a half million years ago, australopithecines were living in a range of habitats," he says. "But if you look at the oldest evidence for bipedalism—Laetoli, and now the *anamensis* sites—these are actually more open habitats."

However, Alan Walker of Penn State, one of the codiscoverers of *anamensis*, disagrees. Though the regional climate may have been as hot and arid back then as it is now, he says, the local habitat of *anamensis* was probably quite different. Back then the lake was much bigger than it is today and would have supported a massive ring of vegetation. Animal fossils found at the *anamensis* sites—everything from little forest monkeys to grass-eating antelope—were lodged in deltalike sediments that must have been deposited by "monstrous great rivers," says Walker, with gallery forests as much as a mile or two wide on both banks.

Even Laetoli is proving to be a mixed blessing for savanna lovers. In the first studies, researchers focused their attention mainly on the abundance of fossils of arid-adapted antelope at the site. Based on these remains, they concluded that Laetoli was a grassland with scattered trees. But according to Kaye Reed of the Institute of Human Origins, these conclusions are debatable. For one thing, the antelope are gregarious herders, so one should expect to find more of their bones than those of more solitary species. Moreover, the original studies underplayed evidence for a more diverse community, which included woodland-dwelling antelope and an assortment of monkeys. In separate studies, Reed and Peter Andrews of the British Natural History Museum took a more thorough look at Laetoli, and both concluded that the original description was far too bleak. "Monkeys have to have trees to eat and sleep in," says Reed. "I don't want to give the impression that this was some kind of deep forest. But certainly Laetoli was more heavily wooded than we thought."

Ardipithecus ramidus poses a potentially more devastating blow to the savanna hypothesis. The 4.4-million-year-old species has a skull and teeth that are even more primitive and chimplike than *anamensis*. Other traits of its anatomy, however, align it with later hominids such as *afarensis*. It remains to be seen whether *ramidus* is an early cousin of our direct ancestors or is indeed at the very base of the hominid lineage (its name perhaps reveals the hope of its discoverers, deriving from the Afar word for "root"). What we do know is that the species lived in a densely wooded habitat along with forest-dwelling monkey species and the kudu, an antelope that prefers a bushy habitat.

"We interpret these initial results as evidence that the hominids lived and died in a wooded setting," says Tim White.

A. ramidus would be the final nail in the coffin of the savanna hypothesis, except for one crucial bit of missing information: none of its discoverers will yet say if it was bipedal. Researchers have unearthed a partial skeleton consisting of over 100 fragments of dozens of bones, including hand bones, foot bones, wrist bones—more than enough to determine whether the creature walked like a human or like a chimp. Unfortunately the fragile bones are encased in sediment that must be laboriously chipped away before a proper analysis can begin. Neither White nor any member of his team will comment on what the bones say about locomotion until a thorough study can be completed—which at this point won't be until 1998 at the earliest.

There are plenty of other puzzling bones to ponder in the meantime. Back in South Africa, where Dart found the first australopithecine, researchers have begun analyzing a horde of some 500 new and previously collected specimens from a site called Sterkfontein. Some of them may have a potent impact on the savanna hypothesis.

The most widely publicized is "Little Foot," a tantalizing string of four connected foot bones running from the ankle to the base of the big toe. Little Foot is between 3 and 3.5 million years old, hundreds of thousands of years younger than the *ramidus* and *anamensis* specimens. Yet according to

Phillip Tobias and his Witwatersrand colleague Ronald Clarke, the fossil demonstrates that this early species of australopithecine—quite probably *A. africanus*—still spent time in the trees. While the anklebone, Tobias and Clarke say, is built to take the weight of a bipedal stride, the foot is also surprisingly primitive. This is especially true of the big toe, which they contend splayed out to the side like a chimpanzee's, all the better to grasp tree branches when climbing.

ALTHOUGH not everyone agrees that Little Foot's foot is so apelike, a host of other fossils from Sterkfontein also speak of the trees. Lee Berger, also at Witwatersrand, has analyzed several shoulder girdles from the collection and found them even better suited to climbing and suspending behavior than those of *afarensis*. He and Tobias have analyzed a shinbone from the same site and concluded it was more chimplike than human. And in what is perhaps the most compelling finding so far, Berger and Henry McHenry of the University of California at Davis have analyzed the proportions of arms and legs of the Sterkfontein *africanus* specimens and found that they were closer to chimps than to humans.

The tree-climbing anatomy of *africanus* has prompted Tobias, long a savanna loyalist, to wonder why such an animal would be out on the South African veld, where today there aren't any trees big enough to climb around in. A possible answer came from recent reevaluations of the environments at several *africanus* sites. As in East Africa, the ancient fauna and pollen suggest that conditions there were warmer, wetter, and more wooded than previously thought. Tobias's colleague Marian Bamford has even recovered traces at Sterkfontein of liana vines, which grow primarily in dense forests. The climate at the site did become drier and more open 2.5 million years ago, but by then hominids had been walking on two legs for a least a million and a half years.

If they live up to their discovers' initial claims, the Sterkfontein fossils make hominid history more complicated than we thought. Lucy and her fellow *afarensis* were traipsing through Ethiopia at about the same time Little Foot and the other *africanus* hominids were in South Africa—and the two hominids were using different ways of moving around. While Lucy was more committed to life on the ground, Little Foot went for a mixed strategy, sometimes scuttling up trees. "What this suggests," says Berger, "is that bipedalism may have evolved not once but twice."

And in both cases, the South African researchers argue, bipedalism was not associated with the savanna. "The idea that bipedalism evolved as an adaptation to the savanna," declares Tobias, "can be thrown out the window."

If so, it shall be missed. For all its shortcomings—a shortage of evidence being the first among them—the savanna hypothesis provided a tidy, plausible explanation for a profound mystery: What set human beings apart from the rest of creation? If not the savanna, what did cause the first hominids to become bipedal? Why develop anatomy good only for walking on the ground when you are still living among the trees? For all the effort it has taken to bring down the savanna hypothesis, it will take much more to build up something else in its place.

"I don't really know why we became bipedal," says Andrew Hill. "It's such an unusual thing."

"We're back to square one," says Tobias.

"Square one" is not completely empty. Kevin Hunt of Indiana University, for instance, has recently revived the idea that bipedalism was initially an adaptation for woodland feeding rather than a new way of getting around. Chimps often stand while feeding in small trees and bushes, stabilizing themselves by hanging onto an overhanging branch. Hunt suggests that the earliest australopithecines made an anatomic commitment to this specialized way of obtaining food.

Nina Jablonski of the California Academy of Science in San Francisco sees the beginnings of our bipedalism mirrored instead in the upright threat displays of great apes. Perhaps our ancestors resorted to this behavior more

than their ape cousins to maintain the social hierarchy. Originally Jablonski and her colleague George Chaplin of the University of Western Australia linked the increase in bipedal threat displays to a move to the savanna, where there would be more competition for resources. But in light of the new evidence for wooded habitats, she now concedes that this "essentially primary cause" of bipedalism could have emerged in a forest.

There is a stubborn paradox in such models, based as they are on living primates: since chimps and gorillas have presumably been performing these behaviors for millennia without the evolution of bipedalism, how could the same behaviors have driven just such an evolution in hominids? In the early 1980's, Owen Lovejoy of Kent State in Ohio proposed an elegant explanation for bipedalism that bypassed this logical difficulty and had nothing to do with moving about on the savanna.

"Bipedality is a lousy form of locomotion," says Lovejoy. "It's slower and more awkward, and it puts the animal at greater risk of injury. The advantage must come from some other motivating selective force." To Lovejoy, the force is reproduction itself. In his view, what separated the protohominids from their ape contemporaries was a wholly new reproductive strategy, in which males provided food to females and their mutual offspring. With the males' assistance, the females could forage less and give birth more frequently than their ape counterparts because they could care for more than one child at a time. In return a male gained continual sexual access to a particular female, ensuring that the children he provisioned were most likely his own.

This monogamous arrangement would have provided an enormous evolutionary advantage, says Lovejoy, since it would directly affect the number of offspring an individual female could bear and raise to maturity. But the males would need the anatomic apparatus to carry food back to be shared in the first place. Bipedalism may have been a poor way of getting around, but by freeing the hands for carrying, it would have been an excellent way to bear more offspring.

And in evolution, of course, more offspring is the name of the game.

If the savanna hypothesis is barely kicking, it's worth remembering that it isn't dead. While scientists no longer believe in the classic portrait of protohominids loping about on a treeless gray plain, there's treeless, and then there's treeless. Despite the revisionism of recent years, the fact remains that Africa as a whole has gradually cooled and dried over the past 5 million years. Even if the trees did not disappear completely, early hominids may still have faced sparser forests than their ancestors. Bipedalism may still have been an important part of their adjustment to this new setting.

"We have to be careful about what we call savanna," says Peter Wheeler. "Most savanna is a range of habitats, including bushland and quite dense trees. My arguments for bipedalism being a thermo-regulatory adaptation would still apply, unless there was continuous shade cover, as in a closed-canopy forest. Nobody is saying that about *anamensis*."

Unfortunately they *are* saying that about *anamensis's* older cousin, *Ardipithecus ramidus*, in its deep forest home. Which brings us full circle to the business of storytelling. Every reader knows that a good story depends on having strongly drawn characters, ones the author understands well. Rather suddenly, two new protagonists have been added to the opening chapters of human evolution. About *amamensis* we still know very little. But about *ramidus* we know next to nothing. Once upon a time, nearly four and a half million years ago, there was a hominid that lived in the middle of a dark forest. Did it walk on two legs or four? If it walked on two, why?

"The locomotor habit of *ramidus* is crucial," says Lovejoy, who is among those charged with the enviable task of analyzing its bones. "If it is bipedal, then the savanna hypothesis in all its mundane glory would be dead. But if it is quadrupedal, then the old idea, even though I think it is inherently illogical, would not be disproved."

East Side Story:
The Origin of Humankind

The Rift Valley in Africa holds the secret to the divergence of hominids from the great apes and to the emergence of human beings

Yves Coppens

Yves Coppens specializes in the study of human evolution and prehistory. He received his degrees from the Sorbonne, where he studied vertebrate and human paleontology. A member of many organizations, including the French Academy of Sciences and the National Academy of Medicine, Coppens is currently chair of paleoanthropology and prehistory at the College of France in Paris. He is also known for having done 20 years of extensive fieldwork in Africa, particularly in Chad and Ethiopia.

Humans are creatures whose roots lie in the animals. Accordingly, we find ourselves at the tip of one of the branches of an immense tree of life, a tree that has been developing and growing ever more diverse over a period of four billion years. From an evolutionary standpoint, it is important to locate the place and the time that our branch separated from the rest of the tree. It is these questions that the present article attempts to answer. When, where and why did the branch that led to us, the genus *Homo,* diverge from the branch that led to our closest cousin, the genus *Pan,* or the chimpanzee? Because this parting of the ways seems to unfold several million years before *Homo,* properly speaking, was born, the issue of our precise origin also needs to be

addressed. When, where and why did *Homo* appear in the bosom of a family, Hominidae, that was well planted in its ecosystem and well adapted to its environment?

I first realized in 1981 that it might be possible to find answers to these questions. The occasion was an international conference in Paris organized by UNESCO to celebrate the 100-year anniversary of the birth of Pierre Teilhard de Chardin. As an invited speaker, I gave a talk on the French paleontologist and philosopher's scientific work. Although this aspect of Teilhard's writing is often forgotten by biographers, who are essentially interested in his philosophical texts, he produced more than 250 scientific reports over the course of 40 years. His opus includes articles on the structural geology of Jersey, Somalia, Ethiopia and China; on the Paleocene and Eocene mammals of Europe; on the Tertiary and Quaternary mammals of the Far East; on the fossil men of China and Java; on the southern African australopithecines (a kind of prehuman, one that was already hominid, but not yet *Homo*); as well as on the Paleolithic and Neolithic tools of all those countries.

A member of the audience, whom I did not know at the time, came up to me after my talk and congratulated me very courteously, admitting that he had not known about this technical aspect of Father Teilhard's work. He asked me several questions about this science

of evolution that I practiced and about its state of development. My visitor ended this short interview with a precise question: Is there at present an important issue that is still being debated in your field?

Yes, I responded, there is a problem of chronology, as is often the case in historical sciences. Biochemists, struck by the great molecular proximity between humans and chimpanzees, place the beginning of the divergence of these two groups some three million years ago. This discipline also assigns a strictly African origin to humanity. In contrast, the field of paleontology describes a divergence that dates as far back as 15 million years ago. Paleontologists also postulate a broad origin, that is, one radiating from both the Asian and the African tropics.

The gentleman seemed interested, thanked me and left. Several months later I received a letter of invitation to a conference in Rome that he proposed to hold in May 1982. My questioner had been none other than Carlos Chagas, president of the Papal Academy of Sciences! In search of subjects that would have both current interest and important philosophical implications, he had considered what I had said and had organized, under the aegis of his institution, a confrontation between paleontologists and biochemists.

That meeting did take place and, although discreet, its influence on scientific thought was considerable. Two

Common ancestor of *Pan* and *Homo*

Deinotherium

Hippopotamus Struthionidae Giraffidae
Crocodilus

Gomphotherium Hipparion

The Omo River Sequence | LATE MIOCENE | AROUND EIGHT MILLION YEARS AGO

significant facts, one paleontological and one biochemical, were presented to the participants. The first was the announcement by David Pilbeam, professor of paleontology at Harvard University, that his research group had discovered, in the Upper Miocene levels of the Potwar Plateau in Pakistan, the first known face of a ramapithecid. This face resembles an orangutan's much more closely than it does a chimpanzee's face. Pilbeam's data were particularly important because the ramapithecids had for many years been considered by some paleoanthropologists to be the first members of the human family.

The second fact presented was a statement by Jerold M. Lowenstein of the University of California at San Francisco that active proteins had been discovered in the dental material of a ramapithecid. He had determined that activity by injecting extract from the ramapithecid teeth into a rabbit, where it brought on the formation of antibodies. Lowenstein then told us of the indisputable reaction of these antibodies to the antigens of orangutans. This strong reaction made it clear that some of the ramapithecid proteins were still preserved and that the creature seemed related to orangutans.

Before the discovery of the ramapithecid face, scientists had procured only some of this genus's teeth and jaw fragments. Although these features were certainly interesting, it is necessary to know that all the bones of a skeleton do not carry information of equal value. These pieces were less significant than the orbit area and the nose and upper jaw region found in the new Pakistani piece. Paleontologists use such facial fossils to draw anatomical comparisons with similar or contemporary fossils. A simple comparison of the face of this ramapithecid, an orangutan and a chimpanzee clearly revealed the similarities between the ramapithecid and the orangutan.

Rather than comparing anatomical attributes, biochemists examine molecular details. They look at DNA, at the proteins and chromosomal maps of current species—elements that are not usually conserved in fossils. Their work helps paleontologists, who can then arrange species in order of complexity and compare their protein maps. The progression from simple to complex and the sequence that emerges

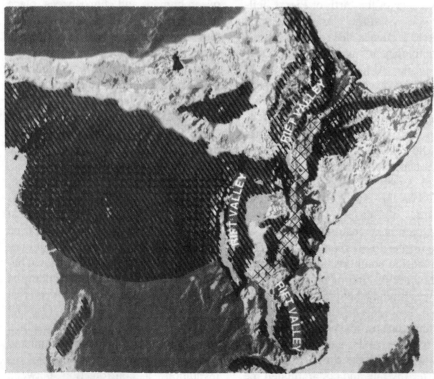

VEGETATION AND CLIMATE vary dramatically on either side of the Rift Valley: wet western woods (*striped areas*) give way to eastern grasslands (*light areas*). Reflecting these ecological differences, which arose millions of years ago, chimpanzees are distributed only to the west (*dotted area*), whereas hominid fossils are found only to the east (*cross-hatching*).

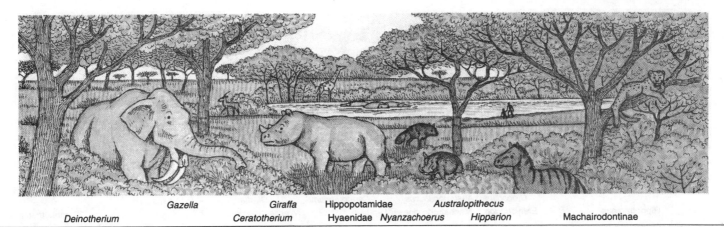

| | Gazella | | Giraffa | Hippopotamidae | | Australopithecus | |
| Deinotherium | | | Ceratotherium | Hyaenidae | Nyanzachoerus | Hipparion | Machairodontinae |

AROUND SIX MILLION YEARS AGO	LOWER LOTHAGAMIAN (LOWER PLIOCENE)	FIVE MILLION YEARS AGO

reproduces, in some fashion, the evolution of creatures in the fossil record. In the case of the ramapithecid, however, biochemistry had made, as never before, a foray back in time by examining fossil proteins.

Circumstances had come together in such a way that we could finally put the ramapithecid in his place. This hominoid had been known to be Eurasiatic, and he remained so. Now that his relationship to the great ape of Asia, the orangutan, had been brought to light, the geographic picture became clear. Indeed, it made complete sense, as so often happens when one has found the solution to a problem. The origin of humanity, as the molecular biologists had suspected, appeared to be Africa, and Africa alone. The question of our family's place of birth seemed settled.

But the question of the date of this birth remained to be addressed. Several paleontologists present at this congress continued to defend the great antiquity of the hominids, whereas the molecular biologists extolled the extraordinary brevity of the independent part of our branch. The most generous of the paleontologists had arrived in Rome convinced of the 15-million-year history of our family. The most extreme of the molecular biologists were sure that three million years, at most, would measure the length of existence of the human family. Both sides came to the conclusion—made, of course, with only the most serious considerations possible—that seven and a half million years was a good span. I dubbed this conclusion "the prehistoric compromise."

The two paleontological and biochemical announcements of the Rome meeting were not the only crucial items that came to light in the early 1980s. Another set of results further clarified our understanding of human origins. Twenty years of excavations in eastern Africa (between 1960 and 1980) had finally yielded a mass of information in which could be sought evolutionary sequences and patterns. This extensive material had not been looked at in such a way before because it takes time to study and identify fossils. Its

COMPARISON OF THREE HOMINOID SKULLS illustrates the proximity between two of the creatures. The ramapithecid (*center*) found in Pakistan resembles the great ape of Asia, the orangutan (*left*), much more closely than it does one of the African apes, the chimpanzee (*right*). Indeed, this very comparison led paleontologists to reject the Eurasiatic ramapithecids as close ancestors of humans and to focus on an African origin.

5. THE FOSSIL EVIDENCE

Hippotraginae	Crocodilus Australopithecus Enhydriodon Machairodontinae	Ceratotherium Lepus	Giraffa Nyanzachoerus	Hyaena	Gomphotheriidae

FIVE MILLION YEARS AGO	UPPER LOTHAGAMIAN (LOWER PLIOCENE)	3.5 MILLION YEARS AGO

implications were vast, particularly when coupled with the information from the ramapithecid and the new-found consensus on dates.

The entry of paleoanthropologists into eastern Africa was actually an ancient affair. In 1935 Louis Leakey's expedition to Olduvai Gorge in Tanzania discovered remains attributed to *Homo erectus.* In 1939 the German team of Ludwig Kohl-Larsen found fossils that were named *Praeanthropus africanus*—later considered to be *Australopithecus*—near Lake Garusi, an area also called Laetoli, in Tanzania. In 1955 another Olduvai expedition led by Leakey revealed a single australopithecine tooth. These modest discoveries, however, did not command much interest.

It was not until the 1960s that the world eagerly turned its attention to eastern Africa. In 1959 Mary Leakey found at Olduvai an australopithecine skull equipped with all its upper teeth. This skull could be absolutely dated to about two million years ago by the volcanic tuff below which it had been enveloped. The new hominid was named *Zinjanthropus;* it was a small-brained bipedal hominid species that went extinct about one million years ago. After that significant finding, expeditions started to arrive in abundance: a new team came each year for the first 12 years, and each one excavated for 10 or 20 seasons. Never before had such an effort been de-ployed by paleontologists or paleo-anthropologists.

The results reflected the investment. Hundreds of thousands of fossils were discovered, of which about 2,000 were hominid remains. Yet, despite the constant work of preparation, analysis and identification of these fossils as they were unearthed, it is understandable that it was not until the 1980s that the first complete inventory of these thousands of finds was published. It is precisely this new information that, when added to the data received at the Rome conference, became essential to solving the mystery.

What emerged so clearly was that there was absolutely no sign of *Pan,* or one of its direct ancestors, in eastern Africa during the time of the australopithecines. Molecular biology, biochemistry and cytogenetics continued to demonstrate that humans and chimpanzees were molecularly extremely close, which meant, in evolutionary terms, that they had shared a common ancestor not very far back in time, geologically speaking. And field-workers had just revealed that Hominidae, as of seven or eight million years ago, were present in Ethiopia, Kenya and Tanzania. But during the same period, this region had not seen the least sign of the family Panidae, no precursor of the chimpanzee and no precursor of the gorilla. Even though one cannot base a hypothesis on a lack of evidence, the striking absence of these Panidae where Hominidae were abundant represented a sufficient contrast to cause concern—all the more so because the 200,000 to 250,000 vertebrate fossils that had been collected constituted a statistical base with a certain authority.

I had been thinking about this puzzle during the conference in Rome. A quite simple explanation came to mind when I opened an atlas marking the distribution of vertebrates. The map devoted to chimpanzees and gorillas showed a significant group of territories, including all the large forested regions of tropical Africa, but stopped, almost without overflow, at the great furrow that cuts perpendicularly across the equator from north to south: the Rift Valley. All the hominid sites that dated to more than three million years ago were found, without exception, on the eastern side of this furrow. Only one solution could explain how, at one and the same time, Hominidae and Panidae were close in molecular terms but never side by side in the fossil record. Hominidae and Panidae had never been together.

I therefore suggested the following model. Before Hominidae and Panidae had separated, the Rift Valley did not constitute an irregularity sufficient to divide equatorial Africa. From the Atlantic to the Indian Ocean, the African continent constituted one homogeneous biogeographical province in which the common ancestors of the future Hominidae and Panidae lived. Then, about eight million years ago, a tectonic

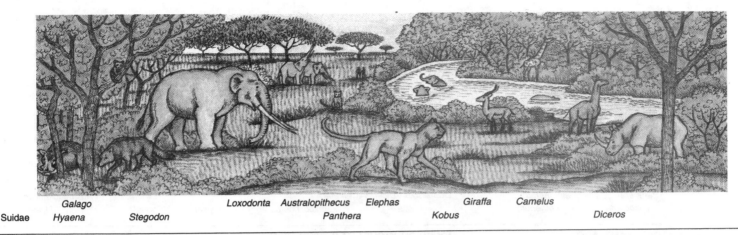

| Galago | | | Loxodonta | Australopithecus | Elephas | | Giraffa | Camelus | |
| Suidae | Hyaena | Stegodon | | Panthera | | Kobus | | | Diceros |

3.5 MILLION YEARS AGO	LOWER SHUNGURIAN (UPPER PLIOCENE)	2.5 MILLION YEARS AGO

crisis arose that entailed two distinct movements: sinking produced the Rift Valley, and rising gave birth to the line of peaks forming the western rim of the valley.

The breach and the barrier obviously disturbed the circulation of air. The air masses of the west maintained, thanks to the Atlantic, a generous amount of precipitation. Those of the east, coming into collision with the barrier of the western rim of the Tibetan plateau, which also was rising, became organized into a seasonal system, today called the monsoon. Thus, the original extensive region was divided into two, each possessed of a different climate and vegetation. The west remained humid; the east became ever less so. The west kept its forests and its woodlands; the east evolved into open savanna.

By force of circumstance, the population of the common ancestor of the Hominidae and the Panidae families also found itself divided. A large western population existed, as did a smaller eastern one. It is extremely tempting to imagine that we have here, quite simply, the reason for the divergence. The western descendants of these common ancestors pursued their adaptation to life in a humid, arboreal milieu: these are the Panidae. The eastern descendants of these same common ancestors, in contrast, invented a completely new repertoire in order to adapt to their new life in an open environment: these are the Hominidae.

This uncomplicated model has the advantage of explaining why Hominidae and Panidae are so close in a genetic sense and yet never together geographically. It also has the advantage of offering, by means of a situation that is at first tectonic and then ecological, a variant of the situation found on islands. Compared to complex solutions about the movements of Hominidae from the forest to the savanna or about the movements of Panidae from the savanna to the forest, the Rift Valley theory is quite straightforward.

It was only later, when I was reading the work of geophysicists, that I learned that the activity of the Rift Valley some eight million years ago was well known. Reading the studies of paleoclimatologists fortified me with the knowledge that the progressive desiccation of eastern Africa was also a well-known event, whose start-

MARY AND LOUIS LEAKEY examine the *Zinjanthropus* skull and upper jaw at Olduvai Gorge in Tanzania in 1959. Their discovery of a hominid fossil at this site led to a bone rush: paleontologists flooded in, and hundreds of thousands of fossils were excavated in subsequent decades.

5. THE FOSSIL EVIDENCE

| Panthera | Deinotherium | | Damaliscus Diceros | Giraffa | | | Dinofelis | Australopithecus |
| Loxodonta | Homotherium | Lepus | | Phacochoerus | Hipparion | Equus | | Hyaena |

| 2.5 MILLION YEARS AGO | UPPER SHUNGURIAN (UPPER PLIOCENE) | 1.8 MILLION YEARS AGO |

ing point had been placed at about eight million years ago. Finally, reading the declarations of paleontologists further reassured me, because they placed the emergence of eastern African animal life—a fauna labeled Ethiopian, to which the australopithecines belong—at about eight or 10 million years ago. Each discipline knew this date and in one way or another was familiar with the event or its consequences, but no interdisciplinary effort had brought them all into a synthesis. Adrian Kortlandt, a famous ethologist from the University of Amsterdam, had thought about such a possible scenario, but without any paleontological support, some years before.

The hypothesis lacked only a name. Three years later I was invited by the American Museum of Natural History

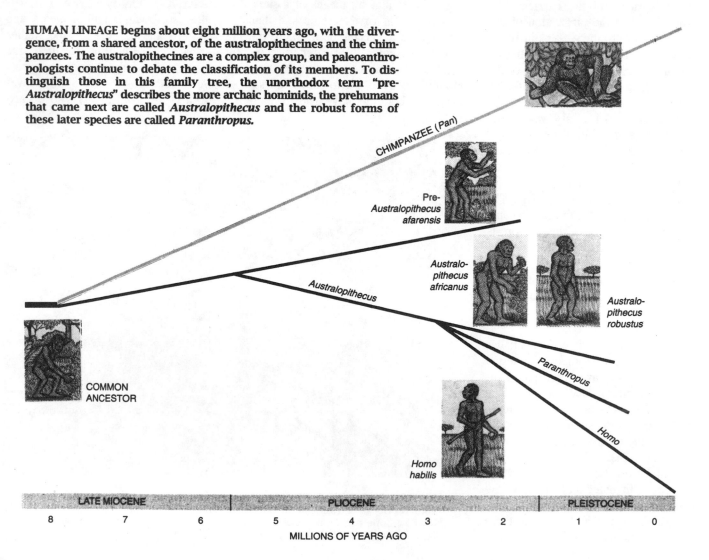

HUMAN LINEAGE begins about eight million years ago, with the divergence, from a shared ancestor, of the australopithecines and the chimpanzees. The australopithecines are a complex group, and paleoanthropologists continue to debate the classification of its members. To distinguish those in this family tree, the unorthodox term "pre-*Australopithecus*" describes the more archaic hominids, the prehumans that came next are called *Australopithecus* and the robust forms of these later species are called *Paranthropus*.

CHIMPANZEE (Pan)

Pre-*Australopithecus afarensis*

Australopithecus

Australopithecus africanus

Australopithecus robustus

Paranthropus

Homo

COMMON ANCESTOR

Homo habilis

| LATE MIOCENE | PLIOCENE | PLEISTOCENE |

8   7   6   5   4   3   2   1   0

MILLIONS OF YEARS AGO

158

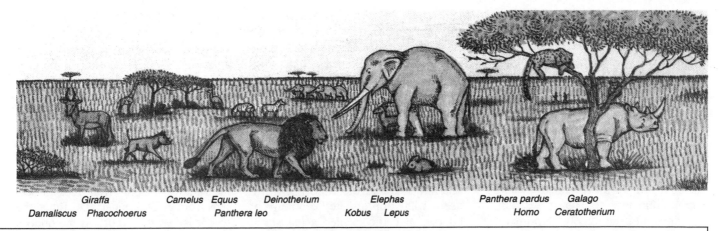

| *Giraffa* | *Camelus* *Equus* | *Deinotherium* | *Elephas* | *Panthera pardus* *Galago* |
| *Damaliscus* *Phacochoerus* | *Panthera leo* | *Kobus* *Lepus* | | *Homo* *Ceratotherium* |

| **1.8 MILLION YEARS AGO** | **PLEISTOCENE** | **ONE MILLION YEARS AGO** |

in New York City to present the 55th James Arthur Lecture on the Evolution of the Human Brain. I also assumed a visiting professorship at the Mount Sinai School of Medicine of the City University of New York. The idea of giving this model a title that would be easy to remember and that would honor my hosts came to me then. I called it the East Side Story.

It is possible that the East Side Story has answered the first volley of questions: the when, where and why of our divergence from Panidae. Our phyletic branch, the one that now bears us, was marked off from the rest of the genealogical tree of living creatures eight million years ago in eastern Africa by reason of geographic isolation. The need for adaptation to the new habitat of the savanna, one that was drier and more bare than the preceding one, promoted further genetic divergence.

The second series of questions is more intricate: the when, where and why of the appearance of the genus *Homo* in the family Hominidae. The past eight million years during which our branch of the tree has grown have revealed themselves to be more complex than one might have imagined. The story begins with the diversification of a subfamily, the australopithecines. These creatures made very modest movements from eastern Africa to southern Africa. The story then continues from about three million years ago to today, with the emergence of another subfamily, the hominines. The hominines moved ex-

tensively, from eastern Africa across the entire planet. The last of the australopithecines coexisted for about two million years with the first of these hominines, which have only one genus, *Homo*.

The emergence of this hominine subfamily can be seen in a remarkable series of geologic beds and fossils found along the banks of the Omo River in Ethiopia. And, not surprisingly, because this is the second part of the East Side Story, the role of climate proves to be as powerful a force for change three million years ago as it did eight million years ago.

The Omo River tale began at the turn of this century, when a French geographic expedition proposed to cross Africa diagonally, from the Red Sea to the Atlantic. The Viscount du Bourg de Bozas directed the expedition. Having departed from Djibouti in 1901, the exploration was to end dramatically in the death of its leader from malaria on the banks of the Congo. The team nonetheless brought back from the journey, which followed the original itinerary, a fine harvest of fossils. Among the collection was a group of vertebrate remains gathered in what was then Abyssinia, on the eastern bank of the lower valley of the Omo River. The Omo lies on the eastern side of the Rift Valley.

Intrigued by this yield, which was described in two or three articles and in Émile Haug's geologic treatise in 1911, Camille Arambourg decided at the beginning of the 1930s to conduct a new expedition. Arambourg, future

professor of paleontology at the National Museum of Natural History in Paris, reached the Omo and stayed eight months in 1932. He returned to Paris with four tons of vertebrate fossils.

The next major operation—the Omo Research Expedition—was undertaken between 1967 and 1977. It was catalyzed, in part, by the bone rush of the 1960s and 1970s, described earlier, which had followed the 1959 find by Mary Leakey at Olduvai. A series of researchers conducted the 10-year Omo expedition in stages. In 1967 Arambourg and I worked on the site with Louis and Richard Leakey and Francis Clark Howell. Between 1968 and 1969 Richard Leakey left the expedition, and Arambourg, Howell and I continued the work. Finally, from 1970 until 1976, Howell and I dug there alone (Arambourg died in 1969).

From the very first expedition, the stratigraphy of this site was eminently visible, a superb column more than 1,000 meters deep. The fauna contained in these beds appeared to change so markedly as it progressed from base to summit that the site was obviously capable, even at mere glance, of telling a story. When dating by potassium-argon and by paleomagnetism finally became available, so that a chronological grid could be placed on this sequence, the history became clear.

Starting four million years ago (the age of the oldest Omo level, the Mursi formation) and ending one million years ago (the age of the most recent level, the top of the Shungura formation), the climate had clearly changed from humid to distinctly less humid. As a consequence, the vegetation had evolved from plants adapted to humidity to those capable of thriving in a drier climate. The fauna had also changed from one suited to a brushwood assemblage to one characteristic of a grassy savanna. And the Hominidae, subject like the other vertebrates to these climate fluctuations, had changed from so-called gracile australopithecines to robust australopithecines and, ultimately, to humans.

In 1975 I informed the international paleontological community of this clear correlation between the evolution of the climate and the evolution of the hominines. I did so in a note to the *Proceedings of the Academy of Sciences* in Paris and in a communication to a congress in London at the Royal Geological Society. The reaction was very skeptical.

Of all the great eastern African paleontological sites, the strata of Omo were the only ones that could have permitted such observations. This site alone offered a continuous sedimentary column that ran from four million years ago to one million years ago. It is precisely between three and two million years ago, or to be very exact between 3.3 to 2.4 million years ago, that the whole earth cooled and that eastern Africa became dry. (Laetoli and Hadar were too old, Olduvai was too young and East Turkana presented a stratigraphic gap at that point, so they could not offer the same demonstration.) We know this fact through several other tests conducted in various regions of the world.

This climatic crisis appears clearly in the fauna and flora records of the Omo sequence. By indexing, both qualitatively and quantitatively, the animals and plants gathered in the various levels, we can interpret the differences that emerge from these species, with regard to changes in the environment.

We know, for example, that the cheek teeth—that is, the premolars and molars—of herbivore vertebrates have a tendency to develop and become more complex when the diet becomes more grassy and less leafy. This change

takes place because grass wears down the teeth more than leaves do. We know also that the locomotion of these same herbivores becomes more digitigrade in open habitats in which they are more vulnerable: one runs better on tiptoe than in boots. A certain number of anatomical features corresponding to very precise functions can also be good indicators: the tree-dwelling feet of some rodents or the feet of others that are adapted to digging. We use, with appropriate caution, of course, a method called actualist; in other words, we believe that the varieties of animals or plants we are considering acted then as they act today.

Many examples demonstrate this transition to a drier environment, and they are extraordinary in their agreement. As one moves from the older strata on the bottom to the younger strata on the top, there is an increase in the hypsodonty—that is, in a tooth's height-to-width ratio—among Elephantidae (elephants close to the ones living in Asia today), Rhinocerotidae (specifically the white rhinoceros), *Hipparion* (ancestors of the horse), Hippopotamidae (precursors of the hippopotamus) and some pigs and antelopes. In other words, these groups exhibited the increasing complexity that we associate with a shift from a diet of leaves to a diet of grass. The Suidae, or precursors to swine, also show an increase in the number of cusps on their molars as they evolved.

On the lower strata are many antelopes—including Tragelaphinae and Reduncinae, which live among shrubs. All these creatures must have lived in an environment of wooded savanna close to water. On the top levels the true horse, *Equus,* appears, as do the high-toothed warthogs, *Phacochoerus* and *Stylochoeras.* We also see the development of the swift antelopes, *Megalotragus, Beatragus* and *Parmularius,* animals found on open grasslands.

On the bottom, three species of small *Galago,* or monkey, and the two Chiroptera, *Eidolon* and *Taphozous,* indicate a well-developed forest and a dense savanna. This conclusion is supported by the large number of Muridae rodents, such as *Mastomys,* as well as

the rodents *Grammomys, Paraxerus, Thryonomys* and *Golunda.* At the top, the rodents *Aethomys, Thallomys, Coleura* and *Gerbillurus* in conjunction with *Jaculus* and *Heterocephalus,* the Chiroptera, and the *Lepus,* or hare, replace the previous inhabitants. All the later rodents inhabit dry savanna.

Pollen specimens on the bottom indicate 24 taxa of trees, whereas the top is characterized by 11. At the bottom, the ratio of pollens from trees to pollens from grasses equals 0.4. But at the top, it is less than 0.01. At the bottom pollens from species that grow in humid conditions are abundant—they include *Celtis, Acalypha, Olea* and *Typha.* In the more recent strata, however, these pollens diminish considerably or even disappear from the record, whereas pollens from Myrica, a plant typical of dry climates, appear. The number of pollens transported by the wind, called allochtone pollens, dwindles from 21 percent at the bottom, where the forest edge is near the Omo River, to 2 percent at the top, where the Omo was low and the forest edge far away.

The story with the hominids is similar. They are clearly represented by *Australopithecus afarensis* on the lower strata. But the younger strata on the top reveal *A. aethiopicus, A. boisei* and *Homo habilis.* The oldest species of australopithecines, the graciles, are more ensconced in tree-filled habitats than are the more recent species, those called robust. As for humans, we are unquestionably a pure product of a certain aridity.

I called this climatic crisis "the (H)Omo event" using the simple play on words of Omo and *Homo,* because it permitted the emergence of humans—an event that affects us quite specifically—and because it was the Omo sequence that revealed it for the first time. Some years later the same data were reported from South Africa.

Thus, it appears strikingly clear that the history of the human family, like that of any other family of vertebrates, was born from one event, as it happens a tectonic one, and progressed under the pressure of another event, this one climatic.

These changes can be but quickly summarized here. Essentially, the first adaptation changed the structure of the brain but did not increase its volume, as suggested by the interpretation of endocasts, latex rubber casts of fossil skulls, done by Ralph L. Holloway of Columbia University. At the same time, the changes caused Hominidae to retain an upright stance as the most advantageous and to diversify the diet while keeping it essentially vegetarian. The second adaptation led in two directions: a strong physique and a narrow, specialized vegetarian diet for the large australopithecines and a large brain and a broad-ranging, opportunistic diet for humans.

Some hundreds of thousands of years later, it was the latter development that proved to be the more fruitful, and it is this one that prevailed. With a larger brain came a higher degree of reflection, a new curiosity. Accompanying the necessity of catching meat came greater mobility. For the first time in the history of the hominids, humanity spread out from its origin. And this mobility is the reason that in less than three million years, humanity has conquered this planet and begun the exploration of other worlds in the solar system.

FURTHER READING

EVOLUTION DES HOMINIDÉS ET DE LEUR ENVIRONNEMENT AU COURS DU PLIO-PLÉISTOCÈNE DANS LA BASSE VALLÉE DE L'OMO EN ETHIOPIE. Yves Coppens in *Comptes Rendus Hebdomadaires des Séances de l'Académie des Sciences,* Vol. 281, Series D, pages 1693–1696; December 3, 1975.

EARLIEST MAN AND ENVIRONMENTS IN THE LAKE RUDOLF BASIN: STRATIGRAPHY, PALEOECOLOGY AND EVOLUTION. Edited by Yves Coppens, F. Clark Howell, Glynn Ll. Isaac and Richard E. F. Leakey. University of Chicago Press, 1976.

RECENT ADVANCES IN THE EVOLUTION OF PRIMATES. Edited by Carlos Chagas. Pontificia Academia Scientiarum, 1983.

L'ENVIRONNEMENT DES HOMINIDÉS AU PLIO-PLÉISTOCÈNE. Edited by Fondation Singer-Polignac. Masson, Paris, 1985.

Asian Hominids Grow Older

Fossils from China could alter the picture of human dispersal and evolution—and they're just one of several findings, described on the following pages, that challenge the textbooks.

Wanderlust has been a potent factor in human evolution, spurring early members of our lineage to leave their African homes and spread throughout the world. But exactly when this itinerant urge struck has become a hotly debated issue, especially since it has major consequences for scenarios of later human evolution. For years, the majority view held that the first footloose hominid was *Homo erectus,* thought to have left Africa about 1 million years ago. In the past year, however, new data from Java and the republic of Georgia have suggested that *H. erectus* was already present in those Asian locales as early as 1.8 million years ago.

Now, in this week's issue of *Nature,* the idea of an earlier migration gets additional support from a team of Chinese and Western scientists. Based on a three-part package of hominid fossils, dating methods, and primitive tools, they argue that early *Homo* reached central China between 1.7 million and 1.9 million years ago—nearly 800,000 years earlier than had been thought.

And the team's claims go beyond dating. They suggest that the ancient wanderer was not *H. erectus* itself, but an even earlier hominid with ties to more primitive African forms. "Our work shows that there was an early dispersal of [primitive] hominids with basic stone tools out of Africa," says paleoanthropologist Russell Ciochon of the University of Iowa, who led the collaboration with the Chinese. Their results could strengthen a minority view that *H. erectus* evolved not in Africa but in Asia, from primitive hominids like the newly reported Chinese finds. If so, *H. erectus*

could be an Asian side branch of the hominid evolutionary tree, rather than part of the African lineage that led to modern humans. But judging from early reaction, the fragmentary new evidence may not be enough to sway researchers who have long held a more classical view of human evolution and dispersal.

The provocative new fossils and stone tools were unearthed in the late 1980s by Chinese scientists, led by Huang Wanpo of the Institute of Vertebrate Paleontology and Paleoanthropology in Beijing. They excavated Longgupo Cave in Sichuan province, known as Dragon Bone Slope in Chinese because the cave's roof and walls have collapsed. They found a rich collection of bones, including prized evidence of hominids: a jaw fragment with two teeth, an upper incisor, and two crude stone tools. The Chinese scientists also analyzed traces of Earth's ancient magnetic field left in sediments associated with the fossils.

Because the field sporadically reverses over time, researchers can date fossils by tying them to a particular period of normal or reversed field. The hominid fossils were determined to have been deposited during a period of normal magnetic polarity, and the Chinese correlated this to a normal polarity event dated at Africa's Olduvai Gorge to 1.77 million to 1.95 million years ago. But their work was not widely known because it was published in a Chinese journal.

Then in 1992 the Chinese invited Ciochon and his colleagues to visit the site to explore its geology and confirm the dates. For this, the researchers used a relatively new method called electron spin resonance (ESR), which measures

the electric charges induced in tooth enamel over time by naturally occurring radioactive materials in the surrounding sediments. They weren't able to excavate the cave and so couldn't date the hominid levels directly. But Henry Schwarcz of McMaster University in Hamilton, Ontario, applied ESR dating to a deer tooth from one of the cave's upper levels; he estimated a minimum age of 750,000 years and a most likely age of 1 million years. Together with associated animal fossils, the ESR date "calibrates and constrains the paleomagnetics," indirectly confirming the hominid ages, says Ciochon.

The Chinese fossils may have a significance even beyond their advanced age: their primitive form. Ciochon says that characters such as a double-rooted premolar and the pattern of cusps on the molar resemble those of early African *Homo* species that predated *H. erectus.* He points to either *H. habilis,* the most ancient member of our genus, known only from Africa, or *H. ergaster,* a species recognized by some researchers as the precursor to *H. erectus* in Africa. The Longgupo Cave fossils provide a link between Asian and early African forms, agrees paleoanthropologist Bernard Wood of the University of Liverpool, noting that the new Asian finds resemble 1.6 million-year-old fossils found in East Africa.

The tools—rounded pieces of igneous rock that show signs of repeated battering—provide additional support for the idea that the fossils belong to a primitive *Homo.* These crude implements recall the basic choppers found with early hominids at Olduvai Gorge,

African Origins: West Side Story

Ask most paleoanthropologists where an ancestral ape took its first humanlike steps, and they're likely to point to East Africa. After all, the oldest known bipedal hominid, 4.1-million-year-old *Australopithecine anamensis,* was found in Kenya, while the slightly younger *Australopithecus afarensis,* typified by the famous skeleton "Lucy," was found in Ethiopia. But the discovery of a 3- to 3.5-million-year-old australopithecine fossil in Chad, some 5400 kilometers to the west in the heart of the African continent, has upset that East African–centric view. "Human origins is not just an east-side story," says Michel Brunet, a paleoanthropologist at the University of Poitiers, who found the partial lower jawbone in January. "It's a west-side story, too."

Brunet's find, preliminarily assigned to *Australopithecus afarensis*—although Brunet himself thinks it may be a new species—is reported in the 16 November issue of *Nature.* It already has scientists backpedaling about previous declarations labeling East Africa the cradle of humankind. "I think that's been a very naive view," says Alan Walker, a paleoanthropologist at Pennsylvania State University, "and so we're going to have to rethink things, which is good for the field."

One idea being heavily rethought is the notion that East Africa's long Rift Valley acted as a geographical barrier to ape populations in the late Miocene, 5 million to 7 million years ago, separating those that became hominids in the savannas of the east from forest-dwelling apes in the west. "Now we have early australopithecines all around Africa," says Brunet, "which makes it impossible to tell the exact place of origin."

Or the cause of that origin. Previously, scientists such as Yves Coppens, a paleoanthropologist at the College de France in Paris and co-author of the new paper, had suggested that hominids had evolved in the eastern part of the continent because of habitat changes associated with the development of the Rift Valley. "The rise of the western Rift has been linked to the development of the more open savanna country one finds in East Africa," explains David Pilbeam, director of Harvard University's Peabody Museum and another co-author of the new paper. Open country, in theory, created selective pressure driving apelike creatures out of the trees and onto the ground. "And that, in turn, was seen as causing the origin of the hominids. But I don't think the Rift Valley was the mechanism," he says. The habitat of the Chad hominids seems to have been a dry, grassy woodland, according to animal fossils from the site. "We have rhinoceroses, giraffe, and hipparion [horse], which suggest grasslands, and pigs and elephants, which are more adapted to woodlands," Brunet says.

The finds will focus more attention on Central and West Africa as potential hotbeds of hominid activity. "We've always thought of the [current] West and Central African tropical rain forest as being around forever, while the east became a savanna," explains Rick Potts, a paleoanthropologist at the Smithsonian Institution's Natural History Museum. "That ecological distinction was thought to be the critical marker of the human-ape split. Now it's clear that we don't really know a single thing about what was going on in West Africa at that time."

Brunet hopes that future discoveries at his site will give scientists a clearer view of ancient West Africa—and help him nail down the precise species of the Chad specimen. Currently, Brunet says, the two australopithecines known from that period are *A. afarensis* and *A. anamensis.* "I think there were more than just two australopithecines 3.5 million years ago—it was more complicated than that, as we know now their origins were, too."

—**Virginia Morell**

rather than the more complicated tools associated with *H. erectus,* says the team's archaeologist, Roy Larick of the University of Massachusetts, Amherst.

All this adds up to a coherent picture of a pre-*erectus* hominid that left Africa perhaps 2 million years ago, says Ciochon. And the simple stone tools show that hominids were able to conquer new territory before they developed the more complex hand axes once thought to be a prerequisite for long-distance dispersal. "This shows that very soon after the origin of *Homo,* hominids became mobile and were able to disperse rapidly over huge distances," agrees Peter Andrews of the Natural History Museum

in London. Furthermore, in this scenario, these first travelers evolved into *H. erectus* while in Asia. And because everyone agrees that our own species arose in Africa, this implies that *erectus* itself was an Asian creature and an evolutionary side branch not directly ancestral to modern humans.

Parts of this theory are extremely controversial, but the early dates are in accord with two recent observations. Last year, Carl Swisher of the Berkeley Geochronology Center and colleagues redated *H. erectus* skulls from Java to 1.6 million and 1.8 million years old (*Science,* 25 February 1994, pp. 1087 and 1118). And earlier this year, scien-

tists in the Republic of Georgia published an *H. erectus* jawbone, estimated to be 1.6 million to 1.8 million years old, from the site of Dmanisi, Georgia.

But many researchers remain skeptical of all three of the earlier dates. The geology of cave deposits such as Longgupo is notoriously complex, because material falling from above may become jumbled with rocks of different ages, says hominid expert Philip Rightmire of the State University of New York, Binghamton. He's not convinced that the hominids are truly older than the deer tooth dated by ESR. And the Chinese team remains leery of the Java dates. The problem, says Ciochon, is that the

Chinese hominid looks more primitive than the Javanese ones. Unless there were two ancient hominids in Asia, it doesn't make sense to have a pre-*erectus* hominid in China at the same time as true *erectus* in Java. Meanwhile, Swisher and Georgian colleagues are now redoing the paleomagnetics for the Dmanisi site.

But even more contentious than the date is the notion that the travelers were "pre-*erectus*." That conclusion is based on "pretty scrappy evidence," says Rightmire. The hominid fossils are incomplete, and the stone tools are so simple that Rightmire and others wonder if they are really artifacts. "This is not the material on which I'd choose to erect bold new scenarios of Chinese pre-history,"

he says. As F. Clark Howell of the University of California, Berkeley, points out, partial jawbones of early hominids are difficult to classify. To paleoanthropologist Alan Walker of Pennsylvania State University, who supports the more classical idea that *H. erectus* led to *H. sapiens,* the Chinese hominid is "just early *erectus.*" If so, *H. erectus* could have evolved in Africa, then dispersed to Asia, albeit earlier than had been thought. But the link between *erectus* and *H. sapiens* would be intact.

Other anthropologists have more fundamental concerns about the fossils' identity. Milford Wolpoff of the University of Michigan, who saw the specimens on a trip to China several years ago, isn't even convinced that the partial

jaw is a hominid. "I believe it is a piece of an orangutan or other *Pongo,*" he says. He bases that conclusion on a wear facet on the preserved premolar, which to him suggests that the missing neighboring tooth is shaped more like an orang's than a human's.

Yet despite the murmurs of doubt, the evidence is mounting in favor of an early excursion out of Africa, accomplished with only crude stone technology. Whether the first travelers are properly called *H. erectus* or something else, the newest work all points to the same conclusion: The urge to wander is an ancient trait that evolved near the dawn of our lineage.

—Elizabeth Culotta

Do Kenya Tools Root Birth of Modern Thought in Africa?

Modern human history, textbooks have it, began with a French revolution. It was heralded not by a mass uprising, but by the carved bone points and elaborate stone blades that suddenly appeared in southwest France between 40,000 and 50,000 years ago; the archaeological record before that held little but crude axes and simple rock flakes. The abrupt shift—known as the Upper Paleolithic transition, which soon spread across Europe—seemed to mark a critical mental advance, a mastery of abstract thought and form lacking in our older ancestors.

But now a group of archaeological sites some 5000 miles away from Europe, and with tools perhaps five times older, seems to be telling a different tale—one of evolution, not revolution. In Kenya, University of Connecticut anthropologist Sally McBrearty has found evi-

dence that our African forebears took the first steps toward modernity as early as 240,000 years ago. This week at the annual meeting of the American Anthropological Association in Washington, D.C., and in a forthcoming paper in the *Journal of Human Evolution,* McBrearty reports the discovery of blades of that age made from long, thin slivers of stone. An extraordinary feature of these tools is that they seem to have been carefully shaped even before they were knocked off a rock core—and that, McBrearty contends, indicates a solid form of abstract thought. Moreover, she argues that similar sites show signs of modern behaviors such as carefully planned resource use and hunting. "Look," she says, "I admire the Paleolithic archaeology of Europe. It's very wonderful. But if you want to learn

about the origins of modern human behavior, you don't look there."

Alison Brooks of George Washington University, who has found modern-looking bone points in Zaire dated to 90,000 years ago (*Science,* 28 April, pp. 495, 548, and 553), says that McBrearty's finds do indeed look modern and that this period in Africa, known as the Middle Stone Age, has long been overlooked by archaeologists with a "Eurocentric" bias. Michael Mehlman, a Middle Stone Age archaeologist at the University of California, Santa Cruz, agrees that "this bias absolutely needs to be challenged" and calls McBrearty's arguments "provocative." Adds Brooks: "What we hope most fervently is that people will understand two things: that the human revolution wasn't a revolution, and that it wasn't European."

Yet the true meaning of these pointed stones is itself a point of dispute. Other researchers say the African artifact record shows no signs that such tools later developed into enhanced forms and thus can't compare with what happened in Europe. And Richard Klein of Stanford University in Palo Alto, California, says the Kenya blades seem almost casually flaked, while the blades made in the European Upper Paleolithic show signs of more precise geometry. "I've never seen anything like that in the Middle Stone Age of Africa, he says, "and I'm not Eurocentric."

The tools engendering this controversy come from the Kapthurin formation, near Kenya's Lake Baringo. The formation, a 125-meter-thick stack of sediments, is interleaved with volcanic deposits, or tuffs. These tuffs have been dated with potassium/argon dating, a well-established and reliable method, and the tuff that caps the formation is about 250,000 years old. McBrearty's artifacts came from just below it.

The Kapthurin blades are long and thin (averaging 10 centimeters in length and less than 1 centimeter in thickness), and in many cases were struck from cores whose surface had been preshaped by flaking. McBrearty has also found the cores to which they can be refitted— as many as a dozen blades to one core— as well as hand axes. And the appearance isn't the only clue to their sophistication, she argues. The blades are made from phonolite, a tough lava that is "a very difficult material to work," she says. It is so tricky, in fact, that Pierre-Jean Texier, a specialist in stone tool manufacture whom she brought to Kenya to try to recreate the fabrication methods, was at first hard-pressed to replicate them at all, although he eventually mastered the preshaped cores.

Other African Middle Stone Age sites, McBrearty says, also contain signs of sophistication. Mehlman, for example, has found evidence at cave sites in Mumba, Tanzania, that the raw materials for tool-making were imported over great distances, signaling advanced planning on the part of the inhabitants. And Brooks has surveyed campsites in Botswana with bones of large and danger-ous animals, such as giant zebra, warthog, and water buffalo, indicating the campers were skilled hunters. Individually these don't mean much, McBrearty admits, but taken together they add up to a signal that the ground of East Africa holds the early stirrings of modernity.

Just who might have been doing that stirring is still debatable. Human fossils were found well below McBrearty's blade levels in the late 1970s, but the remains—a few scraps of jaw, an arm bone, and some toes—have proved difficult to pigeonhole taxonomically. Over the years, they've been called everything from *Homo erectus* (an ancient member of our genus, dating back to 1.5 million years ago) to a transitional form called archaic *Homo sapiens*. McBrearty suspects, "on intuition," that the Kapthurin toolmakers may actually have been anatomically modern *Homo sapiens,* because early members of the species have been found in Ethiopia and in South Africa at least 130,000 years ago. And Klein agrees they could easily have been around 100,000 years before that.

He concurs on little else, however. Comparisons between European Upper Paleolithic blades and McBrearty's finds are like night and day, he says. "The European blades are sort of peeled off the core," he explains, adding that the complex technique used to make them involves not merely knocking one stone against another, but using a piece of hide or antler to distribute the force. In addition, manufacture of European blades results in a so-called prismatic core, a distinctive and sharply geometric form not found in Africa.

The argument for a continuum of technology starting in Africa also has a troubling gap, others point out. There are many younger sites in the Later Stone Age of Africa, says University of Arizona anthropologist Steve Kuhn, but neither these blades nor enhanced forms appear in any of them. If the blades really do provide the "early glimmerings" of the modern mental condition, he says, "it would be more comforting if they had spread, instead of popping up and disappearing."

That spread, which does occur in Europe, signals that something dramatic was happening which hasn't been found in Africa, according to J. Desmond Clark of the University of California, Berkeley: a change not just in technology but communication. "There's no doubt that blades go back for a very long time in Africa, and that's very interesting," he says. But the African tradition seems isolated; the appearance of the sophisticated European tool kit within a few thousand years all over the map, in contrast, means that toolmakers must have also developed language. And that cultural advance, he says, is truly revolutionary.

But McBrearty doesn't buy the idea that blade tools and language had to come in a revolutionary package. "You can point to a lot of so-called revolutions that happened over a long period of time," she says. "As an undergrad I was taught that bipedality, the big brain, and tool use all happened at once; now we know bipedality preceded the big brain by 2 million years." Likewise, she believes, language acquisition could have been a stepwise process, with the simpler African blades the product of its early stages.

She and Brooks both note that archaeologists really have no idea about the spread of technology in the African Middle Stone Age because they haven't looked in many places. Brooks notes that the European record is more elaborate because it's been extensively studied for 150 years, beginning with some of the founding fathers of archaeology, such as Edouard Lartet. "One problem with standing on the shoulders of giants," she observes, "is that you don't have that much choice about what direction you're walking in."

The direction McBrearty plans to go is back to Kenya, where she and her colleagues plan to search for more sites and more tools. "You know how textbooks say that early evolution happened in Africa, but the minute there's any behavior that smacks of being sophisticated, suddenly you're in France?" she says. "If I can change the last chapter in those textbooks in my lifetime, I'll be satisfied."

—JoAnn Gutin

JoAnn Gutin is a science writer in Berkeley, California.

Scavenger Hunt

As paleoanthropologists close in on their quarry, it may turn out to be a different beast from what they imagined

Pat Shipman

Pat Shipman is an assistant professor in the Department of Cell Biology and Anatomy at The Johns Hopkins University School of Medicine.

In both textbooks and films, ancestral humans (hominids) have been portrayed as hunters. Small-brained, big-browed, upright, and usually mildly furry, early hominid males gaze with keen eyes across the gold savanna, searching for prey. Skillfully wielding a few crude stone tools, they kill and dismember everything from small gazelles to elephants, while females care for young and gather roots, tubers, and berries. The food is shared by group members at temporary camps. This familiar image of Man the Hunter has been bolstered by the finding of stone tools in association with fossil animal bones. But the role of hunting in early hominid life cannot be determined in the absence of more direct evidence.

I discovered one means of testing the hunting hypothesis almost by accident. In 1978, I began documenting the microscopic damage produced on bones by different events. I hoped to develop a diagnostic key for identifying the post-mortem history of specific fossil bones, useful for understanding how fossil assemblages were formed. Using a scanning electron microscope (SEM) because of its excellent resolution and superb depth of field, I inspected high-fidelity replicas of modern bones that had been subjected to known events or conditions. (I had to use replicas, rather than real bones, because specimens must fit into the SEM's small vacuum chamber.) I soon established that such common events as weathering, root etching, sedimentary abrasion, and carnivore chewing produced microscopically distinctive features.

In 1980, my SEM study took an unexpected turn. Richard Potts (now of Yale University), Henry Bunn (now of the University of Wisconsin at Madison), and I almost simultaneously found what appeared to be stone-tool cut marks on fossils from Olduvai Gorge, Tanzania, and Koobi Fora, Kenya. We were working almost side by side at the National Museums of Kenya, in Nairobi, where the fossils are stored. The possibility of cut marks was exciting, since both sites preserve some of the oldest known archaeological materials. Potts and I returned to the United States, manufactured some stone tools, and started "butchering" bones and joints begged from our local butchers. Under the SEM, replicas of these cut marks looked very different from replicas of carnivore tooth scratches, regardless of the species of carnivore or the type of tool involved. By comparing the marks on the fossils with our hundreds of modern bones of known history, we were able to demonstrate convincingly that hominids using stone tools had processed carcasses of many different animals nearly two million years ago. For the first time, there was a firm link between stone tools and at least some of the early fossil animal bones.

This initial discovery persuaded some paleoanthropologists that the hominid hunter scenario was correct. Potts and I were not so sure. Our study had shown that many of the cut-marked fossils also bore carnivore tooth marks and that some of the cut marks were in places we hadn't expected—on bones that bore little meat in life. More work was needed.

In addition to more data about the Olduvai cut marks and tooth marks, I needed specific information about the patterns of cut marks left by known hunters performing typical activities associated with hunting. If similar patterns occurred on the fossils, then the early hominids probably behaved similarly to more modern hunters; if the patterns were different, then the behavior was probably also different. Three activities related to hunting occur often enough in peoples around the world and leave consistent enough traces to be used for such a test.

First, human hunters systematically disarticulate their kills, unless the animals are small enough to be eaten on the spot. Disarticulation leaves cut marks in a predictable pattern on the skeleton. Such marks cluster near the major joints of the limbs: shoulder, elbow, carpal joint (wrist), hip, knee, and hock (ankle). Taking a carcass apart at the joints is much easier than breaking or cutting through bones. Disarticulation enables hunters to carry food back to a central place or camp, so that they can share it with others or cook it or even store it by placing portions in trees, away from the reach of carnivores. If early hominids were hunters who transported and shared their kills, disarticulation marks would occur near joints in frequencies comparable to those produced by modern human hunters.

Second, human hunters often butcher carcasses, in the sense of removing meat from the bones. Butchery marks are usually found on the shafts of bones from the upper part of the front or hind limb, since this is where the big muscle masses lie. Butchery may be carried out at the kill site—especially if the animal is very large and its bones very heavy—or it may take place at the base camp, during the process of sharing food with others. Compared with disarticulation, butchery leaves relatively few marks. It is hard for a hunter to locate an animal's joints without leaving cut marks on the bone. In contrast, it is easier to cut the meat away from the midshaft of the bone without making such marks. If early hominids shared their food, however, there ought to be a number of cut marks located on the midshaft of some fossil bones.

Finally, human hunters often remove skin or tendons from carcasses, to be used for clothing, bags, thongs, and so on. Hide or tendon must be separated from the bones in many areas where there is little flesh, such as the lower limb bones of pigs, giraffes, antelopes, and zebras. In such cases, it is difficult to cut the skin without leaving a cut mark on the bone. Therefore, one expects to find many more cut marks on such bones than on the flesh-covered bones of the upper part of the limbs.

Unfortunately, although accounts of butchery and disarticulation by modern human hunters are remarkably consistent, quantitative studies are rare. Further, virtually all modern hunter-gatherers use metal tools, which leave more cut marks than stone tools. For these reasons I hesitated to compare the fossil evidence with data on modern hunters. Fortunately, Diane Gifford of the University of California,

Santa Cruz, and her colleagues had recently completed a quantitative study of marks and damage on thousands of antelope bones processed by Neolithic (Stone Age) hunters in Kenya some 2,300 years ago. The data from Prolonged Drift, as the site is called, were perfect for comparison with the Olduvai material.

Assisted by my technician, Jennie Rose, I carefully inspected more than 2,500 antelope bones from Bed I at Olduvai Gorge, which is dated to between 1.9 and 1.7 million years ago. We made high-fidelity replicas of every mark that we thought might be either a cut mark or a carnivore tooth mark. Back in the United States, we used the SEM to make positive identifications of the marks. (The replication and SEM inspection was time consuming, but necessary: only about half of the marks were correctly identified by eye or by light microscope.) I then compared the patterns of cut mark and tooth mark distributions on Olduvai fossils with those made by Stone Age hunters at Prolonged Drift.

By their location, I identified marks caused either by disarticulation or meat removal and then compared their frequencies with those from Prolonged Drift. More than 90 percent of the Neolithic marks in these two categories were from disarticulation, but to my surprise, only about 45 percent of the corresponding Olduvai cut marks were from disarticulation. This difference is too great to have occurred by chance; the Olduvai bones did not show the predicted pattern. In fact, the Olduvai cut marks attributable to meat removal and disarticulation showed essentially the same pattern of distribution as the carnivore tooth marks. Apparently, the early hominids were not regularly disarticulating carcasses. This finding casts serious doubt on the idea that early hominids carried their kills back to camp to share with others, since both transport and sharing are difficult unless carcasses are cut up.

When I looked for cut marks attributable to skinning or tendon removal, a more modern pattern emerged. On both the Neolithic and Olduvai bones, nearly 75 percent of all cut marks

occurred on bones that bore little meat; these cut marks probably came from skinning. Carnivore tooth marks were much less common on such bones. Hominids were using carcasses as a source of skin and tendon. This made it seem more surprising that they disarticulated carcasses so rarely.

A third line of evidence provided the most tantalizing clue. Occasionally, sets of overlapping marks occur on the Olduvai fossils. Sometimes, these sets include both cut marks and carnivore tooth marks. Still more rarely, I could see under the SEM which mark had been made first, because its features were overlaid by those of the later mark, in much the same way as old tire tracks on a dirt road are obscured by fresh ones. Although only thirteen such sets of marks were found, in eight cases the hominids made the cut marks *after* the carnivores made their tooth marks. This finding suggested a new hypothesis. Instead of hunting for prey and leaving the remains behind for carnivores to scavenge, perhaps hominids were scavenging from the carnivores. This might explain the hominids' apparently unsystematic use of carcasses: they took what they could get, be it skin, tendon, or meat.

Man the Scavenger is not nearly as attractive an image as Man the Hunter, but it is worth examining. Actually, although hunting and scavenging are different ecological strategies, many mammals do both. The only pure scavengers alive in Africa today are vultures; not one of the modern African mammalian carnivores is a pure scavenger. Even spotted hyenas, which have massive, bone-crushing teeth well adapted for eating the bones left behind by others, only scavenge about 33 percent of their food. Other carnivores that scavenge when there are enough carcasses around include lions, leopards, striped hyenas, and jackals. Long-term behavioral studies suggest that these carnivores scavenge when they can and kill when they must. There are only two nearly pure predators, or hunters—the cheetah and the wild dog—that rarely, if ever, scavenge.

What are the costs and benefits of scavenging compared with those of predation? First of all, the scavenger avoids the task of making sure its meal is dead: a predator has already endured the energetically costly business of chasing or stalking animal after animal until one is killed. But while scavenging may be cheap, it's risky. Predators rarely give up their prey to scavengers without defending it. In such disputes, the larger animal, whether a scavenger or a predator, usually wins, although smaller animals in a pack may defeat a lone, larger animal. Both predators and scavengers suffer the dangers inherent in fighting for possession of a carcass. Smaller scavengers such as jackals or striped hyenas avoid disputes to some extent by specializing in darting in and removing a piece of a carcass without trying to take possession of the whole thing. These two strategies can be characterized as that of the bully or that of the sneak: bullies need to be large to be successful, sneaks need to be small and quick.

Because carcasses are almost always much rarer than live prey, the major cost peculiar to scavenging is that scavengers must survey much larger areas than predators to find food. They can travel slowly, since their "prey" is already dead, but endurance is important. Many predators specialize in speed at the expense of endurance, while scavengers do the opposite.

The more committed predators among the East African carnivores (wild dogs and cheetahs) can achieve great top speeds when running, although not for long. Perhaps as a consequence, these "pure" hunters enjoy a much higher success rate in hunting (about three-fourths of their chases end in kills) than any of the scavenger-hunters do (less than half of their chases are successful). Wild dogs and cheetahs are efficient hunters, but they are neither big enough nor efficient enough in their locomotion to make good scavengers. In fact, the cheetah's teeth are so specialized for meat slicing that they probably cannot withstand the stresses of bone crunching and carcass dismembering carried out by scavengers. Other carnivores are less successful at hunting, but have specializations of size, endurance, or

(in the case of the hyenas) dentition that make successful scavenging possible. The small carnivores seem to have a somewhat higher hunting success rate than the large ones, which balances out their difficulties in asserting possession of carcasses.

In addition to endurance, scavengers need an efficient means of locating carcasses, which, unlike live animals, don't move or make noises. Vultures, for example, solve both problems by flying. The soaring, gliding flight of vultures expends much less energy than walking or cantering as performed by the part-time mammalian scavengers. Flight enables vultures to maintain a foraging radius two to three times larger than that of spotted hyenas, while providing a better vantage point. This explains why vultures can scavenge all of their food in the same habitat in which it is impossible for any mammal to be a pure scavenger. (In fact, many mammals learn where carcasses are located from the presence of vultures.)

Since mammals can't succeed as fulltime scavengers, they must have another source of food to provide the bulk of their diet. The large carnivores rely on hunting large animals to obtain food when scavenging doesn't work. Their size enables them to defend a carcass against others. Since the small carnivores—jackals and striped hyenas—often can't defend carcasses successfully, most of their diet is composed of fruit and insects. When they do hunt, they usually prey on very small animals, such as rats or hares, that can be consumed in their entirety before the larger competitors arrive.

The ancient habitat associated with the fossils of Olduvai and Koobi Fora would have supported many herbivores and carnivores. Among the latter were two species of large saber-toothed cats, whose teeth show extreme adaptations for meat slicing. These were predators with primary access to carcasses. Since their teeth were unsuitable for bone crushing, the saber-toothed cats must have left behind many bones covered with scraps of meat, skin, and tendon. Were early hominids among the scavengers that exploited such carcasses?

All three hominid species that were present in Bed I times (*Homo habilis, Australopithecus africanus, A. robustus*) were adapted for habitual, upright bipedalism. Many anatomists see evidence that these hominids were agile tree climbers as well. Although upright bipedalism is a notoriously peculiar mode of locomotion, the adaptive value of which has been argued for years (see Matt Cartmill's article, "Four Legs Good, Two Legs Bad," *Natural History,* November 1983), there are three general points of agreement.

First, bipedal running is neither fast nor efficient compared to quadrupedal gaits. However, at moderate speeds of 2.5 to 3.5 miles per hour, bipedal *walking* is more energetically efficient than quadrupedal walking. Thus, bipedal walking is an excellent means of covering large areas slowly, making it an unlikely adaptation for a hunter but an appropriate and useful adaptation for a scavenger. Second, bipedalism elevates the head, thus improving the hominid's ability to spot items on the ground—an advantage both to scavengers and to those trying to avoid becoming a carcass. Combining bipedalism with agile tree climbing improves the vantage point still further. Third, bipedalism frees the hands from locomotive duties, making it possible to carry items. What would early hominids have carried? Meat makes a nutritious, easy-to-carry package; the problem is that carrying meat attracts scavengers. Richard Potts suggests that carrying stone tools or unworked stones for toolmaking to caches would be a more efficient and less dangerous activity under many circumstances.

In short, bipedalism is compatible with a scavenging strategy. I am tempted to argue that bipedalism evolved because it provided a substantial advantage to scavenging hominids. But I doubt hominids could scavenge effectively without tools, and bipedalism predates the oldest known stone tools by more than a million years.

Is there evidence that, like modern mammalian scavengers, early hominids had an alternative food source, such as either hunting or eating fruits and insects? My husband, Alan Walker, has shown that the microscopic wear on an animal's teeth reflects its diet. Early hominid teeth wear more like that of chimpanzees and other modern fruit eaters than that of carnivores. Apparently, early hominids ate mostly fruit, as the smaller, modern scavengers do. This accords with the estimated body weight of early hominids, which was only about forty to eighty pounds—less than that of any of the modern carnivores that combine scavenging and hunting but comparable to the striped hyena, which eats fruits and insects as well as meat.

Would early hominids have been able to compete for carcasses with other carnivores? They were too small to use a bully strategy, but if they scavenged in groups, a combined bully-sneak strategy might have been possible. Perhaps they were able to drive off a primary predator long enough to grab some meat, skin, or marrow-filled bone before relinquishing the carcass. The effectiveness of this strategy would have been vastly improved by using tools to remove meat or parts of limbs, a task at which hominid teeth are poor. As agile climbers, early hominids may have retreated into the trees to eat their scavenged trophies, thus avoiding competition from large terrestrial carnivores.

In sum, the evidence on cut marks, tooth wear, and bipedalism, together with our knowledge of scavenger adaptation in general, is consistent with the hypothesis that two million years ago hominids were scavengers rather than accomplished hunters. Animal carcasses, which contributed relatively little to the hominid diet, were not systematically cut up and transported for sharing at base camps. Man the Hunter may not have appeared until 1.5 to 0.7 million years ago, when we do see a shift toward omnivory, with a greater proportion of meat in the diet. This more heroic ancestor may have been *Homo erectus,* equipped with Acheulean-style stone tools and, increasingly, fire. If we wish to look further back, we may have to become accustomed to a less flattering image of our heritage.

Late Hominid Evolution

The most important aspect of human evolution is also the most difficult to decipher from the fossil evidence: our development as sentient, social beings, capable of communicating by means of language. We detect hints of incipient humanity in the form of crudely chipped tools, the telltale signs of a home base, or the artistic achievements of ornaments and cave art, as in "The Dawn of Creativity" by William Allman. Yet none of these indicators of a distinctly hominid way of life can provide us with the nuances of the everyday lives of these creatures, their social relations, or their supernatural beliefs, if any. Most of what remains is the rubble of bones and stones from which we interpret what we can of their lifestyle, thought processes, and communicating ability. Our ability to glean from the fossil record is not completely without hope, however. In fact, informed speculation is what makes possible such essays as "Did Neandertals Lose an Evolutionary 'Arms' Race?" by Ann Gibbons and "Old Masters" by Pat Shipman. Each is a fine example of careful, systematic, and thought-provoking work that is based upon an increased understanding of hominid fossil sites as well as the more general environmental circumstances in which our predecessors lived.

Beyond the technological and anatomical adaptations, questions have arisen as to how our hominid forebears organized themselves socially and whether or not modern-day human behavior is inherited as a legacy of our evolutionary past or is a learned product of contemporary circumstances. Attempts to address these questions have given rise to the technique referred to as the "ethnographic analogy." This is a method whereby anthropologists use "ethnographies" or field studies of modern-day hunters and gatherers whose lives we take to be the best approximations we have to what life might have been like for our ancestors. Granted, these contemporary foragers have been living under conditions of environmental

and social change just as industrial peoples have. Nevertheless, it seems that, at least in some aspects of their lives, they have not changed as much as we have. So, if we are to make any enlightened assessments of prehistoric behavior patterns, we are better off looking at them than at ourselves.

As if to show that controversy over lineages is not limited to the earlier hominid period (see unit 5), in this unit we see how long-held beliefs about *Homo erectus* are being threatened by new fossil evidence (see James Shreeve's essay "*Erectus* Rising" and "The First Europeans" by Jean-Jacques Hublin). We also consider new evidence bearing upon the "Eve hypothesis," as addressed in "The Dating Game" and "The Neanderthal Peace," both articles by James Shreeve. In the case of the "Eve hypothesis," the issue of when and where the family tree of modern humans actually began has pitted the bone experts, on the one hand, against a new type of anthropologist specializing in molecular biology, on the other. Granted, for some scientists, the new evidence fits in quite comfortably with previously held positions; for others it seems that reputations, as well as theories, are at stake.

Looking Ahead: Challenge Questions

When, where, and why did *Homo erectus* evolve? Is it one species or two?

What were Cro-Magnons trying to say or do with their cave art?

How do we measure evolutionary time?

Explain whether the Cro-Magnons were the first or the last modern humans to appear on Earth.

What are the strengths and weaknesses of the "Eve hypothesis"?

What happened to the Neanderthals?

How would you draw the late hominid family tree?

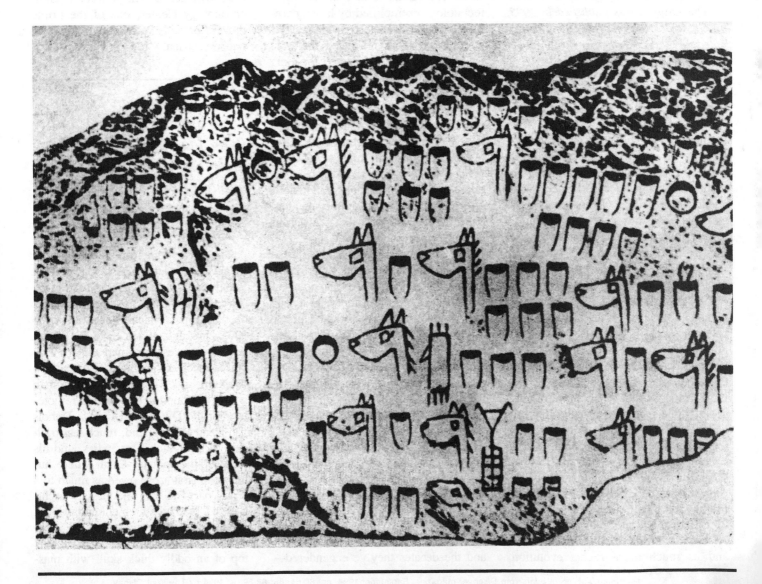

Erectus Rising

Oh No. Not This. The Hominids Are Acting Up Again . . .

James Shreeve

James Shreeve is the coauthor, with anthropologist Donald Johanson, of Lucy's Child: The Discovery of a Human Ancestor. *His book,* The Neandertal Enigma: Solving the Mystery of Modern Human Origins, *was published in 1995, and he is at work on a novel that a reliable source calls "a murder thriller about the species question."*

Just when it seemed that the recent monumental fuss over the origins of modern human beings was beginning to quiet down, an ancient ancestor is once more running wild. Trampling on theories. Appearing in odd places, way ahead of schedule. Demanding new explanations. And shamelessly flaunting its contempt for conventional wisdom in the public press.

The uppity ancestor this time is *Homo erectus*—alias Java man, alias Peking man, alias a mouthful of formal names known only to the paleontological cognoscenti. Whatever you call it, *erectus* has traditionally been a quiet, average sort of hominid: low of brow, thick of bone, endowed with a brain larger than that of previous hominids but smaller than those that followed, a face less apelike and projecting than that of its ancestors but decidedly more simian than its descendants'. In most scenarios of human evolution, *erectus*'s role was essentially to mark time—a million and a half years of it—between its obscure, presumed origins in East Africa just under 2 million years ago and its much more recent evolution

into something deserving the name *sapiens.*

Erectus accomplished only two noteworthy deeds during its long tenure on Earth. First, some 1.5 million years ago, it developed what is known as the Acheulean stone tool culture, a technology exemplified by large, carefully crafted tear-shaped hand axes that were much more advanced than the bashed rocks that had passed for tools in the hands of earlier hominids. Then, half a million years later, and aided by those Acheulean tools, the species carved its way out of Africa and established a human presence in other parts of the Old World. But most of the time, *Homo erectus* merely existed, banging out the same stone tools millennium after millennium, over a time span that one archeologist has called "a period of unimaginable monotony."

Or so read the old script. These days, *erectus* has begun to ad-lib a more vigorous, controversial identity for itself. Research within the past year has revealed that rather than being 1 million years old, several *erectus* fossils from Southeast Asia are in fact almost 2 million years old. That is as old as the oldest African members of the species, and it would mean that *erectus* emerged from its home continent much earlier than has been thought—in fact, almost immediately after it first appeared. There's also a jawbone, found in 1991 near the Georgian city of Tbilisi, that resembles *erectus* fossils from Africa and may be as old as 1.8 million years, though that age is still in doubt. These new dates—and the debates they've engendered—

have shaken *Homo erectus* out of its interpretive stupor, bringing into sharp relief just how little agreement there is on the rise and demise of the last human species on Earth, save one.

"Everything now is in flux," says Carl Swisher of the Berkeley Geochronology Center, one of the prime movers behind the redating of *erectus* outside Africa. "It's all a mess."

Asian and African fossils were lumped into one far-flung taxon, a creature not quite like us but human enough to be welcomed into our genus: Homo erectus.

The focal point for the flux is the locale where the species was first found: Java. The rich but frustration-soaked history of paleoanthropology on that tropical island began just over 100 years ago, when a young Dutch anatomy professor named Eugène Dubois conceived the idée fixe that the "missing link" between ape and man was to be found in the jungled remoteness of the Dutch East Indies. Dubois had never left Holland, much less traveled to the Dutch East Indies, and his pick for the spot on Earth where humankind first arose owed as much to a large part of the Indonesian archipelago's being a Dutch colony as it did to any scientific evidence. He nevertheless found his missing link—the top of an oddly thick skull with massive browridges—in 1891 on the banks of the Solo River, near a community called Trinil in central Java. About a year later a thighbone that Dubois thought might belong to the same individual was found nearby; it looked so much like a modern human thighbone that Dubois assumed this ancient primate had walked upright. He christened the creature *Pithecanthropus erectus*—"erect ape-man"—and returned home in triumph.

Finding the fossil proved to be the easy part. Though Dubois won popular acclaim, neither he nor his "Java man" received the full approbation of the anatomists of the day, who considered his ape-man either merely an ape or merely a man. In an apparent pique, Dubois cloistered away the fossils for a quarter-century, refusing others the chance to view his prized possessions. Later, other similarly primitive human remains began to turn up in China and East Africa. All shared a collection of anatomical traits, including a long, low braincase with prominent browridges and a flattened forehead; a sharp angle to the back of the skull when viewed in profile; and a deep, robustly built jaw showing no hint of a chin. Though initially given separate regional names, the fossils were eventually lumped together into one far-flung taxon, a creature not quite like us but human enough to be welcomed into our genus: *Homo erectus*.

Over the decades the most generous source of new *erectus* fossils has been the sites on or near the Solo River in Java. The harvest continues: two more skulls, including one of the most complete *erectus* skulls yet known, were found at a famous fossil site called Sangiran just in the past year. Though the Javan yield of ancient humans has been rich, something has always been missing—the crucial element of time. Unless the age of a fossil can be determined, it hangs in limbo, its importance and place in the larger scheme of human evolution forever undercut with doubt. Until researchers can devise better methods for dating bone directly—right now there are no tech-

niques that can reliably date fossilized, calcified bone more than 50,000 years old—a specimen's age has to be inferred from the geology that surrounds it. Unfortunately, most of the discoveries made on the densely populated and cultivated island of Java have been made not by trained excavators but by sharp-eyed local farmers who spot the bones as they wash out with the annual rains and later sell them. As a result, the original location of many a prized specimen, and thus all hopes of knowing its age, are a matter of memory and word of mouth.

Despite the problems, scientists continue to try to pin down dates for Java's fossils. Most have come up with an upper limit of around 1 million years. Along with the dates for the Peking man skulls found in China and the Acheulean tools from Europe, the Javan evidence has come to be seen as confirmation that *erectus* first left Africa at about that time.

By the early 1970s most paleontologists were firmly wedded to the idea that Africa was the only human-inhabited part of the world until one million years ago.

There are those, however, who have wondered about these dates for quite some time. Chief among them is Garniss Curtis, the founder of the Berkeley Geochronology Center. In 1971 Curtis, who was then at the University of California at Berkeley, attempted to determine the age of a child's skull from a site called Mojokerto, in eastern Java, by using the potassium-argon method to date volcanic minerals in the sediments from which the skull was purportedly removed. Potassium-argon dating had been in use since the 1950s, and Curtis had been enormously successful with it in dating ancient African hominids—including Louis Leakey's famous hominid finds at Olduvai Gorge in Tanzania. The method takes advantage of the fact that a radioactive

isotope of potassium found in volcanic ash slowly and predictably decays over time into argon gas, which becomes trapped in the crystalline structure of the mineral. The amount of argon contained in a given sample, measured against the amount of the potassium isotope, serves as a kind of clock that tells how much time has passed since a volcano exploded and its ash fell to earth and buried the bone in question.

Applying the technique to the volcanic pumice associated with the skull from Mojokerto, Curtis got an extraordinary age of 1.9 million years. The wildly anomalous date was all too easy to dismiss, however. Unlike the ash deposits of East Africa, the volcanic pumices in Java are poor in potassium. Also, not unexpectedly, a heavy veil of uncertainty obscured the collector's memories of precisely where he had found the fossil some 35 years earlier. Besides, most paleontologists were by this time firmly wedded to the idea that Africa was the only human-inhabited part of the world until 1 million years ago. Curtis's date was thus deemed wrong for the most stubbornly cherished of reasons: because it couldn't possibly be right.

In 1992 Curtis—under the auspices of the Institute for Human Origins in Berkeley—returned to Java with his colleague Carl Swisher. This time he was backed up by far more sensitive equipment and a powerful refinement in the dating technique. In conventional potassium-argon dating, several grams' worth of volcanic crystals gleaned from a site are needed to run a single experiment. While the bulk of these crystals are probably from the eruption that covered the fossil, there's always the possibility that other materials, from volcanoes millions of years older, have gotten mixed in and will thus make the fossil appear to be much older than it actually is. The potassium-argon method also requires that the researcher divide the sample of crystals in two. One half is dissolved in acid and passed through a flame; the wavelengths of light emitted tell how much potassium is in the sample. The other half is used to measure the amount of argon gas that's released when the crystals are heated. This two-step process further increases the chance of error, simply by giving the experiment twice as much opportunity to go wrong.

The refined technique, called argon-argon dating, neatly sidesteps most of these difficulties. The volcanic crystals are first placed in a reactor and bombarded with neutrons; when one of these neutrons penetrates the potassium nucleus, it displaces a proton, converting the potassium into an isotope of argon that doesn't occur in nature. Then the artificially created argon and the naturally occurring argon are measured in a single experiment. Because the equipment used to measure the isotopes can look for both types of argon at the same time, there's no need to divide the sample, and so the argon-argon method can produce clear results from tiny amounts of material.

In some cases—when the volcanic material is fairly rich in potassium—all the atoms of argon from a single volcanic crystal can be quick-released by the heat from a laser beam and then counted. By doing a number of such single-crystal experiments, the researchers can easily pick out and discard any data from older, contaminant crystals. But even when the researchers are forced to sample more than one potassium-poor crystal to get any reading at all—as was the case at Mojokerto—the argon-argon method can still produce a highly reliable age. In this case, the researchers carefully heat a few crystals at a time to higher and higher temperatures, using a precisely controlled laser. If all the crystals in a sample are the same age, then the amount of argon released at each temperature will be the same. But if contaminants are mixed in, or if severe weathering has altered the crystal's chemical composition, the argon measurements will be erratic, and the researchers will know to throw out the results.

Curtis and Swisher knew that in the argon-argon step-heating method they had the technical means to date the potassium-poor deposits at Mojokerto

accurately. But they had no way to prove that those deposits were the ones in which the skull had been buried: all they had was the word of the local man who had found it. Then, during a visit to the museum in the regional capital, where the fossil was being housed, Swisher noticed something odd. The hardened sediments that filled the inside of the fossil's braincase looked black. But back at the site, the deposits of volcanic pumice that had supposedly sheltered the infant's skull were whitish in color. How could a skull come to be filled with black sediments if it had been buried in white ones? Was it possible that the site and the skull had nothing to do with each other after all? Swisher suspected something was wrong. He borrowed a penknife, picked up the precious skull, and nicked off a bit of the matrix inside.

"I almost got kicked out of the country at that point," he says. "These fossils in Java are like the crown jewels."

Luckily, his impulsiveness paid off. The knife's nick revealed white pumice under a thin skin of dark pigment: years earlier, someone had apparently painted the surface of the hardened sediments black. Since there were no other deposits within miles of the purported site that contained a white pumice visually or chemically resembling the matrix in the skull, its tie to the site was suddenly much stronger. Curtis and Swisher returned to Berkeley with pumice from that site and within a few weeks proclaimed the fossil to be 1.8 million years old, give or take some 40,000 years. At the same time, the geochronologists ran tests on pumice from the lower part of the Sangiran area, where *erectus* facial and cranial bone fragments had been found. The tests yielded an age of around 1.6 million years. Both numbers obviously shatter the 1-million-year barrier for *erectus* outside Africa, and they are a stunning vindication of Curtis's work at Mojokerto 20 years ago. "That was very rewarding," he says, "after having been told what a fool I was by my colleagues."

While no one takes Curtis or Swisher for a fool now, some of their colleagues won't be fully convinced by the new

dates until the matrix inside the Mojokerto skull itself can be tested. Even then, the possibility will remain that the skull may have drifted down over the years into deposits containing older volcanic crystals that have nothing to do with its original burial site, or that it was carried by a river to another, older site. But Swisher contends that the chance of such an occurrence is remote: it would have to have happened at both Mojokerto and Sangiran for the fossils' ages to be refuted. "I feel really good about the dates," he says. "But it has taken me a while to understand their implications."

The implications that can be spun out from the Javan dates depend on how one chooses to interpret the body of fossil evidence commonly embraced under the name *Homo erectus*. The earliest African fossils traditionally attributed to *erectus* are two nearly complete skulls from the site of Koobi Fora in Kenya, dated between 1.8 and 1.7 million years old. In the conventional view, these early specimens evolved from a more primitive, smaller-brained ancestor called *Homo habilis,* well represented by bones from Koobi Fora, Olduvai Gorge, and sites in South Africa.

If this conventional view is correct, then the new dates mean that *erectus* must have migrated out of Africa very soon after it evolved, quickly reaching deep into the farthest corner of Southeast Asia. This is certainly possible: at the time, Indonesia was connected to Asia by lower sea levels—thus providing an overland route from Africa—and Java is just 10,000 to 15,000 miles from Kenya, depending on the route. Even if *erectus* traveled just one mile a year, it would still take no more than 15,000 years to reach Java—a negligible amount of evolutionary time.

If *erectus* did indeed reach Asia almost a million years earlier than thought, then other, more controversial theories become much more plausible. Although many anthropologists believe that the African and Asian *erectus* fossils all represent a single species, other investigators have recently argued that

MEANWHILE, IN SIBERIA . . .

The presence of *Homo erectus* in Asia twice as long ago as previously thought has some people asking whether the human lineage might have originated in Asia instead of Africa. This long-dormant theory runs contrary to all current thinking about human evolution and lacks an important element: evidence. Although the new Javan dates do place the species in Asia at around the same time it evolved in Africa, all confirmed specimens of other, earlier hominids—the first members of the genus *Homo,* for instance, and the australopithecines, like Lucy—have been found exclusively in Africa. Given such an overwhelming argument, most investigators continue to believe that the hominid line began in Africa.

Most, but not all. Some have begun to cock an ear to the claims of Russian archeologist Yuri Mochanov. For over a decade Mochanov has been excavating a huge site on the Lena River in eastern Siberia—far from Africa, Java, or anywhere else on Earth an ancient hominid bone has ever turned up. Though he hasn't found any hominid fossils in Siberia, he stubbornly believes he's uncovered the next best thing: a trove of some 4,000 stone artifacts—crudely made flaked tools, but tools nonetheless—that he maintains are at least 2 million years old, and possibly 3 million. This, he says, would mean that the human lineage arose not in tropical Africa but in the cold northern latitudes of Asia.

"For evolutionary progress to occur, there had to be the appearance of new conditions: winter, snow, and, accompanying them, hunger," writes Mochanov. "[The ancestral primates] had to learn to walk on the ground, to change their carriage, and to become accustomed to meat—that is, to become 'clever animals of prey.' " And to become clever animals of prey, they'd need tools.

Although he is a well-respected investigator, Mochanov has been unable to convince either Western anthropologists or his Russian colleagues of the age of his site. Until recently the chipped rocks he was holding up as human artifacts were simply dismissed as stones broken by natural processes, or else his estimate of the age of the site was thought to be wincingly wrong. After all, no other signs of human occupation of Siberia appear until some 35,000 years ago.

But after a lecture swing through the United States earlier this year—in which he brought more data and a few prime examples of the tools for people to examine and pass around—many archeologists concede that it is difficult to explain the particular pattern of breakage of the rocks by any known natural process. "Everything I have heard or seen about the context of these things suggests that they are most likely tools," says anthropologist Rick Potts of the Smithsonian Institution, which was host to Mochanov last January.

They're even willing to concede that the site might be considerably older than they'd thought, though not nearly as old as Mochanov estimates. (To date the site, Mochanov compared the tools with artifacts found early in Africa; he also employed an arcane dating technique little known outside Russia.) Preliminary results from an experimental dating technique performed on soil samples from the site by Michael Waters of Texas A&M and Steve Forman of Ohio State suggest that the layer of sediment bearing the artifacts is some 400,000 years old. That's a long way from 2 million, certainly, but it's still vastly older than anything else found in Siberia—and the site is 1,500 miles farther north than the famous Peking man site in China, previously considered the most northerly home of *erectus.*

"If this does turn out to be 400,000 years old, it's very exciting," says Waters. "If people were able to cope and survive in such a rigorous Arctic environment at such an early time, we would have to completely change our perception of the evolution of human adaptation."

"I have no problem with hominids being almost anywhere at that age—they were certainly traveling around," says Potts. "But the environment is the critical thing. If it was really cold up there"—temperatures in the region now often reach −50 degrees in deep winter—"we'd all have to scratch our heads over how these early hominids were making it in Siberia. There is no evidence that Neanderthals, who were better equipped for cold than anyone, were living in such climates. But who knows? Maybe a population got trapped up there, went extinct, and Mochanov managed to find it." He shrugs. "But that's just arm waving."

—J. S.

the two groups are too different to be so casually lumped together. According to paleoanthropologist Ian Tattersall of the American Museum of Natural History in New York, the African skulls traditionally assigned to *erectus* often lack many of the specialized traits that were originally used to define the species in Asia, including the long, low cranial structure, thick skull bones, and robustly built faces. In his view, the African group deserves to be placed in a separate species, which he calls *Homo ergaster.*

Most anthropologists believe that the only way to distinguish between species in the fossil record is to look at the similarities and differences between bones; the age of the fossil should not play a part. But age is often hard to ignore, and Tattersall believes that the new evidence for what he sees as two distinct populations living at the same time in widely separate parts of the Old World is highly suggestive. "The new dates help confirm that these were indeed two different species," he says.

"In my view, *erectus* is a separate variant that evolved only in Asia."

Other investigators still contend that the differences between the African and Asian forms of *erectus* are too minimal to merit placing them in separate species. But if Tattersall is right, his theory raises the question of who the original emigrant out of Africa really was. *Homo ergaster* may have been the one to make the trek, evolving into *erectus* once it was established in Asia. Or perhaps a population of some even more primitive, as-yet-unidentified common ancestor ventured forth, giving rise to *erectus* in Asia while a sister population evolved into *ergaster* on the home continent.

Furthermore, no matter who left Africa first, there's the question of what precipitated the migration, a question made even more confounding by the new dates. The old explanation, that the primal human expansion across the hem of the Old World was triggered by the sophisticated Acheulean tools, is no longer tenable with these dates, simply because the tools had not yet been invented when the earliest populations would have moved out. In hindsight, that notion seems a bit shopworn anyway. Acheulean tools first appear in Africa around 1.5 million years ago, and soon after at a site in the nearby Middle East. But they've never been found in the Far East, in spite of the abundant fossil evidence for *Homo erectus* in the region.

Until now, that absence has best been explained by the "bamboo line." According to paleoanthropologist Geoffrey Pope of William Paterson College in New Jersey, *erectus* populations venturing from Africa into the Far East found the land rich in bamboo, a raw material more easily worked into cutting and butchering tools than recalcitrant stone. Sensibly, they abandoned their less efficient stone industry for one based on the pliable plant, which leaves no trace of itself in the archeological record. This is still a viable theory, but the new dates from Java add an even simpler dimension to it: there are no Acheulean tools in the Far East because the first wave of *erectus*

to leave Africa didn't have any to bring with them.

So what *did* fuel the quick-step migration out of Africa? Some researchers say the crucial development was not cultural but physical. Earlier hominids like *Homo habilis* were small-bodied creatures with more apelike limb proportions, notes paleoanthropologist Bernard Wood of the University of Liverpool, while African *erectus* was built along more modern lines. Tall, relatively slender, with long legs better able to range over distance and a body better able to dissipate heat, the species was endowed with the physiology needed to free it from the tropical shaded woodlands of Africa that sheltered earlier hominids. In fact, the larger-bodied *erectus* would have required a bigger feeding range to sustain itself, so it makes perfect sense that the expansion out of Africa should begin soon after the species appeared. "Until now, one was always having to account for what kept *erectus* in Africa so long after it evolved," says Wood. "So rather than raising a problem, in some ways the new dates in Java solve one."

Of course, if those dates are right, the accepted time frame for human evolution outside the home continent is nearly doubled, and that has implications for the ongoing debate over the origins of modern human beings. There are two opposing theories. The "out of Africa" hypothesis says that *Homo sapiens* evolved from *erectus* in Africa, and then—sometime in the last 100,000 years—spread out and replaced the more archaic residents of Eurasia. The "multiregional continuity" hypothesis says that modern humans evolved from *erectus* stock in various parts of the Old World, more or less simultaneously and independently. According to this scenario, living peoples outside Africa should look for their most recent ancestors not in African fossils but in the anatomy of ancient fossils within their own region of origin.

As it happens, the multiregionalists have long claimed that the best evidence for their theory lies in Australia,

which is generally thought to have become inhabited around 50,000 years ago, by humans crossing over from Indonesia. There are certain facial and cranial characteristics in modern Australian aborigines, the multiregionalists say, that can be traced all the way back to the earliest specimens of *erectus* at Sangiran—characteristics that differ from and precede those of any more recent, *Homo sapiens* arrival from Africa. But if the new Javan dates are right, then these unique characteristics, and thus the aborigines' Asian *erectus* ancestors, must have been evolving separately from the rest of humankind for almost 2 million years. Many anthropologists, already skeptical of the multiregionalists' potential 1-million-year-long isolation for Asian *erectus,* find a 2-million-year-long isolation exceedingly difficult to swallow. "Can anyone seriously propose that the lineage of Australian aborigines could go back that far?" wonders paleoanthropologist Chris Stringer of the Natural History Museum in London, a leading advocate of the out-of-Africa theory.

The multiregionalists counter that they've never argued for *complete* isolation—that there's always been some flow of genes between populations, enough interbreeding to ensure that clearly beneficial *sapiens* characteristics would quickly be conferred on peoples throughout the Old World. "Just as genes flow now from Johannesburg to Beijing and from Melbourne to Paris, they have been flowing that way ever since humanity evolved," says Alan Thorne of the Australian National University in Canberra, an outspoken multiregionalist.

Stanford archeologist Richard Klein, another out-of-Africa supporter, believes the evidence actually *does* point to just such a long, deep isolation of Asian populations from African ones. The fossil record, he says, shows that while archaic forms of *Homo sapiens* were developing in Africa, *erectus* was remaining much the same in Asia. In fact, if some *erectus* fossils from a site called Ngandong in Java turn out to be as young as 100,000 years, as some researchers believe, then *erectus* was still alive on Java at the same time that fully modern human beings were living in Africa and the Middle East. Even more important, Klein says, is the cultural evidence. That Acheulean tools never reached East Asia, even after their invention in Africa, could mean the inventors never reached East Asia either. "You could argue that the new dates show that until very recently there was a long biological and cultural division between Asia on one hand, and Africa and Europe on the other," says Klein. In other words, there must have been two separate lineages of *erectus,* and since there aren't two separate lineages of modern humans, one

of those must have gone extinct: presumably the Asian lineage, hastened into oblivion by the arrival of the more culturally adept, tool-laden *Homo sapiens.*

Naturally this argument is anathema to the multiregionalists. But this tenacious debate is unlikely to be resolved without basketfuls of new fossils, new ways of interpreting old ones—and new dates. In Berkeley, Curtis and Swisher are already busy applying the argon-argon method to the Ngandong fossils, which could represent some of the last surviving *Homo erectus* populations on Earth. They also hope to work their radiometric magic on a key

erectus skull from Olduvai Gorge. In the meantime, at least one thing has become clear: *Homo erectus,* for so long the humdrum hominid, is just as fascinating, contentious, and elusive a character as any other in the human evolutionary story.

FURTHER READING

Eugène Dubois & the Ape-Man from Java. Bert Theunissen. Kluwer Academic, 1989. When a Dutch army surgeon, determined to prove Darwin right, traveled to Java in search of the missing link between apes and humans, he inadvertently opened a paleontological Pandora's box. This is Dubois's story, the story of the discovery of *Homo erectus.*

THE FIRST EUROPEANS

*Isolated from populations in Africa and the Near East, archaic
Homo sapiens in Europe evolved into Neandertals.*

Jean-Jacques Hublin

JEAN-JACQUES HUBLIN, *a director of research in France's Centre National de Récherche Scientifique in Paris, works in the Musée de l'Homme. He studies the evolution of archaic Homo sapiens around the Mediterranean basin.*

Jean-Jacques Hublin

Late 19th-century savants examine a fossil jawbone.

In the late 1850's the academic world and readers of daily newspapers were fascinated by Darwin's *On The Origin of Species, stone tools discovered along the Somme by Jacques Boucher de Perthe, and a skullcap and long bones of a Neandertal unearthed near Düsseldorf. The following decades witnessed a virtual "flint-rush" during which adventurers, university professors, and country priests excavated (and*

often destroyed) many major European Ice Age sites. Nineteenth-century Europe was economically, politically, and scientifically the leader of Western civilization. For an intellect of the time, an "antediluvian" man, if he ever existed, had to have been European. In reality, during hundreds of thousands of years of human evolution, while the first hominids were emerging in Africa, Europe remained uninhabited.

The peopling of Europe took place in at least two waves. The best known involved modern humans, who moved into central and, later, western Europe from the Middle East more than 40,000 years ago. These people, popularly known as the Cro-Magnons, created the paintings found at Lascaux (France), Altamira (Spain), and the recently explored Chauvet Cave in southeastern France. When the caves were first painted, ca. 30,000 years ago, the last survivors of an older wave of colonization, the Neandertals, still existed in the western extreme of Europe. The Neandertals eventually became extinct, but recent discoveries demonstrate that their ancestors had occupied Europe for hundreds of thousands of years, evolving from archaic *Homo sapiens* as they became isolated from the mother population in Africa and the Near East.

Neandertals are generally assigned to the same biological species as modern humans, *Homo sapiens,* although some scientists consider them a separate species, *Homo neanderthalensis.* While

they represent an essentially European group, they have close relatives in the Near and Middle East, from Israel to Uzbekistan. Neandertal skulls and mandibles display a singular morphology. Although some of their features can occasionally be found in other hominids, the combination of characteristic traits is unique. European Neandertals are rather short and sturdy, with a long trunk and short legs. The skeleton is robust, and the muscle attachments imply a powerful body. The head is remarkable. It is big, enclosing a brain comparable in volume to that of modern humans, but the braincase and face are very long, the forehead is low, and the browridges protrude. The mandible is strong and lacks a projecting chin. Seen from the rear, the braincase has a rounded, almost circular shape in contrast to the pentagonal shape observed in both the earlier *Homo erectus* and modern humans. The face is structured around a large nasal cavity, and its middle part projects forward. The bone below the eye sockets is flat or even convex, receding laterally. The cheekbones are weak and oriented obliquely. This projecting face contrasts strongly with the short, flat face of the first modern humans. Other anatomical details of the ear area and rear and base of the skull are unique to Neandertals. The occipital bone, at the back of the skull, is marked by a conspicuous depression called the suprainiac fossa.

We know that the Neandertal lineage lasted from at least 127,000 to 30,000 years before present (B.P.), but during the last two decades a number of fossil discoveries and the reexamination of previously known specimens have fed discussion about the peopling of Europe before the Neandertals. One of the main issues has been the relationship of these first Europeans to their successors, Neandertals and modern humans. Another involves *Homo erectus*. Was this older species, known in Africa and Asia before 400,000 B.P., present in Europe before these two groups of *Homo sapiens*? The date of the first peopling of Europe has been hotly disputed. Hominids dating more than four million years old have been found in Africa, but the bulk of Palaeolithic remains in Europe are much later, from the Middle Pleistocene (780,000 to 127,000 B.P.) to the end of the Ice Age (10,000 B.P.). The debate has divided scholars into "short chronology" and "long chronology" camps.

Supporters of the short chronology dispute the human origin of stone tools from the earliest sites in Europe as well as the dating of the sites. Wil Roebroeks from the University of Leiden, The Netherlands, has, for example, even claimed that none of the European archaeological sites dated before 500,000 B.P. is valid. Until recently they could also point to a lack of human remains before 500,000 B.P. Rich fossil sites have been known for more than a century, especially in France's Massif Central, but amidst the thousands of large mammal remains not one single human bone or tooth was ever found. Why would humans have stayed out of Europe for such a long time? Climate seemed the most obvious reason, but other environmental factors were proposed to explain this limited expansion. Clive Gamble of the University of Southampton has suggested that hominids could not have successfully adapted to the temperate forest environment before 500,000 B.P. and were restricted to open areas to the south. According to Alan Turner of the University of Liverpool, the archaic humans who subsisted by hunting and scavenging could not compete with the numerous large carnivores then found in Europe.

Hominids, however, were at the gates of Europe long before 500,000 B.P. In North Africa, just beyond the Strait of Gibraltar, their artifacts are at least one million years old, and in the Near East, at Ubeidiya in Israel, stone artifacts and fragmentary human remains have been found with animal bones dated to ca. 1.4 million years ago. But the most striking evidence was found in 1991 on the southern side of the Caucasus mountains. A team from the Georgian Academy of Sciences and the Römisch-Germanisches Zentralmuseum of Mainz, Germany, excavating at Dmanisi, a medieval site near Tbilisi, unexpectedly found underlying strata with a well-preserved human mandible, stone artifacts, and animal remains. A basalt layer beneath these strata has been dated to about 1.8 million years ago. The paleomagnetism—the orientation of the earth's magnetic poles as reflected by the orientation of some particles in the deposits—of both the basalt and the bone-bearing layer confirms a date for the hominid just after 1.8 million years ago. Anatomically, the specimen is remi-

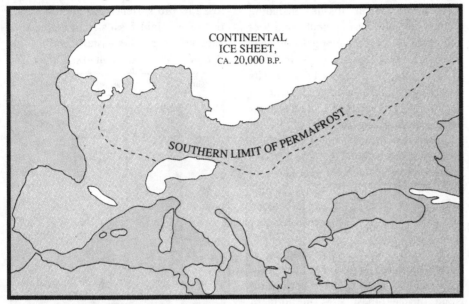

Neandertals evolved from archaic Homo sapiens *populations in Europe that were isolated from Africa and Asia, especially during periods of glacial advance. Map, top, shows Europe at the height of the last glacial peak, ca. 20,000 B.P. Though by this time Neandertals had been replaced by modern humans, map does indicate how glacial advances could have isolated earlier European populations. Humans were able to move northward only during temperate interglacial or moderately cold periods. Key Neandertal and pre-Neandertal sites are on map above.*

niscent of the contemporary early *Homo erectus* of Africa. The Dmanisi find suggests that early waves of people in the lowest latitudes of Europe are not unlikely, but before extended colonization was successful various attempts must have been made that left few archaeological traces.

Supporters of a long chronology believe that people reached Europe long before 500,000 B.P., and that traces of them are scarce because they were few in number and their sites were later destroyed by advancing glaciers. Still, some sites have been assigned to the Lower Pleistocene or even to the end of the Pliocene, between 5.5 and 1.7 million years ago. For example, in the French Massif Central, Saint-Eble has been dated between 2.2 and 2.5 million years ago and Blassac could be as old as 2.0 million years. But the flaked stones found at these sites are never flint but rather quartz or microgranites, and they are frequently interpreted as teph-rofacts, stones naturally fragmented in the course of volcanic eruptions. In the same area, on the shores of an ancient lake, a more reliable archaelogical site with flint tools and fossil remains of large mammals has been excavated by Eugene Bonifay of the Centre National de Récherche Scientifique. It is, however, more recent, no earlier than about one million years. Other sites roughly of the same age have been reported—Le Vallonet in the French Riviera and Stránská Skála in Moravia are the best known—but their dates and the human origin of their tools have been criticized.

The recent discovery of fossil hominids at the site of Gran Dolina at Atapuerca in northern Spain has not conclusively demonstrated that humans were present in Europe as early as 800,000 B.P. Atapuerca was already famous for the Sima De los Huesos, or Pit of Bones, which has yielded the largest number of Middle Pleistocene hominids ever found. A few hundred years away, the Gran Dolina site, under investigation by Eudald Carbonell of Rovira y Virgili University, displays an extraordinary geological profile 60 feet deep that includes several archaeological layers. In 1994 a small area of a stratum known as TD6 was excavated and yielded 36 human fossils, including teeth, skull and jaw fragments, and foot and hand bones, belonging to at least four individuals; excavations in 1995 yielded more remains. The TD6 deposits immediately predate a paleomagnetic inversion (a reversal of the earth's magnetic poles) dating to 780,000 B.P. In addition, the site has two older archaeological layers below TD6. The TD6 remains are too fragmentary to be securely assigned to a species. However, the new evidence can hardly be rejected by even the most vigorous supporters of the short chronology.

It is likely that the early peopling of Europe occurred in the Mediterranean zone and only later extended northward. None of the oldest sites in northern Europe dates to a period of glacial expansion. These include Mauer, Miesenheim I, Kärlich, and Bilzingsleben in Germany; Cagny-l'Epinette in France; and Boxgrove, Hoxne, and Clacton-on-Sea in England. Three have yielded human remains. For almost a century, a human mandible discovered in a sand quarry at Mauer, near Heidelberg, in 1907 was said to be the oldest European human fossil. Study of the associated fauna indicates an age slightly before 500,000 B.P. A tibia found at Boxgrove in 1994 was declared, after a resounding announcement in the press and publication in the prestigious journal *Nature*, to be between 485,000 and 515,000 years old. It was only a matter of weeks, however, before the excavation of stratum TD6 at Atapuerca took the title away from Boxgrove and Mauer. The fragmentary remains found at Bilzingsleben are Middle Pleistocene in date, but more recent than Mauer and Boxgrove.

In the south the Middle Pleistocene fossil record is richer. An impressive series of hominid fossils from the cave of Tautavel, near the eastern end of the Pyrénées, is dated about 400,000 B.P. The

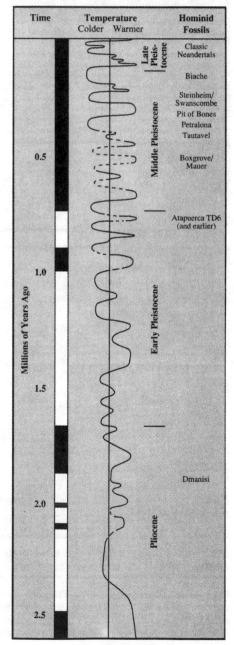

Periodic reversals of the earth's magnetic poles (indicated by black-and-white bar), recorded in cooling volcanic rocks, provide chronological markers.

Fossil pollen from sites reflects temperature changes over the millennia. Oxygen isotopes in marine shells reflect the advance and retreat of ice caps and confirm the pollen data.

Evidence from Dmanisi, Georgia, shows that humans were at the gates of Europe by ca. 1.8 million years ago. Artifacts and human bones from Atapuerca, Spain, indicate hominids reached Europe nearly one million years ago or earlier, but their remains are too fragmentary to be securely assigned to a species, as are those from Boxgrove and Mauer, dated ca. 500,000 B.P. By 400,000 years ago Europe was inhabited by archaic Homo sapiens *whose remains have been found at Tautavel and elsewhere. As intense glacial periods isolated them from populations in Africa and Asia, these European populations evolved into classic Neandertals through pre-Neandertal forms such as Steinheim and Swanscombe, ca. 250,000 years ago.*

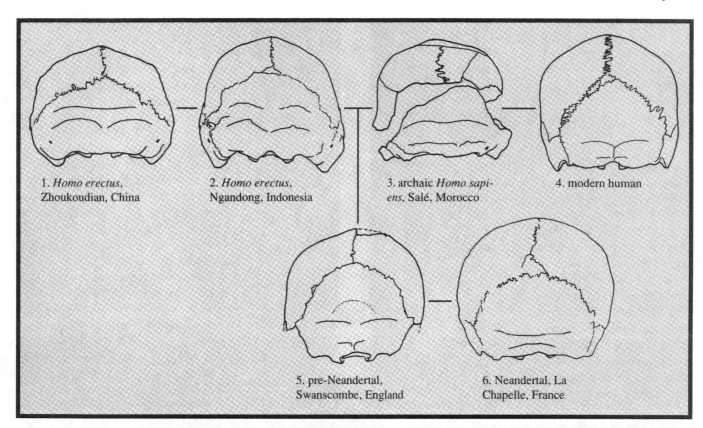

1. *Homo erectus*, Zhoukoudian, China

2. *Homo erectus*, Ngandong, Indonesia

3. archaic *Homo sapiens*, Salé, Morocco

4. modern human

5. pre-Neandertal, Swanscombe, England

6. Neandertal, La Chapelle, France

Evolution of the rear of the human skull shows progression from low pentagonal shape in Homo erectus *(1 and 2), through archaic* Homo sapiens *(3), to high pentagonal shape of modern humans (4). European pre-Neandertal shows moderately high pentagonal skull (5) from which rounded skull of classic Neandertals developed (6).*

Tautavel remains, unearthed by Henri de Lumley of the Institut de Paleontologie Humaine in Paris and his colleagues, includes a spectacular human face, an isolated parietal, two mandibles, a pelvic bone, many teeth, and fragments of post-cranial bones. Fragmentary human remains were also found at Vertésszöllös in Hungary.

What do these finds tell us about the peopling of Europe before the Neandertals? Was the older hominid, *Homo erectus*, present in Europe? Milford Wolpoff of the University of Michigan has claimed that the Mauer mandible, robust and chinless, is a European *Homo erectus*, comparable to the African and Asian representatives of this species. In terms of chronology this is not impossible, as 500,000 B.P. belongs to the transitional period between *Homo erectus* and archaic *Homo sapiens*. But there are no clear anatomical differences between the mandibles of the two species, so the identification of Mauer is very difficult. The development of clear Neandertal features begins only with later specimens.

The Tautavel-Vertésszöllös-Bilzingsleben remains provide us with a clear portrait of a 400,000-year-old European. (A complete skull found at Petralona Cave in northern Greece shows many similarities to this group, but it is difficult to date accurately.) According to Christopher Stringer of the British Museum they have a rather advanced expansion of the cranial vault, indicating that we are already dealing with archaic *Homo sapiens* rather than *Homo erectus*. The affinity of this group to the later Neandertals is obvious in that the specimens already display some Neandertal facial characteristics. The surface below the eye sockets is flat or even convex and the cheekbones are obliquely oriented, giving the face a laterally receding pattern. The large nasal aperture of the Petralona skull also evokes the Neandertal morphology, as do some mandibular features in Tautavel. All this suggests that the period between 500,000 and 400,000 B.P. could well be the beginning of Neandertal history.

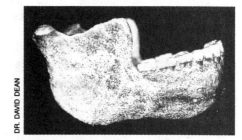

DR. DAVID DEAN

The Mauer mandible, ca. 500,000 years old, was the oldest-known European human fossil for nearly a century.

The interpretation of later specimens was based, in part, on the hypothesis of a European origin of modern humans. The decade following 1910 was the golden age of Neandertal discoveries in southwestern France. The sites of La Chapelle-aux-Saints, La Quina, La Ferrassie, and Le Moustier yielded nearly complete skeletons on which our knowledge of European Neandertal anatomy is largely based. The description of the La Chapelle-aux-Saints remains by French paleontologist Marcellin Boule deeply influenced Neandertal studies,

6. LATE HOMINID EVOLUTION

leading scholars to portray Neandertals as primitive, even ape-like, compared to the succeeding Cro-Magnons. As a result, they were excluded from the ancestry of modern humans, while anthropologists hypothesized connections between the Cro-Magnon inhabitants of France and remove living populations. Late prehistoric dwellers of the Perigord were seriously considered as ancestors of the Eskimos, and inhabitants of the Riviera as forefathers of African people. Obviously, scholars at the time could not imagine that the ancestors of modern Europeans were to be found in southwestern Asia and Africa.

The rejection of Neandertals and the psychological need to find European roots of modern humankind led to the search for a new character in the Palaeolithic landscape, a European forerunner of the handsome Cro-Magnon hunters and a much more honorable and acceptable ancestor than the brutish Neandertal. So-called "Pre-sapiens" finds were for the most part dubious, fragmentary, and poorly dated. The acceptance of the Piltdown forgery—a recent skull and the mandible of an immature orangutan discovered in Sussex between 1912 and 1915—resulted mainly from this quest for a European Pre-sapiens.

The Pre-sapiens lineage did include two well-preserved, authentic and reasonably dated hominid fossils. One, found at Swanscombe in England, was an occipital and two parietal bones; the other was a more complete skull from Steinheim, Germany. These specimens are usually dated to ca. 250,000 B.P., but may be older. Both display some Neandertal features. In the late 1970s I argued that one could observe an incipient supra-iniac fossa, the depression on the back of the skull that is characteristic of Neandertal lineage, on both specimens. This observation was rather well ac-

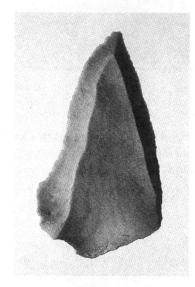

Artifacts found with both pre-Neandertals and Neandertals in Europe include, from top, Mousterian point, Levallois point, and Acheulean hand ax. The discovery of artifacts usually associated with Neandertals, such as the Mousterian point, in pre-Neandertal levels reflects the cultural continuity in the evolution of Neandertals from preceding archaic populations of Europe.

cepted for Swanscombe, but less so for Steinheim. The problem was that the general shape of the skulls was more similar to those of modern humans than those of Neandertals. Both display a pentagonal shape in rear view, somewhat reminiscent of a modern head and quite different from the rounded shape of the Neandertal skull. More detailed analysis of the evolution of the European hominids has finally resolved this contradiction. The pentagonal shape of the skull is a primitive trait shared by many archaic hominids, including *Homo erectus*, but still observed in modern humans. The ancestors of the Neandertals passed through a similar stage, but lost this pentagonal shape in the course of their evolution (in a sense modern humans are more primitive than Neandertals in this regard). The supra-iniac fossa suggests that Swanscombe and Steinheim are in the Neandertal ancestry, and the pentagonal occipital does not exclude them from the early stages of this lineage.

New discoveries have fully confirmed this interpretation. A skull found at Biache, in northern France, dating to ca. 175,000 B.P. provides a good intermediate between Swanscombe-Steinheim and later Neandertals in the shape of its braincase. On the other hand, the extraordinary specimens—hundreds of bones and three complete skulls—found in the Pit of Bones at Atapuerca by Juan Luis Arsuaga of Madrid's Complutense University and his colleagues fill the gap between the Tautavel-Vertésszöllös-Bilzingsleben group (400,000 B.P.) and Swanscombe-Steinheim (250,000 B.P.). The Pit of Bones skulls have some midfacial projection but show an earlier stage of development of the supra-iniac fossa than the Swanscombe skull. In the rear view the Pit of Bones skulls also retain a pentagonal outline.

The evidence indicates that Neandertal traits developed continuously over the course of at least 300,000 years. In the beginning these characteristics were rare, but they became more common, more consistent, and more pronounced over the millennia. Different anatomical areas evolved at different rates. First the face changed, developing an incipient midfacial projection and receding cheeks.

Then the occipital area developed a more Neandertal look. Finally the vault and the temporal bone took on Neandertal proportions. By 127,000 B.P. the Neandertal morphology was established.

This long biological continuity is echoed in the archaeological record. For nearly a century, archaeologists separated the Middle Palaeolithic (then thought to be limited to between 80,000 and 40,000 B.P.) from an earlier, more primitive Lower Palaeolithic. With the excavation and study of new sites, this boundary has become more and more difficult to draw. In contrast, the arrival of modern humans in Europe is reflected in the sharp division between the Middle Palaeolithic and the beginning of the Upper Palaeolithic (ca. 40,000 B.P.). The former, mainly represented by the Mousterian assemblages, are basically composed of side-scrapers, points, notches, and denticulates shaped from flint flakes. Bone objects are rare and very simple. The dwelling sites do not seem very structured. In the Upper Palaeolithic there is a large variety of flint tools, many shaped from elongated blades and bladelets. There are numerous bone and antler objects, especially spear points. It is the time of prehistoric art, of symbolism, of well-structured dwellings. It is quite likely the time of a new social organization.

Why and how did the Neandertals evolve? The evolution of living beings is mainly the result of two phenomena. One is natural selection, which, by favoring the transmission of characteristics, leads to an adaptation of the species to its environment. Others act at random. In the phenomenon called the founder effect, individual gene combinations of the few individuals that colonize a new territory strongly influence the genetic fate of their descendants. If a descendant population remains or becomes sparse, a haphazard phenomenon known as genetic drift can act on small groups. Relatively rare genes carried by some individuals can become rather frequent in their descendants.

To maintain its particular features, and eventually to become a new species, a diverging group needs to be isolated from its mother population. This segregation can happen in different ways, but it usually results from physical isolation created by a geographical barrier. European hominids evolving into Neandertals had to be somewhat isolated from nearby archaic *Homo sapiens* populations. Europe is a peninsula in the far west of Eurasia, and for hundreds of thousands of years the Mediterranean Sea seems to have been a nearly impassable barrier. The Strait of Gibraltar, in spite of its narrowness, was not a place of intense exchanges of people. The fossil hominids of northwestern Africa contemporary with the Neandertals followed a quite different evolutionary line. This does not mean that no human ever crossed the Strait during the second half of the Middle Pleistocene, just that such movements were not numerous and did not have a strong biological influence on the evolution of the resident population.

The gate of Europe was to the east, and the main path of colonization likely remained, for a long time, across the Bosphorus and along the northern Mediterranean coast. The first arrival of small groups of hominids may have created some founding effect. During temperate stages in the middle of the Middle Pleistocene, ca. 500,000 B.P., the descendants of these populations expanded to the north. However, pre-Neandertal bands may never have been able to survive in higher latitudes during the glacial maxima, when immense tundras developed hundreds of miles south of the European ice sheet.

Archaic *Homo sapiens* began evolving into Neandertals at a time of major climatic change. After 400,000 B.P. there was an increase in the severity of climatic swings. Intense glaciations occurred after this date. Between 300,000 and 172,000 B.P., two glacial advances were separated only by a cool interglacial stage during which an extensive ice sheet persisted in northwestern Europe. That classic Neandertals were adapted to a cold environment is demonstrated by their body proportions, which Erik Trinkaus of the University of New Mexico has shown were similar to those of modern Eskimos or Lapps. Such physical adaptations may have been influenced by climate, abetted by a limited ability to make warm clothing, build protective dwellings, and use fire. The principal impact of cold or cool periods, however, was the isolation of the European population. During these times the bulk of the pre-Neandertals lived in the south. The frozen grounds to the north, the Mediterranean Sea, the ice sheets covering the mountains (especially the Caucasus), and the dramatic extension of the Caspian Sea to the northwest severely reduced the already limited genetic exchanges between Europe and nearby areas.

A climatic and geographic explanation of the Neandertal evolution was proposed in the 1950s by paleoanthropologist F. Clark Howell, then of the University of Chicago, who emphasized the role of the glaciation in isolating them. But the fact that Neandertals existed during the last interglacial provoked criticism of Howell's idea by other scholars. Recent discoveries and a better interpretation of the Neandertal features have shown the lineage to be much older than was thought. While Howell's proposed timing was wrong, his linking of human evolution with environmental changes during the Pleistocene in Europe appears to have merit. The astounding findings of the last few years, at Atapuerca and other sites, will allow us to test this theory. Undoubtedly the earliest Europeans and the origins of the Neandertals will remain matters of heated scientific discussion for years to come.

Did Neandertals Lose an Evolutionary "Arms" Race?

Ann Gibbons

It has been called "The Mystery of Mount Carmel," a riddle embodied in the ancient humans who lived in caves around this Israeli mountain 40,000 to 100,000 years ago. They had a great deal in common: Not only where they in the same place at about the same time, but they hunted the same prey, used similar tools, and buried their dead in the same manner. What they didn't share, however, was their species: Some were short, stocky Neandertals, while others were tall, slender early modern humans. Only one group survived—and it wasn't the Neandertals. That, says University of New Mexico paleoanthropologist Erik Trinkaus, is the puzzle: "What is it about the early modern humans that made them ultimately evolutionarily successful?"

Now clues to that success are beginning to emerge, not from the stone tools, but from the bones of the toolmakers. By analyzing the shape and inner structure of fossil skeletons, Trinkaus and his colleagues can show how the bones were altered by behavior during an individual's lifetime. And Neandertal bones seem to reflect activity patterns that differed slightly from those of early moderns. At the Paleoanthropology Society meeting in April in New Orleans, Trinkaus and his colleagues reported that Neandertals' upper arm bones record more vigorous use, perhaps because they were less efficient at processing food than were early modern humans. And their hip bones indicate that Neandertals were more active as

children, possibly because they had to follow along with the adults as the group hunted and foraged; early modern society, in contrast, might have been organized differently, allowing youngsters to stay safely in camp with baby-sitters. Both differences, says Trinkaus, might have given the evolutionary edge to modern humans.

The findings stand out amid the current ambiguity of the archaeological record in the Near East, says archaeologist John Speth of the University of Michigan: "This work is very worthwhile. There's an awful lot of behavior that is reflected in skeletal morphology. . . . We need more attempts like this to link morphology and anatomy with behavior." But, like many researchers, he also cautions that the leap from activity patterns to changes in group social structure is a stretch and needs further support.

It isn't clear whether the two groups were using the sides around Mount Car-

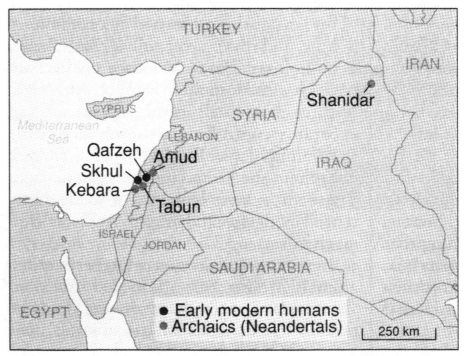

SOURCE: O. BAR-YOSEF

Middle East muddle. Neandertals and early modern humans lived in the same area and used similar tools—yet only one group survived.

mel at precisely the same time or alternately. But they left behind similar tool kits, which date to 80,000 and perhaps as early as 150,000 years ago for modern humans in caves at Qafzeh and Skhul, and 40,000 to 75,000 years ago (and perhaps more than 100,000 years ago) for Neandertals in caves at Tabun, Amud, Kebara, and, in Iraq, Shanidar. "Right now, there is nothing clear in the archaeology that indicates these humans differ in terms of behavior," says Harvard University archaeologist Ofer Bar-Yosef.

That's why Trinkaus turned to the bones. "Some bones are developmentally plastic," he explains. This means, says Johns Hopkins University biological anthropologist Chris Ruff, who collaborates with Trinkaus, that "if you don't use it, you lose it. If you do use it, you gain more bone." In the racquet arms of professional tennis players, for example, the upper arm bone, or humerus, is 60% thicker than in the other arm. For those of us who don't compete at Wimbledon, the asymmetry between arms is closer to 5%.

The best gauge of use, Trinkaus and Ruff have found, is a bone's internal architecture. "If you think of the shaft of a bone as a structural beam, like a hollow tube, you can measure the amount and distribution of bone in a cross section," says Trinkaus, and that indicates how much weight it bore and what kind of torsional or twisting forces it was subjected to on a regular basis. So the two scientists take x-rays—and computerized tomography scans, when possible—of fossil bones. They use software designed for orthopedic labs to calculate the area of the bone within each cross section. After factoring in the body shape and weight of the bone's owner—short, stocky Neandertals needed stronger leg bones to support their greater weight—Ruff can use mechanical engineering equations to figure out how much additional weight or force the bone could have sustained.

When Ruff and Trinkaus did this for major arm or leg bones from 17 Neandertals and early modern humans from the Middle East (and then added European specimens, for a total of 30), they were surprised. "I used to think there was a major difference in the strength of their legs," says Trinkaus, implying that Neandertals were ranging further afield in their foraging expeditions—a notion that had made its way into anthropology texts. "But if you take into account their body proportions, those differences in robusticity disappear." Other work, including recent experimental studies of animals by Daniel Lieberman of Rutgers University in New Jersey, supports this. He found that different-sized animals, when exercised equally on a treadmill, laid down proportionally similar amounts of new bone.

But if the foraging theory lost its legs, the bones did point to another area of divergence: the arms. "There was a major and highly significant difference in the strength of the upper arm," says Trinkaus. Working with Duke University paleoanthropologist Steven Churchill, he found that Neandertal humeri had much more bone than did moderns.

The implication, Churchill says, is that because the tool kits of the groups were similar, the moderns were probably more efficient at picking food that was easier to process, for example: "Reduced robusticity in the context of the same technology usually means they're working smarter."

The Neandertals may have pursued a wide range of foods, which required more work to consume, Churchill says. Lieberman has proposed that Neandertals relied more heavily on hunting for their food than on foraging because they may have used the environment in a different way—perhaps at different seasons or when it was colder than when the moderns were there. Or Neandertals may have been less efficient at using their tools to butcher or prepare their food: Work by Trinkaus and others has shown that their front teeth have more wear than do moderns' teeth, as if they needed to use them as vices to hold things.

This picture of inefficient Neandertals, says Trinkaus, fits with evidence from their hip joints. In studies published in 1993, Trinkaus examined a knob at the top of the femur or thigh bone, which fits into the hip socket. At birth this knob, known as the femoral neck, is vertical like the shaft of the femur. But studies of modern humans have demonstrated that the more active the child, the more the femoral neck bends inward and downward as the youngster grows. And Trinkaus has found that the femoral necks of the Neandertals are more sharply bent than those of the early moderns, which indicates that the Neandertals were more active as children. Perhaps, he says, the Neandertals moved as a group in pursuit of their food, while the moderns could split up, leaving their young with baby-sitters while they hunted and bringing back food to them at the end of the day.

If Trinkaus is right, it would mean that a different way of life was emerging among the early moderns in the Skhul and Qafzeh caves. Better food procurement and a more elaborate social system that protected children to a greater degree could have produced changes in population health and survival that gave these people the means to evolutionary success. "It must be something else beyond the tools," Trinkaus says.

But while many of Trinkaus's colleagues do embrace his findings about arms and efficiency, some worry that going from hip structure to social structure is venturing too far. Bar-Yosef, for instance, grumbles that "I have my reservations about simplistic interpretations of the functional morphology. I see no evidence for different social structures."

Still, this summer Bar-Yosef and his colleague Mary Stiner of the University of Arizona plan to test Trinkaus's ideas about differences in behavior near Mount Carmel by looking for differences in the type of small animals that the two groups hunted or the cut marks that show how they butchered them: "We will try to test the hypothesis that they were using their arms differently. Minor differences can make a major difference in determining what population is more successful than another."

THE DAWN OF CREATIVITY

New discoveries show how humans created art, tools and richly thriving communities eons ago

WILLIAM F. ALLMAN

Peering out from the dim past, the cave images still have the power to haunt: Reindeer ford a stream, their necks stretched taut; a massive bison glares, its flanks shimmering; an antler-bedecked half-man, half-beast stares out in expectation of some long lost ritual.

These ancient examples of expression lie at the center of human evolution's deepest mystery: Who were the first "creators"—the toolmakers, craft workers and artisans—who propelled the human species from a run-of-the-mill primate to the tool-wielding, art-making, cultural beings that exist today? At the heart of it lies an evolutionary disconnect. The modern, biological form of Homo sapiens appeared in the fossil record about 100,000 to 130,000 years ago. These were creatures who looked pretty much like us. But as most textbooks still have it, signs of truly human behavior—spirituality, artwork, sophisticated use of the environment and the dense network of family and friends that make up society—did not appear until around 40,000 years ago, when a "cultural explosion" occurred in Europe rather than Africa, where many exerts believe Homo sapiens first arose.

Now several recent archaeological finds hint that the dawning of human creativity may have occurred earlier and was far more widespread than previously thought. In a paper soon to be published in the *Journal of Human Evolution,* Sally McBrearty of the University of Connecticut reports the finding of skillfully crafted stone blades in Africa that date back hundreds of thousands of years before blade tools appear in Europe. In another paper published

last year, researchers announced the finding of delicate bone harpoons in Africa that precede the famed cave paintings of Ice Age Europe by more than 40,000 years. Scientists dating newly discovered cave paintings in France report the artwork is 30,000 years old—nearly twice the age of similarly rendered paintings in Lascaux, France. And the recent find of a bizarre structure apparently made by Neanderthals deep within a cavern indicates that "human behavior" might not have been confined to our species.

The new finds suggest that the traditional scenario that says human creativity suddenly burst forth in Europe is wrong. "The beginning of what we call human behavior was far older than 40,000 years ago," says Alison Brooks of George Washington University, who with National Science Foundation anthropologist John Yellen found the harpoons in Zaire. "And it began in Africa, not Europe." More important, the new research suggests that the way our ancestors' minds were most creative was not in their work on tools and cave walls but in the manner in which they created the fabric of society itself.

NEW CANVASES

Dating back tens of thousands of years before the Ice Age paintings in Europe, Brooks and Yellen's ancient harpoons testify to a crucial leap in human creative thinking: looking beyond stone and wood for raw material for tools. To ancient humans, bone, antler and ivory were Paleolithic plastic: tough, flexible,

durable and capable of being shaped into everything from deadly spear points to sewing needles to flutes. To make the harpoons, the ancient humans ground the tip of the bone into a point, then cut into the side with a sharp stone to create a triangular "tooth." Circular grooves at the tail end helped fasten the tool to a cord or pole. While not as sophisticated as later harpoon designs, these weapons apparently did the trick: The archaeological site is littered with the remains of a species of giant catfish whose descendants still swim in nearby waters.

The new site suggests that the humans who made the harpoons understood the lives of their prey—and were capable of planning ahead to take advantage of it. The ancient humans, it seems, knew that the catfish spawn at only certain times in the year, during which they are plentiful and easy prey, and they timed their visits to the area accordingly. Brooks notes that there is similar evidence of advanced planning at another site found near a large water hole. The ancient humans apparently staked out the site during the dry season, when they knew that other water holes nearby would be empty, and ambushed animals as they came to drink.

THE NEANDERTHAL ENIGMA

This creative approach to exploiting the environment may have been a crucial divider between ancient humans and their evolutionary cousins, the Neanderthals. These powerfully built, large-brained

From *U.S. News & World Report,* May 20, 1996, pp. 53-54, 56-58. © 1996 by U.S. News & World Report. Reprinted by permission.

creatures arose in Europe some 300,000 years ago and disappeared about 30,000 years ago. Scientists long thought Neanderthals were direct ancestors of modern humans, but fossil finds of ancient humans that predate some Neanderthal fossils reveal the two species co-existed for tens of thousands of years. A recent reconstruction of fragments of a 20,000-year-old Neanderthal skull retains the classic features of the species, suggesting they did not interbreed with humans.

Neanderthals have long had an image of being dim-witted, hulking brutes, but recent findings indicate they were capable of sophisticated behaviors such as making complex stone and bone tools, using fire, burying their dead and possibly even speaking. Scientists recently discovered, for instance, a four-walled structure built out of rock by Neanderthals deep within a cave. The ability to create such a structure, which may have had symbolic importance, in the pitch-black cave reveals the Neanderthals had mastered the ability to use torches and to coordinate their activities, perhaps verbally.

For all their intellectual abilities, however, Neanderthals appear to have been subtly different from humans in how they negotiated the world. For instance, scientists have long been puzzled by the fact that in Europe tools used by humans are far more complex than those used by Neanderthals. But in the Near East, the tools of one species are virtually indistinguishable from the other's. New research reveals that *how* these tools were used may have made the difference. Looking at the growth pattern of the teeth of prey found at the sites, which indicates whether the animals were killed during a particular season, Dan Lieberman of Rutgers University and John Shea of the State University of New York at Stony Brook conclude that the Neanderthals stayed at the sites for long periods and at many different times of the year. Humans, on the other hand, were more deliberate and used a site as one of several areas where they took refuge as they followed the weather or their food sources on their migrations. Occupying a single site for a long time inevitably leads to a depletion of the food, argue Lieberman and Shea,

which means Neanderthals had to work harder and harder for each meal.

This way of life seems to have led to severe hardship for Neanderthals. Erik Trinkaus of the University of New Mexico examined the teeth of Neanderthals and humans for telltale defects in the growth of the enamel, which indicate bouts of starvation. He found that more than 70 percent of Neanderthal fossils studied showed at least one defect, whereas "teeth of the ancient humans were clean as a whistle." Equally telling is that defects in Neanderthal teeth dramatically increased after childhood, suggesting that Neanderthals' lives became harder after they had been weaned and had to get food on their own.

The Neanderthals' bones, too, hint of a life that emphasized brawn as much as brain. Trinkaus found that the bones of the Neanderthals, which are thicker and heavier, also are riddled with multiple minor fractures. And these injuries are reflected in the Neanderthals' tool kit: The stone points they made are best suited to being held and thrust, rather than thrown, says Trinkaus. That presumably exposed them to nasty kicks from their prey—and those broken bones.

SOCIAL GLUE

The biggest difference between humans and Neanderthals—one that may have made all the difference in their creative cultures—was how members of each species interacted among themselves. Humans were using long-distance trading networks for the exchange of quality stone and other goods in Africa at least 100,000 years ago, says Brooks. Similar trade networks might have existed among humans in Ice Age Europe, where shells, for instance, are found at sites hundreds of miles from the sea. As people began to live in larger groups, supported by cooperative group hunting, the need for expressing group identity and individuality intensified, argue Steve Kuhn and Mary Stiner of the University of Arizona.

This link between human creativity and sociability is evident in one of the earliest examples of ancient art: bead-

work. As Randall White of New York University points out, the Lascaux cave paintings "are only the midway point of human art history." The dawn of art, he says, began tens of thousands of years earlier and used the human body as a canvas. Ivory beads were sewn into clothing and pierced carnivore teeth were used in belts and headbands.

It probably took about an hour, White says, to make a bead. Such a time-consuming process would never be undertaken unless personal adornment was a vital part of human existence. "We have this image of art being the result of people having lots of free time," says White. "But that's totally contrary to what we see. For these people, art was a necessity." Such an idea should hardly be surprising in a modern culture where wedding bands, high-priced watches and T-shirt slogans all shape how people perceive one another's status and world view.

THE ICE AGE CINEPLEX

Even the magnificent examples of Ice Age cave art—bolstered by the recent discovery in France of two caverns filled with a menagerie of animal paintings—appear to have played a crucial role in the workings of society. Once thought to represent "hunting magic" designed to help get game, the Ice Age galleries may have been part of elaborate ceremonies that perhaps rivaled the best modern-day multimedia displays.

Flutes made of bird bone that play notes in a scale similar to those of today suggest music may have accompanied viewing of the paintings. In one experiment, researchers walked through three ancient caves while whistling through several octaves and mapping where the sounds resonated off the walls best. They found that those places in the caves with the best acoustics nearly always had art nearby, whereas places where sound was dampened typically did not have art. In another experiment, researchers found that near the front of the famed Lascaux cave, where the cave art is dominated by horses, bison and other hoofed animals, a clapping noise gets echoed back and forth among the walls, producing a sound not unlike a stampede. Near the rear of the cave, however, where the images are dominated by panthers and other stealthy creatures, the walls reflect sound in such a way that it is muted.

Nor was the ancient flowering of art confined to Ice Age Europe. Cave paintings also occur in Africa in the same era, and a recent study by Australian archaeologist Rhys Jones suggests that some Australian rock paintings date back 60,000 years. Not only would this be the oldest artwork known, its presence in Australia implies humans could build sophisticated boats 60,000 years ago.

ANCIENT CATHEDRALS

Some of the most mysterious markings in ancient artwork—the strange circles, dots and chevrons on the animals or by themselves—may be the result of the painter being in a hallucinatory trance. South African archaeologist David Lewis-Williams studied the rock art of Africa's !Kung San people, who continue to create rock paintings today. He found the painter in many cases is a shaman using drawing as part of a trancelike state to ward off spirits. The markings reflect the spots and shimmerings that appear to him during the trance.

More evidence that the caves were used in rituals comes from the paintings themselves. In one Ice Age work there are two humanlike figures who seem to be wearing headdresses of antlers. One stares cross-eyed toward the viewer; the other might be playing a musical instrument. In another cave, some bison seem to have been laboriously painted on a vaulted rock ceiling known as the "Sanctuary." A recent chemical analysis found that although the animals were rendered in a very similar style, the paintings were done hundreds of years apart—in the same way that a modern painting of a cross might reflect art symbolism from the Middle Ages.

Evidence our ancestors were moved by spiritual concerns goes back to the oldest human skeletons: At one Israeli cave site dating back nearly 100,000 years, a man is buried with an antler placed in his hands—an ancient precursor, perhaps, to the tomb offerings of King Tut or rosary beads in modern-day burials. At another ancient site, a woman is buried with her legs deliberately pulled up beneath her and a small child lying at her feet, a pairing that suggests a spiritual concern about being joined after death.

All members in a society appear to have participated in the rituals revealed in Ice Age art. Tiny hand prints of children are found as much as a mile deep within some Ice Age caves, notes NYU's White, and in one cave a side chamber with only a 4-foot-high ceiling is covered with the footprints of children.

CULTURAL EXPLOSION

The social setting implied by the cave art may be what made the so-called cultural explosion 40,000 years ago possible in the first place. Throughout human history, a few people may have worked a piece of bone or done an engraving on a rock. But without that artifact's playing a role in maintaining the function of a larger group, suggests White, it would have been discarded.

Sophisticated toolmaking and culture became prevalent only when human society grew to the point that such practices became important for survival—and social networks spread each new innovation like wildfire. A similar sea change in human existence occurred some 8,000 years ago, for instance, when humans around the globe abandoned their hunting-and-gathering way of life and settled into farming communities. In that instance, the ancients knew about domesticating plants long before they began to farm, imply recent studies, and only took up agriculture in response to changing climate, a growing population and a dwindling food supply.

The newly discovered examples of sophisticated tools indicate that ancient humans may have always had the capacity for creative thinking, even if they didn't always express it. The University of Connecticut's McBrearty has found stone blades that appear to date back more than 250,000 years, yet examples of more-rudimentary tool-making exist elsewhere up until some 40,000 years ago—suggesting that there wasn't an important social network to perpetuate the tool design or pass it along to others.

Clearly, it was profoundly rich relations that inspired and reinforced creativity among ancient humans. The artists who created the cave images were people of spirituality and grace; they loved painting, music and the beauty as well as the function of their technology. Mostly, they were people whose creativity connected them with the members of their community—those alongside them in the cave or thousands of years in the future.

OLD MASTERS

Brilliant paintings brightened the caves of our early ancestors. But were the artists picturing their mythic beliefs or simply showing what they ate for dinner?

Pat Shipman

Pat Shipman wrote about killer bamboo in [Discover,] *February [1990].*

Fifty years ago, in a green valley of the Dordogne region of southwest France, a group of teenage boys made the first claustrophobic descent into the labyrinthine caverns of Lascaux. When they reached the main chamber and held their lamps aloft, the sight that flickered into view astonished them. There were animals everywhere. A frieze of wild horses, with chunky bodies and fuzzy, crew-cut manes, galloped across the domed walls and ceiling past the massive figure of a white bull-like creature (the extinct aurochs). Running helter-skelter in the opposing direction were three little stags with delicately drawn antlers. They were followed by more bulls, cows, and calves rounding the corner of the chamber.

Thousands have since admired these paintings in Lascaux's Hall of the Bulls, probably the most magnificent example of Ice Age art known to us today. In fact, by 1963 so many tourists wanted to view the cave that officials were forced to close Lascaux to the general public; the paintings were being threatened as the huge influx of visitors warmed the air in the cave and brought in corrosive algae and pollen. (Fortunately, a nearby exhibit called Lascaux II faithfully reproduces the

The image of Paleolithic humans moving by flickering lamps, singing, chanting, and drawing their knowledge of their world is hard to resist.

paintings.) After I first saw these powerful images, they haunted me for several months. I had looked at photographs of Lascaux in books, of course, so I knew that the paintings were beautiful; but what I didn't know was that they would reach across 17,000 years to grab my soul.

Lascaux is not an Ice Age anomaly. Other animal paintings, many exquisitely crafted, adorn hundreds of caves throughout the Dordogne and the French Pyrenees and the region known as Cantabria on the northern coast of Spain. All these images were created by the people we commonly call Cro-Magnons, who lived during the Upper Paleolithic Period, between 10,000 and 30,000 years ago, when Europe lay in the harsh grip of the Ice Age.

What did this wonderful art mean, and what does it tell us about the prehistoric humans who created it? These questions have been asked since the turn of the century, when cave paintings in Spain were first definitively attributed to Paleolithic humans. Until recently the dominant answers were based on rather sweeping symbolic interpretations—attempts, as it were, to read the Paleolithic psyche.

These days some anthropologists are adopting a more literal-minded approach. They are not trying to empathize with the artists' collective soul—a perilous exercise in imagination, considering how remote Cro-Magnon life must have been from ours. Rather, armed with the tools of the late twentieth century—statistics, maps, computer analyses of the art's distribution patterns—the researchers are trying to make sense of the paintings by piecing together their cultural context.

This is a far cry from earlier attempts at interpretation. At the beginning of the century, with little more to go on than his intuition, the French amateur archeologist Abbé Henri Breuil suggested that the pictures were a form of hunting magic. Painting animals, in other words, was a magical way of capturing them, in the hope that it would make the beasts vulnerable to hunters. Abstract symbols painted on the walls were interpreted as hunting paraphernalia. Straight lines drawn to the animals' sides represented spears, and V and O shapes on their hides were seen as wounds. Rectangular grids, some observers thought, might have been fences or animal traps.

In the 1960s this view was brushed aside for a much more complex, somewhat Freudian approach that was brought into fashion by anthropologist André Leroi-Gourhan. He saw the cave paintings as a series of mythograms, or symbolic depictions, of how Paleolithic people viewed their world—a world split between things male and female. Femaleness was represented by animals such as the bison and aurochs (which were sometimes juxtaposed with human female figures in the paintings), and maleness was embodied by such animals as the horse and ibex (which, when accompanied by human figures, were shown only with males). Female images, Leroi-Gourhan suggested, were clustered in the central parts of the dark, womblike caves, while male images either consorted with the female ones or encircled them in the more peripheral areas.

Leroi-Gourhan also ascribed sex to the geometric designs on the cave walls. Thin shapes such as straight lines, which often make up barbed, arrowlike structures, were seen as male (phallic) signs. Full shapes such as ovals, V shapes, triangles, and rectangles were female (vulval) symbols. Thus, an arrow stuck into a V-shaped wound on an animal's hide was a male symbol entering a complementary female one.

Leroi-Gourhan was the first to look for structure in the paintings system-

atically, and his work reinforced the notion that these cave paintings had underlying designs and were not simply idle graffiti or random doodles. Still, some scholars considered his *"perspective sexomaniaque"* rather farfetched; eventually even he played down some of the sexual interpretations. However, a far bigger problem with both his theory and Breuil's was their sheer monolithic scope: a single explanation was assumed to account for 20,000 years of paintings produced by quite widely scattered groups of people.

Yet it is at least as likely that the paintings carried a number of different messages. The images' meaning may have varied depending on who painted them and where. Increasingly, therefore, researchers have tried to relate the content of the paintings to their context—their distribution within a particular cave, the cave's location within a particular region, and the presence of other nearby dwelling sites, tools, and animal bones in the area.

Anthropologist Patricia Rice and sociologist Ann Paterson, both from West Virginia University, made good use of this principle in their study of a single river valley in the Dordogne

---

. . . animals may have been depicted more or less frequently depending on how aggressive they were to humans.

---

region, an area that yielded 90 different caves containing 1,955 animal portrayals and 151 dwelling sites with animal bone deposits. They wanted to find out whether the number of times an animal was painted simply reflected how common it was or whether it revealed further information about the animals or the human artists.

By comparing the bone counts of the various animals—horses, reindeer, red deer, ibex, mammoths, bison, and au-

rochs—Rice and Paterson were able to score the animals according to their abundance. When they related this number to the number of times a species turned up in the art, they found an interesting relationship: Pictures of the smaller animals, such as deer, were proportionate to their bone counts. But the bigger species, such as horses and bison, were portrayed more often than you'd expect from the faunal remains. In fact, it turned out that to predict how often an animal would appear, you had to factor in not just its relative abundance but its weight as well.

A commonsense explanation of this finding was that an animal was depicted according to its usefulness as food, with the larger, meatier animals shown more often. This "grocery store" explanation of the art worked well, except for the ibex, which was portrayed as often as the red deer yet was only half its size and, according to the bone counts, not as numerous. The discrepancy led Rice and Paterson to explore the hypothesis that animals may have also been depicted more or less frequently depending on how aggressive they were to humans.

To test this idea, the researchers asked wildlife-management specialists to score the animals according to a "danger index." The feisty ibex, like the big animals, was rated as highly aggressive; and like these other dangerous animals, it was painted more often than just the numbers of its remains would suggest. Milder-tempered red deer and reindeer, on the other hand, were painted only about as often as you'd expect from their bones. Rice and Paterson concluded that the local artists may have portrayed the animals for both "grocery store" and "danger index" reasons. Maybe such art was used to impress important information on the minds of young hunters—drawing attention to the animals that were the most worthwhile to kill, yet balancing the rewards of dinner with the risks of attacking a fearsome animal.

One thing is certain: Paleolithic artists knew their animals well. Subtle physical details, characteristic poses, even seasonal changes in coat color or texture, were deftly observed. At Lascaux bison are pictured shedding their dark winter pelts. Five stags are shown swimming across a river, heads held high above the swirling tide. A stallion is depicted with its lip curled back, responding to a mare in heat. The reddish coats, stiff black manes, short legs, and potbellies of the Lascaux horses are so well recorded that they look unmistakably like the modern Przhevalsky's horses from Mongolia.

New findings at Solutré, in east-central France, the most famous horse-hunting site from the Upper Paleolithic, show how intimate knowledge of the animal's habits was used to the early hunter's advantage. The study, by archeologist Sandra Olsen of the Virginia Museum of Natural History, set out to reexamine how vast numbers of horses—from tens to hundreds of thousands, according to fossil records—came to be killed in the same, isolated spot. The archeological deposits at Solutré are 27 feet thick, span 20,000 years, and provide a record of stone tools and artifacts as well as faunal remains.

The traditional interpretation of this site was lots of fun but unlikely. The Roche de Solutré is one of several high limestone ridges running east-west from the Saône River to the Massif Central plateau; narrow valleys run between the ridges. When the piles of bones were discovered, in 1866, it was proposed that the site was a "horse jump" similar to the buffalo jumps in the American West, where whole herds of bison were driven off cliffs to their death. Several nineteenth-century paintings depict Cro-Magnon hunters driving a massive herd of wild horses up and off the steep rock of Solutré. But Olsen's bone analysis has shown that the horse jump scenario is almost certainly wrong.

For one thing, the horse bones are not at the foot of the steep western end of Solutré, but in a natural cul-de-sac along the southern face of the ridge. For the horse jump hypothesis to work, one of two fairly incredible events had to occur. Either the hunters drove the horses off the western end and then dragged all the carcasses around to the southern face to butcher

them or the hunters herded the animals up the steep slope and then forced them to veer off the southern side of the ridge. But behavioral studies show that, unlike bison, wild horses travel not in herds but in small, independent bands. So it would have been extremely difficult for our Cro-Magnons on foot to force lots of horses together and persuade them to jump en masse.

Instead the horse behavior studies suggested to Olsen a new hypothesis. Wild horses commonly winter in the lowlands and summer in the highlands. This migration pattern preserves their forage and lets them avoid the lowland's biting flies and heat in summer and the highland's cold and snow in winter. The Solutré horses, then, would likely have wintered in the Saône's floodplain to the east and summered in the mountains to the west, migrating through the valleys between the ridges. The kill site at Solutré, Olsen notes, lies in the widest of these valleys, the one offering the easiest passage to the horses. What's more, from the hunters' point of view, the valley has a convenient cul-de-sac running off to one side. The hunters, she proposes, used a drive lane of brush, twigs, and rocks to divert the horses from their migratory path into the cul-de-sac and then speared the animals to death. In deed, spear points found at the site support this scenario.

The bottom line in all this is that Olsen's detailed studies of this prehistoric hunting site confirm what the cave art implies: these early humans used their understanding of animals' habits, mating, and migration patterns to come up with extremely successful hunting strategies. Obviously this knowledge must have been vital to the survival of the group and essential to hand down to successive generations. Perhaps the animal friezes in the cave were used as a mnemonic device or as a visual teaching aid in rites of initiation—a means for people to recall or rehearse epic hunts, preserve information, and school their young. The emotional power of the art certainly suggests that this information was cru-

cial to their lives and could not be forgotten.

The transmission of this knowledge may well have been assisted by more than illustration. French researchers Iégor Reznikoff and Michel Dauvois have recently shown that cave art may well have been used in rituals accompanied by songs or chants. The two studied the acoustic resonances of three caves in the French Pyrenees by singing and whistling through almost five octaves as they walked slowly through each cave. At certain points the caves resonated in response to a particular note, and these points were carefully mapped.

For all the finely observed animal pictures, we catch only the sketchiest glimpses of humans, in the form of stick figures or stylized line drawings.

When Reznikoff and Dauvois compared their acoustic map with a map of the cave paintings, they found an astonishing relationship. The best resonance points were all well marked with images, while those with poor acoustics had very few pictures. Even if a resonance point offered little room for a full painting, it was marked in some way—by a set of red dots, for example. It remains to be seen if this intriguing correlation holds true for other caves. In the meantime, the image of Paleolithic humans moving by flickering lamps, singing, chanting, and drawing their knowledge of their world indelibly into their memories is so appealing that I find it hard to resist.

Yet the humans in this mental image of mine are shadowy, strangely elusive people. For all the finely observed animal pictures, we catch only the sketchiest glimpses of humans, in the form of stick figures or stylized line drawings. Still, when Rice and Paterson turned to study these human images in French and Spanish caves, a few striking patterns did emerge. Of the 67 images studied, 52 were male and a mere 15 were female. Only men were depicted as engaged in active behavior, a category that included walking, running, carrying spears, being speared, or falling. Females were a picture of passivity; they stood, sat, or lay prone. Most women were shown in close proximity to another human figure or group of figures, which were always other women. Seldom were men featured in social groups; they were much more likely to be shown facing off with an animal.

These images offer tantalizing clues to Paleolithic life. They suggest a society where males and females led very separate lives. (Male-female couples do not figure at all in Paleolithic art, for all the sexual obsessions of earlier researchers.) Males carried out the only physical activities—or at least the only ones deemed worthy of recording. Their chief preoccupation was hunting, and from all appearances, what counted most was the moment of truth between man and his prey. What women did in Paleolithic society (other than bear children and gather food) remains more obscure. But whatever they did, they mostly did it in the company of other women, which would seem to imply that social interaction, cooperation, and oral communication played an important role in female lives.

If we could learn the sex of the artists, perhaps interpreting the social significance of the art would be easier. Were women's lives so mysterious because the artists were male and chauvinistically showed only men's activities in their paintings? Or perhaps the artists were all female. Is their passive group activity the recording and encoding of the information vital to the group's survival in paintings and carvings? Did they spend their time with other women, learning the songs and chants and the artistic techniques that transmitted and preserved their knowledge? The art that brightened the caves of the Ice Age endures. But the artists who might shed light on its meaning remain as enigmatic as ever.

The Dating Game

By tracking changes in ancient atoms, archeologists are establishing the astonishing antiquity of modern humanity.

James Shreeve

James Shreeve wrote fiction before turning to science writing. He is the coauthor (with anthropologist Donald Johanson) of Lucy's Child: The Discovery of a Human Ancestor *and the author of* Nature: The Other Earthlings.

Four years ago archeologists Alison Brooks and John Yellen discovered what might be the earliest traces of modern human culture in the world. The only trouble is, nobody believes them. Sometimes they can't quite believe it themselves.

Their discovery came on a sun-soaked hillside called Katanda, in a remote corner of Zaire near the Ugandan border. Thirty yards below, the Semliki River runs so clear and cool the submerged hippos look like giant lumps of jade. But in the excavation itself, the heat is enough to make anyone doubt his eyes.

Katanda is a long way from the plains of Ice Age Europe, which archeologists have long believed to be the setting for the first appearance of truly modern culture: the flourish of new tool technologies, art, and body ornamentation known as the Upper Paleolithic, which began about 40,000 years ago. For several years Brooks, an archeologist at George Washington University, had been pursuing the heretical hypothesis that humans in Africa had invented sophisticated technologies even earlier, while their European counterparts were still getting by with

the same sorts of tools they'd been using for hundreds of thousands of years. If conclusive evidence hadn't turned up, it was only because nobody had really bothered to look for it.

"In France alone there must be three hundred well-excavated sites dating from the period we call the Middle Paleolithic," Brooks says. "In Africa there are barely two dozen on the whole continent."

One of those two dozen is Katanda. On an afternoon in 1988 John Yellen—archeology program director at the National Science Foundation and Brooks's husband—was digging in a densely packed litter of giant catfish bones, river stones, and Middle Paleolithic stone tools. From the rubble he extricated a beautifully crafted, fossilized bone harpoon point. Eventually two more whole points and fragments of five others turned up, all of them elaborately barbed and polished. A few feet away, the scientists uncovered pieces of an equally well crafted daggerlike tool. In design and workmanship the harpoons were not unlike those at the very end of the Upper Paleolithic, some 14,000 years ago. But there was one important difference. Brooks and Yellen believe the deposits John was standing in were at least five times that old. To put this in perspective, imagine discovering a prototypical Pontiac in Leonardo da Vinci's attic.

"If the site is as old as we think it is," says Brooks, "it could clinch the

argument that modern humans evolved in Africa."

Ever since the discovery the couple have devoted themselves to chopping away at that stubborn little word *if.* In the face of the entrenched skepticism of their colleagues, it is an uphill task. But they do have some leverage. In those same four years since the first harpoon was found at Katanda, a breakthrough has revived the question of modern human origins. The breakthrough is not some new skeleton pulled out of the ground. Nor is it the highly publicized Eve hypothesis, put forth by geneticists, suggesting that all humans on Earth today share a common female ancestor who lived in Africa 200,000 years ago. The real advance, abiding quietly in the shadows while Eve draws the limelight, is simply a new way of telling time.

To be precise, it is a whole smorgasbord of new ways of telling time. Lately they have all converged on the same exhilarating, mortifying revelation: what little we thought we knew about the origins of our own species was hopelessly wrong. From Africa to the Middle East to Australia, the new dating methods are overturning conventional wisdom with insolent abandon, leaving the anthropological community dazed amid a rubble of collapsed certitudes. It is in this shell-shocked climate that Alison Brooks's Pontiac in Leonardo's attic might actually find a hearing.

"Ten years ago I would have said it was impossible for harpoons like these to be so old," says archeologist Michael Mehlman of the Smithsonian's National Museum of Natural History. "Now I'm reserving judgment. Anything can happen."

An archeologist with a freshly uncovered skull, stone tool, or bone Pontiac in hand can take two general approaches to determine its age. The first is called relative dating. Essentially the archeologist places the find in the context of the surrounding geological deposits. If the new discovery is found in a brown sediment lying beneath a yellowish layer of sand, then, all things being equal, it is older than the yellow sand layer or any other deposit higher up. The fossilized remains of extinct animals found near the object also provide a "biostratigraphic" record that can offer clues to a new find's relative age. (If a stone tool is found alongside an extinct species of horse, then it's a fair bet the tool was made while that kind of horse was still running around.) Sometimes the tools themselves can be used as a guide, if they match up in character and style with tools from other, better-known sites. Relative dating methods like these can tell you whether a find is older or younger than something else, but they cannot pin an age on the object in calendar years.

The most celebrated *absolute* method of telling archeological time, radiocarbon dating, came along in the 1940s. Plants take in carbon from the atmosphere to build tissues, and other organisms take in plants, so carbon ends up in everything from wood to woodchucks. Most carbon exists in the stable form of carbon 12. But some is made up of the unstable, radioactive form carbon 14. When an organism dies, it contains about the same ratio of carbon 12 to carbon 14 that exists in the atmosphere. After death the radioactive carbon 14 atoms begin to decay, changing into stable atoms of nitrogen. The amount of carbon 12, however, stays the same. Scientists can look at the amount of carbon 12 and—based on the ratio—deduce how much carbon 14 was originally present. Since the decay rate of carbon 14 is constant and steady (half of it disappears every 5,730 years), the difference between the

amount of carbon 14 originally in a charred bit of wood or bone and the amount present now can be used as a clock to determine the age of the object.

Conventional radiocarbon dates are extremely accurate up to about 40,000 years. This is far and away the best method to date a find—as long as it is younger than this cutoff point. (In older materials, the amount of carbon 14 still left undecayed is so small that even the slightest amount of contamination in the experimental process leads to highly inaccurate results.) Another dating technique, relying on the decay of radioactive potassium rather than carbon, is available to date volcanic deposits *older* than half a million years. When it was discovered in the late 1950s, radiopotassium dating threw open a window on the emergence of the first members of the human family—the australopithecines, like the famous Lucy, and her more advanced descendants, *Homo habilis* and *Homo erectus*. Until now, however, the period between half a million and 40,000 years—a stretch of time that just happens to embrace the origin of *Homo sapiens*—was practically unknowable by absolute dating techniques. It was as if a geochronological curtain were drawn across the mystery of our species' birth. Behind that curtain the hominid lineage underwent an astonishing metamorphosis, entering the dateless, dark centuries a somewhat precocious bipedal ape and emerging into the range of radiocarbon dating as the culturally resplendent, silver-tongued piece of work we call a modern human being.

Fifteen years ago there was some general agreement about how this change took place. First, what is thought of as an *anatomically modern human being*—with the rounded cranium, vertical forehead, and lightly built skeleton of people today—made its presence known in Europe about 35,000 years ago. Second, along with those first modern-looking people, popularly known as the Cro-Magnons, came the first signs of complex human

behavior, including tools made of bone and antler as well as of stone, and art, symbolism, social status, ethnic identity, and probably true human language too. Finally, in any one region there was no overlap in time between the appearance of modern humans and the disappearance of "archaic" humans such as the classic Neanderthals, supporting the idea that one group had evolved from the other.

"Thanks to the efforts of the new dating methods," says Fred Smith, an anthropologist at Northern Illinois University, "we now know that each of these ideas was wrong."

The technique doing the most damage to conventional wisdom is called thermoluminescence, TL for short. (Reader take heed: the terrain of geochronology is full of terms long enough to tie between two trees and trip over, so acronyms are a must.) Unlike radiocarbon dating, which works on organic matter, TL pulls time out of stone.

If you were to pick an ordinary rock up off the ground and try to describe its essential rockness, phrases like "frenetically animated" would probably not leap to mind. But in fact minerals are in a state of constant inner turmoil. Minute amounts of radioactive elements, both within the rock itself and in the surrounding soil and atmosphere, are constantly bombarding its atoms, knocking electrons out of their normal orbits. All this is perfectly normal rock behavior, and after gallivanting around for a hundredth of a second or two, most electrons dutifully return to their normal positions. A few, however, become trapped en route—physically captured within crystal impurities or electronic aberrations in the mineral structure itself. These tiny prisons hold on to their electrons until the mineral is heated, whereupon the traps spring open and the electrons return to their more stable position. As they escape, they release energy in the form of light—a photon for every homeward-bound electron.

Thermoluminescence was observed way back in 1663 by the great English physicist Robert Boyle. One night Boyle took a borrowed diamond to bed with him, for reasons that remain obscure. Resting the diamond "upon a warm part of my Naked Body," Boyle noticed that it soon emitted a warm glow. So taken was he with the responsive gem that the next day he delivered a paper on the subject at the Royal Society, noting his surprise at the glow since his "constitution," he felt, was "not of the hottest."

Three hundred years later another Englishman, Martin Aitken of Oxford University, developed the methods to turn thermoluminescence into a geophysical timepiece. The clock works because the radioactivity bombarding a mineral is fairly constant, so electrons become trapped in those crystalline prisons at a steady rate through time. If you crush the mineral you want to date and heat a few grains to a high enough temperature—about 900 degrees, which is more body heat than Robert Boyle's constitution could ever have produced—all the electron traps will release their captive electrons at once, creating a brilliant puff of light. In a laboratory the intensity of that burst of luminescence can easily be measured with a device called a photomultiplier. The higher the spike of light, the more trapped electrons have accumulated in the sample, and thus the more time has elapsed since it was last exposed to heat. Once a mineral is heated and all the electrons have returned "home," the clock is set back to zero.

Now, our lineage has been making flint tools for hundreds of thousands of years, and somewhere in that long stretch of prehistory we began to use fire as well. Inevitably, some of our less careful ancestors kicked discarded tools into burning hearths, setting their electron clocks back to zero and opening up a ripe opportunity for TL timekeepers in the present. After the fire went out, those flints lay in the ground, pummeled by radioactivity, and each trapped electron was another tick of the clock. Released by laboratory heat, the electrons flash out photons that reveal time gone by.

In the late 1980s Hélène Valladas, an archeologist at the Center for Low-Level Radioactivity of the French Atomic Energy Commission near Paris, along with her father, physicist Georges Valladas, stunned the anthropological community with some TL dates on burned flints taken from two archeological sites in Israel. The first was a cave called Kebara, which had already yielded an astonishingly complete Neanderthal skeleton. Valladas dated flints from the Neanderthal's level at 60,000 years before the present.

In itself this was no surprise, since the date falls well within the known range of the Neanderthals' time on Earth. The shock came a year later, when she used the same technique to pin a date on flints from a nearby cave called Qafzeh, which contained the buried remains of early modern human beings. This time, the spikes of luminescence translated into an age of around 92,000 years. In other words, the more "advanced" human types were a full 30,000 years *older* than the Neanderthals they were supposed to have descended from.

If Valladas's TL dates are accurate, they completely confound the notion that modern humans evolved from Neanderthals in any neat and tidy way. Instead, these two kinds of human, equally endowed culturally but distinctly different in appearance, might have shared the same little nook of the Middle East for tens of thousands of years. To some, this simply does not make sense.

"If these dates are correct, what does this do to what else we know, to the stratigraphy, to fossil man, to the archeology?" worries Anthony Marks, an archeologist at Southern Methodist University. "It's all a mess. Not that the dates are necessarily wrong. But you want to know more about them."

Marks's skepticism is not entirely unfounded. While simple in theory, in practice TL has to overcome some devilish complications. ("If these new techniques were easy, we would have thought of them a long time ago," says geoclironologist Gifford Miller of the University of Colorado.) To convert into calendar years the burst of luminescence when a flint is heated, one has to know both the sensitivity of that particular flint to radiation and the dose of radioactive rays it has received each year since it was "zeroed" by fire. The sensitivity of the sample can be determined by assaulting it with artificial radiation in the lab. And the annual dose of radiation received from *within* the sample itself can be calculated fairly easily by measuring how much uranium or other radioactive elements the sample contains. But determining the annual dose from the environment *around* the sample—the radioactivity in the surrounding soil, and cosmic rays from the atmosphere itself—is an iffier proposition. At some sites fluctuations in this environmental dose through the millennia can turn the "absolute" date derived from TL into an absolute nightmare.

Fortunately for Valladas and her colleagues, most of the radiation dose for the Qafzeh flints came from within the flints themselves. The date there of 92,000 years for the modern human skeletons is thus not only the most sensational number so far produced by TL, it is also one of the surest.

"The strong date at Qafzeh was just good luck," says Valladas. "It was just by chance that the internal dose was high and the environmental dose was low."

More recently Valladas and her colleague Norbert Mercier turned their TL techniques to the French site of Saint-Césaire. Last summer they confirmed that a Neanderthal found at Saint-Césaire was only 36,000 years old. This new date, combined with a fresh radiocarbon date of about 40,000 years tagged on some Cro-Magnon sites in northern Spain, strongly suggests that the two types of humans shared the same corner of Europe for several thousand years as the glaciers advanced from the north.

While Valladas has been busy in Europe and the Middle East, other TL timekeepers have produced some astonishing new dates for the first human occupation of Australia. As recently as the 1950s, it was widely believed that Australia had been colonized only some five thousand years ago. The reasoning was typically Eurocentric:

since the Australian aborigines were still using stone tools when the first white settlers arrived, they must have just recently developed the capacity to make the difficult sea crossing from Indonesia in the first place. A decade later archeologists grudgingly conceded that the date of first entry might have been closer to the beginning of the Holocene period, 10,000 years ago. In the 1970s radiocarbon dates on human occupation sites pushed the date back again, as far as 32,000 years ago. And now TL studies at two sites in northern Australia drop that first human footstep on the continent—and the sea voyage that preceded it—all the way back to 60,000 years before the present. If these dates stand up, then the once-maligned ancestors of modern aborigines were building ocean-worthy craft some 20,000 years *before* the first signs of sophisticated culture appeared in Europe.

"Luminescence has revolutionized the whole period I work in," says Australian National University archeologist Rhys Jones, a member of the team responsible for the new TL dates. "In effect, we have at our disposal a new machine—a new time machine."

With so much at stake, however, nobody looks to TL—or to any of the other new "time machines"—as a geochronological panacea. Reputations have been too badly singed in the past by dating methods that claimed more than they could deliver. In the 1970s a flush of excitement over a technique called amino acid racemization led many workers to believe that another continent—North America—had been occupied by humans fully 70,000 years ago. Further testing at the same American sites proved that the magical new method was off by one complete goose egg. The real age of the sites was closer to 7,000 years.

"To work with wrong dates is a luxury we cannot afford," British archeologist Paul Mellars intoned ominously earlier this year, at the beginning of a London meeting of the Royal Society to showcase the new dating technologies. "A wrong date does not simply inhibit research. It could conceivably throw it into reverse."

Fear of just such a catastrophe—not to mention the risk that her own reputation could go up in a puff of light—is what keeps Alison Brooks from declaring outright that she has found exquisitely crafted bone harpoons in Zaire that are more than 40,000 years older than such creations are supposed to be. So far the main support for her argument has been her redating of another site, called Ishango, four miles down the Semliki River from the Katanda site. In the 1950s the Belgian geologist Jean de Heinzelin excavated a harpoon-rich "aquatic civilization" at Ishango that he thought was 8,000 years old. Brooks's radiocarbon dating of the site in the mid-1980s pushed the age back to 25,000. By tracing the layers of sediment shared between Ishango and Katanda, Brooks and her colleagues are convinced that Katanda is much farther down in the stratigraphy—twice as old as Ishango, or perhaps even more. But even though Brooks and Yellen talk freely about their harpoons at meetings, they have yet to utter such unbelievable numbers in the unforgiving forum of an academic journal.

"It is precisely because no one believes us that we want to make our case airtight before we publish," says Brooks. "We want dates confirming dates confirming dates."

Soon after the harpoons were discovered, the team went to work with thermoluminescence. Unfortunately, no burned flints have been found at the site. Nevertheless, while TL works best on materials that have been completely zeroed by such extreme heat as a campfire, even a strong dose of sunlight can spring some of the electron traps. Thus even ordinary sediments surrounding an archeological find might harbor a readable clock: bleached out by sunlight when they were on the surface, their TL timers started ticking as soon as they were buried by natural processes. Brooks and Yellen have taken soil samples from Katanda for TL, and so far the results are tantalizing—but that's all.

"At this point we think the site is quite old," says geophysicist Allen Franklin of the University of Mary-

land, who with his Maryland colleague Bill Hornyak is conducting the work. "But we don't want to put a number on it."

As Franklin explains, the problem with dating sediments with TL is that while some of the electron traps might be quickly bleached out by sunlight, others hold on to their electrons more stubbornly. When the sample is then heated in a conventional TL apparatus, these stubborn traps release electrons that were captured perhaps millions of years before the sediments were last exposed to sunlight-teasing date-hungry archeologists with a deceptively old age for the sample.

Brooks does have other irons in the dating fire. The most promising is called electron spin resonance—or ESR, among friends. Like TL, electron spin resonance fashions a clock out of the steadily accumulating electrons caught in traps. But whereas TL measures that accumulation by the strength of the light given off when the traps open, ESR literally counts the captive electrons themselves while they still rest undisturbed in their prisons.

All electrons "spin" in one of two opposite directions—physicists call them up and down. (Metaphors are a must here because the nature of this "spinning" is quantum mechanical and can be accurately described only in huge mathematical equations.) The spin of each electron creates a tiny magnetic force pointing in one direction, something like a compass needle. Under normal circumstances, the electrons are paired so that their opposing spins and magnetic forces cancel each other out. But trapped electrons are unpaired. By manipulating an external magnetic field placed around the sample to be dated, the captive electrons can be induced to "resonate"—that is, to flip around and spin the other way. When they flip, each electron absorbs a finite amount of energy from a microwave field that is also applied to the sample. This loss of microwave energy can be measured with a detector, and it is a direct count of the number of electrons caught in the traps.

ESR works particularly well on tooth

enamel, with an effective range from a thousand to 2 million years. Luckily for Brooks and Yellen, some nice fat hippo teeth have been recovered from Katanda in the layer that also held the harpoons. To date the teeth, they have called in Henry Schwarcz of McMaster University in Ontario, a ubiquitous, veteran geochronologist. In the last ten years Schwarcz has journeyed to some 50 sites throughout Europe, Africa, and western Asia, wherever his precious and arcane services are demanded.

Schwarcz also turned up at the Royal Society meeting, where he explained both the power and the problems of the ESR method. On the plus side is that teeth are hardy remains, found at nearly every archeological site in the world, and that ESR can test a tiny sample again and again—with the luminescence techniques, it's a one-shot deal. ESR can also home in on certain kinds of electron traps, offering some refinement over TL, which lumps them all together.

On the minus side, ESR is subject to the same uncertainties as TL concerning the annual soaking of radiation a sample has received from the environment. What's more, even the radiation from *within* a tooth cannot be relied on to be constant through time. Tooth enamel has the annoying habit of sucking up uranium from its surroundings while it sits in the ground. The more uranium the tooth contains, the more electrons are being bombarded out of their normal positions, and the faster the electron traps will fill up. Remember: you cannot know how old something is by counting filled traps unless you know the rate at which the traps were filled, year by year. If the tooth had a small amount of internal uranium for 50,000 years but took in a big gulp of the hot stuff 10,000 years ago, calculations based on the tooth's current high uranium level would indicate the electron traps were filled at a much faster rate than they really were. "The big question is, When did the uranium get there?" Schwarcz says. "Did the tooth slurp it all up in three days, or did the uranium accumulate gradually through time?"

One factor muddying the "big ques-

tion" is the amount of moisture present around the sample during its centuries of burial: a wetter tooth will absorb uranium faster. For this reason, the best ESR sites are those where conditions are driest. Middle Eastern and African deserts are good bets. As far as modern human origins go, the technique has already tagged a date of about 100,000 years on some human fossils from an Israeli cave called Skhul, neatly supporting the TL date of 92,000 from Qafzeh, a few miles away. If a new ESR date from a Neanderthal cave just around the corner from Skhul is right, then Neanderthals were also in the Middle East at about the same time. Meanwhile, in South Africa, a human jawbone from the site of Border Cave— "so modern it boggles the mind," as one researcher puts it—has now been dated with ESR at 60,000 years, nearly twice as old as any fossil like it in Europe.

But what of the cultural change to modern human behavior—such as the sophisticated technological development expressed by the Katanda harpoons? Schwarcz's dating job at Katanda is not yet finished, and given how much is at stake, he too is understandably reluctant to discuss it. "The site has good potential for ESR," he says guardedly. "Let's put it this way: if the initial results had indicated that the harpoons were not very old after all, we would have said 'So what?' to them and backed off. Well, we haven't backed off."

There are other dating techniques being developed that may, in the future, add more certainty to claims of African modernity. One of them, called uranium-series dating, measures the steady decay of uranium into various daughter elements inside anything formed from carbonates (limestone and cave stalactites, for instance). The principle is very similar to radiocarbon dating—the amount of daughter elements in a stalactite, for example, indicates how long that stalactite has been around—with the advantage that uranium-series dates can stretch back half a million years. Even amino acid racemization, scorned for the last 15 years, is making a comeback, thanks to the

discovery that the technique, unreliable when applied to porous bone, is quite accurate when used on hard ostrich eggshells.

In the best of all possible worlds, an archeological site will offer an opportunity for two or more of these dating techniques to be called in so they can be tested against each other. When asked to describe the ideal site, Schwarcz gets a dreamy look on his face. "I see a beautiful human skull sandwiched between two layers of very pure flowstone," he says, imagining uranium-series dating turning those cave limestones into time brackets. "A couple of big, chunky hippo teeth are lying next to it, and a little ways off, a bunch of burned flints."

Even without Schwarcz's dream site, the dating methods used separately are pointing to a common theme: the alarming antiquity of modern human events where they are not supposed to be happening in the first place. Brooks sees suggestive traces of complexity not just at Katanda but scattered all over the African continent, as early as 100,000 years before the present. A classic stone tool type called the blade, long considered a trademark of the European Upper Paleolithic, appears in abundance in some South African sites 40,000 to 50,000 years before the Upper Paleolithic begins. The continent may even harbor the earliest hints of art and a symbolic side to human society: tools designed with stylistic meaning; colorful, incandescent minerals, valueless but for their beauty, found hundreds of miles away from their source. More and more, the Cro-Magnons of Europe are beginning to look like the last modern humans to show themselves and start acting "human" rather than the first.

That's not an easy notion for anthropologists and archeologists to swallow. "It just doesn't fit the pattern that those harpoons of Alison's should be so old," says Richard Klein, a paleoanthropologist at the University of Chicago. Then he shrugs. "Of course, if she's right, she has made a remarkable discovery indeed."

Only time will tell.

The Neanderthal Peace

For perhaps 50,000 years, two radically different types of human lived side by side in the same small land. And for all those millennia, the two apparently had nothing whatsoever to do with each other. Why in the world not?

James Shreeve

James Shreeve is a contributing editor of Discover. *His previous book,* Lucy's Child: The Discovery of a Human Ancestor, *was coauthored with anthropologist Donald Johansen.*

I met my first Neanderthal in a café in Paris, just across the street from the Jussieu metro stop. It was a wet afternoon in May, and I was sitting on a banquette with my back to the window. The café was smoky and charmless. Near the entrance a couple of students were thumping on a pinball machine called Genesis, which beeped approval every time they scored. The place was packed with people—foreign students, professors, young professionals, French workers, Arabs, Africans, and even a couple of Japanese tourists, all thrown together by the rain. Our coffee had just arrived, and I found that if I tucked my elbow down when raising my cup, I could drink it without poking the ribs of a bearded man sitting at the table next to me, who was deep into an argument.

Above the noise of the pinball game and the din of private conversations, a French anthropologist named Jean-Jacques Hublin was telling me about the anatomical unity of man. It was he who had brought along the Neanderthal. When we had come into the cafe, he had placed an object wrapped in a soft rag on the table and had ignored it ever since. Like anything so carefully neglected, it was beginning to monopolize my attention.

"Perhaps you would be interested in this," he said at last, whisking away the rag. There, amid the clutter of demitasses and empty sugar wrappers, was a large human lower jawbone. The teeth, worn and yellowed by time, were all in place. Around us, I felt the café raise a collective eyebrow. The hubbub of talk sank audibly. The bearded man next to me stopped in midsentence, looked at the jaw, looked at Hublin, and resumed his argument. Hublin gently nudged the fossil to the center of the table and leaned back.

"What is it? " I asked.

"It is a Neanderthal from a site called Zafarraya, in the south of Spain," he said. "We have only this mandible and an isolated femur. But as you can see, the jaw is almost complete. We are not sure yet, but it may be that this fossil is only 30,000 years old."

"Only" 30,000 years may seem an odd way of expressing time, but coming from a paleoanthropologist, it is like saying that a professional basketball player is *only* 6 foot 4. Hominids—members of the exclusively human family tree—have been on Earth for at least 4 million years. Measured against the earliest members of our lineage, the mineralized piece of bone on the table was a mewling newborn. Even compared with others of its kind, the jaw was astonishingly young. Neanderthals were supposed to have disappeared fully 5,000 years before this one was born, and I had come to France to find out what might have happened to them.

The Neanderthals are the best known and least understood of all human ancestors. To most people, the name instantly brings to mind the image of a hulking brute, dragging his mate around by her coif. This stereotype, born almost as soon as the first skeleton was found in a German cave in the middle of the last century, has been refluffed in comic books, novels, and movies so often that it has successfully passed from cliché to common parlance. But what actually makes a Neanderthal a Neanderthal is not its size or its strength or any measure of its native intelligence but a suite of exquisitely distinct physical traits, most of them in the face and cranium.

Like all Neanderthal mandibles, for example, the one on the table lacked the bony protrusion on the rim of the jaw called a mental eminence—better known as a chin. The places on the outside of the jaw where chewing muscles had once been attached were grossly enlarged, indicating tremendous torque in the bite. Between the last two molars and the upward thrust of the rear of the jaw, Hublin pointed to gaps of almost a quarter inch, an architectural nicety shifting the business of chewing farther toward the front.

In these and in several other features the jaw was uniquely, quintessentially Neanderthal; no other member of the human family before or since shows the same pattern. With a little instruction the Neanderthal pattern is recognizable even to a layman like me. But unlike Hublin, whose expertise allowed him to sit there calmly sipping coffee while the jaw of a 30,000-year-old man rested within biting distance of his free hand, I felt like stooping down and paying homage.

Several years before, based on a comparison of DNA found in the mitochondria of modern human cells, a team of biochemists in Berkeley, California, had concluded that all humans on Earth could trace their ancestry back to a woman who had lived in Africa only 200,000 years earlier. Every living branch and twig of the human family tree had shot up from this "mitochondrial Eve" and spread like kudzu over the face of the globe, binding all humans in an intimate web of relatedness.

To me the Eve hypothesis sounded almost too good to be true. If all living people can be traced back to a common ancestor just 200,000 years ago, then the entire human population of the globe is really just one grand brother-and-sisterhood, despite the confounding embellishments of culture and race. Thus on a May afternoon, a café in Paris could play host to clientele from three or four continents, but the scene still amounted to a sort of ad hoc family reunion.

But Eve bore a darker message too. The Berkeley study suggested that at some point between 100,000 and 50,000 years ago, people from Africa began to disperse across Europe and Asia, eventually populating the Americas as well. These people, and these alone, became the ancestors of all future human generations. When they arrived in Eurasia, however, there were thousands, perhaps millions, of other human beings already living there—including the Neanderthals. What happened to them all? Eve's answer was cruelly unequivocal: the Neanderthals—including the Zafarraya population represented by the jaw on the table—were pushed aside, outcompeted, or otherwise driven extinct by the new arrivals from the south.

What fascinates me about the fate of the Neanderthals is the paradox of their promise. Appearing first in Europe about 150,000 years ago, the Neanderthals flourished throughout the increasing cold of an approaching ice age; by 70,000 years ago they had spread throughout Europe and western Asia. As for Neanderthal appearance, the stereotype of a muscled thug is not completely off the mark. Thick-boned, barrel-chested, a healthy Neanderthal male could lift an average NFL linebacker over his head and throw him through the goalposts. But despite the

Neanderthal's reputation for dim-wittedness, there is nothing that clearly distinguishes its brain from that of a modern human except that, on average, the Neanderthal version was slightly *larger*. There is no trace of the thoughts that animated those brains, so we do not know how much they resembled our own. But a big brain is an expensive piece of adaptive equipment. You don't evolve one if you don't use it. Combining enormous physical strength with manifest intelligence, the Neanderthals appear to have been outfitted to face any obstacle the environment could put in their path. They could not lose.

And then, somehow, they lost. Just when the Neanderthals reached their most advanced expression, they suddenly vanished. Their demise coincides suspiciously with the arrival in western Europe of a new kind of human: taller, thinner, more modern-looking. The collision of these two human populations—us and the other, the destined parvenu and the doomed caretaker of a continent—is as potent and marvelous a part of the human story as anything that has happened since.

By itself, the half-jaw on the table in front of me had its own tale to tell. Hublin had said that it was perhaps as young as 30,000 years old. A few months before, an American archeologist named James Bischoff and his colleagues had also announced astonishing ages for some objects from Spanish caves. After applying a new technique to date some modern human-style artifacts, they declared them to be 40,000 years old. This was 6,000 years *before* there were supposed to have been modern humans in Europe. If both Bischoff's and Hublin's dates were right, it meant that Neanderthals and modern humans had been sharing Spanish soil for 10,000 years. That didn't make sense to me.

"At 30,000 years," I asked Hublin, "wouldn't this jaw be the last Neanderthal known?"

"If we are right about the date, yes," he said. "But there is still much work to be done before we can say how old the jaw is for certain."

"But Bischoff says modern humans were in Spain 10,000 years before then," I persisted. "I can understand how a population with a superior technology might come into an area and quickly dominate a less sophisticated people already there. But 10,000 years doesn't sound very quick even in evolutionary terms. How can two kinds of human being exist side by side for that long without sharing their cultures? Without sharing their genes?"

Hublin shrugged in the classically cryptic French manner that means either "The answer is obvious" or "How should I know?"

Among all the events and transformations in human evolution, the origins of modern humans were, until recently, the easiest to account for. Around 35,000 years ago, signs of a new, explosively energetic culture in Europe marked the beginning of the period known as the Upper Paleolithic. They included a highly sophisticated variety of tools, made out of bone and antler as well as stone. Even more important, the people making these tools—usually known as Cro-Magnons, a name borrowed from a tiny rock shelter in southern France where their skeletons were first found, in 1868—had discovered a symbolic plane of existence, evident in their gorgeously painted caves, carved animal figurines, and the beads and pendants adorning their bodies. The Neanderthals who had inhabited Europe for tens of thousands of years had never produced anything remotely as elaborate. Coinciding with this cultural explosion were the first signs of the kind of anatomy that distinguishes modern human beings: a well-defined chin; a vertical forehead lacking pronounced brow-ridges; a domed braincase; and a slender, lightly built frame, among other, more esoteric features.

The skeletons in the Cro-Magnon cave, believed to be between 32,000 and 30,000 years old, provided an exquisite microcosm of the joined emergence of culture and anatomy. Five skeletons, including one of an infant, were found buried in a communal grave, and all exhibited the anatomical characteristics of modern human beings. Scattered in the grave with them were hundreds of artificially pierced seashells and animal teeth, clearly the vestiges of necklaces, bracelets, and other body ornaments. The nearly simultaneous appearance of modern culture and modern anatomy provided a readymade explanation for the final step in the human journey. Since they happened at the same time, the reasoning went, obviously one had caused the other. It all made good Darwinian sense. A more efficient technology emerged to take over the survival role previously provided by brute strength, relaxing the need for the robust physiques and powerful chewing apparatus of the Neanderthals. Voilà. Suddenly there was clever, slender Cro-Magnon man. That this first truly modern human should be indigenous to Europe tightened the evolutionary narrative: modern man appeared in precisely the region of the world where culture—according to Europeans—later reached its zenith. Prehistory foreshadowed history. The only issue to sort out was whether the Cro-Magnons had come from somewhere else or whether the Neanderthals had evolved into them.

The latter scenario, of course, assumes that modern humans and Neanderthals didn't coexist, at least not for any appreciable amount of time. But the jawbone from Zafarraya challenged that neat supposition. Even more damaging were some strange findings in the Mideast. Recent discoveries there too suggest that Neanderthals and modern humans may have inhabited the same land at the same time, and for far, far longer than in Spain.

In Israel, on the southern edge of the Neanderthal range, a wooded rise of limestone issues abruptly out of the Mediterranean below Haifa, ascending in an undulation of hills. This is the Mount Carmel of the Song of Solomon, where Elijah brought down the false priests of Baal, and Deborah laid rout to the Canaanites. In subsequent centuries, armies, tribes, and whole cultures tramped through its rocky passes and over its fertile flanks, bringing Hittites, Persians, Jews, Romans, Mongols, Muslims, Crusaders, Turks, the modern meddling of Europeans—one people slaughtered or swallowed by the next but somehow springing up again and gaining strength enough to slaughter or swallow in its turn.

The story in the Levant never made sense. Here, how modern a hominid looked said nothing about how modernly it behaved.

My interest here is in more ancient confrontations. Mount Carmel lies in the Levant, a tiny hinge of habitability between the sea and the desert, linking the two great landmasses of Africa and Eurasia. A million years ago a massive radiation of large mammals moved through the Levant from Africa toward the temperate latitudes to the north. Among these mammals were some ancestral humans. Time passed. The humans evolved, diversified. The ones in Europe came to look very different from their now-distant relatives who had remained in Africa. The Europeans became the Neanderthals. Then, still long before history began to scar the Levant with its sieges and slaughters, some Neanderthals from Europe and other humans from Africa wandered into this link between their homelands, leaving their bones on Mount Carmel. What happened when they met? How did two kinds of human respond to each other?

Reaching the Stone Age in Israel is easy; I simply rented a car in Tel Aviv and drove a couple of hours up the coastal road. My destination was the cave of Kebara, an excavation hunched above a banana plantation on the sea-weathered western slope of the mountain.

Inside the cave the present Mideast, with all its political complexities, disappeared—here there was only a cool, sheltered emptiness, greatly enlarged by decades of archeological probing. Scattered through the excavation were a dozen or so scientists and students; an equal number were working at tables along the rim. The atmosphere was one of hushed, almost monkish concentration, like that of a reading room in a great library.

The Kebara excavation began ten years ago, picking up on the previous work of Moshe Stekelis of Hebrew University in the 1950s and early 1960s. Stekelis exposed a sequence of Paleolithic deposits and, before his sudden death, discovered the skeleton of an infant Neanderthal. A greater treasure emerged in 1983. After Stekelis's time, the sharp vertical profiles of the excavation crumbled under the feet of a generation of kibbutz children and assorted other slow ravages. A graduate student named Lynne Schepartz was assigned the mundane task of cleaning up the deteriorated exposures by cutting them a little deeper. One afternoon she noticed what appeared to be a human toe bone peeking out of a fused clod of sediments. The next morning her whisk broom exposed a pearly array of human teeth: the lower jaw of an adult Neanderthal skeleton. Stekelis's team had missed it by two inches.

Lynne Schepartz was no longer a graduate student, but she was still spending her summers at Kebara. I found her and asked her how it felt to uncover the fossil. "Unprintable," she said. "I was jumping up and down and screaming."

She had reason to react unprintably. Her discovery turned out to be not just any Neanderthal but the most complete skeleton ever found: the first complete Neanderthal spinal column, the first complete Neanderthal rib cage, the first complete pelvis of any early hominid known. She showed me a plaster cast of the fossil—affectionately known as Moshe—lying on an adjacent table. The bones were arranged exactly as they had been found. Moshe was resting on his back, his right arm folded over his chest, his left hand on his stomach, in a classic attitude of burial. The only missing parts were the right leg, the extremity of the left, and except for the lower jaw, the skull.

Schepartz led me down ladders to Moshe's burial site, a deep rectangular pit near the center of the excavation. On this July morning, the Neanderthal's grave was occupied by a modern human named Ofer Bar-Yosef, who peered back up at me from behind thick glasses, magnifying my sense that I had disturbed the happy toil of a cavernicolous hobbit. He seemed evolved to the task, nimble and gnomishly compact, the better to fit into cramped quarters.

Bar-Yosef told me that he had directed his first archeological excavation at the age of 11, rounding up a crew of his friends in his Jerusalem neighborhood to help him unearth a Byzantine water system. He had not stopped digging since. Kebara was the latest of three major excavations under his direction. "My daughter has been coming to this site since she was a fetus," he told me. "She used to have a playpen set up right over there."

Throughout his career, Bar-Yosef has dug for answers to two personal obsessions: the origins of Neolithic agricultural societies and—the point where our obsessions converge—the twisting conundrum of modern human origins.

The story in the Levant never really made much sense. In the old days, back when everybody "knew" that modern humans first appeared in western Europe, where the really modern folks still live, you could identify a hominid by the kind of tools he left behind. Bulky Neanderthals made bulky flakes, while svelte Cro-Magnons made slim "blades." Narrowness is, in fact, the very definition of a blade, which in paleoarcheology means nothing more than a stone tool twice as long as it is wide. In Europe a new, efficient way of producing blades from a flint core appeared as part of the "cultural explosion" that coincided with the appearance of the Cro-Magnon people. Here in the Levant, however, the arrival of anatomically modern humans was marked by no fancy new tools not to mention no painted caves, beaded necklaces, or other evidence of exploding Cro-Magnon couture. In this part of the world, how modern a hominid looked in its body said nothing about how modernly it behaved.

Just a couple of bus stops up the coastal road from Kebara is the cave of Tabun, with over 80 vertical feet of deposits spanning more than 100,000

years of human occupation. The treasures of Tabun, like those of Kebara, are Neanderthals. Literally around the corner from Tabun is another cave, called Skhul, where some fairly modern-looking humans were found in the 1930s. And a few miles inland from Kebara on a hill in lower Galilee is Qafzeh, where in 1965 a young French anthropologist named Bernard Vandermeersch found a veritable Middle Paleolithic cemetery of distinctly modern humans. But though the bones in these caves include both Neanderthals and modern humans, the tools found with the bones are all pretty much the same.

In 1982, Arthur Jelinek of the University of Arizona made an inspired attempt to massage some sense into the nagging paradox of Mount Carmel. As in Europe later on, he argued, tools get thinner along with the bodies of the people who make them. Only in this case, the reduction is front to back rather than side to side.

The fattest flakes, he showed, came from a layer near the bottom of Tabun cave, where a partial skeleton of a Neanderthal woman had turned up; if flake thickness was indeed a true measure of time, then she was the oldest in the group. The next oldest would be the Neanderthal infant that Stekelis had found at Kebara. The modern humans of Skhul yielded flake tools that were flatter. And the flattest of all belonged to the moderns of Qafzeh cave. Although the physically modern Skhul-Qafzeh people might not have crossed the line into full-fledged, blade-based humanness, they appeared, as Jelinek wrote, "on the threshold of breaking away.

"Our current evidence from Tabun suggests an orderly and continuous progress of industries in the southern Levant," he went on, "paralleled by a morphological progression from Neanderthal to modern man." According to this scenario, the Neanderthals simply evolved into modern humans. There was no collision of peoples or cultures; two kinds of human never met, because there was really only one kind, changing through time.

If Jelinek's conventional chronology based on slimming tool forms was right, the fossils found in Qafzeh could

be "proto-Cro-Magnons," the evolutionary link between a Neanderthal past and a Cro-Magnon future—and thence to the present moment. But the dating methods he used were *relative,* merely inferring an age for the skeletons by where they fell in an overall chronological scheme. What was needed was a new way of measuring time, preferably an *absolute* dating technique that could label the Mount Carmel hominids with an age in actual calendar years.

The most celebrated absolute dating method is radiocarbon dating, which measures time by the constant, steady decay of radioactive carbon atoms. Developed in the 1940s, radiocarbon dating is still one of the most accurate ways to pin an age on a site, so long as it is younger than around 40,000 years. In older materials the amount of radioactive carbon still left undecayed is so small that even the slightest amount of contamination leads to highly inaccurate results. Another technique, relying on the decay of radioactive potassium instead of carbon, has been used since the late 1950s to date volcanic deposits older than half a million years. Radiopotassium was the method of choice for dating the famous East African early hominids like Lucy, as well as the new "root hominid," *Australopithecus ramidus,* announced in 1994. Until recently, though, everything that lived between the ranges of these two techniques—including the moderns at Qafzeh and the Neanderthals at Kebara—fell into a chronological black hole.

In the early 1980s, however, Hélène Valladas, a French archeologist, used a new technique called thermoluminescence, or TL, to date flints from the Kebara and Qafzeh caves. As applied to these flints, the technique is based on the fact that minerals give off a burst of light when heated to about 900 degrees. It is also based on the certainty that past humans, like present ones, were sometimes careless. In the Middle Paleolithic, some flint tools happened to lie around in the path of careless feet, and some tools got kicked into fires, opening up an exquisite opportunity for absolute dating. When a flint tool was heated sufficiently by the fire, it gave up its thermoluminescent energy. Over thousands of years, that energy slowly built up again. The dating of fire-charred tools is thus, in principle, straightforward: the brighter a bit of flint glows when heated today, the longer since the time it was last used.

By 1987, Valladas and her physicist father, Georges, had squeezed an age of 60,000 years out of the burnt tools found beside Moshe at Kebara. That number pleased everybody, since it agreed with time schemes arrived at through relative dating methods. The shocker came the following year, when Valladas and her colleagues announced the results of their work at Qafzeh: the "modern" skeletons were 92,000 years old, give or take a few thousand.

Several other Neanderthal and modern human sites have since been dated with TL, and the one at Qafzeh remains not only the most sensational but the surest. Key sites in the Levant have also been dated by a "sister" technique called electron-spin resonance (ESR). Large mammal teeth found near the Qafzeh skeletons came back with an ESR date even older than Valladas's thermoluninescent surprise. The skeletons were at least 100,000 and perhaps 115,000 years old. "People said that TL had too many uncertainties," Bernard Vandermeersch told me. "So we gave them ESR. By now it is very difficult to dispute that the first modern humans in the Levant were here by 100,000 years ago."

Clearly, if modern humans were inhabiting the Levant 40,000 years before the Neanderthals, they could hardly have evolved from them. If the dates are indeed correct, it is hard to see what else one can do with the venerated belief in our Neanderthal ancestry but chuck it, once and for all.

Case closed? On the contrary, the dates only twist the mystery on Mount Carmel even tighter. Presuming that the moderns did not just come for a visit 100,000 years ago and then politely withdraw, they must have been around when the Neanderthals arrived 40,000 years later—if the Neanderthals as well weren't there to begin with: the latest ESR dates for the Tabun Neanderthal woman place her there 110,000 years ago. Either way, two distinct

kinds of human were apparently squeezed together in an area not much larger than the state of New Jersey, and for a long time—at least 25,000 years and perhaps 50,000 or more.

Rather than resolving the paradox, the new dating techniques only teased out its riddles. If two kinds of human were behaving the same way in the same place at the same time, how can we call them different? If modern humans did not descend from the Neanderthals but replaced them instead, why did it take them so long to get the job done?

At Kebara, I took the paradox with me to mull over outside, on a still summer afternoon, where the horizon manifested the present moment in the silhouette of an oil tanker, far out at sea. If the names "Neanderthal" and "modern human" are meaningful distinctions, if they have as much reality, say, as the oil tanker pasted onto the horizon, then they cannot be blended, any more than one can blend the sea and the sky. But what if they are mere edges after all, edges that might have had firm content in France and Spain but not here, not in this past; edges whose contents spilled over and leaked into each other so profusely that no true edges can be said to exist at all?

In that case, there would be no more mystery. The Levantine paradox would be a trick knot; pull gently from both ends and it unravels on its own. Think of one end of the rope as cultural. Every species has its own ecological niche, its unique set of adaptations to local habitats. The "principle of competitive exclusion" states that two species cannot squeeze into the same niche: the slightly better adapted one will eventually drive the other one out. Traditionally, the human niche has been defined by culture, so it would be impossible for two kinds of human to coexist using the same stone tools to compete for the same plant and animal resources. One would drive the other into extinction, or never allow it to gain a foothold.

"Competitive exclusion would preclude the coexistence of two different kinds of hominid in a small area over a

40,000- or 50,000-year period unless they had different adaptations," says Geoffrey Clark of Arizona State University. "But as far as we can tell, the adaptations were identical at Kebara and Qafzeh." Clark adds to the list of common adaptations the use of symbols—or lack of it. Perhaps Neanderthals lacked complex social symbols like beads, artwork, and elaborate burial. But so, he believes, did their skinny contemporaries down the road at Qafzeh. If neither was littering the landscape with signs of some new mental capacity, by what right do we favor the skinny one with a brilliant future and doom the other to dull extinction?

This leads to the morphological end of the rope. If the two human types cannot be distinguished on the basis of their tools, then the only valid way of telling a Neanderthal from a modern human is to declare that one looks "Neanderthalish" and the other doesn't. If you were to take all the relevant fossils and line them up, could you really separate them into two mutually exclusive groups, with no overlap? A replacement advocate might think so, but a believer in continuity like Geoffrey Clark insists that you could not. He thinks the lineup might better be characterized as one widely variable population, running the gamut from the most Neanderthal to the most modern. The early excavators at Tabun and Skhul saw the fossils there as an intermediate grade between archaic and modern Homo sapiens. Perhaps they were right. "The skeletal material is anything but clearly 'Neanderthal' and clearly 'modern,'" Clark maintains, "whatever those terms mean in the first place, which I don't think is much."

This view preserves the traditional idea of continuity but abandons the process: there was no evolution from one kind of human to another—from Neanderthal to modern—because there was, in fact, no "other." But for all its appeal, the "oneness" solution to the Levantine paradox is fundamentally flawed. Nobody disputes that the tool kits of the two human types are virtually identical. But it does not logically follow that the toolmakers must be identical as well. Middle Paleolithic tool kits are associated in our minds with Neanderthals because they are the best known human occupants of the Middle Paleolithic. But if people with modern anatomy turn out to have been living back then, too, why *wouldn't* they be using the same culture as the Neanderthals?

"If you ask me, forget about the stone tools," Ofer Bar-Yosef told me. "They can tell you nothing, zero. At most they say something about how they were preparing food. But is what you do in the kitchen all of your life? Of course not. Being positive people, we are not willing to admit that some of the missing evidence might be the crucial evidence we need to solve this problem."

Whatever the tools suggest, the skeletons of moderns and Neanderthals look different, and the pattern of their differences is too consistent to dismiss. As anthropologist Erik Trinkaus of the University of New Mexico has shown, those skeletal differences clearly reflect two distinct patterns of behavior, however alike the archeological leavings may be. Furthermore, the two physical types do not follow one from the other, nor do they meet in a fleeting moment before one triumphs and the other fades. They just keep on going, side by side but never mingling. In his behavioral approach to bones, Trinkaus purposely disregards the features that might best discriminate Neanderthals and moderns from each other genetically. By definition, these traits are poor indicators of the effects of lifestyle on bone, since their shape and size are decided by heredity, not by use. But there is one profoundly important aspect of human life where behavior and heredity converge: the act that allows human lineages to continue in the first place.

Humans love to mate. They mate all the time, by night and by day, through all the phases of the female's reproductive cycle. Given the opportunity, humans throughout the world will mate with any other human. The barriers between races and cultures, so cruelly evident in other respects, melt away when sex is at stake. Cortés began the systematic annihilation of the Aztec

people—but that did not stop him from taking an Aztec princess for his wife. Blacks have been treated with contempt by whites in America since they were first forced into slavery, but some 20 percent of the genes in a typical African American are "white." Consider James Cook's voyages in the Pacific in the eighteenth century. "Cook's men would come to some distant land, and lining the shore were all these very bizarre-looking human beings with spears, long jaws, browridges," archeologist Clive Gamble of Southampton University in England told me. "God, how odd it must have seemed to them. But that didn't stop the Cook crew from making a lot of little Cooklets."

Project this universal human behavior back into the Middle Paleolithic. When Neanderthals and modern humans came into contact in the Levant, they would have interbred, no matter how "strange" they might initially have seemed to each other. If their cohabitation stretched over tens of thousands of years, the fossils should show a convergence through time toward a single morphological pattern, or at least some swapping of traits back and forth.

But the evidence just isn't there, not if the TL and ESR dates are correct. Instead the Neanderthals stay staunchly themselves. In fact, according to some recent ESR dates, the least "Neanderthalish" among them is also the oldest. The full Neanderthal pattern is carved deep at the Kebara cave, around 60,000 years ago. The moderns, meanwhile, arrive very early at Qafzeh and Skhul and never lose their modern aspect. Certainly, it is possible that at any moment new fossils will be revealed that conclusively demonstrate the emergence of a "Neandermod" lineage. From the evidence in hand, however, the most likely conclusion is that Neanderthals and modern humans were not interbreeding in the Levant.

Of course, to interbreed, you first have to meet. Some researchers have contended that the coexistence on the slopes of Mount Carmel for tens of thousands of years is merely an illusion created by the poor archeological record. If moderns and Neanderthals were physically isolated from each

other, then there is nothing mysterious about their failure to interbreed. The most obvious form of isolation is geographic. But imagine an isolation in time as well. The climate of the Levant fluctuated throughout the Middle Paleolithic—now warm and dry, now cold and wet. Perhaps modern humans migrated up into the region from Africa during the warm periods, when the climate was better suited to their lighter, taller, warm-adapted physiques. Neanderthals, on the other hand, might have arrived in the Levant only when advancing glaciers cooled their European range more than even their cold-adapted physiques could stand. Then the two did not so much cohabit as "time-share" the same pocket of landscape between their separate continental ranges.

Humans love to mate. The barriers between races, so cruelly evident in other respects, melt away when sex is at stake.

While the solution is intriguing, there are problems with it. Hominids are remarkably adaptable creatures. Even the ancient *Homo erectus*—who lacked the large brain, hafted spear points, and other cultural accoutrements of its descendants—managed to thrive in a range of regions and under diverse climatic conditions. And while hominids adapt quickly, glaciers move very, very slowly, coming and going. Even if one or the other kind of human gained sole possession of the Levant during climatic extremes, what about all those millennia that were neither the hottest nor the coldest? There must have been long stretches of time—perhaps enduring as long as the whole of recorded human history—when the Levant climate was perfectly suited to both Neanderthals and modern humans. What part do these in-between periods play in the time-sharing scenario? It doesn't make sense that one human population should politely vacate Mount Carmel just before the other moved in.

If these humans were isolated in neither space nor time but were truly contemporaneous, then how on earth did they fail to mate? Only one solution to the mystery is left. Neanderthals and moderns did not interbreed in the Levant because they *could* not. They were reproductively incompatible, separate species—equally human, perhaps, but biologically distinct. Two separate species, who both just happened to be human at the same time, in the same place.

Cohabitation in the Levant in the last ice age conjures up a chilling possibility. It forces you to imagine two equally gifted, resourceful, emotionally rich human entities weaving through one tapestry of landscape—yet so different from each other as to make the racial diversity of present-day humans seem like nothing. Take away the sexual bridge and you end up with two fully sentient human species pressed into one place, as mindless of each other as two kinds of bird sharing the same feeder in your backyard.

When paleoanthropologists bicker over whether Neanderthal anatomy is divergent enough to justify calling Neanderthals a separate species from us, they are using a *morphological* definition of a species. This is a useful pretense for the paleoanthropologists, who have nothing but the shapes of bone to work with in the first place. But they admit that in the real, vibrantly unruly natural world, bone morphology is a pitifully poor indication of where one species leaves off and another begins. Ian Tattersall, an evolutionary biologist at the American Museum of Natural History, points out that if you stripped the skin and muscle off 20 New World monkey species, their skeletons would be virtually indistinguishable. Many other species look the same even with their skins still on.

The most common definition of *biological* species, as opposed to the morphological make-believes paleontologists have to work with, is a succinct ut-

terance of the esteemed evolutionary biologist Ernst Mayr: "Species are groups of actually or potentially interbreeding natural populations that are reproductively isolated from other such groups." The key phrase is *reproductively isolated:* a species is something that doesn't mate with anything but itself. The evolutionary barriers that prevent species from wantonly interbreeding and producing a sort of organismic soup on the landscape are called isolating mechanisms. These can be any obstructions that prevent otherwise closely related species from mating to produce fertile off-spring. The obstructions may be anatomical. Two species of hyrax in East Africa share the same sleeping holes, make use of common latrines, and raise their young in communal "play groups." But they cannot interbreed, at least in part because of the radically different shapes of the males' penises. Isolating mechanisms need not be so conspicuous. Two closely related species might have different estrous cycles. Or the barrier might come into play after mating: the chromosomes are incompatible or perhaps recombine into an offspring that is incapable of breeding, an infertile hybrid like a mule.

It is easy to see why paleoanthropologists despair over trying to apply Mayr's biological concept of species to ancient hominids. The characteristics needed to recognize a biological species—the isolating mechanism—are not the kind that usually turn up as fossils. How can an estrous cycle be preserved? What does an infertile hybrid, reduced to a few fragments of its skeleton, look like? How does a chromosomal difference turn into stone?

But there is another way of looking at species that might offer hope. The biological-species concept is a curiously negative one: what makes a species itself is that it doesn't mate with anything else. A few years ago a South African biologist named Hugh Patterson turned the biological-species concept inside out, proposing a view of a species based on not with whom it *doesn't* mate but with whom it *does.* Species, according to Patterson, are groups of individuals in nature that

share "a common system of fertilization mechanisms."

With reproduction at its core, Patterson's concept is just as "biological" as Mayr's. But he turns the focus away from barriers preventing interbreeding and throws into relief the adaptations that together ensure the successful meeting of a sperm and an egg. Obviously, sex and conception are fertilization mechanisms, as is the genetic compatibility of the two parents' chromosomes. But long before a sperm cell gets near a receptive egg, the two sexes must have ways of recognizing each other as potential mates. And therein, perhaps, lies a solution to the mystery of Mount Carmel.

Every mating in nature begins with a message. It may be chemically couched: eggs of the brown alga *Ascophyllum nodosum,* for example, send out a chemical that attracts the sperm of *A. nodosum* and no other. It may be a smell. As any dog owner knows, a bitch in heat lures males from all over the neighborhood. Note that the scent does not draw squirrels, tomcats, or teenage boys. Many birds use vocal signals to attract and recognize the opposite sex, but only of their own species. "A female of one species might *hear* the song of the male of another," explains Judith Masters, a colleague of Patterson's at the University of Witwatersrand, "but she won't make any response. There's no need to talk about what *prevents* her from mating with that male. She just doesn't see what all the fuss is about."

A species' mate-recognition system is extremely stable compared with adaptations to the local habitat. A sparrow born with a slightly too short beak may or may not be able to feed its young as well as another with an average-size beak. But a sparrow who sings an unfamiliar song will not attract a mate and is not going to have any young at all. He will be plucked from the gene pool of the next generation, leaving no evolutionary trace of his idiosyncratic serenade. The same goes, of course, for any sparrow hen who fails to respond to potential mates singing the "correct" tune. With this kind of price for deviance, everybody is a conservative. "The only time a species' mate-recognition system will change is when something really dramatic happens," Masters says.

For the drama to unfold, a population must be geographically isolated from its parent species. If the population is small enough and the habitat radically different from what it was previously, even the powerful evolutionary inertia of the mate-recognition system may be overcome. This change in reproduction may be accompanied by new adaptations to the environment. Or it may not. Either way, the only shift that marks the birth of a new species is the one affecting the recognition of mates. Once the recognition threshold is crossed, there is no going back. Even if individuals from the new population and the old come to live in the same region again—let's say in a well-trafficked corridor of fertile land linking their two continental ranges—they will no longer view each other as potential mates.

The human mate-recognition system is overwhelmingly visual. "Love comes in at the eye," wrote Yeats, and the locus of the human body that lures the eye most of all is the face—a trait our species shares with many other primates. "It is a common Old World anthropoid ploy," says Masters. "Cercopithecoid monkeys have a whole repertoire of eyelid flashes. Forest guenons have brightly painted faces with species-specific patterns, which they wave like flags in the forest gloom. Good old evolution tinkering away, providing new variations on a theme."

Faces are exquisitely expressive instruments. Behind our facial skin lies an intricate web of musculature, concentrated especially around the eyes and mouth, evolved purely for social communication—expressing interest, fear, suspicion, joy, contentment, doubt, surprise, and countless other emotions. Each emotion can be further modified by the raise of an eyebrow or the slight flick of a cheek muscle to express, say, measured surprise, wild surprise, disappointed surprise, feigned surprise, and so on. By one estimate, the 22

expressive muscles on each side of the face can be called on to produce 10,000 different facial actions or expressions.

Among this armory of social signals are stereotyped, formal invitations to potential mates. The mating display we call flirtation plays the same on the face of a New Guinean tribeswoman and a *lycéenne* in a Parisian café: a bashful lowering of the gaze to one side and down, followed by a furtive look at the other's face and a coy retreat of the eyes. A host of other sexual signals are communicated facially—the downward tilt of the chin, the glance over the shoulder, the slight parting of the mouth. The importance of the face as an attractant is underscored by the lengths to which humans in various cultures go to embellish what is already there. But the underlying message is communicated by the anatomy of the face itself. "'Tis not a lip, or eye, we beauty call, / but the joint force and full result of all," wrote Alexander Pope. And it is that "joint force"—over generations—that keeps our species so forcefully joined.

This brings us back to the Levant: two human species in a tight space for a long time. The vortex of anatomy where Neanderthals and early moderns differ most emphatically, where a clear line can be drawn between *them* and *us* by even the most rabid advocate of continuity is, of course, the face. The Neanderthal's "classic" facial pattern—the midfacial thrust picked up and amplified by the great projecting nose, the puffed-up cheekbones, the long jaw with its chinless finish, the large, rounded eye sockets, the extra-thick browridges shading it like twin awnings—is usually explained as a complex of modifications relating to a cold climate, or as a support to heavy chewing forces delivered to the front teeth. Either way it is assumed to be an environmental adaptation. But what if these adaptive functions of the face were not the reason they evolved in the first place? What if the peculiarities evolved instead as the underpinnings of a totally separate, thoroughly Neanderthal mate-recognition system?

Although it is merely a speculation, the idea fits some of the facts and

solves some of the problems. Certainly the Neanderthals' ancestors were geographically cut off from other populations enough to allow some new mate-recognition system to emerge. During glacial periods, contact through Asia was blocked by the polar glaciers and vast uninhabitable tundra. Mountain glaciers between the Black and Caspian Seas all but completed a barrier to the south. "The Neanderthals are a textbook case for how to get a separate species," archeologist John Shea told me. "Isolate them for 100,000 years, then melt the glaciers and let 'em loose."

If mate recognition lay behind a species-level difference between Neanderthals and moderns, the Levantine paradox can finally be put to rest. Their cohabitation with moderns no longer needs explanation. Neanderthals and moderns managed to coexist through long millennia, doing the same human-like things but without interbreeding, simply because the issue never really came up.

The idea seems scarcely imaginable. Continuity believers cannot credit the idea of two human types coexisting in sexual isolation. Replacement advocates cannot conceive of such a long period of coexistence without competition, if not outright violent confrontation. They would rather see Neanderthals and moderns pushing each other in and out of the Levant, in an extended struggle finally won by our own ancestors. Of course, if the Neanderthals were a biologically separate species, something must have happened to cause their extinction. After all, we are still here, and they are not.

Why they faded and we managed to survive is a separate story with its own shocks and surprises. But what happened on Mount Carmel might be more remarkable still. It is something that people today are not prepared to comprehend, especially in places like the Levant. Two human species, with far less in common than any two races or ethnic groups now on the planet, may have shared a small, fertile piece of land for 50,000 years, regarding each other the whole time with steady, untroubled, peaceful indifference.

Living with the Past

Anthroplogy continues to evolve as a discipline, not only in the tools and techniques of the trade but also in the application of whatever knowledge we stand to gain about ourselves. In this context, Patrick Huyghe, in "Profile of an Anthropologist: No Bone Unturned," describes "forensic anthropology," a whole new field involving the use of physical similarities and differences between people in order to identify human remains. In the same practical vein, John Horgan discusses the validity of, and the methods behind, a new field called "behavioral genetics" in his essay "Eugenics Revisited." And Edward Humans, in "The DNA Wars," discusses the legal ramifications of the use of DNA in criminal law. One theme that ties the articles of this section together, then, is that they have as much to do with the present as they have to do with the past.

Sometimes an awareness of our biological and behavioral past may even make the difference between life and death, as Lori Oliwenstein writes in her article "Dr. Darwin." Recent research indicates that the symptoms of disease must first be interpreted as to whether they represent part of the aggressive strategy of microbes or the defensive mechanisms of the patient before treatment can be applied. Meanwhile, in "The Saltshaker's Curse," Jared Diamond argues that a physiological adaptation that once helped American blacks survive slavery may now be predisposing their descendants to an early death from hypertension.

As we reflect upon where we have been and how we came to be as we are in the evolutionary sense, the inevitable question arises as to what will happen next. (See "The Future Evolution of Homo Sapiens.") This is the most difficult issue of all, since our biological future depends so much on long-range ecological trends that no one seems to be able to predict. Some wonder if we will even survive long enough as a species to experience any significant biological changes. Perhaps our capacity for knowledge is outstripping the wisdom to use it wisely, and the consequent destruction of our earthly environments and wildlife is placing us in ever greater danger of creating the circumstances of our own extinction.

Counterbalancing this pessimism is the view that, since it has been our conscious decision making (and not the genetically predetermined behavior that characterizes some species) that has gotten us into this mess, then it will be the conscious will of our generation and future generations that will get us out. But, can we wait much longer for humanity to collectively come to its senses? Or is it already too late?

Looking Ahead: Challenge Questions

What is "forensic anthropology"?

What social policy issues are involved in the nature versus nurture debate?

How reliable is DNA matching in criminal cases?

What relevance does the concept of natural selection have to the treatment of disease?

What is the "saltshaker's curse," and why are some people more affected by it than others?

What do you think the future holds for the evolution of Homo sapiens?

Profile of an Anthropologist
No Bone Unturned

Patrick Huyghe

The research of some physical anthropologists and archaeologists involves the discovery and analysis of old bones (as well as artifacts and other remains). Most often these bones represent only part of a skeleton or maybe the mixture of parts of several skeletons. Often these remains are smashed, burned, or partially destroyed. Over the years, physical anthropologists have developed a remarkable repertoire of skills and techniques for teasing the greatest possible amount of information out of sparse material remains.

Although originally developed for basic research, the methods of physical anthropology can be directly applied to contemporary human problems. . . . In this profile, we look briefly at the career of Clyde C. Snow, a physical anthropologist who has put these skills to work in a number of different settings. . . .

As you read this selection, ask yourself the following questions:

- *Given what you know of physical anthropology, what sort of work would a physical anthropologist do for the Federal Aviation Administration?*
- *What is* anthropometry*? How might anthropometric surveys of pilots and passengers help in the design of aircraft equipment?*
- *What is* forensic anthropology*? How can a biological anthropologist be an expert witness in legal proceedings?*

Clyde Snow is never in a hurry. He knows he's late. He's always late. For Snow, being late is part of the job. In fact, he doesn't usually begin to work until death has stripped some poor individual to the bone, and no one—neither the local homicide detectives nor the pathologists—can figure out who once gave identity to the skeletonized remains. No one, that is, except a shrewd, laconic, 60-year-old forensic anthropologist.

Snow strolls into the Cook County Medical Examiner's Office in Chicago on this brisk October morning wearing a pair of Lucchese cowboy boots and a three-piece pin-striped suit. Waiting for him in autopsy room 160 are a bunch of naked skeletons found in Illinois, Wisconsin, and Minnesota since his last visit. Snow, a native Texan who now lives in rural Oklahoma, makes the trip up to Chicago some six times a year. The first case on his agenda is a pale brown skull found in the garbage of an abandoned building once occupied by a Chicago cosmetics company.

Snow turns the skull over slowly in his hands, a cigarette dangling from his fingers. One often does. Snow does not seem overly concerned about mortality, though its tragedy surrounds him daily.

"There's some trauma here," he says, examining a rough edge at the lower back of the skull. He points out the area to Jim Elliott, a homicide detective with the Chicago police. "This looks like a chopping blow by a heavy bladed instrument. Almost like a decapitation." In a place where the whining of bone saws drifts through hallways and the sweet-sour smell of death hangs in the air, the word surprises no one.

Snow begins thinking aloud. "I think what we're looking at here is a female, or maybe a small male, about thirty to forty years old. Probably Asian." He turns the skull upside down, pointing out the degree of wear on the teeth. "This was somebody who lived on a really rough diet. We don't normally find this kind of dental wear in a modern Western population."

"How long has it been around?" Elliott asks.

Snow raises the skull up to his nose. "It doesn't have any decompositional odors," he says. He pokes a finger in the skull's nooks and crannies. "There's no soft tissue left. It's good and dry. And it doesn't show signs of having been buried. I would say that this has been lying around in an attic or a box for years. It feels like a souvenir skull," says Snow.

Souvenir skulls, usually those of Japanese soldiers, were popular with U.S. troops serving in the Pacific during World War II; there was also a trade in skulls during the Vietnam War years. On closer inspection, though, Snow begins to wonder about the skull's Asian origins—the broad nasal aperture and the jutting forth of the upper-tooth-bearing part of the face suggest Melanesian features. Sifting through the objects found in the abandoned building with the skull, he finds several loose-leaf albums of 35-millimeter transparencies documenting life among the highland tribes of New Guinea. The slides, shot by an anthropologist, include graphic scenes of ritual warfare. The skull, Snow concludes, is more likely to be a trophy from one of these tribal battles than the result of a local Chicago homicide.

"So you'd treat it like found property?" Elliott asks finally. "Like somebody's garage-sale property?"

"Exactly," says Snow.

Clyde Snow is perhaps the world's most sought-after forensic anthropologist. People have been calling upon him to identify skeletons for more than a quarter of a century. Every year he's involved in some 75 cases of identification, most of them without fanfare. "He's an old scudder who doesn't have to blow his own whistle," says Walter Birkby, a forensic anthropologist at the University of Arizona. "He know's he's good."

Yet over the years Snow's work has turned him into something of an unlikely celebrity. He has been called upon to identify the remains of the Nazi war criminal Josef Mengele, reconstruct the face of the Egyptian boyking Tutankhamen, confirm the authenticity of the body autopsied as that of President John F. Kennedy, and examine the skeletal remains of General Custer's men at the battlefield of the Little Bighorn. He has also been involved in the grim task of identifying the bodies in some of the United States' worst airline accidents.

Such is his legend that cases are sometimes attributed to him in which he played no part. He did not, as the *New York Times* reported, identify the remains of the crew of the *Challenger* disaster. But the man is often the equal of his myth. For the past four years, setting his personal safety aside, Snow has spent much of his time in Argentina, searching for the graves and identities of some of the thousands who "disappeared" between 1976 and 1983, during Argentina's military regime.

Snow did not set out to rescue the dead from oblivion. For almost two decades, until 1979, he was a physical anthropologist at the Civil Aeromedical Institute, part of the Federal Aviation Administration in Oklahoma City. Snow's job was to help engineers improve aircraft design and safety features by providing them with data on the human frame.

One study, he recalls, was initiated in response to complaints from a flight attendants' organization. An analysis of accident patterns had revealed that inadequate restraints on flight attendants' jump seats were leading to deaths and injuries and that aircraft doors weighing several hundred pounds were impeding evacuation efforts. Snow points out that ensuring the survival of passengers in emergencies is largely the flight attendants' responsibility. "If they are injured or killed in a crash, you're going to find a lot of dead passengers."

Reasoning that equipment might be improved if engineers had more data on the size and strength of those who use it, Snow undertook a study that required meticulous measurement. When his report was issued in 1975, Senator William Proxmire was outraged that $57,800 of the taxpayers' money had been spent to caliper 423 airline stewardesses from head to toe. Yet the study, which received one of the senator's dubious Golden Fleece Awards, was firmly supported by both the FAA and the Association of Flight Attendants. "I can't imagine," says Snow with obvious delight, "how much coffee Proxmire got spilled on him in the next few months."

It was during his tenure at the FAA that he developed an interest in forensic work. Over the years the Oklahoma police frequently consulted the physical anthropologist for help in identifying crime victims. "The FAA figured it was a kind of community service to let me work on these cases," he says.

The experience also helped to prepare him for the grim task of identifying the victims of air disasters. In December 1972, when a United Airlines plane crashed outside Chicago, killing 43 of the 61 people aboard (including the wife of Watergate conspirator Howard Hunt, who was found with $10,000 in her purse), Snow was brought in to help examine the bodies. That same year, with Snow's help, forensic anthropology was recognized as a specialty by the American Academy of Forensic Sciences. "It got a lot of anthropologists interested in forensics," he says, "and it made a lot of

pathologists out there aware that there were anthropologists who could help them."

Each nameless skeleton poses a unique mystery for Snow. But some, like the second case awaiting him back in the autopsy room at the Cook County morgue, are more challenging than others. This one is a real chiller. In a large cardboard box lies a jumble of bones along with a tattered leg from a pair of blue jeans, a sock shrunk tightly around the bones of a foot, a pair of Nike running shoes without shoelaces, and, inside the hood of a blue windbreaker, a mass of stringy, blood-caked hair. The remains were discovered frozen in ice about 20 miles outside Milwaukee. A rusted bicycle was found lying close by. Paul Hibbard, chief deputy medical examiner for Waukesha County, who brought the skeleton to Chicago, says no one has been reported missing.

Snow lifts the bones out of the box and begins reconstructing the skeleton on an autopsy table. "There are two hundred six bones and thirty-two teeth in the human body," he says, "and each has a story to tell." Because bone is dynamic, living tissue, many of life's significant events—injuries, illness, childbearing—leave their mark on the body's internal framework. Put together the stories told by these bones, he says, and what you have is a person's "osteobiography."

Snow begins by determining the sex of the skeleton, which is not always obvious. He tells the story of a skeleton that was brought to his FAA office in the late 1970s. It had been found along with some women's clothes and a purse in a local back lot, and the police had assumed that it was female. But when Snow examined the bones, he realized that "at six foot three, she would have probably have been the tallest female in Oklahoma."

Then Snow recalled that six months earlier the custodian in his building had suddenly not shown up for work. The man's supervisor later mentioned to Snow, "You know, one of these days when they find Ronnie, he's going to be dressed as a woman." Ronnie, it turned out, was a weekend transvestite. A copy of his dental records later confirmed that the skeleton in women's clothing was indeed Snow's janitor.

The Wisconsin bike rider is also male. Snow picks out two large bones that look something like twisted oysters—the innominates, or hipbones, which along with the sacrum, or lower backbone, form the pelvis. This pelvis is narrow and steep-walled like a male's, not broad and shallow like a female's. And the sciatic notch (the V-shaped space where the sciatic nerve passes through the hipbone) is narrow, as is normal in a male. Snow can also determine a skeleton's sex by checking the size of the mastoid processes (the bony knobs at the base of the skull) and the prominence of the brow ridge, or by measuring the head of an available limb bone, which is typically broader in males.

From an examination of the skull he concludes that the bike rider is "predominantly Caucasoid." A score of bony traits help the forensic anthropologist assign a skeleton to one of the three major racial groups: Negroid, Caucasoid, or Mongoloid. Snow notes that the ridge of the boy's nose is high and salient, as it is in whites. In Negroids and Mongoloids (which include American Indians as well as most Asians) the nose tends to be broad in relation to its height. However, the boy's nasal margins are somewhat smoothed down, usually a Mongoloid feature. "Possibly a bit of American Indian admixture," says Snow. "Do you have Indians in your area?" Hibbard nods.

Age is next. Snow takes the skull and turns it upside down, pointing out the basilar joint, the junction between the two major bones that form the underside of the skull. In a child the joint would still be open to allow room for growth, but here the joint has fused—something that usually happens in the late teen years. On the other hand, he says, pointing to the zigzagging lines on the dome of the skull, the cranial sutures are open. The cranial sutures, which join the bones of the braincase, begin to fuse and disappear in the mid-twenties.

Next Snow picks up a femur and looks for signs of growth at the point where the shaft meets the knobbed end. The thin plates of cartilage—areas of incomplete calcification—that are visible at this point suggest that the boy hadn't yet attained his full height. Snow double-checks with an examination of the pubic symphysis, the joint where the two hipbones meet. The ridges in this area, which fill in and smooth over in adulthood, are still clearly marked. He concludes that the skeleton is that of a boy between 15 and 20 years old.

"One of the things you learn is to be pretty conservative," says Snow. "It's very impressive when you tell the police, 'This person is eighteen years old,' and he turns out to be eighteen. The problem is, if the person is fifteen you've blown it—you probably won't find him. Looking for a missing person is like trying to catch fish. Better get a big net and do your own sorting."

Snow then picks up a leg bone, measures it with a set of calipers, and enters the data into a portable computer. Using the known correlation between the height and length of the long limb bones, he quickly estimates the boy's height. "He's five foot six and a half to five foot eleven," says Snow. "Medium build, not excessively muscular, judging from the muscle attachments that we see." He points to the grainy ridges that appear where muscle attaches itself to the bone. The most prominent attachments show up on the teenager's right arm bone, indicating right-handedness.

Then Snow examines the ribs one by one for signs of injury. He finds no stab wounds, cuts, or bullet holes, here or elsewhere on the skeleton. He picks up the hyoid bone from the boy's throat and looks for the tell-tale fracture signs that would suggest the boy was strangled. But, to Snow's frustration, he can find no obvious cause of death. In hopes of identifying the missing teenager, he suggests sending the skull, hair, and boy's description to Betty Pat Gatliff, a medical illustrator and sculptor in Oklahoma who does facial reconstructions.

Six weeks later photographs of the boy's likeness appear in the *Milwaukee Sentinel.* "If you persist long enough," says Snow, "eighty-five to ninety percent of the cases eventually get positively identified, but it can take anywhere from a few weeks to a few years."

Snow and Gatliff have collaborated many times, but never with more glitz than in 1983, when Snow was commissioned by Patrick Barry, a Miami orthopedic surgeon and amateur Egyptologist, to reconstruct the face of the Egyptian boy-king Tutankhamen. Normally a facial reconstruction begins with a skull, but since Tutankhamen's 3,000-year-old remains were in Egypt, Snow had to make do with the skull measurements from a 1925 postmortem and X-rays taken in 1975. A plaster model of the skull was made, and on the basis of Snow's report—"his skull is Caucasoid with some Negroid admixtures"—Gatliff put a face on it. What did Tutankhamen look like? Very much like the gold mask on his sarcophagus, says Snow, confirming that it was, indeed, his portrait.

Many cite Snow's use of facial reconstructions as one of his most important contributions to the field. Snow, typically self-effacing, says that Gatliff "does all the work." The identification of skeletal remains, he stresses, is often a collaboration between pathologists, odontologists, radiologists, and medical artists using a variety of forensic techniques.

One of Snow's last tasks at the FAA was to help identify the dead from the worst airline accident in U.S. history. On May 25, 1979, a DC-10 crashed shortly after takeoff from Chicago's O'Hare Airport, killing 273 people. The task facing Snow and more than a dozen forensic specialists was horrific. "No one ever sat down and counted," says Snow, "but we estimated ten thousand to twelve thousand pieces or parts of bodies." Nearly 80 percent of the victims were identified on the basis of dental evidence and fingerprints. Snow and forensic radiologist John Fitzpatrick later managed to identify two dozen others by comparing postmortem X-rays with X-rays taken during the victim's lifetime.

Next to dental records, such X-ray comparisons are the most common way of obtaining positive identifications. In 1978, when a congressional committee reviewed the evidence on John F. Kennedy's assassination, Snow used X-rays to show that the body autopsied at Bethesda Naval Hospital was indeed that of the late president and had not—as some conspiracy theorists believed—been switched.

The issue was resolved on the evidence of Kennedy's "sinus print," the scalloplike pattern on the upper margins of the sinuses that is visible in X-rays of the forehead. So characteristic is a person's sinus print that courts throughout the world accept the matching of antemortem and postmortem X-rays of the sinuses as positive identification.

Yet another technique in the forensic specialist's repertoire is photo superposition. Snow used it in 1977 to help identify the mummy of a famous Oklahoma outlaw named Elmer J. McCurdy, who was killed by a posse after holding up a train in 1911. For years the mummy had been exhibited as a "dummy" in a California funhouse—until it was found to have a real human skeleton inside it. Ownership of the mummy was eventually traced back to a funeral parlor in Oklahoma, where McCurdy had been embalmed and exhibited as "the bandit who wouldn't give up."

Using two video cameras and an image processor, Snow superposed the mummy's profile on a photograph of McCurdy that was taken shortly after his death. When displayed on a single monitor, the two coincided to a remarkable degree. Convinced by the evidence, Thomas Noguchi, then Los Angeles County coroner, signed McCurdy's death certificate ("Last known occupation: Train robber") and allowed the outlaw's bones to be returned to Oklahoma for a decent burial.

It was this technique that also allowed forensic scientists to identify the remains of the Nazi "Angel of Death," Josef Mengele, in the summer of 1985. A team of investigators, including

Snow and West German forensic anthropologist Richard Helmer, flew to Brazil after an Austrian couple claimed that Mengele lay buried in a grave on a São Paulo hillside. Tests revealed that the stature, age, and hair color of the unearthed skeleton were consistent with information in Mengele's SS files; yet without X-rays or dental records, the scientists still lacked conclusive evidence. When an image of the reconstructed skull was superposed on 1930s photographs of Mengele, however, the match was eerily compelling. All doubts were removed a few months later when Mengele's dental X-rays were tracked down.

In 1979 Snow retired from the FAA to the rolling hills of Norman, Oklahoma, where he and his wife, Jerry, live in a sprawling, early-1960s ranch house. Unlike his 50 or so fellow forensic anthropologists, most of whom are tied to academic positions, Snow is free to pursue his consultancy work full-time. Judging from the number of miles that he logs in the average month, Snow is clearly not ready to retire for good.

His recent projects include a reexamination of the skeletal remains found at the site of the Battle of the Little Bighorn, where more than a century ago Custer and his 210 men were killed by Sioux and Cheyenne warriors. Although most of the enlisted men's remains were moved to a mass grave in 1881, an excavation of the battlefield in the past few years uncovered an additional 375 bones and 36 teeth. Snow, teaming up again with Fitzpatrick, determined that these remains belonged to 34 individuals.

The historical accounts of Custer's desperate last stand are vividly confirmed by their findings. Snow identified one skeleton as that of a soldier between the ages of 19 and 23 who weighed around 150 pounds and stood about five foot eight. He'd sustained gunshot wounds to his chest and left forearm. Heavy blows to his head had fractured his skull and sheared off his teeth. Gashed thigh bones indicated that his body was later dismembered with an ax or hatchet.

Given the condition and number of the bodies, Snow seriously questions the accuracy of the identifications made by the original nineteenth-century burial crews. He doubts, for example, that the skeleton buried at West Point is General Custer's.

For the last four years Snow has devoted much of his time to helping two countries come to terms with the horrors of a much more recent past. As part of a group sponsored by the American Association for the Advancement of Science, he has been helping the Argentinian National Commission on Disappeared Persons to determine the fate of some of those who vanished during their country's harsh military rule: between 1976 and 1983 at least 10,000 people were systematically swept off the streets by roving death squads to be tortured, killed, and buried in unmarked graves. In December 1986, at the invitation of the Aquino government's Human Rights Commission, Snow also spent several weeks training Philippine scientists to investigate the disappearances that occurred under the Marcos regime.

But it is in Argentina where Snow has done the bulk of his human-rights work. He has spent more than 27 months in and around Buenos Aires, first training a small group of local medical and anthropology students in the techniques of forensic investigation, and later helping them carefully exhume and examine scores of the *desaparecidos,* or disappeared ones.

Only 25 victims have so far been positively identified. But the evidence has helped convict seven junta members and other high-ranking military and police officers. The idea is not necessarily to identify all 10,000 of the missing, says Snow. "If you have a colonel who ran a detention center where maybe five hundred people were killed, you don't have to nail him with five hundred deaths. Just one or two should be sufficient to get him convicted." Forensic evidence from Snow's team may be used to prosecute several other military officers, including General Suarez Mason. Mason is the former commander of the I Army Corps in Buenos Aires and is believed to be responsible for thousands of disappearances. He was recently extradited from San Francisco back to Argentina, where he is expected to stand trial this winter [1988].

The investigations have been hampered by a frustrating lack of antemortem information. In 1984, when commission lawyers took depositions from relatives and friends of the disappeared, they often failed to obtain such basic information as the victim's height, weight, or hair color. Nor did they ask for the missing person's X-rays (which in Argentina are given to the patient) or the address of the victim's dentist. The problem was compounded by the inexperience of those who carried out the first mass exhumations prior to Snow's arrival. Many of the skeletons were inadvertently destroyed by bulldozers as they were brought up.

Every unearthed skeleton that shows signs of gunfire, however, helps to erode the claim once made by many in the Argentinian military that most of the desaparecidos are alive and well and living in Mexico City, Madrid, or Paris. Snow recalls the case of a 17-year-old boy named Gabriel Dunayavich, who disappeared in the summer of 1976. He was walking home from a movie with his girlfriend when a Ford Falcon with no license plates snatched him off the street. The police later found his body and that of another boy and girl dumped by the roadside on the outskirts of Buenos Aires. The police went through the motions of an investigation, taking photographs and doing an autopsy, then buried the three teenagers in an unmarked grave.

A decade later Snow, with the help of the boy's family, traced the autopsy reports, the police photographs, and the grave of the three youngsters. Each of them had four or five closely spaced bullet wounds in the upper chest—the signature, says Snow, of an automatic weapon. Two also had wounds on their arms from bullets that had entered behind the elbow and exited from the forearm.

"That means they were conscious when they were shot," says Snow. "When a gun was pointed at them, they naturally raised their arm." It's details like these that help to authenticate the last moments of the victims and bring a dimension of reality to the judges and jury.

Each time Snow returns from Argentina he says that this will be the last time. A few months later he is back in Buenos Aires. "There's always more work to do," he says. It is, he admits quietly, "terrible work."

"These were such brutal, cold-blooded crimes," he says. "The people who committed them not only murdered; they had a system to eliminate all trace that their victims even existed."

Snow will not let them obliterate their crimes so conveniently. "There are human-rights violations going on all around the world," he says. "But to me murder is murder, regardless of the motive. I hope that we are sending a message to governments who murder in the name of politics that they can be held to account."

TRENDS IN BEHAVIORAL GENETICS

EUGENICS REVISITED

Scientists are linking genes to a host of complex human disorders and traits, but just how valid—and useful—are these findings?

John Horgan, *senior writer*

"How to Tell If Your Child's a Serial Killer!" That was the sound bite with which the television show *Donahue* sought to entice listeners February 25. On the program, a psychiatrist from the Rochester, N.Y., area noted that some men are born with not one Y chromosome but two. Double-Y men, the psychiatrist said, are "at special risk for antisocial, violent behavior." In fact, the psychiatrist had recently studied such a man. Although he had grown up in a "Norman Rockwell" setting, as an adult he had strangled at least 11 women and two children.

"It is not hysterical or overstating it," Phil Donahue told his horrified audience, "to say that we are moving toward the time when, quite literally, just as we can anticipate . . . genetic predispositions toward various physical diseases, we will also be able to pinpoint mental disorders which include aggression, antisocial behavior and the possibility of very serious criminal activity later on."

Eugenics is back in fashion. The message that genetics can explain, predict and even modify human behavior for the betterment of society is promulgated not just on sensationalistic talk shows but by our most prominent scientists. James D. Watson, co-discoverer of the double-helix structure of DNA and former head of the Human Genome Project, the massive effort to map our entire genetic endowment, said recently, "We used to think that our fate was in our stars. Now we know, in large part, that our fate is in our genes."

Daniel E. Koshland, Jr., a biologist at the University of California at Berkeley and editor of *Science*, the most influential peer-reviewed journal in the U.S., has declared in an editorial that the nature/nurture debate is "basically over," since scientists have shown that genes influence many aspects of human behavior. He has also contended that genetic research may help eliminate society's most intractable problems, including drug abuse, homelessness and, yes, violent crime.

Some studies cited to back this claim are remarkably similar to those conducted over a century ago by scientists such as Francis Galton, known as the father of eugenics. Just as the British polymath studied identical twins in order to show that "nature prevails enormously over nurture," so do modern researchers. But the primary reason behind the revival of eugenics is the astonishing successes of biologists in mapping and manipulating the human genome. Over the past decade, investigators have identified genes underlying such crippling diseases as cystic fibrosis, muscular dystrophy and, this past spring, Huntington's disease. Given these advances, researchers say, it is only a matter of time before they can lay bare the genetic foundation of much more complex traits and disorders.

The political base for eugenics has also become considerably broader in recent years. Spokespersons for the mentally ill believe demonstrating the genetic basis of disorders such as schizophrenia and manic depression—and even alcoholism and drug addiction—will lead not only to better diagnoses and

treatments but also to more compassion toward sufferers and their families. Some homosexuals believe society will become more tolerant toward them if it can be shown that sexual orientation is an innate, biological condition and not a matter of choice.

But critics contend that no good can come of bad science. Far from moving inexorably closer to its goals, they point out, the field of behavioral genetics is mired in the same problems that have always plagued it. Behavioral traits are extraordinarily difficult to define, and practically every claim of a genetic basis can also be explained as an environmental effect. "This has been a huge enterprise, and for the most part the work has been done shoddily. Even careful people get sucked into misinterpreting data," says Jonathan Beckwith, a geneticist at Harvard University. He adds, "There are social consequences to this."

The skeptics also accuse the media of having created an unrealistically optimistic view of the field. Richard C. Lewontin, a biologist at Harvard and a prominent critic of behavioral genetics, contends that the media generally give much more prominent coverage to dramatic reports—such as the discovery of an "alcoholism gene"—than to contradictory results or retractions. "Skepticism doesn't make the news," Lewontin says. "It only makes the news when you find a gene." The result is that spurious findings often become accepted by the public and even by so-called experts.

The claim that men with an extra Y chromosome are predisposed toward violence is a case in point. It stems from a survey in the 1960s that found more extra-Y men in prison than in the general population. Some researchers hypothesized that since the Y chromo-

"EERIE" PARALLELS between identical twins raised apart—such as Jerry Levey (*left*) and Mark Newman, who both became firefighters—are said to support genetic models of human behavior. Yet skeptics say the significance of such coincidences has been exaggerated.

gotic twins reared apart are about as similar as are monozygotic twins reared together." (Identical twins are called monozygotic because they stem from a single fertilized egg, or zygote.)

The researchers have buttressed their statistical findings with anecdotes about "eerie," "bewitching" and "remarkable" parallels between reunited twins. One case involved Oskar, who was raised as a Nazi in Czechoslovakia, and Jack, who was raised as a Jew in Trinidad. Both were reportedly wearing shirts with epaulets when they were reunited by the Minnesota group in 1979. They also both flushed the toilet before as well as after using it and enjoyed deliberately sneezing to startle people in elevators.

Some other celebrated cases involved two British women who wore seven rings and named their firstborn sons Richard Andrew and Andrew Richard; two men who both had been named Jim, named their pet dogs Toy, married women named Linda, divorced them and remarried women named Betty; and two men who had become firefighters and drank Budweiser beer.

Other twin researchers say the significance of these coincidences has been greatly exaggerated. Richard J. Rose of Indiana University, who is collaborating on a study of 16,000 pairs of twins in Finland, points out that "if you bring together strangers who were born on the same day in the same country and ask them to find similarities between them, you may find a lot of seemingly astounding coincidences."

Rose's collaborator, Jaakko Kaprio of the University of Helsinki, notes that the Minnesota twin studies may also be biased by their selection method. Whereas he and Rose gather data by combing birth registries and sending questionnaires to those identified as twins, the Minnesota group relies heavily on media coverage to recruit new twins. The twins then come to Minnesota for a week of study—and, often, further publicity. Twins who are "interested in publicity and willing to support it," Kaprio says, may be atypical. This self-selection effect, he adds, may explain why the Bouchard group's estimates of heritability tend to be higher than those of other studies.

One of the most outspoken critics of

some confers male attributes, men with an extra Y become hyperaggressive "supermales." Follow-up studies indicated that while extra-Y men tend to be taller than other men and score slightly lower on intelligence tests, they are otherwise normal. The National Academy of Sciences concluded in a report published this year that there is no evidence to support the link between the extra Y chromosome and violent behavior.

Minnesota Twins

No research in behavioral genetics has been more eagerly embraced by the press than the identical-twin studies done at the University of Minnesota. Thomas J. Bouchard, Jr., a psychologist, initiated them in the late 1970s, and since then they have been featured in the *Washington Post, Newsweek,* the *New York Times* and other publications worldwide as well as on television. *Science* has favorably described the Minnesota team's work in several news stories and in 1990 published a major article by the group.

The workers have studied more than 50 pairs of identical twins who were separated shortly after birth and raised in different households. The assump-

tion is that any differences between identical twins, who share all each other's genes, are caused by the environment; similarities are attributed to their shared genes. The group estimates the relative contribution of genes to a given trait in a term called "heritability." A trait that stems entirely from genes, such as eye color, is defined as 100 percent heritable. Height is 90 percent heritable; that is, 90 percent of the variation in height is accounted for by genetic variation, and the other 10 percent is accounted for by diet and other environmental factors.

The Minnesota group has reported finding a strong genetic contribution to practically all the traits it has examined. Whereas most previous studies have estimated the heritability of intelligence (as defined by performance on intelligence tests) as roughly 50 percent, Bouchard and his colleagues arrived at a figure of 70 percent. They have also found a genetic component underlying such culturally defined traits as religiosity, political orientation (conservative versus liberal), job satisfaction, leisure-time interests and proneness to divorce. In fact, the group concluded in *Science,* "On multiple measures of personality and temperament...monozy-

the Minnesota twin studies—and indeed all twin studies indicating high heritability of behavioral traits—is Leon J. Kamin, a psychologist at Northeastern University. In the 1970s Kamin helped to expose inconsistencies and possible fraud in studies of separated identical twins conducted by the British psychologist Cyril Burt during the previous two decades. Burt's conclusion that intelligence was mostly inherited had inspired various observers, notably Arthur R. Jensen, a psychologist at the University of California at Berkeley, to argue that socioeconomic stratification in the U.S. is largely a genetic phenomenon.

In his investigations of other twin studies, Kamin has shown that identical twins supposedly raised apart are often raised by members of their family or by unrelated families in the same neighborhood; some twins had extensive contact with each other while growing up. Kamin suspects the same may be true of some Minnesota twins. He notes, for example, that some news accounts suggested Oskar and Jack (the Nazi and the Jew) and the two British women wearing seven rings were reunited for the first time when they arrived in Minnesota to be studied by Bouchard. Actually, both pairs of twins had met previously. Kamin has repeatedly asked the Minnesota group for detailed case histories of its twins to determine whether it has underestimated contact and similarities in upbringing. "They've never responded," he says.

Kamin proposes that the Minnesota twins have particularly strong motives to downplay previous contacts and to exaggerate their similarities. They might want to please researchers, to attract more attention from the media or even to make money. In fact, some twins acquired agents and were paid for appearances on television. Jack and Oskar recently sold their life story to a film producer in Los Angeles (who says Robert Duvall is interested in the roles).

Even the Minnesota researchers caution against overinterpretation of their work. They agree with their critics that high heritability should not be equated with inevitability, since the environment can still drastically affect the expression of a gene. For example, the genetic disease phenylketonuria, which causes profound retardation, has a heritability of 100 percent. Yet eliminating the amino acid phenylalanine from the diet of affected persons prevents retardation from occurring.

Such warnings tend to be minimized in media coverage, however. Writers often make the same inference that Koshland did in an editorial in *Science:* "Bet-

ter schools, a better environment, better counseling and better rehabilitation will help some individuals but not all." The prime minister of Singapore apparently reached the same conclusion. A decade ago he cited popular accounts of the Minnesota research in defending policies that encouraged middle-class Singaporeans to bear children and discouraged childbearing by the poor.

Smart Genes

Twin studies, of course, do not indicate which specific genes contribute to a trait. Early in the 1980s scientists began developing powerful ways to unearth that information. The techniques stem from the fact that certain stretches of human DNA, called polymorphisms, vary in a predictable way. If a polymorphism is consistently inherited together with a given trait—blue eyes, for example—then geneticists assume it either lies near a gene for that trait or actually is the gene. A polymorphism that merely lies near a gene is known as a marker.

In so-called linkage studies, investigators search for polymorphisms co-inherited with a trait in families unusually prone to the trait. In 1983 researchers used this method to find a marker linked to Huntington's disease, a crippling neurological disorder that usually strikes carriers in middle age and kills them within 10 years. Since then, the same technique has pinpointed genes for cystic fibrosis, muscular dystrophy and other diseases. In association studies, researchers compare the relative frequency of polymorphisms in two unrelated populations, one with the trait and one lacking it.

Workers are already using both methods to search for polymorphisms associated with intelligence, defined as the ability to score well on standardized intelligence tests. In 1991 Shelley D. Smith of the Boys Town National Institute for Communication Disorders in Children, in Omaha, and David W. Fulker of the University of Colorado identified polymorphisms associated with dyslexia in a linkage study of 19 families exhibiting high incidence of the reading disorder.

Behavioral Genetics: A Lack-of-Progress Report

CRIME: Family, twin and adoption studies have suggested a heritability of 0 to more than 50 percent for predisposition to crime. (Heritability represents the degree to which a trait stems from genetic factors.) In the 1960s researchers reported an association between an extra Y chromosome and violent crime in males. Follow-up studies found that association to be spurious.

MANIC DEPRESSION: Twin and family studies indicate heritability of 60 to 80 percent for susceptibility to manic depression. In 1987 two groups reported locating different genes linked to manic depression, one in Amish families and the other in Israeli families. Both reports have been retracted.

SCHIZOPHRENIA: Twin studies show heritability of 40 to 90 percent. In 1988 a group reported finding a gene linked to schizophrenia in British and Icelandic families. Other studies documented no linkage, and the initial claim has now been retracted.

ALCOHOLISM: Twin and adoption studies suggest heritability ranging from 0 to 60 percent. In 1990 a group claimed to link a gene—one that produces a receptor for the neurotransmitter dopamine—with alcoholism. A recent review of the evidence concluded it does not support a link.

INTELLIGENCE: Twin and adoption studies show a heritability of performance on intelligence tests of 20 to 80 percent. One group recently unveiled preliminary evidence for genetic markers for high intelligence (an IQ of 130 or higher). The study is unpublished.

HOMOSEXUALITY: In 1991 a researcher cited anatomic differences between the brains of heterosexual and homosexual males. Two recent twin studies have found a heritability of roughly 50 percent for predisposition to male or female homosexuality. These reports have been disputed. Another group claims to have preliminary evidence of genes linked to male homosexuality. The data have not been published.

Two years ago Robert Plomin, a psychologist at Pennsylvania State University who has long been active in behavioral genetics, received a $600,000 grant from the National Institute of Child Health and Human Development to search for genes linked to high intelligence. Plomin is using the association method, which he says is more suited than the linkage technique to identifying genes whose contribution to a trait is relatively small. Plomin is studying a group of 64 schoolchildren 12 to 13 years old who fall into three groups: those who score approximately 130, 100 and 80 on intelligence tests.

Plomin has examined some 25 polymorphisms in each of these three groups, trying to determine whether any occur with greater frequency in the "bright" children. The polymorphisms have been linked to genes thought to have neurological effects. He has uncovered several markers that seem to occur more often in the highest-scoring children. He is now seeking to replicate his results in another group of 60 children; half score above 142 on intelligence tests, and half score less than 74 (yet have no obvious organic deficiencies). Plomin presented his preliminary findings at a meeting, titled "Origins and Development of High Ability," held in London in January.

At the same meeting, however, other workers offered evidence that intelligence tests are actually poor predictors of success in business, the arts or even advanced academic programs. Indeed, even Plomin seems ambivalent about the value of his research. He suggests that someday genetic information on the cognitive abilities of children might help teachers design lessons that are more suited to students' innate strengths and weaknesses.

But he also calls his approach "a fishing expedition," given that a large number of genes may contribute to intelligence. He thinks the heritability of intelligence is not 70 percent, as the Minnesota twin researchers have claimed, but 50 percent, which is the average finding of other studies, and at best he can only find a gene that accounts for a tiny part of variance in intelligence. "If you wanted to select on the basis of this, it would be of no use whatsoever," he remarks. These cautions did not prevent the *Sunday Telegraph*, a London newspaper, from announcing that Plomin had found "evidence that geniuses are born not made."

Evan S. Balaban, a biologist at Harvard, thinks Plomin's fishing expedition is doomed to fail. He grants that there may well be a significant genetic compo-nent to intelligence (while insisting that studies by Bouchard and others have not demonstrated one). But he doubts whether investigators will ever uncover any specific genes related to high intelligence or "genius." "It is very rare to find genes that have a specific effect," he says. "For evolutionary reasons, this just doesn't happen very often."

The history of the search for markers associated with mental illness supports Balaban's view. Over the past few decades, studies of twins, families and adoptees have convinced most investigators that schizophrenia and manic depression are not caused by psychosocial factors—such as the notorious "schizophrenogenic mother" postulated by some Freudian psychiatrists—but by biological and genetic factors. After observing the dramatic success of linkage studies in the early 1980s, researchers immediately began using the technique to isolate polymorphic markers for mental illness. The potential value of such research was enormous, given that schizophrenia and manic depression each affect roughly one percent of the global population.

They seemed to have achieved their first great success in 1987. A group led by Janice A. Egeland of the University of Miami School of Medicine claimed it had linked a genetic marker on chromosome 11 to manic depression in an Amish population. That same year another team, led by Miron Baron of Columbia University, linked a marker on the X chromosome to manic depression in three Israeli families.

The media hailed these announcements as major breakthroughs. Far less attention was paid to the retractions that followed. A more extensive analysis of the Amish in 1989 by a group from the National Institute of Mental Health turned up no link between chromosome 11 and manic depression. This year Baron's team retracted its claim of linkage with the X chromosome after doing a new study of its Israeli families with more sophisticated markers and more extensive diagnoses.

Schizophrenic Results

Studies of schizophrenia have followed a remarkably similar course. In 1988 a group headed by Hugh M. D. Gurling of the University College, London, Medical School announced in *Nature* that it had found linkage in Icelandic and British families between genetic markers on chromosome 5 and schizophrenia. In the same issue, however, researchers led by Kenneth K. Kidd of Yale University reported seeing no such linkage in a Swedish family. Although Gurling defended his result as legitimate for several years, additional research has convinced him that it was probably a false positive. "The new families showed no linkage at all," he says.

These disappointments have highlighted the problems involved in using linkage to study mental illness. Neil Risch, a geneticist at Yale, points out that linkage analysis is ideal for studying diseases, such as Huntington's, that have distinct symptoms and are caused by a single dominant gene. Some researchers had hoped that at least certain subtypes of schizophrenia or manic depression might be single-gene disorders. Single-gene mutations are thought to cause variants of breast cancer and of Alzheimer's disease that run in families and are manifested much earlier than usual. But such diseases are rare, Risch says, because natural selection quickly winnows them out of the population, and no evidence exists for distinct subtypes of manic depression or schizophrenia.

Indeed, all the available evidence suggests that schizophrenia and manic depression are caused by at least several genes—each of which may exert only a tiny influence—acting in concert with environmental influences. Finding such genes with linkage analysis may not be impossible, Risch says, but it will be considerably more difficult than identifying genes that have a one-to-one correspondence to a trait. The difficulty is compounded by the fact that the diagnosis of mental illness is often subjective—all the more so when researchers are relying on family records or recollections.

Some experts now question whether genes play a significant role in mental illness. "Personally, I think we have overestimated the genetic component of schizophrenia," says E. Fuller Torrey, a psychiatrist at St. Elizabeth's Hospital in Washington, D.C. He argues that the evidence supporting genetic models can be explained by other biological factors, such as a virus that strikes in utero. The pattern of incidence of schizophrenia in families often resembles that of other viral diseases, such as polio. "Genes may just create a susceptibility to the virus," Torrey explains.

The Drink Link

Even Kidd, the Yale geneticist who has devoted his career to searching for genes linked to mental illness, acknowledges that "in a rigorous, technical, scientific sense, there is very little proof that schizophrenia, manic depression" and other psychiatric disorders have a genetic origin. "Virtually all the evidence supports a genetic explanation, but there are always other explanations, even if they are convoluted."

The evidence for a genetic basis for alcoholism is even more tentative than that for manic depression and schizophrenia. Although some studies discern a genetic component, especially in males, others have reached the opposite conclusion. Gurling, the University College investigator, found a decade ago that identical twins were slightly *more* likely to be discordant for alcoholism than fraternal twins. The drinking habits of some identical twins were strikingly different. "In some cases, one drank a few bottles a day, and the other didn't drink at all," Gurling says.

Nevertheless, in 1990 a group led by Kenneth Blum of the University of Texas Health Science Center at San Antonio announced it had discovered a genetic marker for alcoholism in an association study comparing 35 alcoholics with a control group of 35 nonalcoholics. A page-one story in the *New York Times* portrayed the research as a potential watershed in the diagnosis and treatment of alcoholism without mentioning the considerable skepticism aroused among other researchers.

The Blum group claimed that its marker, called the A1 allele, was associated with a gene, called the D2 gene, that codes for a receptor for the neurotransmitter dopamine. Skeptics noted that the A1 allele was actually some 10,000 base pairs from the dopamine-receptor gene and was not linked to any detectable variation in its expression.

Since the initial announcement by Blum, three papers, including an additional one by Blum's group, have presented more evidence of an association between the A1 allele and alcoholism. Six groups have found no such evidence (and received virtually no mention in the popular media).

In April, Risch and Joel Gelernter of Yale and David Goldman of the National Institute on Alcohol Abuse and Alcoholism analyzed all these studies on the A1 allele in a paper in the *Journal of the American Medical Association.* They noted that if Blum's two studies are cast aside, the balance of the results shows no association between the D2 receptor and alcoholism, either in the disor-

BRAIN OF SCHIZOPHRENIC (*right*) appears different from the brain of his identical twin in these magnetic resonance images. Such findings suggest that factors that are biological but not genetic—such as viruses—may play a significant role in mental illness.

der's milder or most severe forms. "We therefore conclude that no physiologically significant association" between the A1 allele and alcoholism has been proved, the group stated. "It's a dead issue," Risch says.

Gelernter and his colleagues point out that association studies are prone to spurious results if not properly controlled. They suggest that the positive findings of Blum and his colleagues may have derived from a failure to control for ethnic variation. The limited surveys done so far have shown that the incidence of the A1 allele varies wildly in different ethnic groups, ranging from 10 percent in certain Jewish groups to about 50 percent in Japanese.

Blum insists that the ethnic data, far from undermining his case, support it, since those groups with the highest prevalence of the A1 allele also exhibit the highest rates of "addictive behavior." He contends that the only reason the Japanese do not display higher rates of alcoholism is that many also carry a gene that prevents them from metabolizing alcohol. "They're pretty compulsive," explains Blum, who recently obtained a patent for a genetic test for alcoholism.

These arguments have been rejected even by Irving I. Gottesman of the University of Virginia, who is a strong defender of genetic models of human behavior. He considers the papers cited by Blum to support his case to be ambiguous and even contradictory. Some see an association only with alcoholism that leads to medical complications or even death; others discern no association with alcoholism but only with "polysubstance abuse," including cigarette smoking. "I think it is by and large

garbage," Gottesman says of the alleged A1-alcoholism link.

By far the most controversial area of behavioral genetics is research on crime. Last fall complaints by civil-rights leaders and others led the National Institutes of Health to withdraw its funding from a meeting entitled "Genetic Factors in Crime: Findings, Uses and Implications." The conference brochure had noted the "apparent failure of environmental approaches to crime" and suggested that genetic research might yield methods for identifying and treating potential criminals—and particularly those prone to violence—at an early age.

Critics contend that such investigations inevitably suggest that blacks are predisposed to crime, given that blacks in the U.S. are six times more likely than whites to be arrested for a violent crime. In fact, some prominent scientists, notably Richard J. Herrnstein, a psychologist at Harvard, have made this assertion. Others reject this view but insist biological research on attributes linked to violent crime, such as aggression, may still have some value. "People who are unwilling to address genetic and biochemical factors are just putting their heads in the sand," says Goldman, the alcoholism expert. "It is not fair to say that just because there have been geneticists who have had a very narrow view of this in the past, we shouldn't explore this now."

In fact, investigations of the biology of violent crime continue, albeit quietly. Workers at City of Hope Hospital in Duarte, Calif., claim to have found an association between the A1 allele—the alleged alcoholism marker—and "criminal aggression." Last year a group led

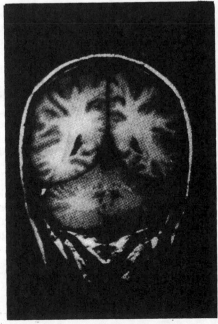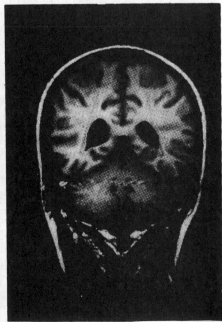

by Markus J. P. Kruesi of the University of Illinois at Chicago presented evidence of an association between low levels of the neurotransmitter serotonin and disruptive-behavior disorders in children. Kruesi concedes there is no way to determine whether the serotonin levels are genetically influenced. In fact, the serotonin levels might be an effect—a reaction to an environmental trauma—rather than a cause. "This might be a scar marker," he says.

One reason such research persists is that studies of families, twins and adoptees have suggested a genetic component to crime. Glenn D. Walters, a psychologist at the Federal Correctional Institution in Schuylkill, Pa., recently reviewed 38 of these studies, conducted from the 1930s to the present, in the journal *Criminology*. His meta-analysis turned up a small genetic effect, "but nothing to get excited about." He observes that "a lot of the research has not been very good" and that the more recent, better-designed studies tended to turn up less evidence. "I don't think we will find any biological markers for crime," he says. "We should put our resources elsewhere."

Gay Genes

The ostensible purpose of investigations of mental illness, alcoholism and even crime is to reduce their incidence. Scientists studying homosexuality have a different goal: simply to test whether homosexuality is innate, as many homosexuals have long professed. That claim was advanced by a report in *Science* in 1991 by Simon LeVay of the Salk Institute for Biological Studies in San Diego. LeVay has acknowledged both that he is gay and that he believes evidence of biological differences between homosexuals and heterosexuals will encourage tolerance toward gays.

LeVay, who recently left the Salk Institute to found the Institute of Gay and Lesbian Education, focused on a tiny neural structure in the hypothalamus, a region of the brain known to control sexual response. He measured this structure, called the interstitial nucleus, in autopsies of the brains of 19 homosexual males, 16 heterosexual males and six heterosexual women. LeVay found that the interstitial nucleus was almost twice as large in the heterosexual males as in the homosexual males or in the women. He postulated that the interstitial nucleus "is large in individuals oriented toward women"—whether male or female.

Of course, LeVay's finding only addresses anatomic differences, not nec-

The Huntington's Disease Saga: A Cautionary Tale

The identification of the gene for Huntington's disease, which was announced in March, was hailed as one of the great success stories of modern genetics. Yet it provides some rather sobering lessons for researchers seeking genes linked to more complex human disorders and traits.

The story begins in the late 1970s, when workers developed novel techniques for identifying polymorphisms, sections of the human genome that come in two or more forms. Investigators realized that by finding polymorphisms linked—always and exclusively—to diseases, they could determine which chromosome the gene resides in. Researchers decided to test the polymorphism technique on Huntington's disease, a devastating neurological disorder that affects roughly one in 10,000 people. Scientists had known for more than a century that Huntington's was caused by a mutant, dominant gene. If one parent has the disease, his or her offspring have a 50 percent chance of inheriting it.

One of the leaders of the Huntington's effort was Nancy Wexler, a neuropsychologist at Columbia University whose mother had died of the disease and who therefore has a 50 percent chance of developing it herself. She and other researchers focused on a poor Venezuelan village whose inhabitants had an unusually high incidence of the disease. In 1983, through what has now become a legendary stroke of good fortune, they found a linkage with one of the first polymorphisms they tested. The linkage indicated that the gene for Huntington's disease was somewhere on chromosome 4.

The finding led quickly to a test for determining whether offspring of carriers—either in utero or already born—have inherited the gene itself. The test requires an analysis of blood samples from several members of a family known to carry the disease. Wexler herself has declined to say whether she has taken the test.

Researchers assumed that they would quickly identify the actual gene in chromosome 4 that causes Huntington's disease. Yet it took 10 years for six teams of workers from 10 institutions to find the gene. It is a so-called expanding gene, which for unknown reasons gains base pairs (the chemical "rungs" binding two strands of DNA) every time it is transmitted. The greater the expansion of the gene, researchers say, the earlier the onset of the disease. The search was complicated by the fact that workers had no physical clues about the course of the disease to guide them. Indeed, Wexler and others emphasize that they still have no idea how the gene actually causes the disease; treatments or cures may be years or decades away.

The most immediate impact of the new discovery will be the development of a better test for Huntington's, one that requires blood only from the person at risk

essarily genetic ones. Various other researchers have tried to establish that homosexuality is not just biological in its origin—caused, perhaps, by hormonal influences in utero—but also genetic. Some have sought evidence in experiments with rats and other animals. A group headed by Angela Pattatucci of the National Cancer Institute is studying a strain of male fruit flies—which wags have dubbed either "fruity" or "fruitless"—that court other males.

In December 1991 J. Michael Bailey of Northwestern University and Richard C. Pillard of Boston University announced they had uncovered evidence of a genetic basis for male homosexuality in humans. They studied 161 gay men, each of whom had at least one identical or fraternal twin or adopted brother. The researchers determined that 52 percent of the identical twins were both

homosexual, as compared with 22 percent of the fraternal twins and 11 percent of the adopted brothers.

Bailey and Pillard derived similar results in a study of lesbians published this year in the *Archives of General Psychiatry*. They compared 147 gay women with identical or fraternal twins or adopted sisters: 48 percent of the identical twins were both gay, versus 16 percent of the fraternal twins (who share only half each other's genes) and 6 percent of the adopted sisters. "Both male and female sexual orientation appeared to be influenced by genetic factors," Bailey and Pillard concluded.

This conclusion has disturbed some of Bailey and Pillard's own subjects. "I have major questions about the validity of some of the assumptions they are making," says Nina Sossen, a gay woman living in Madison, Wis., whose identical twin is heterosexual. Her doubts

"Hierarchy of Worthlessness"

NANCY WEXLER helped to find the gene responsible for Huntington's disease by studying a population in Venezuela that has been ravaged by the disorder.

and not other family members. By measuring the length of the mutant gene, the test might also predict more accurately when carriers will show symptoms.

As difficult as it was to pinpoint the gene for Huntington's, it will be almost infinitely harder to discover genes for behavioral disorders, says Evan S. Balaban, a biologist at Harvard University. Unlike Huntington's disease, he notes, disorders such as schizophrenia and alcoholism cannot be unambiguously diagnosed. Furthermore, they stem not from a single dominant gene but from many genes acting in concert with environmental effects. If researchers do find a statistical association between certain genes and a trait, Balaban says, that knowledge may never be translated into useful therapies or tests. "What does it mean to have a 10 percent increased risk of alcoholism?" he asks.

are shared by William Byne, a psychiatrist at Columbia University. He notes that in their study of male homosexuality Bailey and Pillard found more concordance between unrelated, adopted brothers than related (but non-twin) brothers. The high concordance of the male and female identical twins, moreover, may stem from the fact that such twins are often dressed alike and treated alike—indeed, they are often mistaken for each other—by family members as well as by others.

"The increased concordance for homosexuality among the identical twins could be entirely accounted for by the increased similarity of their developmental experiences," Byne says. "In my opinion, the major finding of that study is that 48 percent of identical twins who were reared together were discordant for sexual orientation."

Byne also criticizes LeVay's conclusion that homosexuality must be biological—although not necessarily genetic—because the brains of male homosexuals resemble the brains of women. That assumption, Byne points out, rests on still another assumption, that there are significant anatomic differences between heterosexual male and female brains. But to date, there have been no replicable studies showing such sexual dimorphism.

Byne notes that he has been suspected of having an antigay motive. Two reviewers of an article he recently wrote criticizing homosexuality research accused him of having a "right-wing agenda," he says. He has also been contacted by conservative groups hoping he will speak out against the admittance of homosexuals to the military. He emphasizes that he supports gay rights and thinks homosexuality, whatever its cause, is not a "choice." He adds that genetic models of behavior are just as likely to foment bigotry as to quell it.

Despite the skepticism of Byne and others, at least one group, led by Dean Hamer of the National Cancer Institute, is searching not merely for anatomic or biochemical differences in homosexuals but for genetic markers. Hamer has done a linkage study of numerous small families, each of which has at least two gay brothers. He says his study has turned up some tentative findings, and he plans to submit his results soon. Hamer's colleague Pattatucci is planning a similar study of lesbians.

What purpose will be served by pinpointing genes linked to homosexuality? In an information sheet for prospective participants in his study, Hamer expresses the hope that his research may "improve understanding between people with different sexual orientations." He adds, "This study is not aimed at developing methods to alter either heterosexual or homosexual orientation, and the results of the study will not allow sexual orientation to be determined by a blood test or amniocentesis."

Yet even Pillard, who is gay and applauds Hamer's work, admits to some concern over the potential uses of a genetic marker for homosexuality. He notes that some parents might choose to abort embryos carrying such a marker. Male and female homosexuals might then retaliate, he says, by conceiving children and aborting fetuses that lacked such a gene.

Balaban, the Harvard biologist, thinks the possible dangers of such research—assuming it is successful—outweigh any benefits. Indeed, he sees behavioral genetics as a "hierarchy of worthlessness," with twin studies at the bottom and linkage studies of mental illness at the top. The best researchers can hope for is to find, say, a gene associated with a slightly elevated risk of schizophrenia. Such information is more likely to lead to discrimination by insurance companies and employers than to therapeutic benefits, Balaban warns.

His colleague Lewontin agrees. In the 1970s, he recalls, insurance companies began requiring black customers to take tests for sickle cell anemia, a genetic disease that primarily affects blacks. Those who refused to take the test or who tested positive were denied coverage. "I feel that this research is a substitute for what is really hard—finding out how to change social conditions," Lewontin remarks. "I think it's the wrong direction for research, given that we have a finite amount of resources."

Paul R. Billings, a geneticist at the California Pacific Medical Center, shares some of these concerns. He agrees that twin studies seem to be inherently ambiguous, and he urges researchers seeking markers for homosexuality to consider what a conservative government—led by Patrick Buchanan, for example—might allow to be done with such information. But he believes some aspects of behavioral genetics, particularly searches for genes underlying mental illness, are worth pursuing.

In an article published in the British journal *Social Science and Medicine* last year, Billings and two other scientists offered some constructive criticism for the field. Researchers engaged in association and linkage studies should establish "strict criteria as to what would constitute meaningful data." Both scientists and the press should emphasize the limitations of such studies, "especially when the mechanism of how a gene acts on a behavior is not known." Billings and his colleagues strive to end their article on a positive note. "Despite the shortcomings of other studies," they say, "there is relatively good evidence for a site on the X chromosome which is associated with [manic depression] in some families." This finding was retracted earlier this year.

"Better Breeding"

Fairly or not, modern genetics research is still haunted by the history of eugenics. "It offers a lot of cautionary lessons," says Daniel J. Kevles, a historian at the California Institute of Technology, who wrote the 1985 book *In the Name of Eugenics*. The British scientist Francis Galton, cousin to Charles Darwin, first proposed that human society could be improved "through better breeding" in 1865 in an article entitled "Hereditary Talent and Character." He coined the term "eugenics," from the Greek for "good birth," in 1883.

Galton's proposal had broad appeal. The American sexual libertarian John Humphrey Noyes bent eugenics into an ingenious argument for polygamy. "While the good man will be limited by his conscience to what the law allows," Noyes said, "the bad man, free from moral check, will distribute his seed beyond the legal limit."

A more serious advocate was the biologist Charles B. Davenport, founder of Cold Spring Harbor Laboratory and of the Eugenics Record Office, which gathered information on thousands of American families for genetic research. After demonstrating the heritability of eye, skin and hair color, Davenport went on to "prove" the heritability of traits such as "pauperism," criminality and "feeble-mindedness." In one monograph, published in 1919, he asserted that the ability to be a naval officer is an inherited trait, composed of subtraits for thalassophilia, or love of the sea, and hyperkineticism, or wanderlust. Noting the paucity of female naval officers, Davenport concluded that the trait is unique to males.

Beginning in the 1920s the American Eugenics Society, founded by Davenport and others, sponsored "Fitter Families Contests" at state fairs around the U.S. Just as cows and sheep were appraised by judges at the fairs, so were human entrants (such as the family shown above at the 1925 Texas State Fair). Less amusingly, eugenicists helped to persuade more than 20 U.S. states to authorize sterilization of men and women in prisons and mental hospitals, and they urged the federal government to restrict the immigration of "undesirable" races.

No nation, of course, practiced eugenics as enthusiastically as Nazi Germany, whose program culminated in "euthanasia" ("good death") of the mentally and physically disabled as well as Jews, Gypsies, Catholics and others. As revelations of these atrocities spread after World War II, popular support for eugenics programs waned in the U.S. and elsewhere.

The DNA Wars

Touted as an infallible method of identifying criminals, DNA matching has mired courts in a vicious battle of expert witnesses.

Edward Humes

Edward Humes is the author of "Murderer With a Badge," published by Dutton. He won a 1989 Pulitzer Prize for specialized reporting while at the Orange County Register.

The ambush was waiting for Prof. Laurence Mueller long before he arrived at the courthouse with his bar charts and slides and other hieroglyphs of science. You could see it in the deputy D.A.'s bear-trap smile, in the pile of dog-eared files he had amassed on Mueller's past courtroom performances, in the patiently highlighted inconsistencies ferreted out of those transcripts. The prosecutor and his experts were even making book—inaccurately, it turned out—that the good professor would wear his trademark tweed blazer, a coat that announces academe faster than any resume.

Mueller, a professor of evolutionary biology at UC Irvine, had been hired on behalf of an accused rapist to testify on the potentially eye-glazing subject of population genetics as it applies to that *Wunderkind* police technology, DNA "fingerprinting." With such a topic, his main concern should have been keeping the jury awake.

He needn't have worried. By the time Mueller left that courtroom in Santa Ana last spring, science, the presumptive subject of the day, had taken a back seat to more personal, and far more heated, topics. In short order, Mueller was branded a hired gun devoid of principles because he has earned more as an expert witness in DNA cases than as a university professor. As for his opinions, Mueller wasn't just proclaimed mistaken, he was accused of attempting to "lie with statistics" and of making deliberately erroneous conclusions with false data, a cardinal sin in the scientific world. (Mueller and his colleagues say his only sin was disagreeing with the government's experts.) Finally, Mueller was mocked for daring to apply lessons from his experiments with fruit flies to human genetics, even though the lowly fruit fly, with its prodigious birth rate, has been a staple of genetic science for most of this century.

LAURENCE MUELLER: The professor's opposition to DNA evidence has prompted vitriolic personal attacks.

The two-fisted, in-your-face cross-examination had jurors on the edges of their chairs and the defense lawyer itching to fling a similar attack at the prosecution's three experts. Which he did, with the same sort of biting accusations Mueller had withstood, charging the experts with ignoring data to reach false conclusions.

Funny thing was, Mueller declared the experience one of his milder courtroom bouts to date. More often than not, the verbal body blows get a lot nastier. "It was heated, yes, but not as bad as some," Mueller said. Later he added, "And I've taken a few other jackets out of mothballs."

Genetic printing, a once seemingly unassailable technology for identifying criminals through their DNA, has been stripped of its luster by a series of furious academic debates, vicious legal battles and highly personal attacks on scientists nationwide. Mueller's experience has become the rule, not the exception, and people on either side of the equation say that both justice and science are the ultimate casualties of this war.

DNA printing, it turns out, though an undeniably powerful tool for the police, is not the nearly infallible magic bullet for solving crimes its proponents once claimed. Bottom line: Much of the time this method of comparing a suspect's DNA to blood, semen or hair found at crime scenes works, sometimes it doesn't. And the close calls—when the spotty, ink-blot-like patterns of deoxyribonucleaic acid are blurred or faint—become so subjective that their utility is questionable. Matching a suspect's DNA to evidence from a crime scene becomes an art, not a science, with lives and liberty hanging in the balance.

This simple realization has created chaos in the courts. Judges and juries with no scientific training are being asked to decide which scientists are right: those who believe DNA printing works well, or those who believe it is fatally flawed. Each side is sincere. Each cites reams of data. Each boasts impeccable resumes, and equally impeccable reasons for slamming those who disagree. So how do jurors know whether a complex technology is being used to convict the guilty, or to railroad the innocent? They can listen to hours of testimony and still not know. Innocents might be sent to Death Row, while killers are set free. It all depends on which side has the cleverest lawyers and the most persuasive scientists. Across the country, the confusion is so great that one judge can allow DNA printing as evidence while another, in the same state or even the same courthouse, could rule that the technology is still too unreliable.

And so, perhaps inevitably, a strange thing has happened as lawyers jockey for the sympathies of juries and jurists. Instead of the science, the scientists themselves are being challenged with unprecedented venom. Reputations have been tarnished, accusations of government conspiracies and defense lawyers' cabals have been thrown onto the fire and bitter rifts among researchers have arisen, making it increasingly difficult for them to work together to resolve the very debate causing the courtroom combat.

The reason for this lies in a long-standing legal principle that judges may admit new scientific evidence in the courtroom only when there is no significant debate within the scientific community over its validity. That means government prosecutors and experts who want to use DNA to convict people cannot simply disagree with scientists who question the reliability of genetic printing. They must discredit and denounce the critics, questioning their motives, ethics and abilities to show that there really is no legitimate debate. Defense lawyers have responded in kind, attacking with equal zest.

Scientists accustomed to the sterility of the laboratory and the decorum of university conferences have been left blinking in anger, and sometimes fear, at the withering attacks. Allegations of perjury and hidden conflicts of interest, smirking remarks about sexual orientation, insinuations about tax audits and immigration status—nothing is off-limits. Many researchers now simply

refuse to testify rather than face being shredded on the stand.

"It's not about science," says a rueful Mueller. "They are interested in dirty laundry."

ROCKNE HARMON: The deputy D.A. is noted for his use of DNA matching—and for discrediting its opponents.

"I'm not ashamed," counters Alameda County Deputy Dist. Atty. Rockne Harmon, a nationally prominent expert at using DNA evidence in court who has been equally prominent in efforts to discredit Mueller and other scientists critical of the technology. During charitable moments, he calls Mueller a "knucklehead." "A dangerous criminal," he says, "could be set free to rape, rob or kill people."

In short, the battle over DNA fingerprinting has become the most entertaining and bewildering legal spectacle around. With jurors and judges unsure just whom to believe, resolution is nowhere in sight.

By the end of last year, the FBI had performed 4,000 DNA comparisons in criminal cases, with private, state and county labs adding thousands more. In about a third of those cases, the tests cleared the suspect. But hundreds of people were prosecuted using DNA evidence in more than 40 states.

Most of these defendants were convicted, a testament to the power of this new scientific evidence. But even when DNA tests were ruled inadmissible, the suspects usually were still found guilty. In part, that is because prosecutors chose slam-dunk cases for their first DNA trials, under the theory that obviously guilty individuals were less likely to pursue appeals of the DNA portion of their cases. Only now are numbers of cases beginning to appear in which DNA is the key piece of evidence, the only link between a suspect and a crime.

In the earlier mold, one case stands out. In New York, a handyman by the name of Joseph Castro was accused of fatally stabbing a pregnant South Bronx woman and her 2-year-old daughter in February, 1987. Lifecodes Corp. of Valhalla, N.Y., one of three private labs in the United States that does forensic genetics testing, was hired by prosecutors to do a DNA analysis. Lifecodes' tests said the DNA from a tiny bloodstain on Castro's wristwatch matched the woman's genes, and that the odds against the blood belonging to someone else were 1 in 100 million.

But the Castro case represented the first full-blown challenge to DNA evidence. Castro's lawyers amassed a team of prestigious scientists, who examined Lifecodes' test results and found numerous errors. Even one of the prosecution experts, after conferring with the defense scientists, reversed himself and said that the vaunted 1 in 100 million match was wrong. The judge hearing Castro's case kicked the DNA evidence out.

Yet prosecutors also proclaimed a victory: Castro later pleaded guilty, admitting that the blood on his watch really had come from the victim. Prosecutors say the Castro case, in which the tests were accurate but ruled inadmissible, raises the specter of criminals' going free. Had there not been other, damning evidence in the case, they say, a double murderer would have walked. (In fact, although no national statistics are kept, experts on both sides of the DNA issue could recall only three or four defendants who won acquittals after DNA evidence was excluded.) Defense attorneys counter that bogus DNA evidence could just as easily convict an innocent person.

The prosecution of David Hicks, meanwhile, demonstrates another extreme—a case where DNA was the pivotal evidence despite problems with the test results. A young, unskilled laborer now on Death Row for the murder-rape of his grandmother in rural Freestone County, Tex., Hicks lacked the expert witnesses and skilled attorneys Castro had. The blurry DNA prints obtained in his case, and the 96 million to 1 probability of guilt announced by prosecution experts, again from Lifecodes, were barely challenged. The one expert hired for Hicks' defense, former UC Irvine geneticist Simon Ford, said he got such limited and late access to lab data that he was unprepared when he testified. Yet the defense attorney appointed to represent Hicks declined to ask for more time.

After Hicks was convicted in early 1989 and sentenced to death, Ford and other scientists, Laurence Mueller among them, reviewed the Lifecodes lab work and concluded that the DNA match was improperly done and vastly overstated. An accurate analysis, Ford estimates, would show not a 96 million to 1 probability that Hicks was the killer, but something on the order of hundreds to 1. Furthermore, in Furney-Richardson, Tex., population 300, where Hicks was born and lived, many residents are genetically related. Families have lived there, married their neighbors and passed around the same genetic patterns, the same DNA, for generations. Therefore, people in the community are far more likely to share DNA prints with Hicks—and the killer. Indeed, relatives of Hicks were suspects early in the case, raising an even greater possibility that the killer's DNA pattern could appear to match Hicks', yet not be his.

Hicks' jury never heard any of this because Ford never had a chance to do his analysis before the trial. Had they known, Hicks might easily have been acquitted, since there was little more than circumstantial evidence against him and no eyewitnesses, says William C. Thompson, the attorney and UC Irvine professor now helping Hicks to find legal representation for his appeal. Because of the certainty the 96 million to 1 finding implied, "the jurors were told that there was no way anyone other than Hicks could have been the killer," Thompson says. "That simply is not true."

Hicks' prosecutor, Robert W. Gage, said in a letter to The Sciences magazine that he was convinced Hicks received a fair trial. Gage also asserted that Ford and Thompson, who together have written articles on DNA printing,

have less than altruistic motives behind their criticism. They "profit by traveling around the country and testifying for desperate defendants about the pitfalls of DNA identification," Gage claimed, noting that Ford and Thompson, thanks to their DNA work for defendants, are known to prosecutors nationwide as the "Combine from Irvine."

Both Ford and Thompson say such remarks are examples of widespread harassment against DNA critics. Thompson, who is married and lives with his wife and children near the UC Irvine campus, says opponents are spreading false rumors among the DNA litigation set that he and Ford are lovers using their DNA legal fees to build a home in Laguna Beach. He has also endured bar complaints filed by prosecutors in New Mexico and California, accusing him of conflict of interest for publishing academic articles on cases in which he had been a lawyer. The complaints were ruled groundless, he says.

Ford also complains of prosecutorial harassment and, though still active as a consultant in DNA cases, now avoids testifying in court. A British national, Ford was on the witness stand in an Arizona murder case two years ago when an FBI lawyer abruptly asked about his visa status. Ford was then trying to obtain a permanent resident "green card." Though the judge quickly silenced the attorney for straying into irrelevant areas, Ford saw it as a veiled threat from a federal official.

"It definitely was chilling," Ford recalls. The judge in the Arizona case eventually dealt a stinging blow to the FBI, by finding that its DNA methods were not generally accepted by the scientific community and so could not be used in court. A state appeals court is currently considering whether to ban DNA evidence throughout Arizona.

In Texas, however, David Hicks remains on Death Row, convicted on similar evidence, insisting he is innocent, but garnering few sympathetic ears. Thompson believes Hicks might well be an innocent man convicted wrongly by his own DNA. But given the current conservative climate in ap-

pellate courts, Thompson concedes, stopping Hicks' date with lethal injection will be an uphill battle. Yet the question his case poses is a haunting one: How can evidence be too unreliable in Arizona, yet be used to put a man to death in Texas?

To keep defense experts at bay, an informal network of prosecutors and scientists who wholeheartedly support DNA printing has sprung up across the country. Rockne Harmon, the Alameda County deputy district attorney, is its unofficial clearinghouse. To help discredit defense experts, network members regularly fax articles, transcripts and other tidbits to one another, including, at times, unpublished papers that DNA experts have tried to keep confidential.

Researchers critical of the technology say they have been pressured to alter or withdraw scientific papers from publication. When pressure on individual scientists has failed, government officials and the experts they employ have lobbied publications, including the prestigious journal Science, to reject articles that challenge DNA fingerprinting. In one highly publicized case last December, Science—after pressure from DNA proponents—published an unprecedented, simultaneous rebuttal to an article criticizing the government's theories on DNA printing.

"I never expected that the government would attempt to interfere in my scholarly activities or publications," wrote Daniel L. Hartl, a co-author of the Science article and a professor of genetics at Washington University School of Medicine in St. Louis, in a sworn affidavit. James Wooley, a federal prosecutor from an organized crime strike force in Ohio, obtained the article before it was published, then telephoned Hartl, who had been a defense witness in a case Wooley prosecuted. According to the professor, Wooley "proceeded to badger me for almost an hour asserting that the article would do incalculable harm to government prosecutions and the criminal justice system. . . . I was particularly disturbed when Mr. Wooley . . . asked me whether I was afraid of having my taxes audited."

Wooley denies any attempt to stifle Hartl's, or anyone else's, academic freedom, saying he may have made some sarcastic remarks toward the end of his telephone conversation with Hartl, but that the scientist was being paranoid.

Perhaps more than anyone, Laurence Mueller has been a favorite whipping boy in the DNA debate. Deputy Dist. Atty. Harmon has written harsh letters questioning the quality and accuracy of his testimony, not only to various scientific journals to which Mueller has submitted articles, but to his department chairman and the UC Irvine chancellor. Editors at Science even informed Harmon by mail that they had rejected an article by Mueller before they told the scientist of their decision. "It sounds like they respect me more than they do him," Harmon says.

The government is not the only side getting down and dirty: Defense lawyers active in DNA cases also have their own network of litigators and scientists, and the government's "pro-DNA" experts have begun to complain that they, too, are being harassed by harsh attacks on their integrity.

In U.S. vs. Yee, an Ohio murder case, both sides of the DNA war pulled out all stops. Last spring, New York defense lawyers Barry Scheck and Peter Neufeld filed a motion for a new trial, charging the government with conducting a national campaign to harass and stifle DNA critics. At the same time, the two attorneys launched an attack of their own: The motion accuses two prominent Texas genetics researchers of warping their scientific opinions in favor of DNA printing in order to garner $500,000 in Justice Department grants.

Scheck and Neufeld's broadside also accuses the FBI's chief expert in DNA printing, Bruce Budowle, of citing a nonexistent study to support convictions in the Yee case. The lawyers have demanded an investigation of this alleged perjury by the FBI, the nation's preeminent purveyor of DNA printing. Budowle did not return repeated phone messages, but his boss, FBI Assistant Director John W. Hicks, defended the DNA expert, saying that Budowle knows that teams of defense lawyers and experts will scrutinize every word he says on the witness stand. "He knows not to say something dumb."

Not even Hicks is immune from attack: Yee lawyers say he tried to destroy evidence that could aid defendants in DNA cases—allegations that Hicks hotly denies, although government memos do show that he wanted to destroy certain files that might have contained information on FBI lab errors. They also criticized Hicks for attempting to influence a recent report by a committee of the prestigious National Academy of Sciences, which in April issued a qualified endorsement of the use of DNA fingerprinting in court, but rejected certain crucial techniques used by the FBI. Hicks calls the allegations raised in the Yee case "a witch hunt."

Finding common ground between the opposing sides has been nearly impossible, in part because of the way DNA evidence was first presented to the courts and the public: as a magic bullet.

The technique first was used in 1985 in Great Britain, when blood was taken from more than 4,000 men to identify a rapist-murderer who had killed two 15-year-old girls in two neighboring villages. The DNA fragments extracted from these blood samples were then compared to DNA from semen found in the victims; police found their man when he tried to get someone else to contribute a blood sample for him.

Proponents of the technique claimed that the genetic comparison was so exact that the odds of a false match were 1 in hundreds of millions. In other words, they said, the odds of a mistake were so minuscule that DNA was as good as a fingerprint.

By 1987, the technology had crossed the Atlantic, with three private laboratories offering DNA analyses in criminal cases. The FBI soon followed with labs in Washington and Quantico, Va., and various state and local government DNA laboratories came on line in the next two years. DNA printing became a growth industry.

Its appeal was enormous, especially in rape and murder cases. Previously, forensic scientists charged with finding evidence at crime scenes could only compare blood types, or a somewhat more refined analysis of blood enzymes. Semen or blood found in the victims or at the crime scenes was collected and compared to blood from suspects. If there was a different blood type, a suspect was exonerated. If there was a match, however, it often proved little; millions of people share the same blood type or enzyme.

"Before DNA, if we got a figure of 1 in 4, maybe 1 in 10, we were lucky. One in 100, we were ecstatic," says John Hartmann of the Orange County sheriff-coroner's DNA lab, one of the most respected in the nation. "It was not very discriminating."

Then, suddenly, a quantum advance occurred. Compare DNA, proponents of the new technology announced, and you could be sure you had the right culprit. To men like Hartmann, who genuinely agonized at the thought of criminals' going free because their lab work was too inexact, DNA printing represented a law-enforcement "home run." Now rapists who might have gone free because of the vagaries of eyewitness testimony or the lack of hard proof could be prosecuted. Serial murderers who left no living witnesses would no longer prey on society with impunity. Their DNA, with the help of the knowing, dispassionate men in white lab coats who analyzed it, would answer the questions of guilt and innocence. And the odds that they might be wrong would be 1 in millions, maybe billions. It almost seemed too good to be true.

It was.

Even the term genetic "fingerprint" turned out to be a misnomer, granting an unwarranted aura of infallibility to the technique. Now many jurisdictions prohibit the term's use in court out of concern that it misleads juries, using instead DNA "profiling" or "typing."

Jurors can look at a projection of conventional fingerprint comparisons and say, yeah, they really do look alike. Furthermore, it is uncontrovertible that no two people possess the same fingerprints, not even twins.

Neither is true with DNA prints.

Unlike fingerprints, a person's entire DNA is not examined in the technique—current technology is too primitive. Instead, scientists analyze only a few areas of the long, complex chains of deoxyribonucleic acid that contain the genetic blueprint for all organisms, bacteria to Homo sapiens. The regions of the DNA molecule that are studied, fragments called VNTRs, serve no known purpose other than linking other portions of DNA, like empty boxcars in the middle of a freight train.

When semen, blood or hair is left by a rapist or killer at a crime scene, the DNA can be chemically removed and broken up into fragments in a solution of organic solvents and enzymes. The purified DNA that results is poured into an electrified gel, which spreads out the fragments according to their length. After several more steps, another chemical solution called a "probe" is added, which mates with a specific group of VNTRs, highlighting it with a radioactive tag. The final step uses radiation-sensitive film to record the VNTR pattern. The resulting "autoradiograph" bears a passing resemblance to a very sloppy bar code like that on most grocery labels. The crime-scene autorad can then be compared to a suspect's autorad. Theoretically, different patterns mean innocence; a match means guilt.

Sounds a lot like fingerprinting. And proponents and critics agree that the technique is theoretically sound. But in practice, the autorads are occasionally hazy and incomplete, due in part to the decomposition that typically occurs before forensic evidence is gathered. The blurry images that can result may lead one expert to see a match where another sees none. Many matches are clear-cut; others are so subjective that the certainty DNA printing is supposed to convey simply vanishes. Worse, jurors and judges at times don't know what they're really looking at when the scientists start passing out the auto-

rads. So one focus of the debate is whether a match can be declared reliably in each and every case.

The hottest battle, though, revolves around what it means when DNA prints do clearly match. Does a match really mean that only one person matches the criminal's DNA print, or are there other people walking around with the same patterns? Proponents of the technology say the odds of more than one person bearing the same VNTR patterns are extremely small. Critics say prosecutors and their witnesses are vastly overstating this certainty level.

To understand this aspect of the debate, you have to understand how the chemical "probes" work. There are several kinds, and each probe examines a different grouping of genetic "boxcars"—the VNTRs. According to FBI population studies of several thousand people who gave blood samples, one VNTR grouping may occur in 1 out of 100 people. Another grouping may occur in 1 out of 17 individuals, another, 1 in 2,000. By using three, four or more such probes, then multiplying the odds together, incredibly small probabilities are generated, anywhere from 1 in 100,000 to 1 in billions. Thus, the image of DNA printing as magic bullet was born.

In the Hicks case in Texas, for example, Lifecodes declared a six-probe match in calculating the 96 million to 1 probability that he raped and murdered his grandmother. In theory, this should be irrefutable evidence. But defense witness Ford's analysis showed that four of the six VNTR patterns contained in a vaginal swab of the victim could have come from the grandmother's cells, not the killer's—there was no way of telling them apart in this case. As relatives, Hicks and his grandmother had identical VNTR groupings in four places. Using only the two VNTR groupings that must have come from the killer and that do match Hicks, the odds of Hicks' being the guilty party, Ford says, are a few hundred to 1, nothing close to the level of certainty represented to the jury that convicted Hicks. By Ford's estimate, millions of others could have commit-

ted the crime, including many people in Hicks' insular hometown.

The possibility that such communities might contain people with common VNTR patterns is another hot point of contention in the DNA wars. Defense attorneys recently sought to embarrass one of the government's most prominent DNA experts, Kenneth Kidd, a professor of genetics, psychiatry and biology at Yale University for claiming that a four-probe match between two different people was virtually impossible. This claim was refuted—with Kidd's own research on an isolated Amazonian Indian tribe called the Karitiana, in which about a third of the 54 people tested had identical DNA patterns for four different probes. (Hicks was sent to Death Row on the basis of two distinct probes.) Once confronted with his own data, Kidd and prosecutors dismissed the inbred Karitiana community as an aberration that does not apply to the much larger gene pool of U.S. populations.

"That's what they say now," Thompson argues. "But before, they said it couldn't happen at all, anywhere."

The DNA critics say such findings dictate that more conservative probabilities should be used to explain to juries what a DNA match means. Prosecutors resist this, saying the FBI and other labs already build in error factors that give defendants the benefit of the doubt. Obviously, they also oppose change because less dramatic odds leave too much room for doubt about a suspect's guilt. The magic bullet turns into a blank.

The recent National Academy of Sciences report, however, urges somewhat more conservative numbers than the FBI and other labs generate—striking a rare compromise between the two DNA factions. It remains to be seen if either side adopts the report's recommendations. So far, the FBI is resisting change, while defense lawyers lobby for probabilities more in their favor. A series of recent appellate court decisions against DNA evidence, including one in Massachusetts and two in California, cite the academy report and side with the critics.

7. LIVING WITH THE PAST

"Really, when you get down to it, there's not much difference between a million to 1 and 100 million to 1, or a million to 1 and a 100,000 to 1, which is really what they're arguing about," Orange County's John Hartmann says. "Either way, we've hit a home run."

If such arguments were all there was to it, the viciousness that marks DNA court battles probably wouldn't have erupted. Each side merely would amass its experts, let them testify, and the judge and jury would decide. But the battle also revolves around a legal precedent set by a federal appeals court in 1923, the Frye Rule, which governs the admissibility of new scientific evidence in many, though not all, jurisdictions. The legal rule was first established to examine (and ultimately bar from court) lie-detector tests, but 70 years later, it provides an uneasy fit with the state-of-the-art technology of DNA printing. In DNA cases, Frye's peculiarities make the intense attacks on scientists' integrity all but inevitable.

According to the Frye decision, scientific discoveries should be admitted only when there is no substantial debate about them within the relevant scientific communities—in this case, geneticists, molecular biologists and statisticians. Under Frye, judges aren't supposed to decide who is right in a scientific debate—they aren't qualified—but merely whether a legitimate debate exists.

That explains the vicious attacks on the scientists, and why so many complaints of harassment have been generated. If government officials can halt articles and paint DNA critics as unethical charlatans, judges may decide there is no legitimate debate.

Naturally, defense lawyers respond in kind, attempting to prove the government experts are the real charlatans whenever they testify that there is no debate about DNA printing. "It's trench warfare, no doubt about that," Hartmann says.

Until recently, prosecutors have won most of the skirmishes: DNA evidence has been admitted in most cases, with decisive results. Only one state's Supreme Court has ruled DNA inadmissible—Massachusetts—while one other, Minnesota, severely limits the way it can be used. Numerous other states, including California, are reconsidering its admissibility. The U.S. Supreme Court has yet to specifically rule on the issue.

At the trial court level, the results are more contradictory. In the same courthouse, one judge may admit a DNA test while another finds a legitimate scientific dispute exists, rendering the DNA evidence inadmissible. Justice, when it comes to DNA evidence, is unequal, varying from state to state and judge to judge. The U.S. 2nd Circuit Court of Appeals (with jurisdiction over New York, Vermont and Connecticut), for instance, recently approved the use of DNA evidence without lengthy admissibility hearings; at the same time, a District of Columbia judge barred prosecutors from using DNA printing in court.

Another example: last year, one division of the California Court of Appeal upheld DNA evidence in a Ventura murder case, which could have set a precedent for all of California. But a month before that ruling, Los Angeles County Superior Court Judge C. Robert Simpson Jr. ruled in a different murder case that DNA evidence was too questionable to be admitted. "There is a profound, significant and honestly held disagreement among these men of science," Simpson wrote. Because that case developed after the Ventura prosecution, Simpson's decision was not undone by the appellate court's embrace of DNA evidence. The two at-odds rulings stand, their illogic intact.

In August, a different division of the state Court of Appeal, citing the national academy report, ruled that DNA evidence was not admissible. Now the state Supreme Court has been asked to rule definitively.

An Orange County rape trial earlier this year, in which Mueller and Kidd were lambasted, shows the kind of balancing act juries must perform to deal with the attacks, disputes and contradictions. Frank Lee Soto was charged with raping his neighbor, an elderly woman in Westminster. The only hard evidence against him was a DNA test of his blood—he provided it voluntarily—that showed a clear match with the rapist's semen. (The victim, too ill to testify, told police Soto didn't do it.)

Soto's attorney, Paul Stark, fought bitterly with Deputy Dist. Atty. Dennis Bauer over what the DNA match meant. Bauer, with Yale's Kidd and several other of the most prominent pro-DNA printing experts in the country, said the odds were 189 million to 1 that Soto was the right man. Equally prominent experts suggested the numbers should be more favorable to Soto; in particular, biology professor William Shields of the State University of New York at Syracuse said the correct number was between 65,000 to 1 and 23,000 to 1. Given Orange County's population, that meant 17 other potential suspects were running around, five times that in all of Southern California.

Both sides sounded certain and sincere. Both had the charts and the numbers and other paraphernalia to support their positions. Both sides impugned the abilities and motives of the opposing experts in typical vitriolic fashion. There was literally no way for the jurors (or the judge, for that matter) to know whom to believe.

Doing what the scientists refused to do, the jurors sought compromises. Their statements after the trial showed that they accepted the smaller defense numbers put forward by Shields, and so convicted Soto of the lesser charge of attempted rape. Had Soto not been a neighbor, they said, they would have acquitted him. But 65,000 to 1, coupled with the fact that he lived next door, was enough to cinch the case. Soto received a three-year prison sentence, but, because of the uncertainties revolving around the DNA issue, he is free on bail while he appeals.

Hartmann, whose Orange County lab performed the tests on Soto's blood, applauded the jury's decision that DNA evidence alone is insufficient to convict. His attitude is among the most reasonable of any expert in the field and though he is ardent in his support for DNA printing, he does not believe that it should on its own determine anyone's fate: "If I was on that jury, I would have had a real problem with it, too. . . . No one should be convicted solely on the basis of DNA evidence, not even a billion to 1. If there was no other evidence, I would vote not guilty, too."

But for most of the combatants, the DNA debate has assumed the aspect of a religious argument, each side certain it is right, each incapable of seeing the other's point of view. Voices of compromise are drowned out.

There is one unique case in Virginia that calls into question the sincerity of these head-butting positions. Joseph R. O'Dell III was convicted of a vicious and sensational murder in Virginia Beach in 1985. O'Dell was seen at the bar where the murder victim had been on the night of her death, though no one saw them arrive or leave together. Later, police seized bloody men's clothes in O'Dell's garage after they received an angry call from his girlfriend. O'Dell, who had a robbery conviction and other legal run-ins behind him, told police his shirt and jacket got bloody from a bar fight that night, but detectives figured they had

found their man. Lab results seemed to confirm their opinion. The case predates DNA printing, but less sophisticated blood enzyme tests showed that the blood on O'Dell's clothes was consistent with the victim's blood. O'Dell was convicted and sentenced to death.

Since then, with the advent of DNA testing, his appellate lawyers petitioned the court for new tests. The result: the more refined DNA technology shows the blood on O'Dell's clothing cannot be matched to the victim's, according to briefs filed in his appeal. Key evidence against O'Dell apparently has been invalidated, a case where DNA printing seemingly exonerates rather than convicts.

But O'Dell remains on Death Row. He has been unable to win a new trial. Prosecutors, who in other courtrooms have championed the use of DNA evidence against defendants, are so convinced of O'Dell's guilt that they have opposed its use in his case, even as defense attorneys want the test results admitted.

The prosecution is winning so far, keeping the Virginia appellate courts from considering the DNA evidence in O'Dell's case by focusing on a technicality. O'Dell's out-of-state lawyers filed an appeal petition that lacked a

table of contents and a memorandum of facts summarizing the case (a formulaic introduction to the appeal). The prosecution seized on this procedural gaffe as a means of throwing out the entire appeal. By the time the defense lawyers noticed the oversight, a filing deadline had passed by three days, and the Virginia Supreme Court refused to consider the case.

Recently, three justices of the U.S. Supreme Court suggested that the Virginia courts ought to think again. Although the full court declined to hear O'Dell's direct appeal, Harry A. Blackmun, John Paul Stevens and Sandra Day O'Connor decided that putting a man to death because his lawyers forgot a table of contents was going too far, especially if he might be innocent. In an unusual letter handed down in the case, they instructed lower federal courts in Virginia to give careful consideration to O'Dell's next appeal, a habeas corpus petition to be filed this year, "because of the gross injustice that would result if an innocent man were sentenced to death." And there it stands.

O'Dell remains on Death Row, waiting for his case to be heard, even as prosecutors across the country continue to use DNA evidence to send him companions.

The Saltshaker's Curse

Physiological adaptations that helped American blacks survive slavery may now be predisposing their descendants to hypertension

Jared Diamond

Jared Diamond is a professor of physiology at UCLA Medical School.

On the walls of the main corridor at UCLA Medical School hang thirty-seven photographs that tell a moving story. They are the portraits of each graduating class, from the year that the school opened (Class of 1955) to the latest crop (Class of 1991). Throughout the 1950s and early 1960s the portraits are overwhelmingly of young white men, diluted by only a few white women and Asian men. The first black student graduated in 1961, an event not repeated for several more years. When I came to UCLA in 1966, I found myself lecturing to seventy-six students, of whom seventy-four were white. Thereafter the numbers of blacks, Hispanics, and Asians exploded, until the most recent photos show the number of white medical students declining toward a minority.

In these changes of racial composition, there is of course nothing unique about UCLA Medical School. While the shifts in its student body mirror those taking place, at varying rates, in other professional groups throughout American society, we still have a long way to go before professional groups truly mirror society itself. But ethnic diversity among physicians is especially important because of the dangers inherent in a profession composed of white practitioners for whom white biology is the norm.

Different ethnic groups face different health problems, for reasons of genes as well as of life style. Familiar examples include the prevalence of skin cancer and cystic fibrosis in whites, stomach cancer and stroke in Japanese, and diabetes in Hispanics and Pacific islanders. Each year, when I teach a seminar course in ethnically varying disease patterns, these by-now-familiar textbook facts assume a gripping reality, as my various students choose to discuss some disease that affects themselves or their relatives. To read about the molecular biology of sickle-cell anemia is one thing. It's quite another thing when one of my students, a black man homozygous for the sickle-cell gene, describes the pain of his own sickling attacks and how they have affected his life.

Sickle-cell anemia is a case in which the evolutionary origins of medically important genetic differences among peoples are well understood. (It evolved only in malarial regions because it confers resistance against malaria.) But in many other cases the evolutionary origins are not nearly so transparent. Why is it, for example, that only some human populations have a high frequency of the Tay-Sachs gene or of diabetes? . . .

Compared with American whites of the same age and sex, American blacks have, on the average, higher blood pressure, double the risk of developing hypertension, and nearly ten times the risk of dying of it. By age fifty, nearly half of U.S. black men are hypertensive. For a given age and blood pressure, hypertension more often causes heart disease and especially kidney failure and strokes in U.S. blacks than whites. Because the frequency of kidney disease in U.S. blacks is eighteen times that in whites, blacks account for about two-thirds of U.S. patients with hypertensive kidney failure, even though they make up only about one-tenth of the population. Around the world, only Japanese exceed U.S. blacks in their risk of dying from stroke. Yet it was not until 1932 that the average difference in blood pressure between U.S. blacks and whites was clearly demonstrated, thereby exposing a major health problem outside the norms of white medicine.

What is it about American blacks that makes them disproportionately likely to develop hypertension and then to die of its consequences? While this question is of course especially "interesting" to black readers, it also concerns all Americans, because other ethnic groups in the United States are not so far behind blacks in their risk of hypertension. If *Natural History* readers are a cross section of the United States, then about one-quarter of you now have high blood pressure, and more than half of you will die of a heart attack or stroke to which high blood pressure predisposes. Thus, we all have valid reasons for being interested in hypertension.

First, some background on what those numbers mean when your doctor inflates a rubber cuff about your arm, listens, deflates the cuff, and finally pronounces, "Your blood pressure is 120 over 80." The cuff device is called a sphygmomanometer, and it measures the pressure in your artery in units of millimeters of mercury (that's the height to which your blood pressure would force up a column of mercury in case, God forbid, your artery were suddenly connected to a vertical mercury column). Naturally, your blood pressure varies with each stroke of your heart, so the first and second numbers refer, respectively, to the peak pressure at each heartbeat (systolic pressure) and to the minimum pressure between beats (diastolic pressure). Blood pressure varies somewhat with position, activity, and anxiety level, so the measurement is usually made while you are resting flat on your back. Under those conditions, 120 over 80 is an average reading for Americans.

There is no magic cutoff between normal blood pressure and high blood pressure. Instead, the higher your blood pressure, the more likely you are to die of a heart attack, stroke, kidney failure, or ruptured aorta. Usually, a pressure reading higher than 140 over 90 is arbitrarily defined as constituting hypertension, but some people with lower readings will die of a stroke at age fifty, while others with higher readings will die in a car accident in good health at age ninety.

Why do some of us have much higher blood pressure than others? In about 5 percent of hypertensive patients there is an identifiable single cause, such as hormonal imbalance or use of oral contraceptives. In 95 percent of such cases, though, there is no such obvious cause. The clinical euphemism for our ignorance in such cases is "essential hypertension."

Nowadays, we know that there is a big genetic component in essential hypertension, although the particular genes involved have not yet been identified. Among people living in the same household, the correlation coefficient for blood pressure is 0.63 between identical twins, who share all of their genes. (A correlation coefficient of 1.00 would mean that the twins share identical blood pressures as well and would suggest that pressure is determined entirely by genes and not at all by environment.) Fraternal twins or ordinary siblings or a parent and child,

who share half their genes and whose blood pressure would therefore show a correlation coefficient of 0.5 if purely determined genetically, actually have a coefficient of about 0.25. Finally, adopted siblings or a parent and adopted child, who have no direct genetic connection, have a correlation coefficient of only 0.05. Despite the shared household environment, their blood pressures are barely more similar than those of two people pulled randomly off the street. In agreement with this evidence for genetic factors underlying blood pressure itself, your risk of actually developing hypertensive disease increases from 4 percent to 20 percent to 35 percent if, respectively, none or one or both of your parents were hypertensive.

But these same facts suggest that environmental factors also contribute to high blood pressure, since identical twins have similar but not identical blood pressures. Many environmental or life style factors contributing to the risk of hypertension have been identified by epidemiological studies that compare hypertension's frequency in groups of people living under different conditions. Such contributing factors include obesity, high intake of salt or alcohol or saturated fats, and low calcium intake. The proof of this approach is that hypertensive patients who modify their life styles so as to minimize these putative factors often succeed in reducing their blood pressure. Patients are especially advised to reduce salt intake and stress, reduce intake of cholesterol and saturated fats and alcohol, lose weight, cut out smoking, and exercise regularly.

Here are some examples of the epidemiological studies pointing to these risk factors. Around the world, comparisons within and between populations show that both blood pressure and the frequency of hypertension increase hand in hand with salt intake. At the one extreme, Brazil's Yanomamö Indians have the world's lowest-known salt consumption (somewhat above 10 milligrams per day!), lowest average blood pressure (95 over 61!), and lowest incidence of hypertension (no cases!). At the opposite extreme, doc-

tors regard Japan as the "land of apoplexy" because of the high frequency of fatal strokes (Japan's leading cause of death, five times more frequent than in the United States), linked with high blood pressure and notoriously salty food. Within Japan itself these factors reach their extremes in Akita Prefecture, famous for its tasty rice, which Akita farmers flavor with salt, wash down with salty miso soup, and alternate with salt pickles between meals. Of 300 Akita adults studied, not one consumed less than five grams of salt daily, the average consumption was twenty-seven grams, and the most salt-loving individual consumed an incredible sixty-one grams—enough to devour the contents of the usual twenty-six-ounce supermarket salt container in a mere twelve days. The *average* blood pressure in Akita by age fifty is 151 over 93, making hypertension (pressure higher than 140 over 90) the norm. Not surprisingly, Akita's frequency of death by stroke is more than double even the Japanese average, and in some Akita villages 99 percent of the population dies before age seventy.

Why salt intake often (in about 60 percent of hypertensive patients) leads to high blood pressure is not fully understood. One possible interpretation is that salt intake triggers thirst, leading to an increase in blood volume. In response, the heart increases its output and blood pressure rises, causing the kidneys to filter more salt and water under that increased pressure. The result is a new steady state, in which salt and water excretion again equals intake, but more salt and water are stored in the body and blood pressure is raised.

At this point, let's contrast hypertension with a simple genetic disease like Tay-Sachs disease. Tay-Sachs is due to a defect in a single gene; every Tay-Sachs patient has a defect in that same gene. Everybody in whom that gene is defective is certain to die of Tay-Sachs, regardless of their life style or environment. In contrast, hypertension involves several different genes whose molecular products remain to be identified. Because there are many causes of raised blood pressure, different hypertensive patients may owe their condition to different gene combinations. Furthermore, whether someone genetically predisposed to hypertension actually develops symptoms depends a lot on life style. Thus, hypertension is not one of those uncommon, homogeneous, and intellectually elegant diseases that geneticists prefer to study. Instead, like diabetes and ulcers, hypertension is a shared set of symptoms produced by heterogeneous causes, all involving an interaction between environmental agents and a susceptible genetic background.

Since U.S. blacks and whites differ on the average in the conditions under which they live, could those differences account for excess hypertension in U.S. blacks? Salt intake, the dietary factor that one thinks of first, turns out on the average not to differ between U.S. blacks and whites. Blacks do consume less potassium and calcium, do experience more stress associated with more difficult socioeconomic conditions, have much less access to medical care, and are therefore much less likely to be diagnosed or treated until it is too late. Those factors surely contribute to the frequency and severity of hypertension in blacks.

However, those factors don't seem to be the whole explanation: hypertensive blacks aren't merely like severely hypertensive whites. Instead, physiological differences seem to contribute as well. On consuming salt, blacks retain it on average far longer before excreting it into the urine, and they experience a greater rise in blood pressure on a high-salt diet. Hypertension is more likely to be "salt-sensitive" in blacks than in whites, meaning that blood pressure is more likely to rise and fall with rises and falls in dietary salt intake. By the same token, black hypertension is more likely to be treated successfully by drugs that cause the kidneys to excrete salt (the so-called thiazide diuretics) and less likely to respond to those drugs that reduce heart rate and cardiac output (so-called beta blockers, such as propanolol). These facts suggest that there are some qualitative differences between the causes of black and white hyperten-

sion, with black hypertension more likely to involve how the kidneys handle salt.

Physicians often refer to this postulated feature as a "defect": for example, "kidneys of blacks have a genetic defect in excreting sodium." As an evolutionary biologist, though, I hear warning bells going off inside me whenever a seemingly harmful trait that occurs frequently in an old and large human population is dismissed as a "defect." Given enough generations, genes that greatly impede survival are extremely unlikely to spread, unless their net effect is to increase survival and reproductive success. Human medicine has furnished the best examples of seemingly defective genes being propelled to high frequency by counterbalancing benefits. For example, sickle-cell hemoglobin protects far more people against malaria than it kills of anemia, while the Tay-Sachs gene may have protected far more Jews against tuberculosis than it killed of neurological disease. Thus, to understand why U.S. blacks now are prone to die as a result of their kidneys' retaining salt, we need to ask under what conditions people might have benefited from kidneys good at retaining salt.

That question is hard to understand from the perspective of modern Western society, where saltshakers are on every dining table, salt (sodium chloride) is cheap, and our bodies' main problem is getting rid of it. But imagine what the world used to be like before saltshakers became ubiquitous. Most plants contain very little sodium, yet animals require sodium at high concentrations in all their extracellular fluids. As a result, carnivores readily obtain their needed sodium by eating herbivores, but herbivores themselves face big problems in acquiring that sodium. That's why the animals that one sees coming to salt licks are deer and antelope, not lions and tigers. Similarly, some human hunter-gatherers obtained enough salt from the meat that they ate. But when we began to take up farming ten thousand years ago, we either had to evolve kidneys superefficient at conserving salt or learn to

extract salt at great effort or trade for it at great expense.

Examples of these various solutions abound. I already mentioned Brazil's Yanomamö Indians, whose staple food is low-sodium bananas and who excrete on the average only 10 milligrams of salt daily—barely one-thousandth the salt excretion of the typical American. A single Big Mac hamburger analyzed by *Consumer Reports* contained 1.5 grams (1,500 milligrams) of salt, representing many weeks of intake for a Yanomamö. The New Guinea highlanders with whom I work, and whose diet consists up to 90 percent of low-sodium sweet potatoes, told me of the efforts to which they went to make salt a few decades ago, before Europeans brought it as trade goods. They gathered leaves of certain plant species, burned them, scraped up the ash, percolated water through it to dissolve the solids, and finally evaporated the water to obtain small amounts of bitter salt.

Thus, salt has been in very short supply for much of recent human evolutionary history. Those of us with efficient kidneys able to retain salt even on a low-sodium diet were better able to survive our inevitable episodes of sodium loss (of which more in a moment). Those kidneys proved to be a detriment only when salt became routinely available, leading to excessive salt retention and hypertension with its fatal consequences. That's why blood pressure and the frequency of hypertension have shot up recently in so many populations around the world as they have made the transition from being self-sufficient subsistence farmers to members of the cash economy and patrons of supermarkets.

This evolutionary argument has been advanced by historian-epidemiologist Thomas Wilson and others to explain the current prevalence of hypertension in American blacks in particular. Many West African blacks, from whom most American blacks originated via the slave trade, must have faced the chronic problem of losing salt through sweating in their hot environment. Yet in West Africa, except on the coast and certain inland areas, salt was traditionally as scarce for African farmers as it has been for Yanomamö and New Guinea farmers. (Ironically, those Africans who sold other Africans as slaves often took payment in salt traded from the Sahara.) By this argument, the genetic basis for hypertension in U.S. blacks was already widespread in many of their West African ancestors. It required only the ubiquity of saltshakers in twentieth-century America for that genetic basis to express itself as hypertension. This argument also predicts that as Africa's life style becomes increasingly Westernized, hypertension could become as prevalent in West Africa as it now is among U.S. blacks. In this view, American blacks would be no different from the many Polynesian, Melanesian, Kenyan, Zulu, and other populations that have recently developed high blood pressure under a Westernized life style.

But there's an intriguing extension to this hypothesis, proposed by Wilson and physician Clarence Grim, collaborators at the Hypertension Research Center of Drew University in Los Angeles. They suggest a scenario in which New World blacks may now be at more risk for hypertension than their African ancestors. That scenario involves very recent selection for superefficient kidneys, driven by massive mortality of black slaves from salt loss.

Grim and Wilson's argument goes as follows. Black slavery in the Americas began about 1517, with the first imports of slaves from West Africa, and did not end until Brazil freed its slaves barely a century ago in 1888. In the course of the slave trade an estimated 12 million Africans were brought to the Americas. But those imports were winnowed by deaths at many stages, from an even larger number of captives and exports.

First, slaves captured by raids in the interior of West Africa were chained together, loaded with heavy burdens, and marched for one or two months, with little food and water, to the coast. About 25 percent of the captives died en route. While awaiting purchase by slave traders, the survivors were held on the coast in hot, crowded buildings called barracoons, where about 12 percent of them died. The traders went up

and down the coast buying and loading slaves for a few weeks or months until a ship's cargo was full (5 percent more died). The dreaded Middle Passage across the Atlantic killed 10 percent of the slaves, chained together in a hot, crowded, unventilated hold without sanitation. (Picture to yourself the result of those toilet "arrangements.") Of those who lived to land in the New World, 5 percent died while awaiting sale, and 12 percent died while being marched or shipped from the sale yard to the plantation. Finally, of those who survived, between 10 and 40 percent died during the first three years of plantation life, in a process euphemistically called seasoning. At that stage, about 70 percent of the slaves initially captured were dead, leaving 30 percent as seasoned survivors.

Even the end of seasoning, however, was not the end of excessive mortality. About half of slave infants died within a year of birth because of the poor nutrition and heavy workload of their mothers. In plantation terminology, slave women were viewed as either "breeding units" or "work units," with a built-in conflict between those uses: "These Negroes breed the best, whose labour is least," as an eighteenth-century observer put it. As a result, many New World slave populations depended on continuing slave imports and couldn't maintain their own numbers because death rates exceeded birth rates. Since buying new slaves cost less than rearing slave children for twenty years until they were adults, slave owners lacked economic incentive to change this state of affairs.

Recall that Darwin discussed natural selection and survival of the fittest with respect to animals. Since many more animals die than survive to produce offspring, each generation becomes enriched in the genes of those of the preceding generation that were among the survivors. It should now be clear that slavery represented a tragedy of unnatural selection in humans on a gigantic scale. From examining accounts of slave mortality, Grim and Wilson argue that death was indeed selective: much of it was related to unbalanced salt loss, which quickly

brings on collapse. We think immediately of salt loss by sweating under hot conditions: while slaves were working, marching, or confined in unventilated barracoons or ships' holds. More body salt may have been spilled with vomiting from seasickness. But the biggest salt loss at every stage was from diarrhea due to crowding and lack of sanitation—ideal conditions for the spread of gastrointestinal infections. Cholera and other bacterial diarrheas kill us by causing sudden massive loss of salt and water. (Picture your most recent bout of *turista*, multiplied to a diarrheal fluid output of twenty quarts in one day, and you'll understand why.) All contemporary accounts of slave ships and plantation life emphasized diarrhea, or "fluxes" in eighteenth-century terminology, as one of the leading killers of slaves.

Grim and Wilson reason, then, that slavery suddenly selected for superefficient kidneys surpassing the efficient kidneys already selected by thousands of years of West African history. Only those slaves who were best able to retain salt could survive the periodic risk of high salt loss to which they were exposed. Salt supersavers would have had the further advantage of building up, under normal conditions, more of a salt reserve in their body fluids and bones, thereby enabling them to survive longer or more frequent bouts of diarrhea. Those superkidneys became a disadvantage only when modern medicine began to reduce diarrhea's lethal impact, thereby transforming a blessing into a curse.

Thus, we have two possible evolutionary explanations for salt retention by New World blacks. One involves slow selection by conditions operating in Africa for millennia; the other, rapid recent selection by slave conditions within the past few centuries. The result in either case would make New World blacks more susceptible than whites to hypertension, but the second explanation would, in addition, make them more susceptible than African blacks. At present, we don't know the relative importance of these two explanations. Grim and Wilson's provocative hypothesis is likely to stimulate medical and physiological comparisons of American blacks with African blacks and thereby to help resolve the question.

While this piece has focused on one medical problem in one human population, it has several larger morals. One, of course, is that our differing genetic heritages predispose us to different diseases, depending on the part of the world where our ancestors lived. Another is that our genetic differences reflect not only ancient conditions in different parts of the world but also recent episodes of migration and mortality. A well-established example is the decrease in the frequency of the sickle-cell hemoglobin gene in U.S. blacks compared with African blacks, because selection for resistance to malaria is now unimportant in the United States. The example of black hypertension that Grim and Wilson discuss opens the door to considering other possible selective effects of the slave experience. They note that occasional periods of starvation might have selected slaves for superefficient sugar metabolism, leading under modern conditions to a propensity for diabetes.

Finally, consider a still more universal moral. Almost all people alive today exist under very different conditions from those under which every human lived 10,000 years ago. It's remarkable that our old genetic heritage now permits us to survive at all under such different circumstances. But our heritage still catches up with most of us, who will die of life style related diseases such as cancer, heart attack, stroke, and diabetes. The risk factors for these diseases are the strange new conditions prevailing in modern Western society. One of the hardest challenges for modern medicine will be to identify for us which among all those strange new features of diet, life style, and environment are the ones getting us into trouble. For each of us, the answers will depend on our particular genes, hence on our ancestry. Only with such individually tailored advice can we hope to reap the benefits of modern living while still housed in bodies designed for life before saltshakers.

Dr. Darwin

With a nod to evolution's god, physicians are looking at illness through the lens of natural selection to find out why we get sick and what we can do about it.

Lori Oliwenstein

Lori Oliwenstein a former DISCOVER senior editor, is now a freelance journalist based in Los Angeles.

PAUL EWALD KNEW FROM THE BEGINNING that the Ebola virus outbreak in Zaire would fizzle out. On May 26, after eight days in which only six new cases were reported, that fizzle became official. The World Health Organization announced it would no longer need to update the Ebola figures daily (though sporadic cases continued to be reported until June 20).

The virus had held Zaire's Bandundu Province in its deadly grip for weeks, infecting some 300 people and killing 80 percent of them. Most of those infected hailed from the town of Kikwit. It was all just as Ewald predicted. "When the Ebola outbreak occurred," he recalls, "I said, as I have before, these things are going to pop up, they're going to smolder, you'll have a bad outbreak of maybe 100 or 200 people in a hospital, maybe you'll have the outbreak slip into another isolated community, but then it will peter out on its own."

Ewald is no soothsayer. He's an evolutionary biologist at Amherst College in Massachusetts and perhaps the world's leading expert on how infectious diseases—and the organisms that cause them—evolve. He's also a force behind what some are touting as the next great medical revolution: the application of Darwin's theory of natural selection to the understanding of human diseases.

> *"If you look at it from an evolutionary point of view, you can sort out the 95 percent of disease organisms that aren't a major threat from the 5 percent that are."*

A Darwinian view can shed some light on how Ebola moves from human to human once it has entered the population. (Between human outbreaks, the virus resides in some as yet unknown living reservoir.) A pathogen can survive in a population, explains Ewald, only if it can easily transmit its progeny from one host to another. One way to do this is to take a long time to disable a host, giving him plenty of time to come into contact with other potential victims. Ebola, however, kills quickly, usually in less than a week. Another way is to survive for a long time outside the human body, so that the pathogen can wait for new hosts to find it. But the Ebola strains encountered thus far are destroyed almost at once by sunlight, and even if no rays reach them, they tend to lose their infectiousness outside the human body within a day. "If you look at it from an evolutionary point of view,

you can sort out the 95 percent of disease organisms that aren't a major threat from the 5 percent that are," says Ewald. "Ebola really isn't one of those 5 percent."

The earliest suggestion of a Darwinian approach to medicine came in 1980, when George Williams, an evolutionary biologist at the State University of New York at Stony Brook, read an article in which Ewald discussed using Darwinian theory to illuminate the origins of certain symptoms of infectious disease—things like fever, low iron counts, diarrhea. Ewald's approach struck a chord in Williams. Twenty-three years earlier he had written a paper proposing an evolutionary framework for senescence, or aging. "Way back in the 1950s I didn't worry about the practical aspects of senescence, the medical aspects," Williams notes. "I was pretty young then." Now, however, he sat up and took notice.

While Williams was discovering Ewald's work, Randolph Nesse was discovering Williams's. Nesse, a psychiatrist and a founder of the University of Michigan Evolution and Human Behavior Program, was exploring his own interest in the aging process, and he and Williams soon got together. "He had wanted to find a physician to work with on medical problems," says Nesse, "and I had long wanted to find an evolutionary biologist, so it was a very natural match for us." Their collaboration led to a 1991 article that most researchers say signaled the real birth of the field.

NESSE AND WILLIAMS DEfine Darwinian medicine as the hunt for evolutionary explanations of vulnerabilities to disease. It can, as Ewald noted, be a way to interpret the body's defenses, to try to figure out, say, the reasons we feel pain or get runny noses when we have a cold, and to determine what we should—or shouldn't—be doing about those defenses. For instance, Darwinian researchers like physiologist Matthew Kluger of the Lovelace Institute in Albuquerque now say that a moderate rise in body temperature is more than just a symptom of disease; it's an evolutionary adaptation the body uses to fight infection by making itself inhospitable to invading microbes. It would seem, then, that if you lower the fever, you may prolong the infection. Yet no one is ready to say whether we should toss out our aspirin bottles. "I would love to see a dozen proper studies of whether it's wise to bring fever down when someone has influenza," says Nesse. "It's never been done, and it's just astounding that it's never been done."

Diarrhea is another common symptom of disease, one that's sometimes the result of a pathogen's manipulating your body for its own good purposes, but it may also be a defense mechanism mounted by your body. Cholera bacteria, for example, once they invade the human body, induce diarrhea by producing toxins that make the intestine's cells leaky. The resultant diarrhea then both flushes competing beneficial bacteria from the gut and gives the cholera bacteria a ride into the world, so that they can find another hapless victim. In the case of cholera, then, it seems clear that stopping the diarrhea can only do good.

But the diarrhea that results from an invasion of shigella bacteria—which cause various forms of dysentery—seems to be more an intestinal defense than a bacterial offense. The infection causes the muscles surrounding the gut to contract more frequently, apparently in an attempt to flush out the bacteria as quickly as possible. Studies done more than a decade ago showed that us-

ing drugs like Lomotil to decrease the gut's contractions and cut down the diarrheal output actually prolong infection. On the other hand, the ingredients in over-the-counter preparations like Pepto Bismol, which don't affect how frequently the gut contracts, can be used to stem the diarrheal flow without prolonging infection.

Seattle biologist Margie Profet points to menstruation as another "symptom" that may be more properly viewed as an evolutionary defense. As Profet points out, there must be a good reason for the body to engage in such costly activities as shedding the uterine lining and letting blood flow away. That reason, she claims, is to rid the uterus of any organisms that might arrive with sperm in the seminal fluid. If an egg is fertilized, infection may be worth risking. But if there is no fertilized egg, says Profet, the body defends itself by ejecting the uterine cells, which might have been infected. Similarly, Profet has theorized that morning sickness during pregnancy causes the mother to avoid foods that might contain chemicals harmful to a developing fetus. If she's right, blocking that nausea with drugs could result in higher miscarriage rates or more birth defects.

DARWINIAN MEDICINE ISN'T simply about which symptoms to treat and which to ignore. It's a way to understand microbes—which, because they evolve so much more quickly than we do, will probably always beat us unless we figure out how to harness their evolutionary power for our own benefit. It's also a way to realize how disease-causing genes that persist in the population are often selected for, not against, in the long run.

Sickle-cell anemia is a classic case of how evolution tallies costs and benefits. Some years ago, researchers discovered that people with one copy of the sickle-cell gene are better able to resist the protozoans that cause malaria than are people with no copies of the gene. People with two copies of the gene may die, but in malaria-plagued regions such as tropical Africa, their numbers will be more than made up for by the offspring left by the disease-resistant kin.

Cystic fibrosis may also persist through such genetic logic. Animal studies indicate that individuals with just one copy of the cystic fibrosis gene may be more resistant to the effects of the cholera bacterium. As is the case with malaria and sickle-cell, cholera is much more prevalent than cystic fibrosis; since there are many more people with a single, resistance-conferring copy of the gene than with a disease-causing double dose, the gene is stably passed from generation to generation.

"With our power to do gene manipulations, there will be temptations to find genes that do things like cause aging, and get rid of them," says Nesse. "If we're sure about everything a gene does, that's fine. But an evolutionary approach cautions us not to go too fast, and to expect that every gene might well have some benefit as well as costs, and maybe some quite unrelated benefit."

"I used to hunt saber-toothed tigers all the time, thousands of years ago. Now I sit in front of a computer and don't get exercise, so I've changed my body chemistry."

Darwinian medicine can also help us understand the problems encountered in the New Age by a body designed for the Stone Age. As evolutionary psychologist Charles Crawford of Simon Fraser University in Burnaby, British Columbia, put it: "I used to hunt saber-toothed tigers all the time, thousands of years ago. I got lots of exercise and all that sort of stuff. Now I sit in front of a computer, and all I do is play with a mouse, and I don't get exercise. So I've changed my body biochemistry in all sorts of unknown ways, and it could affect me in all sorts of ways, and we have no idea what they are."

Radiologist Boyd Eaton of Emory University and his colleagues believe such biochemical changes are behind to-day's breast cancer epidemic. While it's impossible to study a Stone Ager's bio-chemistry, there are still groups of hunter-gathers around—such as the San of Africa—who make admirable stand-ins. A foraging life-style, notes Eaton, also means a life-style in which men-struation begins later, the first child is born earlier, there are more children al-together, they are breast-fed for years rather than months, and menopause comes somewhat earlier. Overall, he says, American women today probably experience 3.5 times more menstrual cy-cles than our ancestors did 10,000 years ago. During each cycle a woman's body is flooded with the hormone estrogen, and breast cancer, a research has found, is very much estrogen related. The more frequently the breasts are exposed to the hormone, the greater the chance that a tumor will take seed.

Depending on which data you choose, women today are somewhere between 10 and 100 times more likely to be stricken with breast cancer than our ancestors were. Eaton's proposed solutions are pretty radical, but he hopes people will at least entertain them; they include de-laying puberty with hormones and using hormones to create pseudopregnancies, which offer a woman the biochemical advantages of pregnancy at an early age without requiring her to bear a child.

In general, Darwinian medicine tells us that the organs and systems that make up our bodies result not from the pursuit of perfection but from millions of years of evolutionary compromises designed to get the greatest reproductive benefit at the lowest cost. We walk upright with a spine that evolved while we scampered on four limbs; balancing on two legs leaves our hands free, but we'll probably always suffer some back pain as well.

"What's really different is that up to now people have used evolutionary the-ory to try to explain why things work, why they're normal," explains Nesse. "The twist—and I don't know if it's simple or profound—is to say we're try-ing to understand the abnormal, the vul-nerability to disease. We're trying to understand why natural selection has not made the body better, why natural se-lection has left the body with vulner-

abilities. For every single disease, there is an answer to that question. And for very few of them is the answer very clear yet."

One reason those answers aren't yet clear is that few physicians or medical researchers have done much serious sur-veying from Darwin's viewpoint. In many cases, that's because evolutionary theories are hard to test. There's no way to watch human evolution in progress—at best it works on a time scale involving hundreds of thousands of years. "Dar-winian medicine is mostly a guessing game about how we think evolution worked in the past on humans, what it designed for us," say evolutionary biolo-gist James Bull of the University of Texas at Austin. "It's almost impossible to test ideas that we evolved to respond to this or that kind of environment. You can make educated guesses, but no one's going to go out and do an experiment to show that yes, in fact humans will evolve this way under these environ-mental conditions."

Yet some say that these experiments can, should, and will be done. Howard Howland, a sensory physiologist at Cor-nell, is setting up just such an evolution-ary experiment, hoping to interfere with the myopia, or nearsightedness, that af-flicts a full quarter of all Americans. Myopia is thought to be the result of a delicate feedback loop that tries to keep images focused on the eye's retina. There's not much room for error: if the length of your eyeball is off by just a tenth of a millimeter, your vision will be blurry. Research has shown that when the eye perceives an image as fuzzy, it compensates by altering its length.

This loop obviously has a genetic component, notes Howland, but what drives it is the environment. During the Stone Age, when we were chasing buf-falo in the field, the images we saw were usually sharp and clear. But with mod-ern civilization came a lot of close work. When your eye focuses on something nearby, the lens has to bend, and since bending that lens is hard work, you do as little bending as you can get away with. That's why, whether you're con-scious of it or not, near objects tend to be a bit blurry. "Blurry image?" says the eye. "Time to grow." And the more it

grows, the fuzzier those buffalo get. Myopia seems to be a disease of indus-trial society.

To prevent that disease, Howland suggests going back to the Stone Age—or at least convincing people's eyes that that's where they are. If you give folks with normal vision glasses that make their eyes think they're looking at an ob-ject in the distance when they're really looking at one nearby, he says, you'll avoid the whole feedback loop in the first place. "The military academies induct young men and women with twenty-twenty vision who then go through four years of college and are trained to fly an airplane or do some dif-ficult visual task. But because they do so much reading, they come out the other end nearsighted, no longer eligible to do what they were hired to do," How-land notes. "I think these folks would very much like not to become near-sighted in the course of their studies." He hopes to be putting glasses on them within a year.

THE NUMBING PACE OF EVO-lution is a much smaller problem for researchers interested in how the bugs that plague us do their dirty work. Bacteria are present in such large numbers (one person can carry around more pathogens than there are people on the planet) and evolve so quickly (a sin-gle bacterium can reproduce a million times in one human lifetime) that ex-periments we couldn't imagine in hu-mans can be carried out in microbes in mere weeks. We might even, says Ewald, be able to use evolutionary the-ory to tame the human immunodefi-ciency virus.

"HIV is mutating so quickly that surely we're going to have plenty of sources of mutants that are mild as well as severe," he notes. "So now the ques-tion is, which of the variants will win?" As in the case of Ebola, he says, it will all come down to how well the virus manages to get from one person to an-other.

"If there's a great potential for sexual transmission to new partners, then the viruses that reproduce quickly will spread," Ewald says. "And since they're reproducing in a cell type that's critical for the well-being of the host—the helper T cell—then that cell type will be decimated, and the host is likely to suffer from it." On the other hand, if you lower the rate of transmission—through abstinence, monogamy, condom use—then the more severe strains might well die out before they have a chance to be passed very far. "The real question," says Ewald, "is, exactly how mild can you make this virus as a result of reducing the rate at which it could be transmitted to new partners, and how long will it take for this change to occur?" There are already strains of HIV in Senegal with such low virulence, he points out, that most people infected will die of old age. "We don't have all the answers. But I think we're going to be living with this virus for a long time, and if we have to live with it, let's live with a really mild virus instead of a severe virus."

Though condoms and monogamy are not a particularly radical treatment, that they might be used not only to stave off the virus but to tame it is a radical notion—and one that some researchers find suspect. "If it becomes too virulent, it will end up cutting off its own transmission by killing its host too quickly," notes James Bull. "But the speculation is that people transmit HIV primarily within one to five months of infection, when they spike a high level of virus in the blood. So with HIV, the main period of transmission occurs a few months into the infection, and yet the virulence—the death from it—occurs years later. The major stage of transmission is decoupled from the virulence." So unless the protective measures are carried out by everyone, all the time, we won't stop most instances of transmission; after all, most people don't even know they're infected when they pass the virus on.

But Ewald thinks these protective measures are worth a shot. After all, he says, pathogen taming has occurred in the past. The forms of dysentery we encounter in the United States are quite mild because our purified water supplies have cut off the main route of transmission for virulent strains of the bacteria. Not only did hygienic changes reduce the number of cases, they selected for the milder shigella organisms, those that leave their victim well enough to get out and about. Diphtheria is another case in point. When the diphtheria vaccine was invented, it targeted only the most severe form of diphtheria toxin, though for economic rather than evolutionary reasons. Over the years, however, that choice has weeded out the most virulent strains of diphtheria, selecting for the ones that cause few or no symptoms. Today those weaker strains act like another level of vaccine to protect us against new, virulent strains.

"We did with diphtheria what we did with wolves. We took an organism that caused harm, and unknowingly, we domesticated it into an organism that protects us."

"You're doing to these organisms what we did to wolves," says Ewald. "Wolves were dangerous to us, we domesticated them into dogs, and then they helped us, they warned us against the wolves that were out there ready to take our babies. And by doing that, we've essentially turned what was a harmful organism into a helpful organism. That's the same thing we did with diphtheria; we took an organism that was causing harm, and without knowing it, we domesticated it into an organism that is protecting us against harmful ones."

Putting together a new scientific discipline—and getting it recognized—is in itself an evolutionary process. Though Williams and Neese say there are hundreds of researchers working (whether they know it or not) within this newly built framework, they realize the field is still in its infancy. It may take some time before *Darwinian medicine* is a household term. Nesse tells how the editor of a prominent medical journal, when asked about the field, replied, "Darwinian medicine? I haven't heard of it, so it can't be very important."

But Darwinian medicine's critics don't deny the field's legitimacy; they point mostly to its lack of hard-and fast answers, its lack of clear clinical guidelines. "I think this idea will eventually establish itself as a basic science for medicine, " answers Nesse. "What did people say, for instance, to the biochemists back in 1900 as they were playing out the Krebs cycle? People would say, 'So what does biochemistry really have to do with medicine? What can you cure now that you couldn't before you knew about the Krebs cycle?' And the biochemists could only say, 'Well, gee, we're not sure, but we know what we're doing is answering important scientific questions, and eventually this will be useful.' And I think exactly the same applies here."

The Future Evolution of *Homo Sapiens*

An eminent biologist speculates about what the future may hold for our species.

Colin Tudge

Visiting research fellow at the Centre for Philosophy of the London School of Economics, Colin Tudge has written for New Scientist *and* Nature. *His books include* Last Animals at the Zoo *and* The Engineer in the Garden, *shortlisted for the Rhone-Poulenc Science Book of the Year award.*

Suppose that our species survives the next 1,000 years and is thus able to embark upon a journey of a million years. The human genus has changed radically over the past million years—from *Homo erectus* into us. Would the next million bring comparable change?

Of course, human beings will continue to change culturally; and this might be considered to be evolution of a kind. After all, if we entertain different ideas and behave differently and have a different attitude to the world, then we would be ecologically quite different. Our impact on our fellow creatures would be altered, and from their point of view at least we would effectively be a different kind of creature.

We can argue, too, (as others have argued) that the infusion of new ideas to some extent occurs by natural selection. The ideas that produce worldly success do indeed tend to spread, just as science and capitalism have spread this past few centuries (and the present growth of Islam is certainly interesting). But we can also point out, as others have done, that the growth of new ideas proceeds according to Lamarck's fourth law: That is, ideas acquired by one generation are passed on to the next. Darwin himself would have had no problem in accepting that Darwinian and Lamarckian mechanisms might proceed in tandem and indeed more than toyed with the notion that Lamarckian systems of inheritance obtain generally.

The issue here, however, is that of bona fide Neo-Darwinian evolution: whether it is possible to change the overall gene pool of human beings to such a significant degree that our descendants can properly be considered a different species.

The short answer, on purely theoretical grounds, is "Yes." Remarkable creatures though we are, we are manifestations of our collective gene pool just like any other animal. And it is possible to subtract alleles [alternative forms of a particular gene as, say, for an eye color] from that pool and add them, and go on doing this until we have a new creature just as it was possible, by these means, to turn a gomphothere [a mastodon with four tusks] into an elephant or *Australopithecus* into *Homo*. But it is not as easy to envisage the circumstances, or at least the natural circumstances, in which such a change could be brought about.

It is obvious, after all, that the human gene pool is changing. To our shame, rare tribes of aboriginal people worldwide continue to disappear, and they must take at least a few recondite alleles with them. Mutation continues, too, and since the human population is now so large, the gene pool as a whole must be accumulating mutations more rapidly than ever before. Some groups, too, are now breeding much faster than others—Kenyans faster than Germans, for example—so some alleles are becoming relatively more common than others. Even so, we cannot argue that such fluctuations truly represent evolution. The loss of alleles, sad though it is, is genetically marginal. At the same time, few of the new mutations contribute significantly to the life of our species. The shift in frequency of alleles within the pool is only the usual "noise" in the system. No one seriously supposes, after all, that Northern Europeans are going to disappear all together.

In short, the fluctuations of the human gene pool are providing raw material for evolution—the genetic variation. But the key ingredient is missing. Natural selection is simply not acting forcefully or consistently upon that variation. The loss of alleles in aboriginal peoples is random; at least, we cannot argue that particular alleles are being

From *Earth* magazine, February 1996, pp. 36-40. Adapted from *The Time Before History* by Colin Tudge. © 1996 by Colin Tudge. Reprinted by permission of Scribner, an imprint of Simon & Schuster, Inc.

lost because they are disadvantageous. At the same time, the people who seem to be successful are not necessarily breeding particularly quickly. Neither can we argue that the success of particular present-day groups is correlated with their genes. The materially successful people of California belong to all conceivable human genotypes.

Finally, and I think crucially, the human population is just too big to change significantly by Neo-Darwinian mechanisms. Natural selection works best on middle-sized, isolated or semi-isolated populations, like those of the australopithecines on the margins of Africa's retreating Pliocene forest. It is hard to see how it could operate on a population of five billion individuals who, by courtesy of the world's airlines, are all in close genetic contact with each other.

There are, however, two feasible sets of circumstances in which we might envisage Neo-Darwinian change. The first is if there is after all some kind of world catastrophe: an ecological crash. In such circumstances we might envisage that the human species would be reduced to patchy populations, separated by various kinds of badlands. Excess ultraviolet radiation, or the still-present specter of nuclear war (or even peacetime nuclear catastrophe) might provide the conditions. Then we could envisage the isolated groups evolving afresh, just as hominid groups of comparable size, with comparable degrees of stress, evolved in the past.

It is interesting to speculate how those isolated groups might evolve. There is absolutely no good reason to assume that the trend of the past few million years would be continued: that intelligent human beings would produce a race of superintelligents. That is actually an unlikely option, since it is hard to see why natural selection should favor geniuses in the straitened circumstances of a devastated future, any more than it would favor the emergence of a literary codfish beneath the Arctic ice. Besides, it seems likely (though nobody knows) that genius is a genetically difficult trick to pull. Overall, indeed, isolated human populations of the future seem far more likely to generate a significantly less-intelligent lineage.

The brain, physiologically, is an expensive item. One way and another it is said to commandeer 20 percent of total metabolic effort. Unless this prodigal brain can be put to good use, it is a luxury which future hominids in straitened circumstances may prefer to do without. In the same way, many island birds from many different groups have abandoned the apparently self-evident advantage of flight. How glorious to fly! But if you do not need to fly because there are no predators to escape from, then you are advised to stay on the ground because it is cheaper.

In the same way, *Australopithecus afarensis* apparently gave rise to the robust paranthropines [apes with thick-boned skulls and jaws and small brains] as well as to the hominines [light-boned apes with larger brains who evolved into *Homo sapiens*]. Of course we cannot tell at this range whether the paranthropines were less bright than *afarensis*.

The point is, though, that the hominine emphasis on brain was only one of the options open to the hominids. Muscularity was another. Neither can we assert that the paranthropines were a failure. They lasted a million years and might be with us still. Neither were the hominines bound to succeed. I believe that we can identify their clear advantages. But advantage does not come with guarantees. Every lineage needs good luck. "Time and chance," as Ecclesiastes reminds us, "happeneth to all men."

We can, however, envisage a quite different set of circumstances in which human beings might continue to evolve; not by natural selection, but by artificial selection. In other words, our descendants could in theory breed a new kind of hominine, probably (to speed things up) with the aid of genetic engineering. However, I mention this only in the interests of logical completeness.

But the theoretical possibility of future evolutionary change does raise an issue that is of universal significance. What, in theory, might human beings evolve into? More broadly, of what is life capable?

The concept of progress in evolution is worth entertaining. Some creatures really are demonstrably better than others at carrying out definable tasks: better both in engineering (for example, extracting more useful calories from a given quantity of food) and logistically. On the whole, too, the "better" creatures evolve from the less good, even though there seem at times to have been some interesting reversals.

But one reason why these somewhat obvious and, I believe, useful ideas have become unfashionable is, I suggest, because people seem to confuse progress with destiny. In fact, progress in general is likely to happen. Better creatures are bound to appear through the chance processes of mutation, and having appeared they are liable to succeed. Furthermore, some lines of development are more likely than others. In short, animals must obey the rules of engineering, which are founded in the laws of physics; and in practice only a limited range of body forms, or ecomorphs, is feasible. Put the two ideas together and it looks as if the gene pools of ancestral creatures are bound to be pushed in particular directions. Indeed as Nobel Prize–winning biochemist Christian de Duve comments in *Vital Dust*, "Should things start all over again, here or elsewhere, the final outcome could not be the same. But how different would it be?"

In other words, we can predict to a large extent that any suite of creatures given the freedom to evolve will eventually fill at least some of a range of the niches that can define, that they will do so in at least some of the ways that we have already observed. But we absolutely cannot predict which particular creatures are liable to evolve. It was never possible to predict that life on Earth would, within the lifetime of the Earth, have produced *Homo sapiens*, or indeed the hominids as a whole.

The story of human evolution may be one of opportunities taken, of the right gene pool in the right place at the right time. What was always absolutely unpredictable, however, was that such a set of circumstances would ever arise. If India had not crunched into Asia, there would be no Himalaya. If there were no Himalaya, then the world might still be covered in tropical forests. [The uplifted Himalaya underwent weathering, which caused chemical reactions that pulled carbon dioxide from the atmosphere. This resulted in cooling, which led to temperate climates.] Given that there were plenty of primates around in the Eocene, we can reasonably predict that there would be plenty around now. But the predominant forms surely would be monkeys, supremely adapted to the trees. There is absolutely no reason to assume that any of them would have developed along hominine lines.

More generally, there is no reason whatever to suppose that intelligence of human proportions would ever have appeared among any lineage. Indeed there is no reason to assume that such a quality would have appeared before the next meteorite collision, which, if paleontologists Dave Raup and Jack Sepkoski are right, is due in another 13 million years. In fact the world might have begun its final cooling, with the death of all life, before circumstances ever arose that could produce creatures like us. In short there is progress but there is no destiny. Time and chance always play their part.

Yet there is another notion that seems to me intriguing and relevant. It is obvious that nothing can happen of any kind, except the things that are possible. To put the point more simply: Everything that does happen must be possible. Perhaps this seems too self-evident to be worth stating. Yet it is interesting. For if it is possible for phenomenon B to result from phenomenon A, then we have to conclude that phenomenon A had the potential to produce phenomenon B. Indeed we might argue, as I think Aristotle essentially did, that the potential to produce B must be numbered among the innate properties of A.

Thus it seems to me entirely unexceptionable to point out that intelligence of the kind we recognize in human beings is a product of living flesh. Intelligence can be seen as emergent property; something that results when some of the molecules contained in flesh are suitably arranged. By the same token, we can argue that the molecules of that flesh must have had the potential to produce intelligence. Look at those molecules more closely and you find they are compounded from carbon, oxygen, hydrogen, nitrogen, phosphorus, sulfur, and a few metals—a pretty commonplace array.

Yet we have to concede that this chemistry-set collection, suitably arranged, has the potential to produce thought. We could go even further back and observe that these elements are themselves compounded of fundamental particles. These particles, then, had the potential, once suitably arranged, to produce the highest flights of genius.

Why is this relevant? Suppose Dr. Who, the extraordinarily accomplished scientist of children's TV capable of traveling through time, traveled back to the time of the Big Bang, when the Universe began and all the fundamental particles were still separate. Could he have predicted just by looking at those particles (and leaving aside for a moment his knowledge of the future) that they had the potential to form themselves into the elements of the periodic table? I am not at all sure how. I feel he would have to have waited a few years, or a few million years, to see how things turned out.

Once the elements had formed, in the depths of successive generations of stars, could he have predicted that they would when suitably combined produce living things? Again I think not. If he visited Earth two billion years ago or so and saw the first organisms with nucleated cells, could he have predicted that from these humble creatures would evolve animals that can think? Again one asks, "How could he?" Where are the clues in those primitive creatures to indicate future possibilities?

So I would like now to make precisely the same point in the context of present-day creatures, and indeed of ourselves. That is, we simply cannot tell, by looking at existing life, exactly what life is capable of. Some eminent biologists of the 20th century have felt that human beings, bright creatures that we are, represent the ultimate in biological evolution. The same point, stated only slightly differently, is made in the Old Testament: We are supposed to have been created "in God's image" and God, by definition, is ultimate. But what reason do we have to assume that this is the case? What do we really know of life? I submit that we are no more able to predict the possibilities—the potential—of life, than Dr. Who would have been in trying to predict the potential of newly emerging elements.

Specifically, we think that our brains are wonderful, and so they are. Yet we can identify areas of extreme feebleness. We are extremely poor at math and feel that those who gain first degree honors in that subject must be very bright indeed. But why then do those "geniuses" use calculators? Because a $5.00 pocket calculator with a tiny microchip can carry out the mechanics of math a hundred times quicker than they can. Of course the calculator lacks imagination but the point stands nonetheless; here is a comparatively simple trick of math that even the greatest mathematicians simply cannot do.

We assume that because in general we are smarter than most other animals, we must be brighter than them in all respects. Yet this too is nonsense. Memory is generally conceded to be a component of intelligence: Certainly there is at least some correlation between measurable memory and measurable IQ. But animals that hide food for the winter sometimes—like nuthatches—have memories of topography that are quite out of our league.

In short, it is very easy to envisage a dozen ways in which we could in theory improve the components of our present-day intelligence by leaps and bounds. There is no a priori reason why a human being should not combine the qualities, say, of Einstein, Shakespeare, Mozart, Darwin, J. M. W. Turner, a nuthatch and a pocket calculator. Indeed there is no a priori reason why such a paragon should not be considered ordinary.

But the qualities of known geniuses could well prove mundane, compared to what might be possible. For example, homing pigeons and many other creatures have been shown to possess some magnetic sense by which they can orient themselves relative to the Earth's field. Could not future lineages—in theory—develop such a sense to a high degree? On the other hand, no creature present or past is known to respond to, or transmit, long-wave electromagnetic radiations, that is, radio waves. Yet there seems to be nothing in those possibilities to offend any known laws of physics.

In theory then, we could envisage future creatures that carry their own two-way radios in their heads. Combine this with a more advanced magnetic sense, and we can envisage creatures that might be able to pinpoint their positions anywhere on the Earth's surface. It is very difficult to envisage the natural circumstances that would favor an emergence of such creatures, of course, but that is not my point. I am simply asking the more fundamental question, Of what is life really capable?

So we can see the future evolution of living things on Earth as a kind of obstacle race. Living things as a whole might have the potential to do a whole range of things that at present are unknowable and unpredictable. The question is, which and how many of those things will in practice be realized? It would be nice if there were creatures in the future that were intelligent enough to take note. Indeed, it would be nice, speaking chauvinistically, if those creatures included future members of our own lineage.

However, the chances that any untapped potential will be realized, by whatever lineage, depend very much on our actions over the next few hundred years. But those actions must not be geared only to the needs of the next few hundred years. The events of this planet can take far longer than that to unfold.

We surely have the potential to survive as a species for at least a million years; there is no reason to doubt that. But if we want our descendants to claim that million years and do so in the company of other creatures, then we must think from the beginning in such terms. In short, we cannot claim to be taking our species and our planet seriously until we acknowledge that a million years is a proper unit of political time.

Index

Credits/Acknowledgments

Cover design by Charles Vitelli

1. Natural Selection
Facing overview—New York Public Library illustration.

2. Primates
Facing overview—United Nations photo by George Love.

3. Sex and Society
Facing overview—National Geographic Society photo by Baron Hugo van Lawick.

4. The Hominid Transition
Facing overview—American Museum of Natural History photo.

5. The Fossil Evidence
Facing overview—Dushkin Publishing Group illustration by Mike Eagle. 143—Illustration from the Lancelyn Green Collection. 148-149—*Discover* graphics by Nenad Jakesevic. © 1996 by The Walt Disney Company. 154-158—*Scientific American* Time Line with illustrations by Patricia J. Wynne. 154 (bottom)—*Scientific American* map by Johnny Johnson. 155 (bottom)—Photo by Yves Coppens. 157 (bottom)—Photo by Des Bartlett/Photo Researchers. 158 (bottom)—*Scientific American* illustration by Patricia J. Wynne.

6. Late Hominid Evolution
Facing overview—WHO photo. 181-182—Illustration and photos by Jean-Jacques Hublin.

7. Living with the Past
Facing overview—Australian Information Service photo. 214—© 1995 by Bob Sacha. 217—Photos courtesy of Drs. E. Fuller Torrey and Daniel R. Weinberger, N.I.M.H. Neuroscience Center, Washington, DC. 219—Photo courtesy of Nick Kelsh. 220—American Philosophical Society photo.

ANNUAL EDITIONS ARTICLE REVIEW FORM

■ NAME: _____ DATE: _____

■ TITLE AND NUMBER OF ARTICLE: _____

■ BRIEFLY STATE THE MAIN IDEA OF THIS ARTICLE: _____

_____

■ LIST THREE IMPORTANT FACTS THAT THE AUTHOR USES TO SUPPORT THE MAIN IDEA:

_____

_____

_____

_____

■ WHAT INFORMATION OR IDEAS DISCUSSED IN THIS ARTICLE ARE ALSO DISCUSSED IN YOUR TEXTBOOK OR OTHER READINGS THAT YOU HAVE DONE? LIST THE TEXTBOOK CHAPTERS AND PAGE NUMBERS:

_____

_____

_____

_____

■ LIST ANY EXAMPLES OF BIAS OR FAULTY REASONING THAT YOU FOUND IN THE ARTICLE:

_____

_____

_____

_____

■ LIST ANY NEW TERMS/CONCEPTS THAT WERE DISCUSSED IN THE ARTICLE, AND WRITE A SHORT DEFINITION:

_____

_____

_____

_____

We Want Your Advice

ANNUAL EDITIONS revisions depend on two major opinion sources: one is our Advisory Board, listed in the front of this volume, which works with us in scanning the thousands of articles published in the public press each year; the other is you—the person actually using the book. Please help us and the users of the next edition by completing the prepaid article rating form on this page and returning it to us. Thank you for your help!

ANNUAL EDITIONS: PHYSICAL ANTHROPOLOGY 97/98
Article Rating Form

Here is an opportunity for you to have direct input into the next revision of this volume. We would like you to rate each of the 41 articles listed below, using the following scale:

1. **Excellent: should definitely be retained**
2. **Above average: should probably be retained**
3. **Below average: should probably be deleted**
4. **Poor: should definitely be deleted**

Your ratings will play a vital part in the next revision. So please mail this prepaid form to us just as soon as you complete it.
Thanks for your help!

Rating	Article	Rating	Article
	1. The Growth of Evolutionary Science		22. Ape Cultures and Missing Links
	2. Evolution's New Heretics		23. Dawson's Dawn Man: The Hoax at Piltdown
	3. Keeping Up Down House		24. The Case of the Missing Link
	4. Curse and Blessing of the Ghetto		25. Sunset on the Savanna
	5. The Future of AIDS		26. East Side Story: The Origin of Humankind
	6. Black, White, Other		27. Asian Hominids Grow Older, Do Kenya Tools Root Birth of Modern Thought in Africa?
	7. Racial Odyssey		28. Scavenger Hunt
	8. Machiavellian Monkeys		29. *Erectus* Rising
	9. What Are Friends For?		30. The First Europeans
	10. Gut Thinking		31. Did Neandertals Lose an Evolutionary "Arms" Race?
	11. The Mind of the Chimpanzee		32. The Dawn of Creativity
	12. Dian Fossey and Digit		33. Old Masters
	13. These Are Real Swinging Primates		34. The Dating Game
	14. Natural-Born Mothers		35. The Neanderthal Peace
	15. Sex and the Female Agenda		36. Profile of an Anthropologist: No Bone Unturned
	16. Why Women Change		37. Eugenics Revisited
	17. What's Love Got to Do with It?		38. The DNA Wars
	18. Apes of Wrath		39. The Saltshaker's Curse
	19. Dim Forest, Bright Chimps		40. Dr. Darwin
	20. To Catch a Colobus		41. The Future Evolution of *Homo Sapiens*
	21. Ape at the Brink		

(Continued on next page)

ABOUT YOU

Name _____ Date _____

Are you a teacher? ☐ Or a student? ☐

Your school name _____

Department _____

Address _____

City _____ State _____ Zip _____

School telephone # _____

YOUR COMMENTS ARE IMPORTANT TO US !

Please fill in the following information:

For which course did you use this book? _____

Did you use a text with this *ANNUAL EDITION*? ☐ yes ☐ no

What was the title of the text? _____

What are your general reactions to the *Annual Editions* concept?

Have you read any particular articles recently that you think should be included in the next edition?

Are there any articles you feel should be replaced in the next edition? Why?

Are there any World Wide Web sites you feel should be included in the next edition? Please annotate.

May we contact you for editorial input?

May we quote your comments?

ANNUAL EDITIONS: PHYSICAL ANTHROPOLOGY 97/98